AF361747

# The Future of Coral Reefs: Research Submitted to ICRS 2020, Bremen, Germany

# The Future of Coral Reefs: Research Submitted to ICRS 2020, Bremen, Germany

Editors

**Rupert Ormond**
**Peter Schupp**

MDPI • Basel • Beijing • Wuhan • Barcelona • Belgrade • Manchester • Tokyo • Cluj • Tianjin

*Editors*
Rupert Ormond
Heriot-Watt University
UK

Peter Schupp
Helmholtz Institute for Functional Marine Biodiversity at
the University of Oldenburg (HIFMB)
Germany

*Editorial Office*
MDPI
St. Alban-Anlage 66
4052 Basel, Switzerland

This is a reprint of articles from the Special Issue published online in the open access journal *Oceans* (ISSN 2673-1924) (available at: https://www.mdpi.com/journal/oceans/special_issues/coral_reefs).

For citation purposes, cite each article independently as indicated on the article page online and as indicated below:

LastName, A.A.; LastName, B.B.; LastName, C.C. Article Title. *Journal Name* **Year**, *Volume Number*, Page Range.

ISBN 978-3-0365-4033-7 (Hbk)
ISBN 978-3-0365-4034-4 (PDF)

Cover image courtesy of Rupert Ormond.

# Contents

# About the Editors

**Rupert Ormond** now lives on the Isle of Mull in Scotland and is currently affiliated with Heriot-Watt University in Edinburgh, having previously worked at the Universities of Cambridge, York, London, and Glasgow, as well as for two conservation organisations—the Save Our Seas Foundation and Marine Conservation International. He has been involved in research and conservation on coral reefs for over 50 years, mainly in the Red Sea and Indian Ocean regions. His research has focused on reef fish communities, ecological interactions between reef organisms, impacts affecting corals and the behaviour of sharks. At the same time, he has played a key role in the establishment and management of Marine Protected Areas and National Parks in a series of coral reef countries. He was Corresponding Secretary of the International Coral Reef Society from 2011 to 2019 and is now Editor-in-Chief of the Society's bulletin *Reef Encounter*.

**Peter Schupp** is head of the Environmental Biochemistry group at the Institute of Chemistry and Biology of the Marine Environment and a founding member of the Helmholtz Institute for Functional Marine Biodiversity, both at the University of Oldenburg, Germany. Previously he has been working as faculty and director at the University of Guam Marine Laboratory, Guam, USA. He has over 30 years of experience in benthic coral reef ecology, mainly on Indo-Pacific coral reefs. His research focuses on the chemical and microbial ecology of marine invertebrates, often using sponges as model organisms to study the microbe-host associations. Another research emphasis is the identification and function of chemical signals from crustose coralline algae and associated biofilms during coral settlement. Recently, research also involves investigations on coral reef benthic community changes, especially the identification of possible drivers towards communities dominated by organisms other than hard corals.

# Preface to "The Future of Coral Reefs: Research Submitted to ICRS 2020, Bremen, Germany"

In connection with the 14th International Coral Reef Symposium (14th ICRS), originally planned to be held in Bremen, Germany, in July 2020, arrangements were made with the MDPI journal Oceans to produce a Special Issue associated with the conference, to which conference delegates could submit papers based on their oral presentations or posters. However, due to the COVID pandemic, we had to postpone the event. A virtual conference with more than 1300 participants was then held as the 14th ICRS in July 2021, organised by University of Bremen, while an in-person meeting will take place as 15th ICRS in July 2022 in Bremen, Germany. We expect about 1000 in-person participants plus at least 500 virtual participants.

The first Special Issue (entitled "The Future of Coral Reefs: Research submitted to 14th ICRS") based on papers submitted by July 2021 has now been concluded. This Special Issue contains 16 very interesting papers from different disciplines that can be downloaded here: https://www.mdpi.com/journal/oceans/special_issues/coral_reefs. I would like to express my sincere thanks to the dedicated Guest Editor of this first Special Issue, Rupert Ormond, for his immense and tireless job in assisting the authors with improving their manuscripts.

Since the follow-up in-person meeting will take place in July 2022, it has been agreed that a second Special Issue will also be published in *Oceans*. This sister volume will retain the same name and continue to reflect the conference themes. Manuscripts for this second volume are to be submitted by 30 September 2022.

*Oceans* is an open access journal that normally charges authors a fee. We are grateful to MDPI and Oceans that they agreed to publish papers related to abstracts presented at the 14th ICRS (virtual) and the 15th ICRS (in person) free of any charge or fees. This is a unique opportunity for participants of both the 14th and the 15th ICRS to publish their work in an international, open access, and peer-reviewed journal without any associated costs. The offer clearly goes beyond previous proceedings of ICRS events.

This second volume will be edited by *Oceans* editors in cooperation with an editorial support team consisting of Leila Chapron, Sebastian Ferse, Rupert Ormond, Ronald Osinga, Peter Schupp, and myself.

We look forward to receiving contributions for this second Special Issue!

**Christian Wild**

Professor for Marine Ecology, University of Bremen, Germany,
and Chair of 14th and 15th ICRS Organising Committee

*Article*

# Investigation into the Presence of *Symbiodiniaceae* in Antipatharians (*Black Corals*)


Erika Gress [1,2,3,*], Igor Eeckhaut [2,4], Mathilde Godefroid [5], Philippe Dubois [5], Jonathan Richir [6] and Lucas Terrana [2]

[1] ARC Centre of Excellence for Coral Reef Studies, James Cook University, Townsville 4811, Australia
[2] Biology of Marine Organisms and Biomimetics, University of Mons, 7000 Mons, Belgium; Igor.EECKHAUT@umons.ac.be (I.E.); lucas.terrana@gmail.com (L.T.)
[3] Nekton Foundation, Oxford OX5 1PF, UK
[4] Institute of Halieutics and Marine Sciences, University of Toliara, Toliara 601, Madagascar
[5] Marine Biology Laboratory, Free University of Brussels, 1050 Brussels, Belgium; mathilde.an.godefroid@ulb.ac.be (M.G.); phdubois@ulb.ac.be (P.D.)
[6] Chemical Oceanography Unit, FOCUS, University of Liege, 4000 Liege, Belgium; jonathan.richir@uliege.be
* Correspondence: gresserika@gmail.com or erika@nektonmission.org



**Abstract:** Here, we report a new broad approach to investigating the presence and density of Symbiodiniaceae cells in corals of the order Antipatharia subclass Hexacorallia, commonly known as black corals. Antipatharians are understudied ecosystem engineers of shallow (<30 m depth), mesophotic (30–150 m) and deep-sea (>200 m) reefs. They provide habitat to numerous organisms, enhancing and supporting coral reef biodiversity globally. Nonetheless, little biological and ecological information exists on antipatharians, including the extent to which global change disturbances are threatening their health. The previous assumption that they were exempted from threats related to the phenomenon known as bleaching was challenged by the recent findings of high densities of dinoflagellates within three antipatharian colonies. Further studies were thus necessary to investigate the broader uniformity of these findings. Here we report results of an integrated methodology combining microscopy and molecular techniques to investigate the presence and estimate the density of Symbiodiniaceae cells within two antipatharians species—*Cupressopathes abies* and *Stichopathes maldivensis*—from both shallow and mesophotic reefs of SW Madagascar. We found that Symbiodiniaceae-like cells were present within samples of both species collected from both shallow and mesophotic reefs, although the overall cell density was very low (0–4 cell mm$^{-3}$). These findings suggest that presence or high abundance of Symbiodiniaceae is not characteristic of all antipatharians, which is relevant considering the bleaching phenomenon affecting other corals. However, the possibility of higher densities of dinoflagellates in other antipatharians or in colonies exposed to higher light irradiance deserves further investigation.

**Keywords:** coral reefs; symbiotic algae; dinoflagellates; Madagascar; symbiosis



**Citation:** Gress, E.; Eeckhaut, I.; Godefroid, M.; Dubois, P.; Richir, J.; Terrana, L. Investigation into the Presence of *Symbiodiniaceae* in Antipatharians (*Black Corals*). *Oceans* **2021**, *2*, 772–784. https://doi.org/10.3390/oceans2040044

Academic Editors: Rupert Ormond, Peter Schupp, Ronald Osinga and Michael W. Lomas

Received: 29 October 2020
Accepted: 18 November 2021
Published: 25 November 2021




## 1. Introduction

Coral reefs are both among the most biodiverse of marine habitats and among the most productive ecosystems on Earth [1]. However, most of our information 'about them is based on our knowledge of shallow reefs (<30 m depth). Light-dependent reefs can thrive in deeper waters though. Mesophotic coral ecosystems (MCEs) are light-dependent coral reefs found typically from 30 to 150 m in tropical and subtropical regions [2]. Hereinafter, MCEs are also referred to as 'mesophotic reefs' for the ease of the reader, a phrase considered analogous with the term 'shallow reefs'. Mesophotic reefs have been estimated to represent about 80% of potential coral reef habitat by area worldwide; however, we know little about them compared to shallow reefs [3]. Scleractinians (hard corals) are present in MCEs, but to a much lesser extent than in shallow reefs [3–5]. Instead, antipatharians, octocorals, sponges and macroalgae provide most of the available habitat structure at these depths [3–5].

The order Antipatharia (subclass Hexacorallia) consists of seven families and around 270 species [6,7]. These corals occur in most oceans at depths ranging from 2 to 8900 m, generally favouring strong currents and low-light environments [8,9]. Antipatharians do not produce a calcium carbonate skeleton; instead, the thorny axial skeleton is composed of a proteinaceous complex called antipathin [10,11]. It has been suggested that antipatharians tend to increase in diversity and abundance with depth, reaching a peak at mesophotic depths (30–150 m) [12]. Nonetheless, dense aggregations of antipatharians have also been observed in shallow (<30 m depth) clear water environments [13,14], Gress pers. obs. Further, they are important habitat-providing corals with which an array of marine fauna associates [9,13,15–18]. For instance, in the Philippines, the average density of invertebrate macrofauna associated with antipatharians ranged from 82 to 8313 individuals $m^{-2}$ [14]. Thus, antipatharians can be considered ecosystem engineers supporting and enhancing marine biodiversity at shallow, mesophotic and deep-sea depths.

However, limited studies have evaluated their current condition under increasing threats, such as those associated with global warming. In the closely related Hexacorallian order of scleractinia, global warming has led to extensive mortality due to temperature-related coral bleaching. Bleaching is the consequence of a disturbance of the symbiotic relationship between dinoflagellates and scleractinians on both shallow and mesophotic reefs as a result of climate change—a causal relationship that is well established [19–22]. Nevertheless, our understanding of the physiological mechanisms underlying the endosymbiotic association between dinoflagellates and their cnidarian host has been frequently revised, and this relationship is still not fully understood. It now appears that four of the seven genera of Symbiodiniaceae are found in symbioses with scleractinian corals (*Symbiodinium*, *Breviolum*, *Cladocopium* and *Durusdinium*) [23,24]. The translocation of photosynthetically fixed carbon from the symbiont to the host is considered the best-known aspect of the coral–algae symbiosis. However, the amount of photosynthetic carbon translocated to the host and the identity of the compounds are not fully known [25].

In the coral–algae symbiotic association, Symbiodiniaceae harvest sunlight for photosynthesis and dissipate excess energy so as to prevent light-induced oxidative stress [26–28]. Under ambient conditions (i.e., not heat and light-stressed), Symbiodiniaceae absorbs light that can be (1) used to drive photochemistry, (2) re-emitted as fluorescence, (3) dissipated as heat or (4) decayed via the chlorophyll triplet state [26–29]. Experiments have shown that Symbiodiniaceae in scleractinians, under typical irradiances at shallow coral reefs (640 $\mu$mol photons $m^{-2}$ $s^{-1}$), dissipate 96% of the energy and use only 4% of the absorbed light energy for photosynthesis [26]. However, over prolonged periods of water temperature alterations, the invertebrate host needs to lower the number of its symbionts because the oxidative stress, which results from the production and accumulation of reactive oxygen species (ROS), can damage lipids, proteins and DNA [27,30]. When corals reduce the number of their symbionts, the main source of ROS production is removed, although the coral host itself may also produce ROS as a result of light and temperature [27]. Oxidative stress on corals results in a lack or low number of dinoflagellates and their photosynthetic pigments, an effect known as 'coral bleaching'. Three potential mechanisms have been suggested for the regulation of symbiont numbers under such stressful conditions: (i) expulsion of excess symbionts; (ii) degradation of symbionts by host cells; and (iii) inhibition of symbiont cell growth and division controlled by the pH of the host cell [25,31].

The earliest suggestion of dinoflagellates being present in antipatharian tissues comes from the *Report on the Antipatharia collected by H.M.S. Challenger* by Brook [32]. A few years later, in a report on *The Antipatharia of the Siboga Expedition*, van Pesch [33] documented six species containing what he referred to as 'symbiotic Algae' ranging from 7–10 $\mu$m in diameter present in the gastrodermis. Significantly, he reported observing these cells in only six out of the thirty species he examined [33]). With no more empirical studies or reports for many decades and given antipatharians ability to thrive at abyssal depths and in low-light environments, they were assumed to lack Symbiodiniaceae, which is commonly referred to as 'being azooxanthellate' [8,34]. Moreover, a few later reports using molecular techniques did not find dinoflagellates in the antipatharian species examined. For instance, after intense morphological studies, dinoflagellates were reported absent in *Antipathes grandis* VERRILL, 1928, from Hawaii [35]. Likewise, dinoflagellate-specific primers and spectrophotometric methods that detect dinoflagellate chlorophyll absorbance patterns did not reveal any found microalgae in the species *Stichopathes luetkeni* (BROOK, 1889) (formerly called *Cirrhipathes lutkeni*) [36].

In contrast, in accordance with the early historical suggestions, two more recent studies have confirmed the presence of Symbiodiniaceae in various antipatharian species. A histological analysis of 14 antipatharian species collected from a depth between 10 and 396 m from Hawaii and Johnston Atoll revealed low densities (0–92 cells mm$^{-3}$) of Symbiodiniaceae cells inside antipatharian gastrodermal tissues, suggesting that the dinoflagellates are endosymbiotic [34]. Additionally, dinoflagellates sequences retrieved from the antipatharians confirmed the presence of Symbiodiniaceae in the genera *Cladocopium*, *Gerakladium* and *Durusdinium*. However, it was concluded that the endosymbiotic dinoflagellates had no significant role in the 'nutrition' of the species examined and suggested more research to determine whether the association might be parasitic [34]. The conclusion was based on the low density of microalgae cells within the antipatharians and their presence in colonies at depths where light penetration does not enable photosynthesis. They did not find any pattern in the types of Symbiodiniaceae present in the different antipatharian species; therefore, they suggested that endosymbiont acquisition might occur opportunistically and not be host-specific. In another recent study conducted on a single species of the genus *Cirrhipathes* from Indonesia, two colonies were sampled at 38 m and one at 15 m and evidence of abundant (~$10^7$ cells cm$^{-2}$) Symbiodiniaceae cells in the gastrodermis of the corals was found [37]. Among these, the authors identified two genera—*Cladocopium* and *Gerakladium*—the latter commonly found in association with clinoid sponges. They concluded that a mutualistic endosymbiosis existed based on the presence of the dinoflagellates inside the antipatharian gastrodermis and a symbiosome surrounding the microalgal cell, combined with evidence of its division inside the host [37]. These findings led us to reconsider our view of the vulnerability of antipatharians to global change and prompted us to further investigate the occurrence of dinoflagellate symbionts in antipatharian species.

Most lineages in the subclass Hexacorallia are believed to have evolved photosymbioses independently [38], with the order Antipatharia being one of the exceptions until evidence of antipatharian species hosting dinoflagellates inside the coral gastrodermis was found [34,37]. However, these two studies presented two contrasting conclusions. In one case, it was concluded that the presence of the Symbiodiniaceae was opportunistic, and in the other, it was concluded that a mutualistic endosymbiosis existed. Those conclusions were based on the presence and abundance of the Symbiodiniaceae cells and their location in the host tissue, although from a very limited number of specimens. The present study was therefore undertaken with the objective of gaining further insight into the possibility of antipatharians hosting high abundances of dinoflagellates, expanding the geographic range of the species studied and the number of colonies examined. We used an integrated methodological approach, combining both microscopy and molecular techniques to investigate the presence, abundance, location and identity of Symbiodiniaceae in two antipatharian species—*Cupressopathes abies* (LINNAEUS, 1758) and *Stichopathes maldivensis* COOPER, 1903. Our samples represent two different morphologies and were collected from both shallow and mesophotic reefs in SW Madagascar.

## 2. Materials and Methods

### 2.1. Site Description

Sample collection took place near the Great Reef of Toliara (GRT), in SW Madagascar, at a barrier reef almost 20 km long and 2 km wide, bordered by freshwater rivers at its northern (Fiherenana River) and southern (Onilahy River) extremities (Figure 1a). Coral communities, including scleractinians and antipatharians, were first documented in shallow (<30 m depth) and mesophotic reef (at depths between 35 m and 55 m) areas in the 1970s [39]. More recent studies have documented severe reef degradation due to fisheries, pollution and heavy sedimentation derived from the adjacent rivers [40–42], as well as coral bleaching episodes [43]. Antipatharian samples were collected in November–December 2018 on shallow (20 m depth, 23°20.978′ S, 43°36.885′ E—Site 1) and mesophotic (40 m depth, 23°21.345′ S, 43°36.348′ E—Site 2) reefs (Figure 1a) during a non-bleaching episode. Figure 1b shows the annual median for 2018 of the diffuse attenuation coefficient for downwelling irradiance at 490 nm in m$^{-1}$ (K$_d$ 490) over the wider region as obtained from satellite data (MODIS-Aqua), a proxy for water turbidity [44].

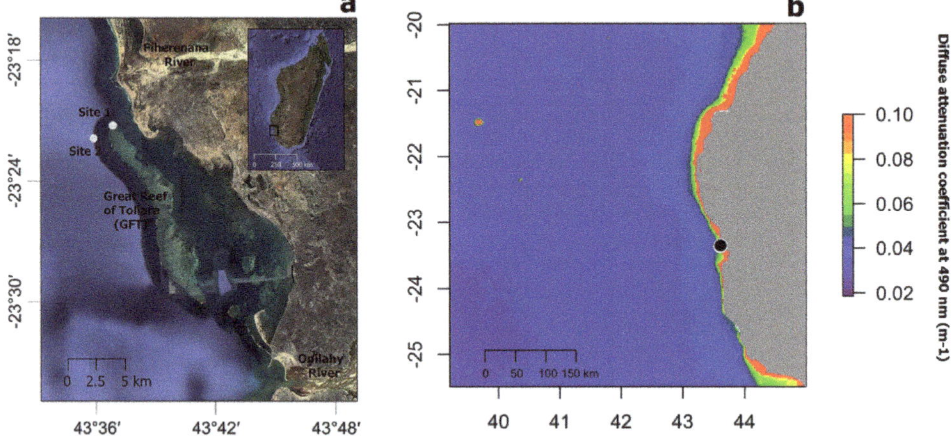


**Figure 1.** (**a**) Map showing the location of study sites 1 and 2 (white circles) near the Great Reef of Toliara (GRT) in SW Madagascar. (**b**) Annual median of the diffuse attenuation coefficient for downwelling irradiance at 490 nm in m$^{-1}$ (K$_d$ 490) for the year 2018 over the wider region (black circle indicated the location of the GRT). Satellite data derived from MODIS-Aqua accessed through OceanColor (https://oceancolor.gsfc.nasa.gov, accessed on 2 April 2019). Increasing coefficient values indicate higher water turbidity.

### 2.2. Sample Collection

Two antipatharian species, which were the most abundant species at the sites, the bottle-brush-like *Cupressopathes abies* (Figure 2a) and the whip-like *Stichopathes maldivensis* (Figure 2b), were sampled. Six colonies of each species were sampled at a 20 m depth and five colonies of each at a 40 m depth in November–December 2018 (permit no. 089/19/MESipReS). Colony height ranged between 25 and 35 cm for *C. abies* and 200 and 250 cm in height for *S. maldivensis*. It has been shown that Symbiodiniaceae abundance and clade can vary at the intra-colony level [24]; therefore, three samples 3–4 cm long were taken from each colony at the top, the middle and the base, making a total of 66 samples from 22 separate colonies. Half of every sample was preserved in 100% ethanol for subsequent molecular analysis. The other half was fixed in 3% glutaraldehyde buffered with 0.1 M sodium cacodylate before rinsing and storing them in 70% ethanol for later morphological study. Due to the minute size of their polyps (~0.7 mm, Figure 2a), prior to fixation, samples of *C. abies* were placed in a 5 g/L magnesium chloride solution buffered in filtered sea water in order to relax the polyps and tentacles so as to facilitate further observation.

**Figure 2.** (**a**) The bottle-brush-like *Cupressopathes abies* and (**b**) the whip-like *Stichopathes maldivensis*, with an image showing a section of two individual polyps in the bottom right corner. Scale bars of polyp images = 1 mm.

*2.3. Isolation of Symbiodiniaceae Cells*

From each of the 66 samples preserved in 100% ethanol, 1 cm fragments were cut, amounting to an estimated 20 mg of tissue for both species. Each fragment included about two polyps in the case of *S. maldivensis or* about eight polyps in the case of *C. abies*. Following a method adapted from [45] to isolate Symbiodiniaceae cells from cnidarians, samples were placed in 1.5 mL tubes with 500 µL of a 2 M sodium hydroxide solution. These were incubated at 37 °C for 1 h and then vortexed at a medium speed (using an Analitik Jena Tmix homogenizer) at the same temperature for about 2 h until complete lysis of the antipatharian tissues had occurred. The tubes were then centrifuged at 8000 rpm for 3 min, and the supernatant was discarded, then the pellet was resuspended in ultra-pure (Milli-Q) water and vortexed. One drop of lugol was added to the solution to re-stain the cells that might have lost pigments after preservation. About 8 µL of the pellet in 7 µL was used per chamber. Counts of visible Symbiodiniaceae were undertaken using a glass haemocytometer and an Axioscope A1 (Zeiss) light microscope. The number of the microalgae cells in all nine squares (1 mm$^2$ each) of four counting chambers per sample was counted in order to estimate the density of cells per µL. To corroborate the efficiency of the procedure, the isolation protocol was also carried out on fresh and preserved (100% ethanol) samples of the scleractinian coral *Seriatopora hystrix* (Figure 3). To test for differences in microalgae density between antipatharian species, depths and colony region (top, middle or bottom part), a quasi-Poisson generalised linear model (GLM) was fitted using the 'stats' package in R [46]. The power to detect differences between depths was calculated using the function 'power.t.test' in the same 'stats' package.

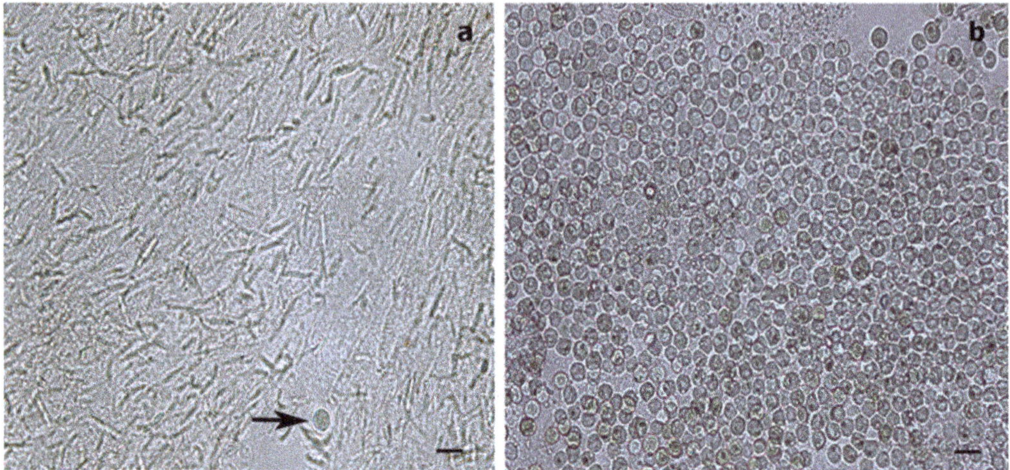

**Figure 3.** Light microscopy images of Symbiodiniaceae-like extracts obtained through the sodium hydroxide isolation method. (**a**) Extract from the antipatharian *Cupressopathes abies* showing only one Symbiodiniaceae-like cell (arrow) but abundant cnidocytes. (**b**) Extract from the scleractinian *Seriatopora hystrix* showing numerous Symbiodiniaceae-like cells, corroborating the efficiency of the isolation method. Scale bars = 10 μm.

### 2.4. Microscopy

(a)  Histology

Histological sections were taken to determine the presence and location of Symbiodiniaceae cells within the coral's tissue. About 1.5 cm of each sample was dehydrated using increasing concentrations of ethanol, then soaked in butanol at 60 °C for 24 h. These samples were then embedded in liquid paraffin. Serial sections of 7 μm were cut using a Microm HM 340E (Zeiss) microtome and stained with a Masson's trichrome. The sections were observed and photographed using an Axioscope A1 (Zeiss) light microscope and Axiocam 305 camera.

(b)  Transmission electron microscope (TEM)

For each species, sub-sample sections were made from the samples of the top, middle and bottom part of one colony from a 20 m depth and one colony from a 40 m depth. For *S. maldivensis*, due to the large size of the polyps (~6 mm, Figure 2b), sections of the tentacles and of the oral cone were analysed separately, making a total of 18 sub-samples in all. These sections, each about 1 cm long, were post-fixed for 1 h at room temperature with 1% osmium tetroxide in a 0.1 M sodium cacodylate and 2.3% sodium chloride buffer. They were rinsed several times in the same buffer and dehydrated in an increasing concentration series of ethanol. Samples were then placed in Spurr resin overnight before polymerisation at 70 °C for 24 h. Then ultrathin sections of 50–70 nm were cut on a Leica Ultracut (UCT) microtome equipped with a diamond knife and collected on formvar-coated copper grids. These were stained with uranyl acetate and lead citrate and observed with an LEO 906E (Zeiss) transmission electron microscope.

(c)  Scanning electron microscope (SEM)

For each species, the samples corresponding to the middle sections of one colony from a 20 m depth and one from a 40 m depth (each about 1.5 cm long) were studied under a scanning electron microscope (SEM). The four samples stored in 70% ethanol were then dehydrated for TEM analysis. Sub-samples were also made in order to be frozen with liquid nitrogen before being cut randomly. The eight sub-samples were dried in a critical-point dryer using $CO_2$ as the transition fluid (Agar Scientific Ltd.) before being mounted on aluminium stubs and coated with gold in a JFC-1100E (JEOL) sputter coater. These samples were observed and photographed with a JSM-7200F (JEOL) scanning electron microscope.

### 2.5. Molecular Analyses

Middle parts of the colonies of both species and from both depths (preserved in 100% ethanol) were cut into 8 mm fragments. The total genomic DNA of each sample was extracted using QIAGEN DNeasy Blood & Tissue Kit using the manufacturer's protocol. The concentration and quality of DNA were examined at 260 nm using a spectrophotometer (DeNovix). The internal transcribed spacer-2 (ITS2) region of the dinoflagellate ribosomal DNA (rDNA) was amplified using the primers ITS-DINO (5′ GTGAATTGCAGAACTCCGTG 3′) and ITS2-REV2 (5′ CCTCCGCTTACTTATATGCTT 3′) following conditions described in [47]. A second set of primers, SYM_VAR_5.8S2 (5′ GAATTGCAGAACTCCGTGAACC 3′) and SYM_VAR_REV (5′ CGGGTTCWCTTGTYTGACTTCATGC 3′), were used for PCR amplification of the ITS2 region, following the protocol described in [48]. This second set of primers was used in addition to the former because it has been found to perform better than other ITS2 primer sets tested on a range of Symbiodiniaceae ITS2 rDNA [48]. The presence of amplicons was checked in a 2% agarose gel in Tris-Borate-EDTA buffer. A DNA extract from a scleractinian coral of the genus *Acropora* from Toliara (Madagascar) was used as positive control.

### 3. Results

### 3.1. Dinoflagellate Cells Count

The isolation of Symbiodiniaceae-like cells with sodium hydroxide should have enabled efficient isolation of the microalgal cells if present within the coral tissues and also isolated cnidocytes (Figure 3). For both antipatharian species and both depths (all 66 samples), dinoflagellate cell density ranged between 0–4 cells mm$^{-3}$. Microalgae density was different between the two species ($t = 2.36$, $p = 0.029$; Table 1), with a mean density of $0.037 \pm 0.013$ cells mm$^{-3}$ (mean $\pm$ SE) for *C. abies* and $0.121 \pm 0.034$ cells mm$^{-3}$ for *S. maldivensis*. The mean density difference between the two species was 0.084 cells mm$^{-3}$ regardless of the depth. No significant difference was detected based on the region of the colony using the quasi-Poisson GLM; therefore, the factor 'region' was removed from the model (the average dinoflagellate cell density from the three regions of each colony was calculated to obtain a mean cell density per colony for further analyses). No significant difference was detected in Symbiodiniaceae-like density between depths ($t = -1.577$, $p = 0.131$; Table 1). However, the statistical power to detect a significant difference in cell density between depths for each species was limited. The results of a power analysis showed a low power of 0.26 (Type II error rate 74%) for *S. maldivensis* and 0.06 (Type II error rate 94%) for *C. abies* to detect differences between depths.

**Table 1.** Results of the quasi-Poisson GLM test for differences between species and depths. The intercept represents *C. abies* dinoflagellate cell density. Significant *p*-value ($p < 0.05$) are shown in bold.

| Factor | Estimate | Standard Error | *t*-Value | *p*-Value |
|---|---|---|---|---|
| Intercept | −2.27 | 0.74 | −3.08 | **0.006** |
| *S. maldivensis* | 1.16 | 0.49 | 2.36 | **0.029** |
| Depth | −0.04 | 0.02 | −1.58 | 0.131 |

### 3.2. Morphological Analyses

Histological sections could only be produced for the *S. maldivensis* samples. Symbiodiniaceae-like cells could be observed in both the polyps and the gastrovascular gastrodermis of the sections examined (Figure 4). From most paraffin blocks containing samples of *C. abies*, sections of the tentacles and oral cone were not obtained because there is only a thin layer of tissue surrounding the skeleton, but cells likely to be Symbiodiniaceae could not be detected in the few successful sections that were made. Ultrastructural

(TEM) observations were possible for all 18 samples examined from the two species, but no Symbiodiniaceae-like cells could be identified in any of them. Mucous cells, zymogen granules, and a large number of cnidocytes, including spirocysts and b-mastigophores were evident (Figure 5). On SEM examination, round cells of different sizes (3–10 µm) were observed, among which the larger round cells inside the gastrodermis of the polyp gastrovascular cavity (~8 µm; Figure 6a) could be dinoflagellates cells. More abundant were smaller round cells (3–4 µm), which are likely to be the vesicular mucous cells also observed on the ultrastructural analysis (Figure 6b).

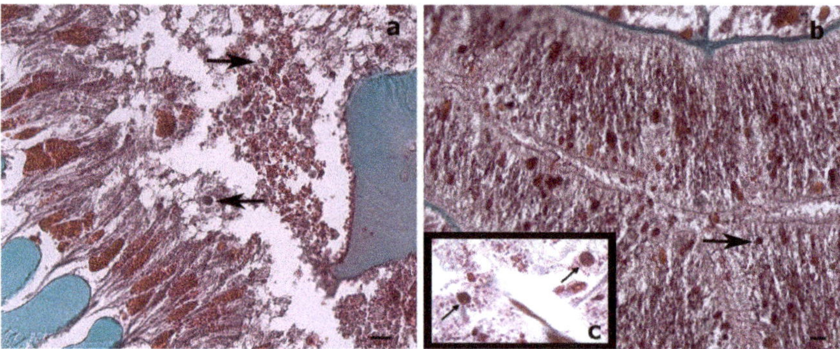

**Figure 4.** Histological cross-sections of a *Stichopathes maldivensis* polyp. Symbiodiniaceae-like cells (arrows) are observed in the gastrodermis of the gastrovascular cavity (**a**) and in the gastrodermis of a lateral tentacle (**b**). (**c**) Inset for comparison is an image from Wagner et al. ([34], Figure 2c) in which Symbiodiniaceae cells were identified through histological examination and molecular analysis. Scale bars = 20 µm.

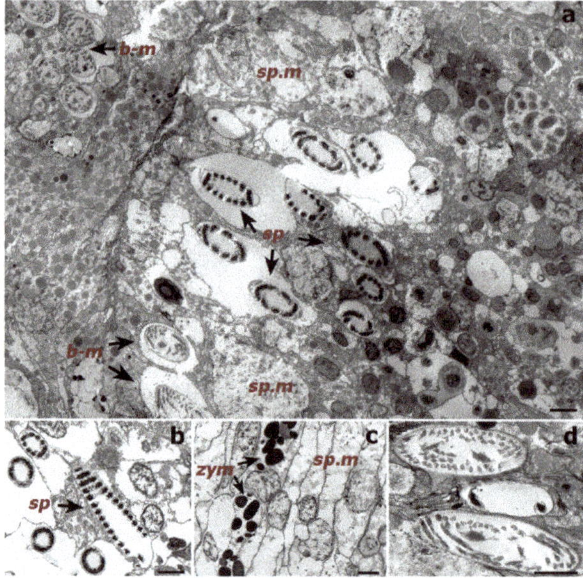

**Figure 5.** Transmission electron microscope (TEM) images of a cross-section of a *Cupressopathes abies* polyp tentacle. (**a**) Ultrastructural view of numerous cnidocytes, including: spirocysts (sp) and b-mastigophores (b-m), as well as spumous mucous cells (sp.m). (**b**) Longitudinal section of mature spirocysts (sp). (**c**) Close view of zymogen granules (zym) and spumous mucous cells (sp.m) inside the gastrodermis. (**d**) Cross-section of mature b-mastigophores. Scale bars = 2 µm.

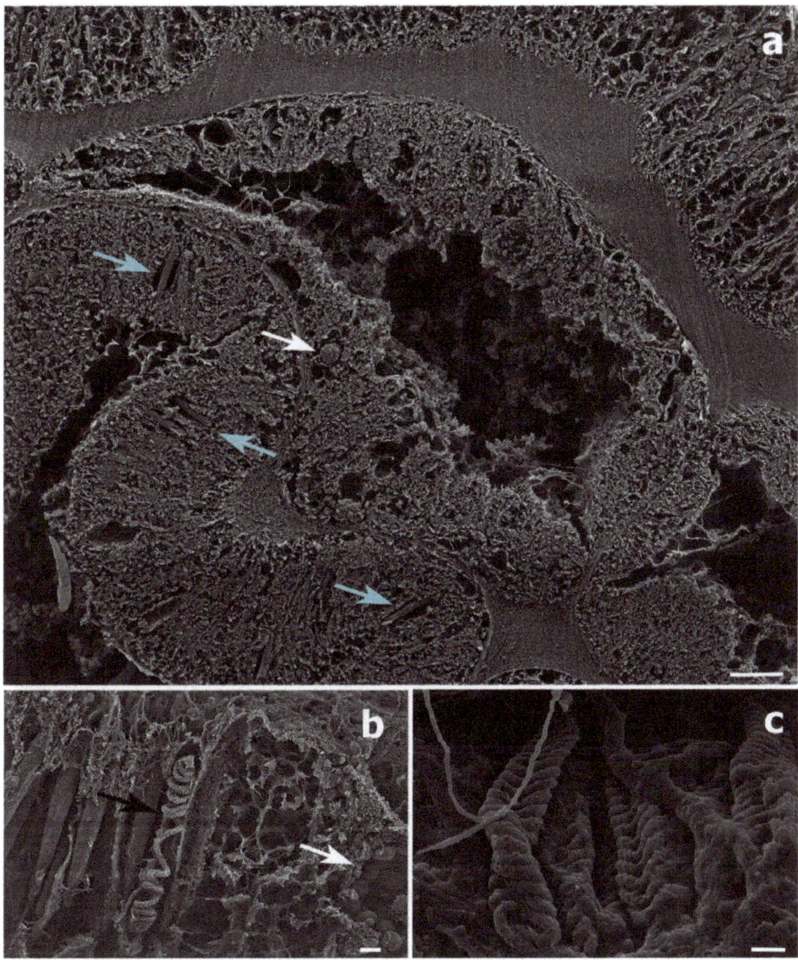

**Figure 6.** Scanning electron microscope (SEM) images of a *Stichopathes maldivensis polyp*. (**a**) A round potentially dinoflagellate-cell of about 8 µm inside the gastrodermis of the polyp gastrovascular cavity (white arrow). Numerous b-mastigophores can be observed in the ectoderm (blue arrows). (**b**) Smaller round cells (3–4 µm) that are likely to be mucous cells (white arrow) in the polyp ectoderm. The image also shows several b-mastigophores aligned among which one has a broken capsule which exposes the cnidocyte tubule (black arrow). (**c**) Closer view of spirocysts in a polyp tentacle ectoderm. Scale bars: a = 10 µm, b = 2 µm, c =1 µm.

### 3.3. Symbiodiniaceae Molecular Analysis

No ITS2 sequences of Symbiodiniaceae were successfully amplified from any of the antipatharian DNA extractions. However, the positive control (scleractinian coral DNA extraction) did produce a band on the agarose gel, implying that the method was being used correctly. This was the case for both sets of primers used.

## 4. Discussion

### 4.1. Overall Analysis Results

Despite photosymbiosis being evolutionarily conserved within most lineages in the subclass Hexacorallia [38], an endosymbiotic association between antipatharians and

Symbiodiniaceae has not been clearly established. High densities of dinoflagellates ($\sim 10^7$ cells cm$^{-2}$) were found in three whip-like colonies (*Cirrhipathes* sp.) [37]. Therefore, it was considered that a study involving a higher number of samples was required in order to assess the microalgal presence and estimate their densities in a greater number of samples from more species and to identify possible patterns regarding species, colony morphology and depth. The present study showed either none or only very low densities (0–4 cells mm$^{-3}$) of Symbiodiniaceae-like cells in both the whip-like *S. maldivensis* and the bushy *C. abies*, both on shallow and on mesophotic reefs. These new findings align with both previous historical observations and more subsequent studies that found few, if any, endosymbiotic algae in antipatharians [32–36]. These studies—ours included—evidence that high abundances of dinoflagellates within antipatharians is not a common finding. Considering the relevance of antipatharians for habitat provision on reefs and the increasing extent of mass bleaching events affecting scleractinian corals, the possibility of most antipatharians being exempt from this threat warrants further investigation.

While, in our study, the dinoflagellates' identity could not be determined by molecular analysis, the Symbiodiniaceae-like cells observed in the histological sections of the whip-like *S. maldivensis* closely resembled those described in the other two studies [34,37] (Figure 4). For future histological examination of antipatharians, where the tissue cannot be separated from the skeleton (such as *C. abies*), we suggest experimenting with softening the skeletons with a lytic polysaccharide monooxygenases (LPMOs) treatment [49]. We applied the same sodium hydroxide isolation protocol—a protocol found to be effective for scleractinians and other cnidarians [45]—to a scleractinian coral fragment, and it was clear that the cells obtained were Symbiodiniaceae (Figure 3). Nevertheless, it is probably safer to refer to the cells in our antipatharian samples only as being Symbiodiniaceae-like.

In the present study, Symbiodiniaceae-like cells were observed in the SEM and histological examination located inside the coral gastrodermis of *S. maldivensis* (Figures 5 and 6). This suggests an endosymbiotic association, which was also suggested from histological sections analysis and from ultrastructural analysis [34,37]. The fact that no Symbiodiniaceae ITS2 sequences were amplified from the DNA extractions in this study seems most likely due to the extremely low density of dinoflagellates. In the two studies where identification was possible, different genera (*Cladocopium*, *Gerakladium* and *Durusdinium*) were reported to be associated with different antipatharian species [34,37]. Such plasticity, even at the intra-colony level, has also been evidenced in more recent studies on scleractinian corals and is believed to be environmentally driven [24]. Moreover, there is evidence to suggest that the same Symbiodiniaceae species may be mutualistic in one host context but opportunistic in another [50]. Therefore, the identity of the dinoflagellates alone is insufficient to determine the type of symbiosis. Further studies will be necessary to determine the nature of the association between Symbiodiniaceae and antipatharians, particularly in cases where microalgae abundance is high, as in [37].

### 4.2. Dinoflagellate Density Difference between Species

Density estimates showed that colonies of the whip-like *S. maldivensis* had significantly more Symbiodiniaceae-like cells within their tissues compared to *C. abies* regardless of depth (0.084 mean cells mm$^{-3}$ density difference, Table 1). However, due to the very low densities of dinoflagellates observed in both species, it is difficult to determine the biological significance of this difference. To date, a single study examined multiple species (14) and different colony morphologies (both whip-like and branching) [34]. The branching species, *Antipathes griggi*, recorded the highest density (0–92 cells mm$^{-3}$), although microalgal cells were observed in only one of the eight colonies of this species examined. Variation in density was also observed in our study, which is the first to have examined several colonies of the same species from two contrasting depths. Of the eleven colonies of each species, only four colonies of *C. abies* and seven of *S. maldivensis* were observed to contain microalgae. On the other hand, no differences were found between the three different regions (top, middle and base) within any single colony. This suggests that the intra-specific variability recorded in [34], as well as ourselves, is unlikely to be caused by variations in the microalgal cell density between different parts of the colonies. From this limited information, it seems plausible that other factors, rather than species or morphology, might account for the greater numbers of microalgal cells found in some antipatharian colonies.

*4.3. Dinoflagellate Density Difference between Depths*

Contrary to expectation, no statistically significant difference in dinoflagellate density was found between shallow and mesophotic depths, although the mean dinoflagellates density at 20 m depth was greater than at 40 m depth in both species (the mean differences between depths being 0.200 and 0.016 cells mm$^{-3}$ for *S. maldivensis* and *C. abies*, respectively). Differences in microalgae density can be inferred from previous studies. For instance, among the 14 samples from different colonies of *A. griggi* from Hawaii for which the cell density was reported, the highest densities were from the two colonies sampled on shallow reefs [34]. The samples of *Cirrhipathes* collected in Bunaken, Indonesia, were from two different depths (15 and 38 m) [37]. While it was not reported whether there were any differences in Symbiodiniaceae density related to depth, the $K_d$ 490 (proxy of turbidity) values at 15 m and 38 m depth in Bunaken have been estimated to be very similar [50]. In Toliara, Madagascar, where the $K_d$ 490 values indicate high turbidity (Figure 1b) due to sedimentation derived from river runoff, light penetration at a 40 m depth is likely to be considerably reduced (as was noticeable during field work). However, it is not inevitable that higher densities of Symbiodiniaceae should be found in association with antipatharians in shallower water since it is believed that the uptake of dinoflagellates in the coral–algae symbiotic association can be controlled by the coral host [25,31]. More studies on antipatharian colonies exposed to higher radiations are therefore necessary to assess the significance of high densities of dinoflagellates in antipatharians (assuming they occur), and to understand the mechanism behind the association.

## 5. Conclusions

This study represents a broader integrative approach to investigating the presence and density of Symbiodiniaceae in antipatharians than has been used in previous studies. We combined microalgal cell extraction, histological examination and transmission and scanning microscopy with attempted sequencing of DNA samples. We found that Symbiodiniaceae-like cells were present within some of the antipatharian samples of *S. maldivensis* and *C. abies* from SW Madagascar, although the overall density of macroalgae cells in both antipatharian species from both shallow and mesophotic reefs was very low. This low density aligns with the majority of previous findings, indicating that high Symbiodiniaceae densities are not characteristic of antipatharians. These findings are significant in the context of extensive 'coral bleaching' events threatening the integrity of coral reefs, suggesting that most antipatharians are likely to be less prone to this phenomenon. Nonetheless, considering that high densities of dinoflagellates have been documented in three colonies of one antipatharian species, more studies are desirable to understand the mechanism and implications of the coral–algae relationship within this coral taxon. Studies on other effects that climate change might have on antipatharians are also desirable.

**Author Contributions:** Conceptualization, E.G. and I.E.; Data curation, E.G., M.G. and J.R.; Formal analysis, E.G.; Funding acquisition, I.E., P.D. and L.T.; Methodology, E.G., I.E. and L.T.; Writing—original draft, E.G.; Writing—review and editing, I.E. All authors have read and agreed to the published version of the manuscript.

**Funding:** This research was funded by the Fonds National de la Recherche Scientifique, Belgium (no. PDR T0083.18), under the 'Conservation Biology of Black Corals' research project co-directed by the University of Mons, the University of Liège, and the Free University of Brussels, in Belgium.

**Informed Consent Statement:** Not applicable.

**Data Availability Statement:** The dataset generated for this study are available on request to the corresponding author. No sequence data was generated on this study.

**Acknowledgments:** We thank: All the members at the Marine Organisms Biology and Biomimetics (BOMB) Laboratory in Belgium for their assistance during sample analyses. Nicolas Sturaro, University of Liege, for help with sample collection. Members at the University of Toliara, Madagascar, for support during field work. Liz Tynan, James Cook University, for editorial comments and Rupert Ormond, Heriot-Watt University, for editorial review. Figure 1b created on R (R Team 2019) with an adapted script provided by J.C. Fischer, University of Bayreuth, and A. Wiefels, University of Reunion Island. We are very grateful to Dennis Opresko, Smithsonian Institute, for continuous support.

**Conflicts of Interest:** We declare that the research was conducted in the absence of any commercial or financial relationships that could be construed as a potential conflict of interest.

## References


1. Rogers, C.S. The effect of shading on coral reef structure and function. *J. Exp. Mar. Biol. Ecol.* **1979**, *41*, 269–288. [CrossRef]
2. Hinderstein, L.M.; Marr, J.C.A.; Martinez, F.A.; Dowgiallo, M.J.; Puglise, K.A.; Pyle, R.L.; Zawada, D.G.; Appeldoorn, R. Theme section on "Mesophotic Coral Ecosystems: Characterization, Ecology, and Management". *Coral Reefs* **2010**, *29*, 247–251. [CrossRef]
3. Pyle, R.L.; Copus, J.M. Mesophotic Coral Ecosystems: Introduction and Overview. *Coral Reefs World* **2019**, *39*, 1469–1482.
4. Baker, E.K.; Puglise, K.A.; Harris, P.T. *Mesophotic Coral Ecosystems—A Lifeboat for Coral Reefs?* The United Nations Environment Programme and GRID-Arendal: Arendal, Norway, 2016.
5. Gress, E.; Arroyo-Gerez, M.J.; Wright, G.; Andradi-Brown, D.A. Assessing mesophotic coral ecosystems inside and outside a Caribbean marine protected area. *R. Soc. Open Sci.* **2018**, *5*, 180835. [CrossRef]
6. Brugler, M.R.; Opresko, D.M.; France, S.C. The evolutionary history of the order Antipatharia (*Cnidaria: Anthozoa: Hexacorallia*) as inferred from mitochondrial and nuclear DNA: Implications for black coral taxonomy and systematics. *Zool. J. Linn. Soc.* **2013**, *169*, 312–361. [CrossRef]
7. Opresko, D.M. New species of black corals (*Cnidaria: Anthozoa: Antipatharia*) from the New Zealand region, part 2. *N. Z. J. Zool.* **2019**, *47*, 149–186. [CrossRef]
8. Grigg, R.W. Ecological studies of Black Coral in Hawaii. *Pac. Sci.* **1965**, *19*, 244–260.
9. Wagner, D.; Luck, D.G.; Toonen, R.J. The Biology and Ecology of Black Corals (*Cnidaria: Anthozoa: Hexacorallia: Antipatharia*). *Adv. Mar. Biol.* **2012**, *63*, 67–132. [CrossRef]
10. Goldberg, W.M. Chemical changes accompanying maturation of the connective tissue skeletons of gorgonian and antipatharian corals. *Mar. Biol.* **1978**, *49*, 203–210. [CrossRef]
11. Goldberg, W.M.; Hopkins, T.L.; Holl, S.M.; Schaefer, J.; Kramer, K.J.; Morgan, T.D.; Kim, K. Chemical composition of the sclerotized black coral skeleton (*Coelenterata: Antipatharia*): A comparison of two species. *Comp. Biochem. Physiol. Part B Biochem.* **1994**, *107*, 633–643. [CrossRef]
12. Bo, M.; Montgomery, A.D.; Opresko, D.M.; Wagner, D.; Bavestrello, G. Antipatharians of the Mesophotic Zone: Four Case Studies. In *Coral Reefs of the World*; Springer: Singapore, 2019; pp. 683–708.
13. Tazioli, S.; Bo, M.; Boyer, M.; Boyer, M.; Rotinsulu, H.; Bavestrello, G. Ecological observations of some common antipatharian corals in the marine park of Bunaken (North Sulawesi, Indonesia). *Zool. Stud.* **2007**, *46*, 227–241.
14. Suarez, H.N.; Dy, D.T.; Violanda, R.R. Density of associated macrofauna of black corals (*Anthozoa: Antipatharia*) in Jagna, Bohol, central Philippines. *Philipp. J. Sci.* **2015**, *144*, 107–115.
15. Parrish, F.A.; Abernathy, K.; Marshall, G.J.; Buhleier, B.M. Hawaiian monk seals (*Monachus schauinslandi*) foraging in deep-water coral beds. *Mar. Mammal. Sci.* **2002**, *18*, 244–258. [CrossRef]
16. Boland, R.C.; Parrish, F.A. Description of Fish Assemblages in the Black Coral Beds off Lahaina, Maui, Hawai'i. *Pac. Sci* **2005**, *59*, 411–420. [CrossRef]
17. Bruckner, A.W. Advances in Management of Precious Corals to Address Unsustainable and Destructive Harvest Techniques. In *The Cnidaria, Past, Present and Future*, 1st ed.; Goffredo, S., Dubinsky, Z., Eds.; Springer International Publishing: Cham, Switzerland, 2016; pp. 747–786.
18. Terrana, L.; Lepoint, G.; Eeckhaut, I. Assessing trophic relationships between shallow-water black corals (*Antipatharia*) and their symbionts using stable isotopes. *Belg. J. Zool.* **2019**, *149*, 107–121. [CrossRef]
19. Baker, A.C.; Glynn, P.W.; Riegl, B. Climate change and coral reef bleaching: An ecological assessment of long-term impacts, recovery trends and future outlook. *Estuar. Coast Shelf. Sci.* **2008**, *80*, 435–471. [CrossRef]
20. Hughes, T.P.; Kerry, J.T.; Simpson, T. Large-scale bleaching of corals on the Great Barrier Reef. *Ecology* **2018**, *99*, 501. [CrossRef]
21. Bongaerts, P.; Smith, T.B. *Beyond the "Deep Reef Refuge" Hypothesis: A Conceptual Framework to Characterize Persistence at Depth*; Springer: Cham, Switzerland, 2019; pp. 881–895.
22. Fisher, R.; Bessell-Browne, P.; Jones, R. Synergistic and antagonistic impacts of suspended sediments and thermal stress on corals. *Nat. Commun.* **2019**, *10*, 2346. [CrossRef]
23. Lajeunesse, T.C.; Parkinson, J.; Gabrielson, P.W.; Jeong, H.J.; Reimer, J.D.; Voolstra, C.R.; Santos, S.R. Systematic Revision of Symbiodiniaceae Highlights the Antiquity and Diversity of Coral Endosymbionts. *Curr. Biol.* **2018**, *28*, 2570–2580.e6. [CrossRef]

24. Meistertzheim, A.-L.; Pochon, X.; Wood, S.A.; Ghiglione, J.-F.; Hédouin, L. Development of a quantitative PCR–high-resolution melting assay for absolute measurement of coral-Symbiodiniaceae associations and its application to investigating variability at three spatial scales. *Mar. Biol.* **2019**, *166*, 13. [CrossRef]
25. Davy, S.K.; Allemand, D.; Weis, V. Cell Biology of Cnidarian-Dinoflagellate Symbiosis. *Microbiol. Mol. Biol Rev.* **2012**, *76*, 229–261. [CrossRef]
26. Brodersen, K.E.; Lichtenberg, M.; Ralph, P.; Kühl, M.; Wangpraseurt, D. Radiative energy budget reveals high photosynthetic efficiency in symbiont-bearing corals. *J. R. Soc. Interface* **2014**, *11*, 20130997. [CrossRef]
27. Roth, M.S. The engine of the reef: Photobiology of the coral-algal symbiosis. *Front. Microbiol.* **2014**, *5*, 422. [CrossRef]
28. Roth, M.; Padilla-Gamiño, J.; Pochon, X.; Bidigare, R.; Gates, R.; Smith, C.; Spalding, H. Fluorescent proteins in dominant mesophotic reef-building corals. *Mar. Ecol. Prog. Ser.* **2015**, *521*, 63–79. [CrossRef]
29. Weis, V.M. Cellular mechanisms of Cnidarian bleaching: Stress causes the collapse of symbiosis. *J. Exp. Biol.* **2008**, *211*, 3059–3066. [CrossRef]
30. Lesser, M.P. Oxidative stress in marine environments: Biochemistry and physiological ecology. *Annu. Rev. Physiol.* **2006**, *68*, 253–278. [CrossRef]
31. Barott, K.L.; Venn, A.A.; Perez, S.O.; Tambutté, S.; Tresguerres, M. Coral host cells acidify symbiotic algal microenvironment to promote photosynthesis. *Proc. Natl. Acad. Sci. USA* **2015**, *112*, 607–612. [CrossRef]
32. Brook, G. Report on the Antipatharia collected by HMS Challenger during the years 1873–1876. Report on the Scientific Results of the Voyage of HMS Challenger During the Years 1873–76. *Zoology* **1889**, *32*, 1–222. [CrossRef]
33. van Pesch, A.J. The Antipatharia of the Siboga Expedition. *Siboga Exped. Monogr.* **1914**, *17*, 1–258.
34. Wagner, D.; Pochon, X.; Irwin, L.; Toonen, R.; Gates, R.D. Azooxanthellate? Most Hawaiian black corals contain *Symbiodinium*. *Proc. R. Soc. B Biol. Sci.* **2010**, *278*, 1323–1328. [CrossRef]
35. Grigg, R.W. A Contribution to the Biology and Ecology of the Black Coral, *Antipathes grandis* in Hawai'i. MS Thesis in Zoology, 1964, p. 74. Hawai'i., Honolulu. Available online: http://cn.deziderkostrec.xyz/read/?id=Qv7PHAAACAAJ&format=pdf&server=1 (accessed on 2 April 2019).
36. Santiago-Vázquez, L.Z.; Brück, T.B.; Brück, W.M.; Duque-Alarcón, A.P.; McCarthy, P.J.; Kerr, R.G.; Br, T.B. The diversity of the bacterial communities associated with the azooxanthellate hexacoral Cirrhipathes lutkeni. *ISME J.* **2007**, *1*, 654–659. [CrossRef]
37. Bo, M.; Baker, A.; Gaino, E.; Wirshing, H.; Scoccia, F.; Bavestrello, G. First description of algal mutualistic endosymbiosis in a black coral (*Anthozoa: Antipatharia*). *Mar. Ecol. Prog. Ser.* **2011**, *435*, 1–11. [CrossRef]
38. McFadden, C.S.; Quattrini, A.M.; Brugler, M.R.; Cowman, P.F.; Dueñas, L.F.; Kitahara, M.V.; A. Paz-García, D.; Reimer, J.D.; Pichon, M. Recherches sur les peuplements à dominance d'anthozoaires dans les récifs coralliens de Tuléar (Madagascar). *Atoll Res. Bull.* **1978**, *222*, 1–490.
39. Harris, A.; Manahira, G.; Sheppard, A.; Gouch, C.; Sheppard, C. Demise of Madagascar's once great barrier reef: Changes in coral reef conditions over 40 years. *Atoll Res. Bull.* **2010**, *574*, 1–16. [CrossRef]
40. Todinanahary, G.; Terrana, L.; Lavitra, T. First records of illegal harvesting and trading of black corals (*Antipatharia*) in Madagascar. *Madag. Conserv. Dev.* **2016**, *11*, 1–6. [CrossRef]
41. Todinanahary, G.G.; Refoty, M.E.; Terrana, L.; Lavitra, T.; Eeckhaut, I. Previously unlisted scleractinian species recorded from the Great Reef of Toliara, southwest Madagascar. *West. Indian Ocean J. Mar. Sci.* **2018**, *17*, 67. [CrossRef]
42. Gudka, M.; Obura, D.; Mwaura, J.; Porter, S.; Yahya, S.; Mabwa, R. *Impact of the 3rd Global Coral Bleaching Event on the Western Indian Ocean in 2016*; Global Coral Reef Monitoring Network (GCRMN)/Indian Ocean Commission: Port Louis, Mauritius, 2016; p. 67.
43. Zhang, T.; Fell, F. An empirical algorithm for determining the diffuse attenuation coefficient $K_d$ in clear and turbid waters from spectral remote sensing reflectance. *Limnol. Oceanogr. Methods* **2007**, *5*, 457–462. [CrossRef]
44. Zamoum, T.; Furla, P. *Symbiodinium* isolation by NaOH treatment. *J. Exp. Biol.* **2012**, *215*, 3875–3880. [CrossRef]
45. R Core Team. *R: A Language and Environment for Statistical Computing*; R Foundation for Statistical Computing: Vienna, Austria, 2019; Available online: https://www.R-project.org/ (accessed on 2 April 2019).
46. Stat, M.; Pochon, X.; Cowie, R.; Gates, R. Specificity in communities of *Symbiodinium* in corals from Johnston Atoll. *Mar. Ecol. Prog. Ser.* **2009**, *386*, 83–96. [CrossRef]
47. Hume, B.C.; Ziegler, M.; Poulain, J.; Pochon, X.; Romac, S.; Boissin, E.; De Vargas, C.; Planes, S.; Wincker, P.; Voolstra, C.R. An improved primer set and amplification protocol with increased specificity and sensitivity targeting the *Symbiodinium* ITS2 region. *PeerJ* **2018**, *6*, e4816. [CrossRef]
48. Mutahir, Z.; Mekasha, S.; Loose, J.S.M.; Abbas, F.; Vaaje-Kolstad, G.; Eijsink, V.G.H.; Forsberg, Z. Characterization and synergistic action of a tetra-modular lytic polysaccharide monooxygenase from *Bacillus cereus*. *FEBS Lett.* **2018**, *592*, 2562–2571. [CrossRef]
49. Pettay, D.T.; Wham, D.C.; Smith, R.T.; Iglesias-Prieto, R.; LaJeunesse, T.C. Microbial invasion of the Caribbean by an Indo-Pacific coral zooxanthella. *Proc. Natl. Acad. Sci. USA* **2015**, *112*, 7513–7518. [CrossRef] [PubMed]
50. Holden, H. Characterisation of Optical Water Quality in Bunaken National Marine Park, Indonesia. *Singap. J. Trop. Geogr.* **2002**, *23*, 23–36. [CrossRef]

*Article*

# After the Fall: The Demographic Destiny of a Gorgonian Population Stricken by Catastrophic Mortality

Simona Ruffaldi Santori [1], Maria Carla Benedetti [1,*], Silvia Cocito [2], Andrea Peirano [2], Roberta Cupido [2], Fabrizio Erra [1] and Giovanni Santangelo [1]

[1] Demography and Conservation of Long-Lived Species Lab., Department of Biology, University of Pisa, via Volta 4, 56126 Pisa, Italy; simonaruffaldi@gmail.com (S.R.S.); fabrizio.erra@unipi.it (F.E.); giovanni.santangelo@unipi.it (G.S.)

[2] ENEA Marine Environment Research Centre, Località Pozzuolo di Lerici, 19032 Lerici (La Spezia), Italy; silvia.cocito@enea.it (S.C.); andrea.peirano@enea.it (A.P.); robi.cupido@gmail.com (R.C.)

* Correspondence: carlottabenedetti88@hotmail.it

**Abstract:** In recent years, the frequency of mass mortality events in marine ecosystems has increased, and several populations of benthic organism have been affected, reducing their density and changing their size and age structure. Few details are known about the dynamics of these populations over long time intervals. In late summer of both 1999 and 2003 two drastic mass mortality events, co-occurring with anomalous temperature increases, affected the northwestern Mediterranean rocky coastal communities. Due to these events the *Paramuricea clavata* population living at the western edge of La Spezia Gulf (Italy) was stricken, and 78% of the colonies died. This population was monitored from 1998 (pre-mortality) until 2013. This paper deals with the photographic sampling of permanent plots carried out in 2013. The findings were compared with those from the previous sampling series. This long-term, non-destructive sampling highlights the demographic trajectory of the octocoral population there after two anomalous mortality events, indicating that some new drop-point between local extinction and complete recovery may be have been reached. Long-term monitoring (including pre-mortality data) could allow evaluating the effects of global climate change on the conservation of impacted populations.

**Keywords:** long-term mortality series; population dynamics; mass mortality; octocorals; habitat forming species; extinction; survival; equilibrium points

**Citation:** Ruffaldi Santori, S.; Benedetti, M.C.; Cocito, S.; Peirano, A.; Cupido, R.; Erra, F.; Santangelo, G. After the Fall: The Demographic Destiny of a Gorgonian Population Stricken by Catastrophic Mortality. *Oceans* **2021**, *2*, 337–350. https://doi.org/10.3390/oceans2020020

Academic Editor: Rupert Ormond

Received: 11 January 2021
Accepted: 9 April 2021
Published: 19 April 2021

**Publisher's Note:** MDPI stays neutral with regard to jurisdictional claims in published maps and institutional affiliations.



## 1. Introduction

Over the last few years, the frequency of mass mortalities in marine ecosystems linked to Global Climate Change (GCC) has increased greatly, and several authors have described the dramatic effects of such anomalous events on benthic communities and populations [1–4] among others. However, few research studies address the delayed effects on, and the long-term growth trends of the affected populations, whose complex dynamics cannot be adequately expressed through the narrow terms of "survival" or "extinction".

In marine ecosystems, temperature is one of the main factors controlling species distribution, and unusual deviations from typical seasonal patterns may represent the greater impacts of global change on such systems [5]. The effects of abnormal warming of the water column are clearly and easily observable not only at shallow depths as environmental degradation and the direct responses of the organisms (e.g., lower population growth rates), but also in the deep sea, where modifications of the water's physicochemical characteristics are impacting biodiversity [6].

Habitat-forming species such as sea grasses, kelps, corals, and oysters are among the main organisms profoundly influenced by thermal stressors [7], although neither all species nor all communities are equally affected or respond in the same way. For example, among tropical hexacorals, differential susceptibility and resistance to bleaching and mass mortalities have been observed due to species peculiarity, the environmental variability between reefs, and even genetic differences between colonies of the same species on the same reef [8–10].

In recent decades, several worldwide populations of octocorals have been affected by stressors linked to GCC, making it essential that a demographic study of these populations is carried out in order to formulate reliable survivorship predictions [11–19]. It has been observed that the frequency of negative events may be a main factor determining the survival of such populations [17]. Even if they do endure, these populations could dramatically change their demographic structure and, depending upon their peculiar life-history traits, respond to anomalous mortality events in unexpected ways [13,20–23]. Octocorals are among the most common suspension feeders dwelling within the so-called animal marine forest *sensu* Rossi [24], where they often act as *habitat engineers*. Thus, the dynamics of their populations may have profound effects on the whole species assemblage. In the Mediterranean Sea, some populations of the temperate red gorgonian *Paramuricea clavata* (Risso 1826) have been affected by mass mortality events following unusual thermal stress [20,25–27] among others. Widely distributed along boulders and cliffs between 12 and 120 m depth in highly hydrodynamic areas [28,29], *P. clavata* is one of the main components of the coralligenous species assemblages [30] and determines the composition of the entire epibenthic community [31]. *P. clavata* is a large sea fan gorgonian (up to 1 m tall) that can form dense patches (up to 59 colonies per $m^2$, [32]) and is characterized by high reproductive output and long life cycle [21]. This species has recently been classified as a "bioindicator" [33] and "vulnerable" by the IUCN (International Union for Conservation of Nature). A recent genetic study has highlighted the strategic role of the *P. clavata* populations in the eastern Ligurian Sea (northwestern Mediterranean, Italy) for the persistence of the species at a regional level [34].

At summer's end of both 1999 and 2003, the benthic suspension feeders living along the coastal shores of the Ligurian Sea were affected by two mass mortality events co-occurring with exceptional warming of the water column [35,36]. Exposure to abnormal temperature increases was likely the cause of an ensuing outbreak of opportunistic pathogens such as bacteria, protozoans, and fungi on the *P. clavata* colonies, leading to their total or partial death [35,37]. The local population of this gorgonian living at the western edge of the Spezia Gulf (Figure 1) was stricken, and 78% of the colonies died or were heavily damaged [32]. This population was monitored from 1998 (pre-mortality) to 2010 [13,21]. In the years following the catastrophic events both adult and recruit density fell drastically, and the population size/age structure was altered due to the mortality disproportionally affecting the larger colonies, thereby shifting it towards the smaller ones [32]. Since 2007, a significant increase in adult and recruit density has occurred [13,20].

This paper analyzes the population size/age structure, the adult and annual recruit densities, and the percentage cover of *P. clavata* and the main organisms living in the area in 2013. The results have been compared with the previous findings to identify long-term variations affecting both the *P. clavata* population and the whole community after the catastrophic mortality events.

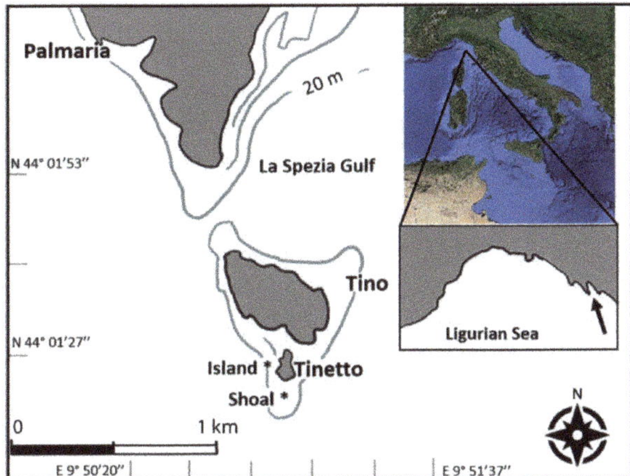



**Figure 1.** Sampling area, Tinetto islet: stars represent the sampling sites (Eastern Liguria Sea, NW Mediterranean, Italy).

## 2. Materials and Methods

### 2.1. Study Area and Sampling Design

The study area is located at Tinetto Islet (Figure 1), part of a small archipelago of the Ligurian Sea at the western edge of the La Spezia Gulf (NW Mediterranean, Italy), an area included in the World Heritage list and under the protection of the United Nations Educational, Scientific and Cultural Organization (UNESCO). Oriented from northwest to southeast, the Gulf is about 5 km wide and 10 km long, and its circulation is affected by several factors: the Ligurian current, seasonal weather changes, and the strength of the so-called 'Sirocco' warm southerly wind. The area is characterized by high turbidity due to the Magra river plume influencing the water and its hydrodynamics, especially during rainy seasons [21,38]. The northwest and western sides of Tinetto Islet are composed of vertical calcareous walls while the southern and the eastern sides are characterized by medium slopes. A more detailed description of the area has been previously published [13,21].

The population of *Paramuricea clavata* living on a sub-vertical rocky cliff between 17 and 25 m depth was monitored between 1998 (pre-mortality) and 2010 by identification and measurement during SCUBA diving [13,21]. This paper deals with the results of the photographic sampling carried out via a housed X 100 G10 Canon (Canon Inc., Tokyo, Japan) digital camera at a 1:1 magnification on the same population in July 2013.

In 2013, 24 of the permanent polyvinyl chloride (PVC) quadrates (1 × 1 m, thus 1 square meter each) previously fixed at 2 sites a few hundred meters from one another were sampled for analysis. Twelve were on the northwestern side (Tinetto Island) of Tinetto Islet (44°01′26″ N, 09°51′03″ E), and 12 at the Tinetto Shoal, on the southern side of the islet (44°01′23″ N, 09°51′05″ E; unfortunately, one of the quadrates fixed at this site was lost). At each site, the squares were randomly arranged at a few meters' distance. Of the 23 quadrates sampled for this study, 12 corresponded to the same ones used for the previous monitoring study. Four photos of a 0.25 m² area in each one-square meter quadrate were taken vertically to the cliff. Overall, 92 photos of 23 quadrates in the whole area have been analyzed for the present study.

The main taxa of the sessile organisms in each photo-sample (Figure 2) were identified, and their percent cover was measured by the graphic program Image J. The "TSH" substrate (including algal turf, *sensu* Connell et al. [39], sediment, and small hydrozoans) was considered as "free surface", putatively suitable for the settlement of *P. clavata*, as recruits and young colonies of this gorgonian have been frequently found emerging from this substrate [20]. The remaining substrate, hereafter named "occupied substrate" was covered by sessile macro-benthos.

**Figure 2.** *P. clavata*: (**a**) a large/old adult and (**b**) a small/young colony.

### 2.2. Size Structure and Density of the P. clavata Population

The number of colonies of *P. clavata* in each plot ($n = 23$) was counted, and the densities (col m$^{-2}$) of adults and annual recruits recorded. Colony size (in width) was measured by a ruler reported on one side of each square in a top view image. Considering an average annual growth rate of 3 cm y$^{-1}$ in width, all colonies were divided into annual age classes according to their size (i.e., size/age classes [12,40]), and all those <4.5 cm were assigned to the 0–1 age class (recruits) according to Cupido et al. [21]. The colonies assigned to classes 2 to 4 were defined as small/young adults, while colonies belonging to classes >5 were considered large/old adults (Figure 2).

Since time series data and the size/age structure of the population between 1998 and 2010 were available [20,21,32], a comparison was made with the data collected in 2013.

### 2.3. Data Analysis

Data normality and homogeneity have been tested using the Shapiro–Wilk test [41] and the Levene test [42], respectively. Non-normal data have been transformed into Log(x) and Log(x + 1), and the remaining non-normally distributed data have been analyzed by non-parametric tests.

The density variability in adult colonies and recruits over the years 1998 to 2013 has been analyzed via the non-parametric Friedman test [43]. Moreover, the population size/age structures over this period have been compared by multidimensional scaling (MDS): in the resulting configuration of points ($n = 9$, years of sampling), the closeness of one year's sample data to another's provides a measure of similarity. The MDS analysis has been performed using the free software Past4.03 and the Euclidean index of similarity has been used.

Differences in percent cover of TSH and occupied substrate between the two sites ($n = 12$ at Tinetto Island and $n = 11$ in Tinetto Shoal) sampled in 2013 have been tested via the *t*-test [44]. Percent cover and density differences in adult *P. clavata* colonies between sites have also been analyzed via the *t*-test, while those of the recruits with the Mann–Whitney test.

The relations between adult and recruit densities and between these data and free substrate have been tested by $R^2$ best fit coefficient, linear Pearson's coefficient and the *t*-test. The relations between recruit densities in the same quadrate on 2 subsequent years have been explored on the scatter plot: the density of one year was plotted against that of the following year in the same fixed plot (2007 vs. 2008 and 2009 vs. 2010), finally a linear trend has been calculated.

## 3. Results

### 3.1. Population Density and Size/Age Structure over the 1998–2013 Period

In 2013, the overall density of *P. clavata* adult colonies reached about half the 'pristine' 1998 values (Figure 3a), while the overall recruit density was 4.5 times higher than that in 1998 (Figure 3b). For both adults and recruits, 2004 was a crucial year in which the population was reduced by 78%, while 2007 was the turning point after which they underwent a rapid increase, remaining stable thereafter (Figure 3a,b), as no differences in adult colony and recruit density have been found over the period 2007–2013 (Friedman tests, adult $\chi^2 = 0.98$, $p > 0.05$ and recruit $\chi^2 = 5.98$, $p > 0.05$).

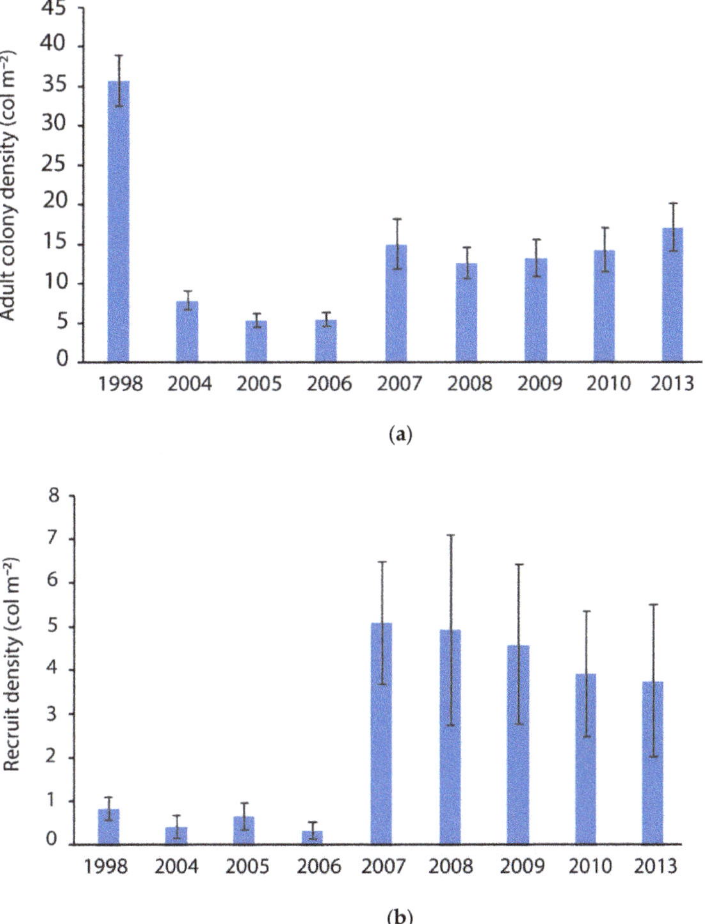

**Figure 3.** (**a**) Adult colony density and (**b**) recruit density during the observational period.

The size/age structures of the *P. clavata* population in 1998 (before the anomalous mortality events) and the following years 2004–2013 are reported in Figure 4a–i. The maximum size found (53.5 cm) corresponded to a maximum life span of about 18 years. The population was thus divided into 18 size/age classes based on the average annual colony growth rate [12,20].

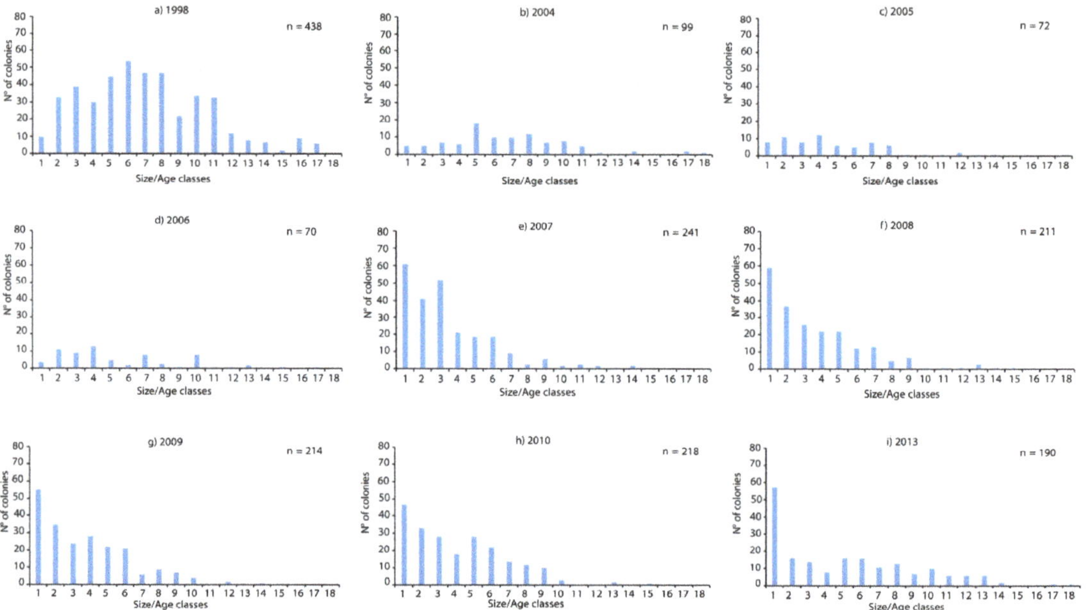

**Figure 4.** (a–i) Population size/age structure during the observational period.

The pre-mortality 1998 population (Figure 4a) exhibited a non-monotonic structure, skewed toward larger/older colonies (reaching class 17), in which classes 5–8 were dominant and recruitment (the first class) was low. In the following years (2004–2006; Figure 4b–d), recruitment, together with all classes, fell, and larger/older colonies nearly disappeared, being heavily impacted by the 1999 and 2003 mortality events. In 2007, and up to 2013 (Figure 4e–i), the population size/age structure changed again, following a monotonic, regularly decreasing pattern with dominant recruitment indicating, according to Caswell [45], a population in steady state. In 2013, the number of larger/older classes increased, and classes 17–18 were again represented (Figure 4i).

These findings are consistent with the results of the MDS analysis (Figure 5), which separated the 1998 pristine structure from the others and grouped together all the years between 2007 and 2013, which exhibited a similar demographic structure, characterized by dominant recruitment and a regularly decreasing abundance of larger/older classes. The population structure in 2004 (immediately after the second mortality event) is not grouped with the other years, but is instead set apart from them. The 2005–2006 cluster likely represents an intermediate point between the pristine structure and the partial recovery occurring in the following years.

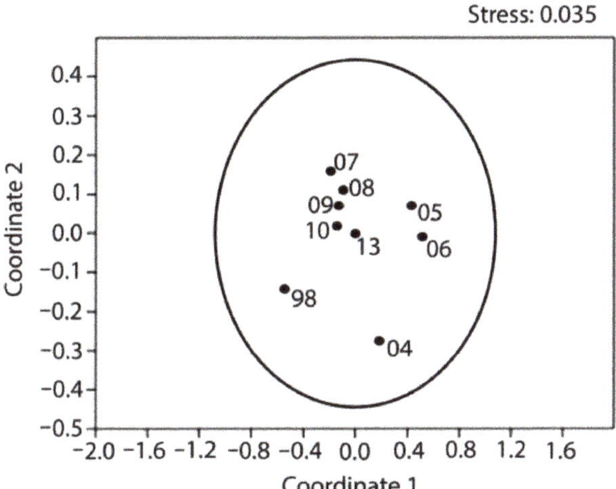


**Figure 5.** Multidimensional scaling (MDS) of population size/age structure in different years. The closeness of one sample to another (belonging to a different year) is a measure of similarity. The elliptical area represents 95% confidence. The 1998 pristine structure is set apart from the others, while all the years between 2007 and 2013 are grouped together. The 2005–2006 cluster likely represents an intermediate point between the pristine structure and the subsequent partial recovery.

*3.2. Percentage Cover of Benthic Organisms and Density of Paramuricea Clavata*

In 2013, the majority of the substrate (65%) was covered by TSH and the remaining 34% by sessile macro-benthonic organism (occupied substrate). It was impossible to read the remaining 1% of the photo-sample (undetermined area, Table 1). *P. clavata* (50%) and Porifera (27.7%) were the dominant organisms. Hexacorallia (mainly *Parazoanthus axinellae* and *Leptosamnia pruvot*i), Phaeophyceae, Rodophyceae, Ascidiacea, Bryozoa, and Polychaeta covered the remaining surface of the occupied cliff (22.3%).

**Table 1.** Percent cover (%) of the whole Tinetto Islet area and two sites (Tinetto Island and Tinetto Shoal) in 2013 and 1998 [46]. The TSH substrate is composed by algal turf, sediment, and small hydrozoans; occupied substrate includes sessile macro-benthonic organisms listed below. * = $p < 0.05$ level. TSH, occupied substrate, density of *P. clavata* adults and recruits (col m$^{-2}$) in the whole Tinetto Islet area in 2013 are compared with those measured in 1998 [46].

| | | 2013 | | | 1998 |
|---|---|---|---|---|---|
| | | **Tinetto Islet Area** | **Tinetto Island** | **Tinetto Shoal** | **Tinetto Islet Area** |
| | | $n = 23$ | $n = 12$ | $n = 11$ | $n = 24$ |
| | TSH | 65.1 ± 1.4 | 65 ± 2.3 | 65.3 ± 1.7 | 39 ± 1 |
| | Undetermined area | 1.3 ± 1.4 | 1.5 ± 0.2 | 1 ± 0.2 | 0 |
| | Occupied substrate | 33.6 ± 1.5 | 33.5 ± 2.5 | 33.7 ± 1.8 | 61 ± 1 |
| | Porifera | 9.3 ± 0.7 | 10.9 ± 0.6 * | 7.6 ± 1.2 * | - |
| **% Cover** | Hexacorallia | 4 ± 0.4 | 5.8 ± 1 * | 2 ± 0.2 * | - |
| $\bar{x}$ ± ES | Algae | 2 ± 0.04 | 1.1 ± 0.1 * | 2.9 ± 0.6 * | - |
| | Ascidiacea, Bryozoa, and Polychetae | 0.29 ± 0.06 | 0.4 ± 0.1 * | 0.2 ± 0.03 * | - |
| | *Paramuricea clavata* | 16.8 ± 1.7 | 14.0 ± 2.3 * | 19.9 ± 2.4 * | 52.0 ± 1 |
| **Colony density (col m$^{-2}$)** | Recruits | 3.75 ± 1.74 | 1.33 ± 0.48 * | 6.18 ± 2.14 * | 0.83 ± 0.27 |
| $\bar{x}$ ± ES | Adults | 17.10 ± 3.01 | 12.07 ± 1.81* | 21.30 ± 4.17* | 35.67 ± 3.20 |

A comparison with the data collected in the same area in 1998 on a similar number of samples highlighted some major changes occurring within the community over the pre-mortality/post mortality transition, showing about a twofold increase in TSH and halving of the cover of benthic organisms (Table 1). The *P. clavata* % cover was reduced threefold, and the adult density decreased twofold, while during the same time interval recruitment increased by four and a half times.

In 2013, no significant difference in TSH % cover or the % of the occupied substrate by sessile organisms between the two sample sites was found ($t = -0.095$, $p > 0.05$ and $t = -0.07$, $p > 0.05$), while the % cover of Porifera was significantly higher at Tinetto Island than at Tinetto Shoal ($t = 2.56$, $p < 0.05$, Table 1).

Overall 355 colonies were counted in 23 m$^2$ of the study area (130 at Tinetto Island and 225 at Tinetto Shoal), reaching an average density of about 15 colonies per m$^2$. Recruits settled only on some of the plots (61.8%): overall, 64 recruits were found on 14 plots (6/12 plots at Tinetto Island and 8/11 plots at Tinetto Shoal), with an average density of 3.75 recruits per m$^2$. A significantly higher recruit density was found at Tinetto Shoal ($6.18 \pm 2.14$ col m$^{-2}$) vs. at Tinetto Island ($1.33 \pm 0.48$ col m$^{-2}$; Mann–Whitney $w = -2.06$, $p < 0.05$, Table 1). Moreover, adult colony cover and density were also significantly higher in this latter area ($14.0 \pm 2.3$ vs. $19.9 \pm 2.4$% cover: $t = -1.80$, $p < 0.05$; $12.07 \pm 1.81$ col m$^{-2}$ vs. $21.30 \pm 4.17$ col m$^{-2}$: $t = -2.08$, $p < 0.05$).

No correlation emerged between larger/older colony and recruit densities ($n = 23$, $r = 0.35$, $p > 0.05$, Figure 6a), but a significant positive linear correlation was found between the density of recruits and that of smaller/younger colonies ($n = 23$, $r = 0.78$, $p < 0.01$, Figure 6b). Some linear trend was also found in the recruitment density measured in the same plot in two consecutive years ($n = 24$, Figure 6c). Both the last findings suggest that samples showing high recruitment in one year show similar values in the following years, as well.

No correlation was found between colony density and Porifera % cover ($r = -0.34$, $p > 0.05$) or between recruit density and TSH % cover measured in the same plots ($r = 0.038$, $p > 0.05$). These results suggest that neither the density of adult colonies nor the kind of substrate (at the gross-grain level examined) likely have any linear effect on *P. clavata* recruitment.

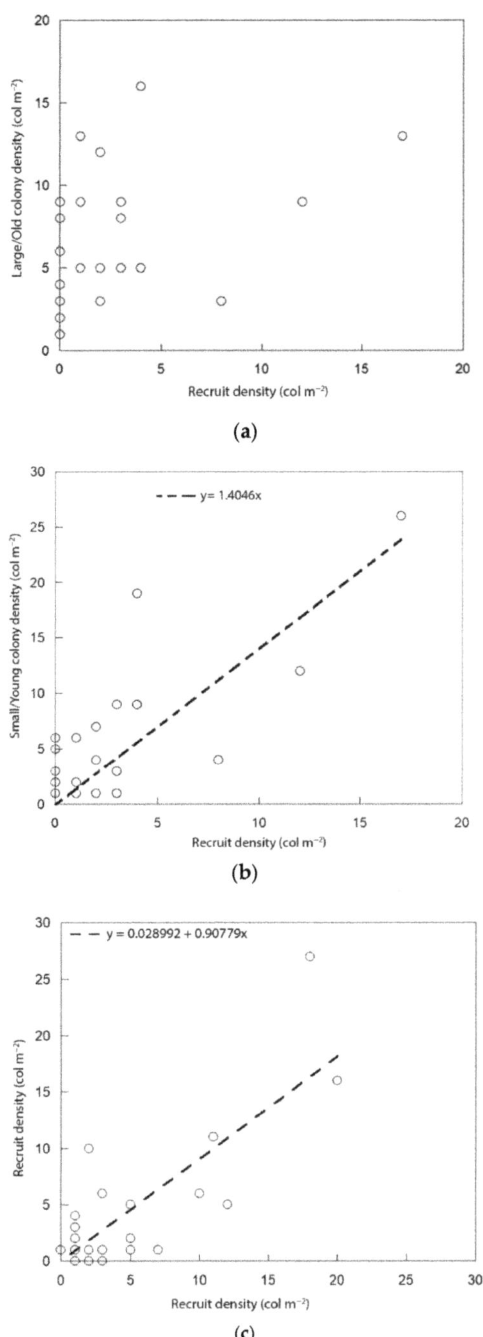

**Figure 6.** Trends of recruit and colony density: (**a**) density of recruits vs. that of larger/older colonies; no linear trend was found; (**b**) density of recruits vs. that of small/young colonies; a linear trend emerged; (**c**) recruit density measured in the same photo sample in two successive years (2007 vs. 2008 and 2009 vs. 2010) and the linear trend.

## 4. Discussion

Our understanding of the long-lasting effects of GCC on long-lived populations of marine benthic communities is quite limited. Generally, simulations try to fill this gap by devising possible scenarios of affected population trends [40,47] among others. An analysis of a dataset from 1998 to 2013 allowed us to follow the main trends of the community under study and the locally dominant gorgonian population before and after two dramatic mass mortality events occurring throughout the northwest Mediterranean during late summer in both 1999 and 2003.

The area where the temperate gorgonian *Paramuricea clavata* population examined here dwells is characterized by a sciaphilic epibenthic community settled on a vertical cliff. The entire area is affected by high sedimentation rates due to the outflow of the Magra river [38]. The high concentration of seston typical of this area is likely a factor favoring passive suspension feeders [32]. Cnidarians, and the gorgonian *P. clavata* in particular, are the dominant group, while the algal cover is poor, limited by the high turbidity of the area.

The *P. clavata* population of Tinetto Islet has been divided into 18 annual age/size classes, according to the mean annual colony growth rate [20]. The distribution of size classes observed in 1998 in the crowded, pre-mortality population with abundant larger (older) colonies could be driven, by the effect of the increasing density of larger/older colonies on recruitment rate, progressively affecting the population over some years, leading to a structure characterized by a low abundance of recruits and smaller/younger classes. After the 1999 and 2003 mortalities, only a few of the largest sized gorgonians survived, and the structure of the population drastically shifted towards the smaller/younger classes. In 2007, colony density significantly increased and remained stable subsequently. As mortality disproportionally affected the larger/older colonies and fecundity is strictly correlated with colony size [21], a drastic reduction in the larger classes could have decreased the population's reproductive output. However, the abundant egg and planulae production of this species could have attenuated the effects of the loss of larger colonies on population resilience [21].

Ten years after the last *P. clavata* anomalous mortality event (2003) a twofold increase was found in the substrate covered by TSH, indicating that a clear-cut change in the whole community occurred. Concerning the *P. clavata* population, both % cover and density of adult colonies were also reduced, albeit to different degrees, as they were respectively one third and half those recorded in 1998. Recruitment, on the other hand, was four and a half times higher than in the pre-mortality population. The difference between the reduction in adult colony cover and the density found could be due to the different size/age structure of the post-mortality population, in which the number of larger/older colonies fell disproportionally [20], thereby affecting colony cover more than density.

Several papers have reported that algal turf coverage and high sedimentation rates negatively affect recruitment and reproduction in benthic communities [48,49] among others. For example, in the gorgonian *Eunicella singularis*, algal turf seems to negatively affect the process of settlement by decreasing the larva's access to suitable substrate [50]. Contrary to this general finding, *P. clavata* recruitment density increased since 2007, despite a significant rise in TSH % cover between 1998 and 2004 [45], and remained substantially unchanged in the following years. However, the lack of any correlation between the TSH and *P. clavata* recruitment or adult colony density likely indicates the absence of any linear effect of TSH cover on the gorgonian population examined. This finding seems to be confirmed by the similarity in TSH % cover between the two sites sampled which, however, differ in *P. clavata* adult and recruit densities. As no correlation between these parameters and TSH cover was observed, other factors likely caused the difference found between the sites. According to Padròn et al. [34], who examined the genetic structure and connectivity of *P. clavata* populations across our sampling area and six other sites in the eastern Ligurian Sea, the locally dominant currents may have fostered a higher larval input in our more external site (Tinetto Shoal), thereby partly driving the dynamics of this population. Indeed, the input of larvae released in other areas may play an important role in the recovery of benthic sessile populations affected by anomalous mortality events. In *P. clavata* the larval stage can span between 6 and 26 days under experimental conditions [45,51], and planulae are able to travel distances of up to tens of kilometers. Recent genetic analyses in the Ligurian Sea suggest that tight connectivity among *P. clavata* populations affected by anomalous mortality can avoid their collapse [34], while highly frequent mortality events may lead isolated octocoral populations even to extinction [12]. Thus, the strategic position of the population under study and the connectivity with the other neighboring populations could be a key factor in fast recovery after disturbing events [51].

There is some hint towards a small-scale spatial constancy of recruitment over time; this is based on two findings: (1) a significant positive linear correlation between the density of recruits and that of the cohort of small/young adult colonies measured in the same plot; (2) some linear trend of recruit density recorded in the same plot in two consecutive years. Both these findings suggest that the plots more intensively recruiting in one year will also recruit more in following years. Thus, some plots should likely, present particular features that make them more suitable for *P. clavata* settlement over time. It can be speculated that micro-scale hydrodynamics, local substrate texture, and/or chemical attractors produced by the local species associations could foster more abundant recruitment in some plots.

The lack of any relation between the density of recruits and that of larger/older colonies suggests some independence of recruitment from the density of these colonies, as well as the lack of any effect of their high fecundity on local recruitment [21]. Although these findings may appear to be in contrast with the initial relation between recruits and small/young adults, the selectively higher mortality affecting the larger/older classes may, over time, lead to an alteration of such relation.

The highly dense canopy made mainly by older colonies in the pre-mortality, overcrowded population may have had some "shadow effect" to avoid larval settlement in their surroundings by limiting the substrate suitable for recruits. After the 1999 and 2003 mortalities, reduction of the canopy by about 2/3 could have driven a clear-cut increase in recruitment. As the increase in recruitment was delayed by four years since the mortality events and co-occurred with the detachment of dead colonies from the cliff in 2007 [20], the shadow effect caused by the dead 'scaffold' of larger colonies could have been maintained until they detached. Only after the dead colonies fell off the cliff and their shadow effect disappeared did recruitment start to rise, remaining at similar values in the following years, as well.

*Paramuricea clavata* has been considered a species with low population turnover and recovery rates [52,53]. Moreover, according to a demographic model aimed at projecting the trends of this population over time by means of simulations based on an ad hoc Leslie-Lewis transition matrix [46], wide fluctuations in density that stabilize only after several decades have been described [13]. However, in the present study, we observed, in the real population, faster density stabilization than predicted by said model.

After 14 and 10 years since the two mortality events, the population of *P. clavata* is still alive and has partially recovered, although with severely reduced density and canopy. Moreover, the population structure, dominated in recent years by recruits, is quite different from the pre-mortality one, dominated by larger/older colonies. As these features have been maintained for 7 years, the population seems to have reached a new equilibrium point.

All the findings presented herein indicate that the effects of drastic mortality events on a long-lived gorgonian population cannot be limited to the narrow definitions of "extinction or recovery", but that a stricken population may reach a new equilibrium. The main concern regards the stability of such a new equilibrium reached at remarkably lower densities: repeated mortality events and undamped oscillations may even devolve the local population to extinction.

## 5. Conclusions

A decade after two mass mortality events in 1999 and 2003, the circalittoral community thriving on a vertical cliff has undergone some profound changes. In subsequent years, the TSH (algal turf, sediment and small hydrozoans) percentage cover increased twofold, and the cover and density of the dominant gorgonian population *P. clavata* was reduced by 2/3 and 1/2, respectively, while recruitment increased two and a half times. A tendency of some quadrates to recruit more intensely over time was found. No direct effect of TSH and *P. clavata* percentage cover on recruit and adult colony density resulted.

In 2013 the still surviving population had not recovered its pristine structure, and had only partially restored its canopy. In addition, although it had recovered, the population seemed to have reached a new equilibrium point, whose stability is as yet disturbingly unknown.

**Author Contributions:** Software, formal and statistical analysis: F.E., S.R.S. and M.C.B.; sampling: S.C., R.C., G.S. and A.P.; writing: G.S., M.C.B., S.C., and A.P.; supervision and coordination: G.S., M.C.B. and S.C. All authors have read and agreed to the published version of the manuscript.

**Funding:** M.C. Benedetti was supported by a doctoral fellowship funded by Enzo Liverino s.r.l., Chii Lih Coral Co., Ltd. of Taiwan and the University of Pisa. She was also supported by a post-doctoral fellowship (Luigi e Francesca Brusarosco) funded by the Italian Ecological Society (S.It.E).

**Data Availability Statement:** The data presented in this study are available on request from the corresponding author. The data are not available because the authors reserve to carry out supplementary analyses and models.

**Acknowledgments:** The authors would like to thank F. Bulleri (Pisa University) for his supervisor contribution to S. Ruffaldi- Santori Master Degree thesis and A. Cafazzo (Pisa University) for the revision of the English text.

**Conflicts of Interest:** The authors declare no conflict of interest.

## References

1. Frölicher, T.L.; Laufkötter, C. Emerging risks from marine heat waves. *Nat. Commun.* **2018**, *9*, 1–4. [CrossRef] [PubMed]
2. Decarlo, T.M.; Cohen, A.L.; Wong, G.T.F.; Davis, K.A.; Lohmann, P.; Soong, K. Mass coral mortality under local amplification of 2 °C ocean warming. *Sci. Rep.* **2017**, *7*, srep44586. [CrossRef]
3. Langangen, Ø.; Ohlberger, J.; Stige, L.C.; Durant, J.M.; Ravagnan, E.; Stenseth, N.C.; Hjermann, D.Ø. Cascading effects of mass mortality events in Arctic marine communities. *Glob. Chang. Biol.* **2016**, *23*, 283–292. [CrossRef]
4. Verdura, J.; Linares, C.; Ballesteros, E.; Coma, R.; Uriz, M.J.; Bensoussan, N.; Cebrian, E. Biodiversity loss in a Mediterranean ecosystem due to an extreme warming event unveils the role of an engineering gorgonian species. *Sci. Rep.* **2019**, *9*, 5911. [CrossRef]
5. Lirman, D.; Schopmeyer, S.; Manzello, D.; Gramer, L.J.; Precht, W.F.; Muller-Karger, F.; Banks, K.; Barnes, B.; Bartels, E.; Bourque, A.; et al. Severe 2010 Cold-Water Event Caused Unprecedented Mortality to Corals of the Florida Reef Tract and Reversed Previous Survivorship Patterns. *PLoS ONE* **2011**, *6*, e23047. [CrossRef] [PubMed]
6. Danovaro, R.; Dell'Anno, A.; Fabiano, M.; Pusceddu, A.; Tselepides, A. Deep-sea ecosystem response to climate changes: The eastern Mediterranean case study. *Trends Ecol. Evol.* **2001**, *16*, 505–510. [CrossRef]
7. Teagle, H.; Smale, D.A. Climate-driven substitution of habitat-forming species leads to reduced biodiversity within a temperate marine community. *Divers. Distrib.* **2018**, *24*, 1367–1380. [CrossRef]
8. Marshall, P.A.; Baird, A.H. Bleaching of corals on the Great Barrier Reef: Differential susceptibilities among taxa. *Coral Reefs* **2000**, *19*, 155–163. [CrossRef]
9. Loya, Y.; Sakai, K.; Yamazato, K.; Nakano, Y.; Sambali, H.; Van Woesik, R. Coral bleaching: The winners and the losers. *Ecol. Lett.* **2001**, *4*, 122–131. [CrossRef]
10. Brown, B.E.; Dunne, R.P.; Goodson, M.S.; Douglas, A.E. Experience shapes the susceptibility of a reef coral to bleaching. *Coral Reefs* **2002**, *21*, 119–126. [CrossRef]
11. Garzón-Ferreira, J.; Zea, S. A mass mortality of *Gorgonia ventalina* (Cnidaria: Gorgoniidae) in the Santa Marta area, Caribbean coast of Colombia. *Bull. Mar. Sci.* **1992**, *50*, 522–526.
12. Santangelo, G.; Bramanti, L.; Iannelli, M. Population dynamics and conservation biology of the over-exploited Mediterranean red coral. *J. Theor. Biol.* **2007**, *244*, 416–423. [CrossRef] [PubMed]
13. Santangelo, G.; Cupido, R.; Cocito, S.; Bramanti, L.; Priori, C.; Erra, F.; Iannelli, M. Effects of increased mortality on gorgonian corals (Cnidaria, Octocorallia): Different demographic features may lead affected populations to unexpected recovery and new equilibrium points. *Hydrobiologia* **2015**, *759*, 171–187. [CrossRef]

14. Linares, C.; Doak, D.F. Forecasting the combined effects of disparate disturbances on the persistence of long-lived gorgonians: A case study of *Paramuricea clavata*. *Mar. Ecol. Prog. Ser.* **2010**, *402*, 59–68. [CrossRef]
15. Prada, C.; Weil, E.; Yoshioka, P.M. Octocoral bleaching during unusual thermal stress. *Coral Reefs* **2010**, *29*, 41–45. [CrossRef]
16. Sánchez, J.A.; Ardila, N.E.; Andrade, J.; Dueñas, L.F.; Navas, R.; Ballesteros, D. Octocoral densities and mortalities in Gorgona Island, Colombia, Tropical Eastern Pacific. *Rev. Biol. Trop.* **2014**, *62*, 209–219. [CrossRef]
17. Arizmendi-Mejía, R.; LeDoux, J.-B.; Civit, S.; Antunes, A.; Thanopoulou, Z.; Garrabou, J.; Linares, C. Demographic responses to warming: Reproductive maturity and sex influence vulnerability in an octocoral. *Coral Reefs* **2015**, *34*, 1207–1216. [CrossRef]
18. Bramanti, L.; Benedetti, M.C.; Cupido, R.; Cocito, S.; Priori, C.; Erra, F.; Iannelli, M.; Santangelo, G. Demography of Animal Forests: The Example of Mediterranean Gorgonians. In *Marine Animal Forests: The Ecology of Benthic Biodiversity Hotspots*; Rossi, S., Bramanti, L., Gori, A., Orejas, C., Eds.; Springer: Cham, Switzerland, 2017; pp. 529–548.
19. Tsounis, G.; Edmunds, P.J. Three decades of coral reef community dynamics in St. John, USVI: A contrast of scleractinians and octocorals. *Ecosphere* **2017**, *8*, e01646. [CrossRef]
20. Cupido, R.; Cocito, S.; Barsanti, M.; Sgorbini, S.; Peirano, A.; Santangelo, G. Unexpected long-term population dynamics in a canopy-forming gorgonian coral following mass mortality. *Mar. Ecol. Prog. Ser.* **2009**, *394*, 195–200. [CrossRef]
21. Cupido, R.; Cocito, S.; Manno, V.; Ferrando, S.; Peirano, A.; Iannelli, M.; Bramanti, L.; Santangelo, G. Sexual structure of a highly reproductive, recovering gorgonian population: Quantifying reproductive output. *Mar. Ecol. Prog. Ser.* **2012**, *469*, 25–36. [CrossRef]
22. Bramanti, L.; Edmunds, P.J. Density-associated recruitment mediates coral population dynamics on a coral reef. *Coral Reefs* **2016**, *35*, 543–553. [CrossRef]
23. Riegl, B.; Johnston, M.; Purkis, S.; Howells, E.; Burt, J.; Steiner, S.C.; Sheppard, C.R.C.; Bauman, A. Population collapse dynamics in *Acropora downingi*, an Arabian/Persian Gulf ecosystem-engineering coral, linked to rising temperature. *Glob. Chang. Biol.* **2018**, *24*, 2447–2462. [CrossRef] [PubMed]
24. Rossi, S. The destruction of the 'animal forests' in the oceans: Towards an over-simplification of the benthic ecosystems. *Ocean Coast. Manag.* **2013**, *84*, 77–85. [CrossRef]
25. Bavestrello, G.; Bertone, S.; Cattaneo-Vietti, R.; Cerrano, C.; Gaino, E.; Zanzi, D. Mass mortality of *Paramuricea clavata* (Anthozoa, Cnidaria) on Portofino Promontory cliffs, Ligurian Sea, Mediterranean Sea. *Mar. Life* **1994**, *4*, 15–19.
26. Huete-Stauffer, C.; Vielmini, I.; Palma, M.; Navone, A.; Panzalis, P.; Vezzulli, L.; Misic, C.; Cerrano, C. *Paramuricea clavata* (Anthozoa, Octocorallia) loss in the Marine Protected Area of Tavolara (Sardinia, Italy) due to a mass mortality event. *Mar. Ecol.* **2011**, *32*, 107–116. [CrossRef]
27. Turicchia, E.; Abbiati, M.; Sweet, M.; Ponti, M. Mass mortality hits gorgonian forests at Montecristo Island. *Dis. Aquat. Org.* **2018**, *131*, 79–85. [CrossRef]
28. Carpine, C.; Grasshoff, M. Les gorgonaires de la Mediterranee. *Bull. Inst. Océanogr. (Monaco)* **1975**, *71*, 1–140.
29. Coma, R.; Zabala, M.; Gili, J.M. Sexual reproductive effort in the Mediterranean gorgonian *Paramuricea clavata*. *Mar. Ecol. Prog. Ser.* **1995**, *117*, 185–192. [CrossRef]
30. Ballesteros, E. Mediterranean coralligenous assemblages: A synthesis of present knowledge. *Oceanogr. Mar. Biol.* **2006**, *48*, 123–195.
31. Ponti, M.; Perlini, R.A.; Ventra, V.; Grech, D.; Abbiati, M.; Cerrano, C. Ecological shifts in Mediterranean coralligenous assemblages related to gorgonian forest loss. *PLoS ONE* **2014**, *9*, e102782.
32. Cupido, R.; Cocito, S.; Sgorbini, S.; Bordone, A.; Santangelo, G. Response of a gorgonian (*Paramuricea clavata*) population to mortality events: Recovery or loss? *Aquat. Conserv. Mar. Freshw. Ecosyst.* **2008**, *18*, 984–992. [CrossRef]
33. García-Gómez, J.C.; González, A.R.; Maestre, M.J.; Espinosa, F. Detect coastal disturbances and climate change effects in coralligenous community through sentinel stations. *PLoS ONE* **2020**, *15*, e0231641. [CrossRef]
34. Padrón, M.; Costantini, F.; Bramanti, L.; Guizien, K.; Abbiati, M. Genetic connectivity supports recovery of gorgonian populations affected by climate change. *Aquat. Conserv. Mar. Freshw. Ecosyst.* **2018**, *28*, 776–787. [CrossRef]
35. Cerrano, C.; Bavestrello, G.; Bianchi, C.N.; Cattaneo-Vietti, R.; Bava, S.; Morganti, C.; Morri, C.; Picco, P.; Sara, G.; Schiaparelli, S.; et al. A catastrophic mass-mortality episode of gorgonians and other organisms in the Ligurian Sea (North-western Mediterranean), summer 1999. *Ecol. Lett.* **2000**, *3*, 284–293. [CrossRef]
36. Garrabou, J.; Coma, R.; Bensoussan, N.; Bally, M.; Chevaldonné, P.; Ciglianos, M.; Díaz, D.; Harmelin, J.G.; Gambi, M.C.; Kersting, D.K.; et al. Mass mortality in Northwestern Mediterranean rocky benthic communities: Effects of the 2003 heat wave. *Glob. Chang. Biol.* **2009**, *15*, 1090–1103. [CrossRef]
37. Bally, M.; Garrabou, J. Thermodependent bacterial pathogens and mass mortalities in temperate benthic communities: A new case of emerging disease linked to climate change. *Glob. Chang. Biol.* **2007**, *13*, 2078–2088. [CrossRef]
38. Astraldi, M.; Gasparini, G.P.; Manzella, G.M.R.; Hopkins, T.S. Temporal variability of currents in the eastern Ligurian Sea. *J. Geophys. Res. Space Phys.* **1990**, *95*, 1515–1522. [CrossRef]
39. Connell, S.; Foster, M.; Airoldi, L. What are algal turfs? Towards a better description of turfs. *Mar. Ecol. Prog. Ser.* **2014**, *495*, 299–307. [CrossRef]
40. Bramanti, L.; Vielmini, I.; Rossi, S.; Tsounis, G.; Iannelli, M.; Cattaneo-Vietti, R.; Priori, C.; Santangelo, G. Demographic parameters of two populations of red coral (*Corallium rubrum* L. 1758) in the North Western Mediterranean. *Mar. Biol.* **2014**, *161*, 1015–1026. [CrossRef]

41. Shapiro, S.S.; Wilk, M.B. An analysis of variance test for normality (complete samples). *Biometrika* **1965**, *52*, 591–611. [CrossRef]
42. Levene, H. Robust Tests for Equality of Variances. In *Contributions to Probability and Statistics: Essays in Honor of Harold Hotelling*; Olkin, I., Ed.; Stanford University Press: Palo Alto, CA, USA, 1960; pp. 278–292.
43. Siegel, S. *Nonparametric Statistics for the Behavioral Sciences*; McGraw-Hill: New York, NY, USA, 1956.
44. Underwood, A.J. *Experiments in Ecology. Their Logical Design and Interpretation Using Analysis of Variance*; Cambridge University Press: Cambridge, UK, 1997.
45. Caswell, H. *Matrix Population Models*; Sinauer Associates: Sunderland, MA, USA, 2001; p. 722.
46. Cupido, R. Demography of the Red Gorgonian *Paramuricea clavata* (Anthozoa, Octocorallia) Damaged by Repeated Anomalous Mortality Events. Ph.D. Thesis, University of Pisa, Pisa, Italy, 2010.
47. Santangelo, G.; Fronzoni, L. *Global Climate Change and the Ecology of the Next Decade*; Edizioni ETS: Pisa, Italy, 2008.
48. Fabricius, K.E. Effects of terrestrial runoff on the ecology of corals and coral reefs: Review and synthesis. *Mar. Pollut. Bull.* **2005**, *50*, 125–146. [CrossRef] [PubMed]
49. Arnold, S.N.; Steneck, R.S.; Mumby, P.J. Running the gauntlet: Inhibitory effects of algal turfs on the processes of coral recruitment. *Mar. Ecol. Prog. Ser.* **2010**, *414*, 91–105. [CrossRef]
50. Linares, C.; Cebrian, E.; Coma, R. Effects of turf algae on recruitment and juvenile survival of gorgonian corals. *Mar. Ecol. Prog. Ser.* **2012**, *452*, 81–88. [CrossRef]
51. Pilczynska, J.; Cocito, S.; Boavida, J.; Serrão, E.; Queiroga, H. Genetic Diversity and Local Connectivity in the Mediterranean Red Gorgonian Coral after Mass Mortality Events. *PLoS ONE* **2016**, *11*, e0150590. [CrossRef]
52. Coma, R.; Gili, J.M.; Zabala, M.; Riera, T. Feeding and prey capture cycles in the aposymbiontic gorgonian *Paramuricea clavata*. *Mar. Ecol. Prog. Ser.* **1994**, *115*, 257–270. [CrossRef]
53. Linares, C.; Coma, R.; Garrabou, J.; Díaz, D.; Zabala, M. Size distribution, density and disturbance in two Mediterranean gorgonians: *Paramuricea clavata* and *Eunicella singularis*. *J. Appl. Ecol.* **2008**, *45*, 688–699. [CrossRef]

*Article*

# Shallow-Water Species Diversity of Common Intertidal Zoantharians (Cnidaria: Hexacorallia: Zoantharia) along the Northeastern Coast of Trinidad, Southern Caribbean

**Stanton Belford** [1,2]

[1] Department of Mathematics & Science, Martin Methodist College, Pulaski, TN 38478, USA; sbelford@martinmethodist.edu; Tel.: +1-931-424-4621

[2] Department of Mathematics and Science, University of Tennessee Southern, Pulaski, TN 38478, USA

**Abstract:** Zoantharians are colonial cnidarians commonly found in shallow tropical Caribbean coral reefs, and are known to be globally distributed. Common species in genera *Zoanthus* and *Palythoa* occur at Toco, Trinidad, where they are more abundant than their Scleractinia counterparts relative to benthic coverage. In this study, distribution, morphological and molecular data were collected to determine species and symbiont identification to provide more insight on zoantharians. The Line Intercept Point (LIT) transect method recorded coverage at three sites: Salybia (SB), Pequelle (PB), and Grande L'Anse (GA) Bays along the northeastern coast. Variations in morphology, such as tentacle count, oral disk color and diameter were collected from colonies in situ. All specimens were zooxanthellate, and molecular and phylogenetic analyses were done by sequencing the cytochrome oxidase subunit I (COI) gene, and the internal transcribed spacer (ITS) region for species and symbiont identification, respectively. Results showed mean Zoantharia percentage cover was $32.4\% \pm 5.1$ (X $\pm$ SE) at SB, $51.3\% \pm 6.5$ (PB), and $72.2\% \pm 6.1$ at GA. Zooxanthellate zoantharians were identified as *Palythoa caribaeorum*, *Palythoa grandiflora*, *Zoanthus pulchellus*, and *Zoanthus sociatus*. Symbiodiniaceae genera were identified as *Cladocopium* and *Symbiodinium* in *Palythoa* and *Zoanthus* spp., respectively. Although this is the first molecular examination of zoantharians, and their symbionts in Trinidad, more research is needed to identify and document species distribution and symbiont biodiversity to understand their ecology in these dynamic ecosystems.

**Keywords:** symbiodiniaceae; zoantharians; Trinidad coral reefs; zooxanthellate; biogeography

**Citation:** Belford, S. Shallow-Water Species Diversity of Common Intertidal Zoantharians (Cnidaria: Hexacorallia: Zoantharia) along the Northeastern Coast of Trinidad, Southern Caribbean. *Oceans* **2021**, *2*, 477–488. https://doi.org/10.3390/oceans2030027

Academic Editor: Rupert Ormond

Received: 30 October 2020
Accepted: 14 July 2021
Published: 20 July 2021

**Publisher's Note:** MDPI stays neutral with regard to jurisdictional claims in published maps and institutional affiliations.



## 1. Introduction

Zoantharians (Anthozoa: Hexacorallia: Zoantharia) make up a considerable benthic component of tropical and sub-tropical shallow-water reefs, similar to Scleractinia (hard corals) and Actiniaria (sea anemones) [1–4]. Zoantharians are sessile colonial anemone-like organisms with two rings of tentacles surrounding circular to polygonal disks, and forming colonies of polyps [5]. They grow in dense mats or small patches throughout the shallow intertidal rocky zones, or coral reef ecosystems, and play an important ecological role in their tropical and sub-tropical habitats [6–12].

Zoantharian coverage on coral reef ecosystems can be extensive [9,10]. For instance, Karlson (1981) noted that two species, *Zoanthus sociatus* (Ellis, 1768) and *Zoanthus solanderi* (LeSueur, 1818) had extensive coverage at a northern intertidal area in Jamaica, which he subsequently named the '*Zoanthus* zone', as originally named by Tom Goreau in the 1950s [13]. Additionally, Lopez et al. (2018) reported a zoantharian zone located at Cabo Verde Islands, where molecular methods and morphological analysis confirmed the presence of two zoantharian species. Zoantharians, such as *Palythoa caribaeorum* (Duchassaing and Michelotti, 1860) and *Z. sociatus* cover large areas in subtidal and intertidal zones [7,9,10,12]. Additionally, many, but not all zoantharian species maintain symbiotic relationships with Symbiodiniaceae (zooxanthellae), hence determining the identity of these holobionts adds more understanding of their ecology and physiological characteristics.

Molecular analyses of zoantharians assist in species identification, where morphological identification is difficult, or impossible. For example, phenotypic plasticity in zoantharians, specifically large variation in morphological characteristics, such as polyp shape, colony shape, size and oral disk color, may cause them to be overlooked in ecological surveys, even though their numbers may be abundant [5,8,14,15]. Given the difficulties in morphological and molecular examination of zoantharians, more priority should be given to determining their distribution and diversity in the Caribbean, especially since zooxanthellate zoantharians play key ecological roles in marine ecosystems [10]. Although success in identifying potential unidentified zoantharian species through morphological characteristics has been proven to be successful [16], the addition of molecular analyses has alleviated issues with morphological ambiguities [16,17]. Additionally, zoantharian species diversity including molecular analyses of zooxanthellate symbionts of family Symbiodiniaceae will continue to add more information about symbiont ecology in light of global climate change [11,18,19].

Although distributed world-wide in tropical and subtropical waters in the Atlantic and Indo-Pacific regions [8,20], zoantharians and Symbiodiniaceae genera identity at the southern-most part of the Caribbean, specifically Trinidad and Tobago, are limited. Two zoantharian genera, *Zoanthus* and *Palythoa* spp., are commonly observed occupying shallow waters, and in different intertidal zones along the northeastern coast of Trinidad [9,10,12,14], but relatively few studies have highlighted morphological and molecular details of these species in this part of the southern Caribbean region.

The purpose of this study is to quantitatively assess zoantharian benthic coverage, and use morphological and genetic analyses to identify zoantharians, and confirm Symbiodiniaceae genera identities. Molecular knowledge of these benthic organisms will be recorded here for the first time at this southernmost part of the Caribbean Sea. Consequently, all ecological aspects of identifying the extent of zoantharian benthic coverage provides more understanding of the dynamic coral reef ecosystem. In general, zoantharian distribution in the Caribbean needs to be continually monitored, and species identification confirmed, especially in the face of increased climate change and anthropogenic activities [9].

## 2. Materials and Methods

### 2.1. Study Site and Sampling

The northeastern coast of Trinidad has undefined patch reefs, and a fringing reef (see [12]). Beaches within close vicinity of these reefs are largely defined as sandy, stony, or rocky. The tropical climate has two distinct seasons, with a dry period from January to mid-June and a rainy period that extends the remainder of the year (June–December). Tides are semidiurnal with maximum high tides reaching 2 m (meters) in open water, and extreme low tides can reach 0.2 m.

Study sites in accessible areas along the northeastern coast are located at Salybia Bay (SB) (located between 10°50.097′ N, 60°55.208′ W and 10°50.100′ N, 60°55.157′ W), which is part of the only fringing reef in Trinidad. This system has been an important site for citizen/volunteer coral reef monitoring focused on cnidarian and invertebrate abundances and distributions, because it is very shallow during spring low tides (~0.2 m) and the intertidal zone extends some 200 m parallel to the shoreline. This fringing reef is affected by sediment discharge from local rivers throughout the year, specifically during the rainy period from June to December.

Pequelle Bay (PB), located to the east of SB (between 10°50.111′ N, 60°55.129′ W and 10°50.181′ N, 60°54.954′ W), has a mixture of rocks and tide pools along the intertidal zone. During spring low tides, this part of the reef becomes fully exposed for a 3-h period until tides return. Patchy undeveloped reefs are present at Grande L′Anse (GA), also known as Toco Bay (between 10°50.107′ N, 60°56.772′ W and 10°50.266′ N, 60°56.674′ W), which has a mixture of rocky outcrops. This study site has a mixture of sand, stony and rocky beaches with village homes within close vicinity. Spring low tides revealed a rocky patch reef with scattered tide pools interspersed throughout the intertidal zone.

Volunteers marked the study area with GPS points, and quantified an area of 500 m$^2$ using a 50-m open reel fiberglass measuring tape during extreme low tides (<0.3 m). Each transect was placed within a 500 m$^2$ study area, and a total area of 1500 m$^2$ was examined at each site. The Line Intercept Transect (LIT) method [9] was used to examine only the lower intertidal areas at SB, PB, and GA. Transect positions within each 500 m$^2$ area were marked using a global positioning system (GPS). A 50 m open reel fiberglass measuring tape was placed parallel to the shoreline in each area, and benthic components were recorded at every 0.5 m interval on the measuring tape. Benthic components, such as reef-building corals, zoantharians, macroalgae, coral rubble, and other invertebrates, such as sea urchins, fireworms, and sea cucumbers were recorded if they touched the 0.5 m intervals along the measuring tape. This was repeated three times within each 500 m$^2$ study area, and for a total of 3 study areas. Water temperature and salinity were measured using a YSI Pro 1030 probe at three random points along each 50 m LIT. Percentage cnidarian and benthic cover was calculated for benthic communities.

*2.2. Morphological Analyses and Specimen Collection*

Morphological data for zoantharian colonies ($n = 30$) were recorded at three field sites, SB, PB, and GA in June 2019 from the northeastern coast of Toco, Trinidad. A hand-held caliper was used to measure oral disk diameter of 3 opened polyps per colony. Physical characteristics of each polyp, such as tentacle count and color, oral disc color, and polyp form (immersae, intermediate, liberae, see 4, 21 were recorded, together with in situ photographs, which were used to assist in identification (see Table 1). Additional samples were also collected at three other sites between December 2019–February 2020, and were primarily collected for genetic analyses. Samples were collected in areas with high wave action (open habitat exposed to waves), and low wave action (rocky habitat protected from waves). A total of 13 specimens (3–5 polyps per specimen) were collected from SB, PB, and GA. Additionally, specimens were collected at inaccessible areas along the northeastern coast, such as Galera Point (GP) ($n = 1$ specimen) located just left of the Keshorn Walcott Toco Lighthouse, which is 1 km east of PB, and has a mixture of rocks and tide pools where zoantharians were observed to have extensive coverage. Additionally, samples were collected from Straight Bay (StB) ($n = 2$ specimens), 1 km west of SB, and from western SB (WSB) ($n = 5$ specimens), which is an accessible area, but less frequented by local visitors.

**Table 1.** Zoantharian specimens from the northeastern coast of Trinidad with collection information, morphological data, GenBank accession numbers, and identification conclusions based on COI sequences.

| Sample # | Collection Site $\alpha$ | Disk Color | Tentacle Color | Tentacle Count (3 Polyps Per Colony) | Disk Diameter (mm) | COI I.D. (GenBank Accession #) |
|---|---|---|---|---|---|---|
| P1-br-GA | GA | Brown | Brown | 26 | 11.0 | *P. caribaeorum* (MZ150796) |
| P2-br-GA | GA | Brown | Brown | 36 | 10.8 | *P. caribaeorum* (MZ150797) |
| P3-br-SB | SB | Brown | Brown | 37 | 10.5 | *P. caribaeorum* (MZ150798) |
| P4-gr-GA | GA | Green | Brown | 46 | 13.8 | *P. grandiflora* (MZ150799) |
| P5-gr-GA | GA | Green | Brown | 45 | 12.6 | *P. grandiflora* (MZ150800) |
| P6-br-GA | GA | Brown | Brown | 33 | 10.7 | *P. caribaeorum* (MZ150801) |
| P7-br-WSB | WSB | Brown | Brown | 26 | 11.0 | *P. caribaeorum* (MZ147090) |
| P8-br-WSB | WSB | Brown | Brown | 26 | 11.0 | *P. caribaeorum* (MZ147091) |

**Table 1.** *Cont.*

| Sample # | Collection Site $\alpha$ | Disk Color | Tentacle Color | Tentacle Count (3 Polyps Per Colony) | Disk Diameter (mm) | COI I.D. (GenBank Accession #) |
|---|---|---|---|---|---|---|
| Z1-gr-StB | StB | Green/Blue | Green | 41 | 10.2 | *Z. sociatus* (MZ147096) |
| Z2-gr-StB | StB | Green/Blue | Green | 42 | 10.3 | *Z. sociatus* (MZ147097) |
| Z3-gr-WSB | WSB | Green/Blue | Green | 42 | 10.3 | *Z. sociatus* (MZ150806) |
| Z4-gr-WSB | WSB | Green/Blue | Green | 42 | 10.3 | *Z. sociatus* (MZ150802) |
| Z5-gr-SB | SB | Green | Green | 42 | 7.5 | *Z. pulchellus* (MZ150803) |
| Z6-gr-SB | SB | Green | Green | 42 | 6.9 | *Z. pulchellus* (MZ156026) |
| Z7-br-gr-SB | SB | Green/Blue | Green | 42 | 7.6 | *Z. sociatus* (MZ150807) |
| Z8-org-SB | SB | Green | Green | 46 | 10.2 | *Z. pulchellus* (MZ150805) |
| Z9-org-GA | GA | Orange | Brown | 46 | 10.3 | *Z. pulchellus* (MZ150804) |
| Z10-gr-TLH | TLH | Green | Green | 42 | 10.3 | *Z. pulchellus* (MZ147092) |
| Z13-grey-GA | SB | Grey | Green | 44 | 5.2 | *Z. pulchellus* (MZ147093) |
| Z14-org-GA | SB | Orange | Brown | 45 | 10.2 | *Z. pulchellus* (MZ147094) |
| Z15-blu-TLH | TLH | Green/Blue | Green | 42 | 10.3 | *Z. sociatus* (MZ147095) |

Abbreviations: PB = Pequelle Bay, SB = Salybia Bay, TLH = Toco Lighthouse, GA = Grande L'Anse, StB = Straight Bay, WSB = Western Salybia Bay. All samples (# represents number) were collected during extreme low tide (~0.31 m depth).

Altogether specimens were collected at six sites by snorkeling along the northeast coast of Toco, Trinidad between June 2019 to February 2020. A total of 3–5 closed individual polyps were excised from colonies (specimen) using scalpel and tweezers, and placed in 1.5 mL collection vials with 95% ethanol (Carolina Biological. Burlington, NC, USA), then stored at −20 °C. Specimens are housed at the Andrews Science Building, University of Tennessee-Southern (formerly Martin Methodist College), Pulaski, Tennessee, United States. Specimen catalog nos. are TRIN-2019-001 and TRIN-2020-001.

*2.3. DNA Extraction, PCR and ITS 2*

A total of 21 zoantharian polyps were analyzed. Deoxyribonucleic acid (DNA) was extracted from specimens (30–50 mg) of zoantharian tissue following the manufacturer's protocol of an E.Z.N.A. Tissue DNA Kit (Omega BIO-TEK. Model no. D3396-02 Norcross, GA, USA). Mitochondrial cytochrome oxidase subunit 1 (COI) was amplified using the following zoantharian-specific primers (LCOant 5′-TTTTCYACTAATCATAAAGATAT 3′, COIantr 5′-GCCCACACAATAAAGCCCAATAYYCCAAT 3′) (see [21]). Polymerase chain reaction (PCR) amplifications from template DNA were carried out in a BIORAD 96-well thermocycler (Model No. MyCycler Thermal Cycler; Series No. 580BR 7657, Hercules, CA, USA) performed under the following conditions: initial set up at 95 °C for 3 min, followed by 40 cycles of denaturation at 94 °C for 30 s, annealing at 52 °C at 1 min, extension at 72 °C at 2 min, and final extension at 72 °C for 5 min (see [22]). Aliquots from PCR amplification were checked by 1.7% agarose gel electrophoresis. Each PCR product was enzymatically purified with 1.8 μL Exonuclease I, and 3.6 μL Shrimp Alkaline Phosphatase (ExoSAP, ThermoFisher Scientific, Santa Clara, CA, USA), and incubated in the PCR thermocycler at 37 °C for 30 min, followed by 95 °C for 5 min (see [20]).

The internal transcribed spacer (ITS) 2 region of Symbiodiniaceae samples were amplified using the following primers: ITSintfor2, 5′-GAATTGCAGAACTCCGTG-3′, ITS2 clamp, 5′-CGCCCGCCGCGCCCCGCGCCCGTCCCGCCGCCCCCGCCGGGATCCATATGCTTA AGTTCAGCGGGT-3′, with a "Touchdown" protocol [22–25]. Products from the PCR were electrophoresed overnight in gradient gels between 45–80% and stained with sybergreen. PCR-denaturation gradient gel electrophoresis (DGGE) gels were photographed, and distinct gel bands were excised using a scapula, then transferred to 1.5 mL Eppendorf tubes. Gel bands were excised to determine Symbiodiniaceae genera identities.

*2.4. Phylogenetic Analysis*

PCR products for COI gene were sequenced in both directions at Eurofins Genomics (Louisville, KY, USA). DNA sequences were initially inspected by eye and manually edited using Molecular Evolutionary Genetics Analysis (MEGA X, version 7.0 see [26]). Sequences were aligned using CLUSTAL W in MEGA X. Sequences were deposited in GenBank (accession numbers MZ147090-097, MZ150796-807) and were used to align with publicly accessible sequences for other zoantharians (GenBank accession JX119160, JX119164, JX119165, JX119167, JX119157, JX119156, JX119154, JX119159, JX119168, KT454365, AB214177, KF499705, KF499712). All alignments were inspected by eye using MEGA X and errors in nucleotide sequences that were low quality were trimmed prior to phylogenetic analysis. Analysis using maximum likelihood with bootstrap trees (1000 replicates) were incorporated according to the MEGA X protocols.

## 3. Results

*3.1. Zoantharian Distribution*

The lower intertidal zone measured an area of 12 km$^2$ at Salybia Bay (SB), which extends to Pequelle Bay (PB), and 7 km$^2$ at Grande L'Anse (GA/TB). SB showed benthic hard coral (Scleractinia) cover at 47.3% ± 4.5 (Mean% ± SE) and zoantharian cover at 32.4% ± 5.1 (Figure 1A), and specifically dominant concerning *Porites porites* (Pallas 1766) 47.2% ± 4.6, and zoantharians (*Palythoa* and *Zoanthus* spp. making up 31.4% ± 5.0 (Figure 1B). PB, which extends towards the eastern end of SB was dominated by zoantharians (51.3% ± 6.5, Figure 1C), with *Zoanthus* sp. and *Palythoa* sp. benthic cover as 34.1% ± 4.44 and 17.3% ± 8.8, respectively (Figure 1D). TB had a high zoantharian cover of 72.2% ± 6.0 (Figure 1E), with *Palythoa* sp. representing most of the cover at 62.6% ± 9.2, and *Zoanthus* sp. covering 8.4% ± 3.6 benthic cover (Figure 1F). Of the three sites in this study, benthic coverages were dominated by *Po. porites* and *Pa. caribaeorum*. However, there seemed to be mixed cover of *Po. porites* (24.0% ± 3.4), *Zoanthus* sp. (34.1% ± 4.1), and *Pa. caribaeorum* (17.3% ± 8.8) at PB.

Other reef benthic components included coral rubble, sand, stone, rock, sea anemones, sea urchins, fireworms, and gorgonians. Cnidaria diversity at each site also showed the presence of *Siderastrea radians* (Pallas, 1766), various octocorals, *Millipora alcicornis* Linnaeus, 1758, *Stichodactyla helianthus* (Ellis, 1768) and *Epicystis crucifer* (Le Sueur, 1817), *Porites astreoides* Lamarck, 1816, and *Diploria clivosa* (Ellis and Solander, 1786). Simpson's Index of Diversity (1–D) was 0.57 (SB), 0.65 (PB), and 0.50 (TB) showing PB with the most cnidaria diversity. All three sites showed mean zoantharian cover between 37–72%, which highlighted a zone that was extensively covered by brown mats (*P. caribaeorum*), interspersed with green covered rocks (*Zoanthus* spp.) making this feature a common characteristic at all sites. In this zone, space for algae growth is almost negligible as seen by the low cover, <10% at all sites.

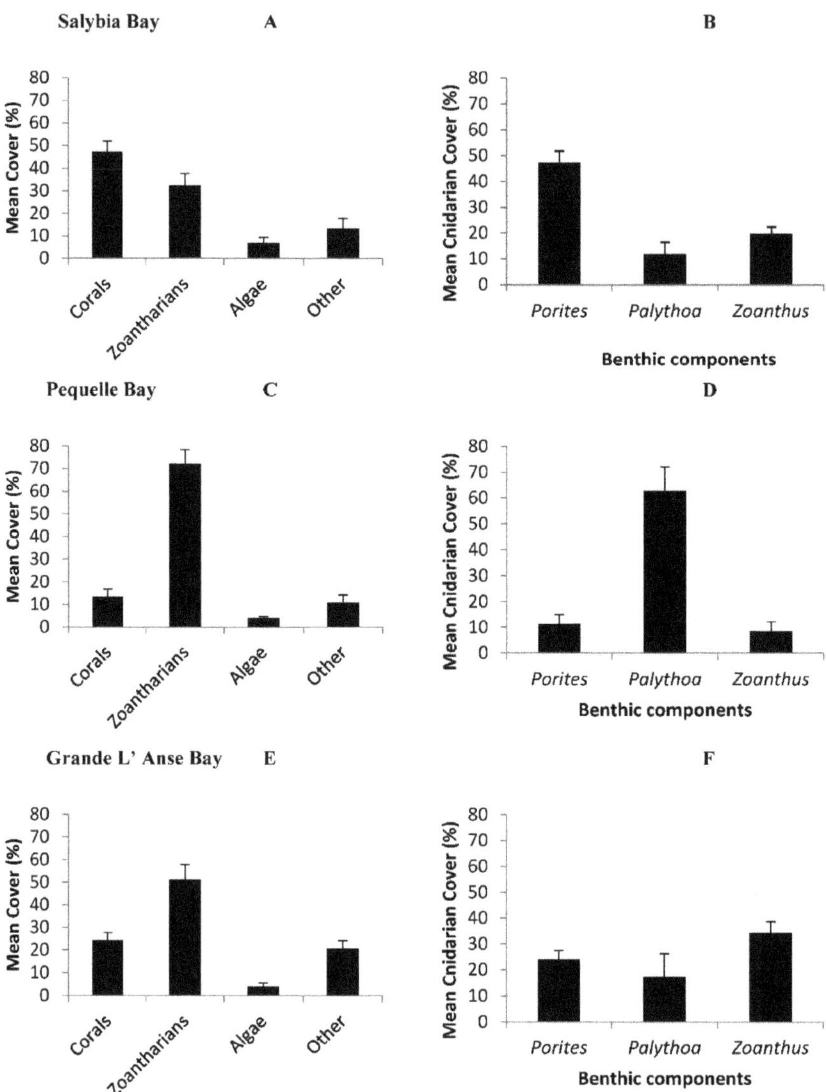


**Figure 1.** Mean benthic coverage (Mean% ± SD) at three sites in the lower intertidal zones illustrating reef-building "corals", zoantharians, algae represent green algae, and "Other" includes all other invertebrates, such as sea urchins, sea cucumbers (**A,C,E**), and mean benthic cover (Mean% ± SD) of *Porites*, *Palythoa*, and *Zoanthus* spp. (**B,D,E**); (**A,B**) Salybia Bay (**C,D**) Pequelle Bay (**E,F**) Grande L'Anse Bay.

*3.2. Specimen Morphological and Molecular Analyses*

Morphological analyses and comparisons (see Table 1) showed specimens of *Palythoa caribaeorum* colonies (Figure 1A) with various levels of coenenchyme thickness (see description in [4,21]). *P. caribaeorum* oral disks and tentacle colors varied between brown and green, in comparison to that observed for *Zoanthus* spp. at all sites, where colors ranged from bright green, orange, grey, blue, and dark green. Maximum mean oral disk diameter for *P. caribaeorum* was 12.8 mm $\pm$ 0.96 (Mean $\pm$ SD, $n$ = 15 polyps; 3 polyps per colony), which was larger than for *Zoanthus* spp. 10.2 mm $\pm$ 0.10, $n$ = 15 polyps (Table 1). Although *P. caribaeorum* and *P. grandiflora* (Figure 2B) could easily be identified through morphological analyses using maximum oral disk size, tentacle color, and tentacle numbers, analyses of *Zoanthus* spp. (Figure 2C–F) were challenging with the aforementioned characteristics. *Zoanthus* spp. colonies showed variation in coenenchyma thickness.

DNA was successfully amplified for 21 COI amplicons, which were approximately 780 bp in length. Phylogenetic analysis of the mitochondrial cytochrome oxidase subunit I (COI) gene identified zoantharians as *Zoanthus pulchellus*, and *Zoanthus sociatus* (see Figure 3) and harboring *Symbiodinium* sp. (formerly Symbiodinium clade A), whereas *Palythoa caribaeorum* and *Palythoa grandiflora* (Verrill, 1900) (see Figure 3) harbored *Cladocopium* sp. (formerly Symbiodinium clade C) in the symbiont family Symbiodiniaceae. The COI tree (Figure 3) distinguished *Palythoa* and *Zoanthus* spp., which belong to families Sphenopidae and Zoanthidae, respectively. Blasted specimens aligned with sequences from Florida (GenBank accession numbers JX119156, JX119154, JX119165) and Brazil (GenBank accession number KT454365). Other Atlantic Ocean sequences (GenBank accession numbers JX119160, JX119164, JX119167, JX119157, JX119159, JX119168 AB214177 KF499705, KF499712) aligned with closely related zoantharian species, such as *P. caribaeorum* and *P. grandiflora*; *Z. pulchellus* and *Z. vietnamensis* (Pax and Müller, 1957); *Z. sociatus* and *Z. sansibaricus* (Carlgren, 1900). For instance, *Palythoa* sp. (MZ150797-801, MZ147090-91) aligned with *Palythoa caribaeorum* (KT454365) and *Palythoa grandiflora* (JX119165). Additionally, *Zoanthus pulchellus* (JX119156) aligned with MZ150802-805, MZ147092-094, MZ156026, and was distinguished from *Zoanthus sociatus* (JX119154) aligned with MZ147095-097 and MZ150806-807, and were well supported by phylogenetic analyses (ML bootstrap% > 60%) (Figure 3).

Results for specimens in family Sphenopidae revealed moderate support for subclade *P. caribaeorum* (KT454365) and *P. grandiflora* (JX119165) (63%). Results for specimens in family Zoanthidae revealed moderate support for *Z. pulchellus* (JX119156) (69%), and high support for *Z. sociatus* (JX119154) (94%) (Figure 3). *Zoanthus pulchellus* specimens (MZ147092-094, MZ150802-805, MZ156026) also matched sequences from *Z. vietnamensis*, which is widely distributed across Indo-Pacific coral reefs, and *Zoanthus sociatus* specimens (MZ147095-097, MZ150806-807) sequences matched *Z. sansibaricus*, which is also widely distributed across Indo-Pacific coral reefs.

**Figure 2.** Zoantharian colonies along the northeastern coast of Toco, Trinidad (**A**) brown color morphotype of *Palythoa caribaeorum* specimen P3-br-SB, (**B**) green color morphotype of *Palythoa grandiflora* specimen P4-gr-GA, (**C–E**) color morphotype variation in *Zoanthus pulchellus* specimen Z9-org-GA, Z5-gr-SB, Z6-gr-SB, respectively (**F**) color morphotype in *Zoanthus sociatus* specimen Z15-blu-GA.

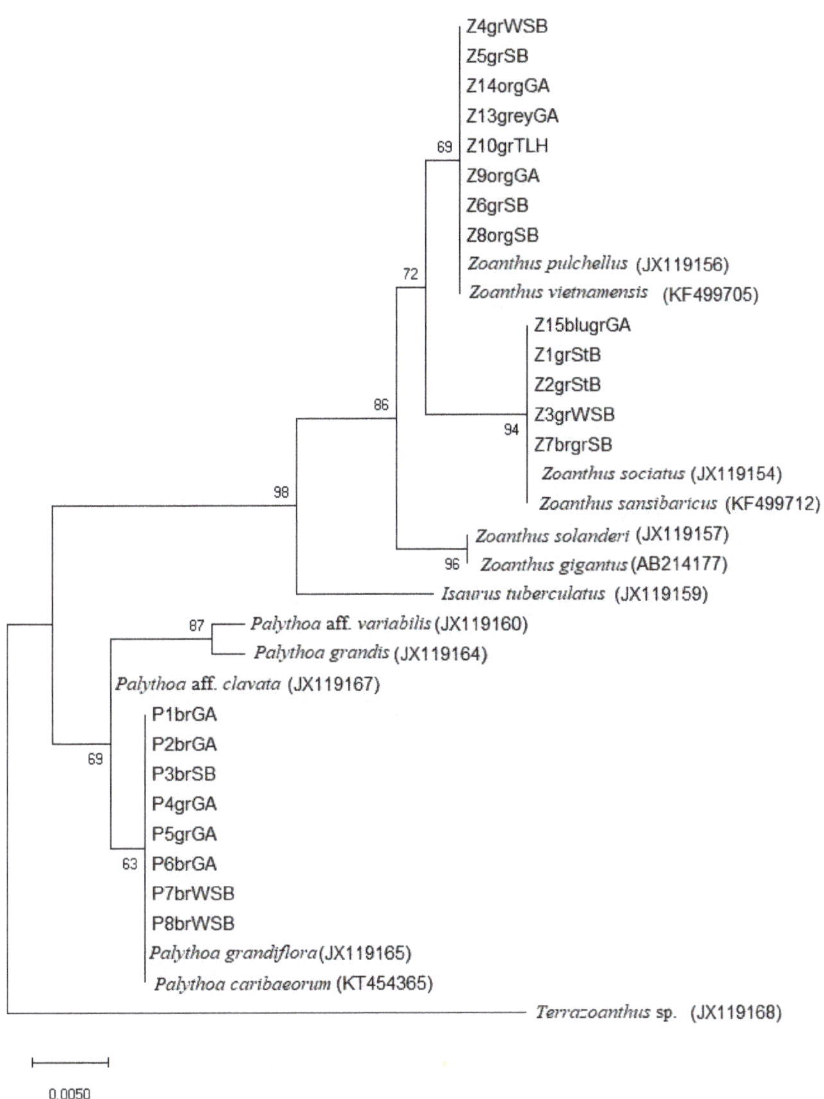



**Figure 3.** Phylogenetic maximum likelihood tree generated from the mitochondrial COI sequence alignment. GenBank accession numbers from other studies of species shown in parentheses. Numbers above branches represent maximum likelihood probabilities.

### 4. Discussion

Results showed the lower intertidal zones at Toco, Trinidad, to be dominated by zoantharians, specifically *Palythoa* and *Zoanthus* spp., similarly reported by Rabelo et al. [27] further south from Trinidad and Tobago on flat sandstone reefs in northeastern Brazil. As well, zoantharians were common in shallow-water habitats, and *Palythoa* and *Zoanthus* spp. were most common in shallow waters (<5 m), as reported at sites along the west coast of Curaçao [11]. In fact, Belford and Phillip [9,10,12,28] highlighted zoantharians being more abundant than their Scleractinia counterparts at this study's main sites. Lopez et al. [29] observed extensive zoantharian coverage for species *Zoanthus solanderi* and *Zoanthus sociatus* at a "zoanthid zone" located at Cabo Verde Islands, central eastern Atlantic,. Additionally, Karlson [6] observed similar extensive *Zoanthus* spp. coverage at Discovery Bay, Jamaica. This study reports a similar presence of a zoantharians covering the majority of the benthic lower intertidal zone, dominated by *Palythoa* and *Zoanthus* spp. in the southern-most part of the Caribbean.

Although zoantharian coverage was extensive, identification of zoantharians was difficult, because color morphotypes varied among sites. Similar observations were mentioned by Reimer et al. [20] related to zoantharian identification in the field, and in other studies [30]. Phylogenetic and morphological analyses using the COI marker revealed *Palythoa* brown and green color morphs were specifically *Palythoa caribaeorum* and *Palythoa grandiflora*, respectively, while green, orange, grey, and blue-green were *Zoanthus pulchellus* and *Zoanthus sociatus*, respectively. However, it is worth mentioning that caution should be taken since only the COI marker was used in this study. For instance, Sinniger et al. [31] noted that although the COI marker is easily amplifiable with universal primers, and was hence used in this study, the addition of the mitochondrial 16S ribosomal DNA marker is useful for comparison and further species identification. In fact, the addition of 16S sequences adds distinct advantages as they are slightly more variable in zoantharians than COI, thereby adding useful phylogenetic information [31].

Closed polyps are often observed during extreme low tides, where desiccation plays a factor in distribution; however, Rabelo et al. [27] reported *Z. sociatus* resisting desiccation better than *P. caribaeorum*. However, in this study, the reverse was observed, where *P. caribaeorum* had a significantly higher benthic coverage than *Zoanthus* spp. at lower intertidal zones, nevertheless zoantharian distribution appears to be related to desiccation tolerance [27]. It is not precisely known why the reverse of [27] was observed in this study, however family Zoanthidae, such as *Z. pulchellus* and *Z. sociatus*, have generally been known to adapt to different environments [10,19]. Reimer et al. [20] reported variation in *Palythoa* sp. polyp form and color as a result of variable environments, such as degree of wave action, and benthic type, which were characteristics similarly observed at Toco, Trinidad [9,10,12,28].

Zoantharian color morphotypes were observed in both high and low wave action habitats, especially at the lower intertidal zones where *P. caribaeorum* carpeted much of the lower intertidal zones. However, *Z. pulchellus* mainly covered individual rocks, or crevasses, and displayed many colors throughout the intertidal zone. *Z. sociatus* also was found within crevasses, however it was not observed to be found in extensive coverage at sites in this study. It should be noted that extreme caution should be taken while surveying zoantharian coverage using color morphotypes to identify zoantharians since species may be conspecifics [8,19]. This study successfully used molecular analyses to assist with identification of *Zoanthus pulchellus* and *Zoanthus sociatus*. *Zoanthus* spp. display phenotypic plasticity in both oral disk color and polyp height [27]. Additionally, molecular analyses assisted in identifying *P. caribaeorum* and *P. grandiflora*; however it should be noted that these two species can be distinguished from morphological characteristics.

Further phylogenetic analyses of zoantharians confirmed the Symbiodiniaceae genera *Cladocopium* and *Symbiodinium* in *Palythoa* and *Zoanthus* spp., respectively. These results are consistent to past analyses of both species at different locations in the Caribbean Sea and Atlantic [31–33]. Similar results for both *Zoanthus* species mentioned in this study hosting *Cladocopium* and *Symbiodinium* were reported at Cape Verde Archipelagos [30]. As global climate will continue to affect oceanic water temperatures [34], identification and distribution of zooxanthellate zoantharians such as in this study will provide important baseline data for future analyses.

**Funding:** This research was funded by contributors to Martin Methodist College Biology Travel fund by Lou Foster and Emily White. Other funding was received from the American Museum of Natural History Lerner-Gray Fund for marine research.

**Institutional Review Board Statement:** Ethical review and approval were waived for this study, because zoantharians were measured in situ, and genetic analyses on polyps did not require colony destruction.

**Data Availability Statement:** Photographs of zoantharian color morphotypes along the northeastern coast of Toco, Trinidad are available online at the Digital Public Library of America and Martin Methodist College Marine Biology Collection websites: https://dp.la/search?q=martin+methodist+college https://www.artstor.org/2016/09/26/case-study-going-underwater-with-shared-shelf-commons/ (accessed on 9 June 2021).

**Acknowledgments:** I am grateful to the University of Tennessee Southern for use of laboratory space and equipment. Field equipment was essential in this study, therefore funding by contributors was appreciated. Thanks to all volunteer assistance during data collection from undergraduate research students and local interested citizens of Trinidad and Tobago. I am extremely grateful for advice and laboratory use from Todd LaJeunesse (Pennsylvania State University), Scott Santos (Auburn University), and Jeff Leblond and Dennis Mullen (Middle Tennessee State University). I am extremely grateful for advice on manuscript improvement by James Reimer (University of Ryukyus), and Douglas Dorer (University of Tennessee Southern) for assistance with molecular analyses of sequences.

**Conflicts of Interest:** There were no conflict of interest for this contribution.

# References

1. Karlson, R.H. Alternate competitive strategies in a periodically disturbed habitat. *Bull. Mar. Sci.* **1980**, *30*, 894–900.
2. Irei, Y.; Nozawa, Y.; Reimer, J.D. Distribution patterns of five zoanthid species in Okinawa Island, Japan. *Zool. Stud.* **2011**, *50*, 426–433.
3. Irei, Y.; Sinniger, F.; Reimer, J.D. Description of two azooxanthellate *Palythoa* species (Subclass Hexacorallia, Order Zoantharia) from Ryukyu Archipelago, southern Japan. *Zookeys* **2015**, *478*, 1–26. [CrossRef] [PubMed]
4. Reimer, J.D.; Shusuke, O.; Takishita, K.; Tsukahara, J.; Maruyama, T. Molecular evidence suggesting species in the zoanthid genera *Palythoa* and *Protopalythoa* (Anthozoa: Hexacorallia) are congeneric. *Zool. Sci.* **2006**, *23*, 87–94. [CrossRef] [PubMed]
5. Sinniger, F.; Montoya-Burgos, J.I.; Chevaldonné, P.; Pawlowski, J. Phylogeny of the order Zoantharia (Anthozoa, Hexacorallia) based on the mitochondrial ribosomal genes. *Mar. Biol.* **2005**, *147*, 1121–1128. [CrossRef]
6. Karlson, R.H. Reproduction patterns on *Zoanthus* spp. from Discovery Bay, Jamaica. Proceedings 4th International Coral Reef Symposium. *Manilla* **1981**, *2*, 699–704.
7. Sebens, P.S. Intertidal distribution of zoanthids on the Caribbean coast of Panama: Effects of predation and desiccation. *Bull. Mar. Sci.* **1982**, *32*, 316–335.
8. Reimer, J.D.; Ono, S.; Fujiwara, Y.; Takishita, K.; Tsukahara, J. Reconsidering *Zoanthus* spp. diversity: Molecular evidence of conspecificity within four previously presumed species. *Zool. Sci.* **2004**, *21*, 517–525. [CrossRef] [PubMed]
9. Belford, S.G.; Phillip, D.A.T. Rapid assessment of a coral reef community in a marginal habitat in the southern Caribbean: S simple way to know what's out there. *Asian J. Biol. Sci.* **2011**, *4*, 520–531. [CrossRef]
10. Belford, S.G.; Phillip, D.A.T. Intertidal distribution patterns of zoanthids compared to their scleractinian counterparts in the southern Caribbean. *Int. J. Oceanogr. Mar. Ecol. Syst.* **2012**, *3*, 67–75. [CrossRef]
11. Reimer, J.D.; Wee, H.B.; García-Hernández, J.E.; Hoeksema, B.W. Zoantharia (Anthozoa: Hexacorallia) abundance and associations with Porifera and Hydrozoa across a depth gradient on the west coast of Curaçao. *Syst. Biodivers.* **2018**, *16*, 820–830. [CrossRef]
12. Belford, S.G.; Phillip, D.A.T.; Rutherford, M.G.; Schmidt, R.S.; Duncan, E.J. Biodiversity of coral reef communities in marginal environments along the north-eastern coast of Trinidad, southern Caribbean. *Prog. Aqu. Farm. Mar. Biol.* **2019**, *2*, 180017.
13. Goreau, T.F. The ecology of Jamaican coral reefs I. Species composition and zonation. *Ecology* **1959**, *40*, 67–90. [CrossRef]
14. Burnett, W.J.; Benzie, J.A.H.; Beardmore, J.A.; Ryland, J.S. Zoanthids (Anthozoa, Hexacorallia) from the Great Barrier Reef and Torres Straights, Australia: Systematics, evolution and a key to species. *Coral Reefs* **1997**, *16*, 55–68. [CrossRef]
15. Ryland, J.S.; Lancaster, J.E. Revision of methods for separating species of *Protopalythoa* (Hexacorallia: Zoanthidea) in the tropical West Pacific. *Invert. Syst.* **2003**, *17*, 407–428. [CrossRef]
16. López, C.; Reimer, J.D.; Brito, A.; Simón, D.; Clemente, S.; Hernández, M. Diversity of zoantharian species and their symbionts from Macaronesian and Cape Verde ecoregions demonstrates their widespread distribution in the Atlantic Ocean. *Coral Reefs* **2019**, *38*, 269–283. [CrossRef]
17. LaJeunesse, T.C.; Parkinson, J.E.; Gabrielson, P.W.; Jeong, H.J.; Reimer, J.D.; Voolstra, C.R.; Santos, S.S. Systematic revision of Symbiodiniaceae highlights the antiquity and diversity of coral endosymbiont. *Curr. Biol.* **2018**, *28*, 2570–2580. [CrossRef] [PubMed]
18. Burnett, W.J. Longitudinal variation in algal symbionts (zooxanthellae) from the Indian Ocean zoanthid *Palythoa caesia*. *Mar. Ecol. Prog. Ser.* **2002**, *234*, 105–109. [CrossRef]

19. Reimer, J.D.; Shusuke, O.; Yasuo, F.; Junzo, T. Seasonal changes in morphological condition of symbiotic dinoflagellates (*Symbiodinium* spp.) in *Zoanthus sansibaricus* (Anthozoa: Hexacorallia) in Southern Japan. *South Pac. Stud.* **2007**, *27*, 2.

20. Reimer, J.D.; Foord, C.; Irei, Y. Species diversity of shallow water zoanthids (Cnidaria: Anthozoa: Hexacorallia) in Florida. *Hindawi Publ. Corp. J. Mar. Biol.* **2012**, *2012*, 856079. [CrossRef]

21. Pax, F. Studien an westindischen Actinien. In *Ergebnisse einer Zoologischen nach Westindien von Prof. W. Kukenthal und Dr. R. Hartmeyer im Jahre, 1907*; Spengel, J.W., Ed.; Zoologische Jahrbucher Supplement: Ann Arbor, MI, USA, 1910; Volume 11, pp. 157–330.

22. Sinniger, F.; Reimer, J.; Pawlowski, J. The Parazoanthidae (Hexacorallia: Zoantharia) DNA taxonomy: Description of two new genera. *Mar. Biodivers.* **2010**, *40*, 57–70. [CrossRef]

23. LaJeunesse, T.C.; Trench, R. Biogeography of two species of *Symbiodinium* (Freudenthal) inhabiting the intertidal sea anemone *Anthopleura elegantissima* (Brandt). *Biol. Bull.* **2000**, *199*, 126–134. [CrossRef] [PubMed]

24. LaJeunesse, T.C. Diversity and community structure of symbiotic dinoflagellates from Caribbean coral reefs. *Mar. Biol.* **2002**, *141*, 387–400.

25. LaJeunesse, T.C.; Thornhill, D.J. Improved resolution of reef-coral endosymbiotic dinoflagellate (*Symbiodinium*) species diversity, ecology, and evolution through psbA non-coding region genotyping. *PLoS ONE* **2011**, *6*, e29013. [CrossRef] [PubMed]

26. Kumar, S.; Strecher, G.; Li, M.; Knyaz, C.; Tamura, K. Molecular evolutionary genetic analysis across computing platforms. *Mol. Biol. Evol.* **2018**, *35*, 1547–1549. [CrossRef] [PubMed]

27. Rabelo, E.F.; Soares, M.D.; Bezerra, L.E.; Matthews-Cascon, H. Distribution patterns of zoanthids (Cnidaria: Zoantharia) on a tropical reef. *Mar. Biol. Res.* **2015**, *11*, 584–592. [CrossRef]

28. Belford, S.G. Spatial abundance and colour morphotype densities of the rock boring sea urchin (*Echinometra lucunter*) at two different habitats. *Thalassas* **2020**, *36*, 157–164. [CrossRef]

29. López, C.; Freitas, R.; Magileviciute, E.; Ratão, S.S.; Brehmer, P.; Reimer, J.D. Report of a *Zoanthus* zone from the Cabo Verde islands (Central eastern Atlantic). *Thalassas* **2018**, *34*, 409–413. [CrossRef]

30. Ong, C.W.; Reimer, J.D.; Todd, P.A. Morphological plastic responses to shading in the zoanthids *Zoanthus sansibaricus* and *Palythoa tuberculosa*. *Mar. Biol.* **2013**, *160*, 1053–1064. [CrossRef]

31. Sinniger, F.; Reimer, J.D.; Pawlowski, J. Potential of DNA sequences to identify zoanthids (Cnidaria: Zoantharia). *Zool. Sci.* **2008**, *25*, 1253–1260. [CrossRef] [PubMed]

32. LaJeunesse, T.C.; Loh, W.; Van Woesik, R.; Hoegh-Guldberg, O.; Schmidt, G.; Fitt, W. Low symbiont diversity in southern Great Barrier Reef corals, relative to those of the Caribbean. *Am. Soc. Limnol. Oceanogr.* **2003**, *48*, 2046–2054. [CrossRef]

33. Kumara, S.; Zacharia, P.U.; Sreenath, K.R.; Kripa, V.; George, G. GIS based mapping of zoanthids along Saurashtra coast, Gujarat, India. *J. Mar. Biol. Assoc. India* **2017**, *59*, 19–25.

34. Hughes, T.P.; Baird, A.H.; Bellwood, D.R.; Card, M.; Connolly, S.R.; Folke, C.; Grosberg, R.; Hoegh-Guldberg, O.; Jackson, J.B.C.; Kleypas, J.; et al. Climate change human impacts and the resilience of coral reefs. *Science* **2003**, *301*, 929–933. [CrossRef] [PubMed]

 *oceans*

*Article*

# Molecular Phylogenetics of *Trapezia* Crabs in the Central Mexican Pacific

**Hazel M. Canizales-Flores, Alma P. Rodríguez-Troncoso *, Eric Bautista-Guerrero and Amílcar L. Cupul-Magaña**

Laboratorio de Ecología Marina, Centro Universitario de la Costa, Universidad de Guadalajara, Av. Universidad No. 203. Puerto Vallarta, Jalisco 48280, Mexico; Hazel_MariaCF@hotmail.com (H.M.C.-F.); ericbguerrero@gmail.com (E.B.-G.); amilcar.cupul@gmail.com (A.L.C.-M.)
* Correspondence: pao.rodriguezt@gmail.com; Tel.: +52-322-226-2319

Received: 15 July 2020; Accepted: 24 August 2020; Published: 26 August 2020

**Abstract:** To date, *Trapezia* spp. crabs have been considered obligate symbionts of pocilloporid corals. They protect their coral hosts from predators and are essential for the health of certain coral species. However, the basic details of this group of crustaceans are lacking, and there is a need for species-level molecular markers. The Tropical Eastern Pacific (TEP) region harbors important coral communities mainly built by corals of the genus *Pocillopora*, with three known *Trapezia* species known to associate with them: *Trapezia bidentata, T. formosa* and *T. corallina*. Both taxonomic and molecular analyses were carried out with samples of all three crab species collected from *Pocillopora* spp. in the Central Mexican Pacific. Analysis of both a mitochondrial and a nuclear gene revealed only two species, *T. corallina* and *T. bidentata. T. formosa* however appears to be a morphotype of *T. bidentata.* The use of integrative taxonomy for this group has increased the knowledge of the biodiversity not only of the study area, but of the whole TEP and will enhance the future study of the *Trapezia–Pocillopora* symbiosis.

**Keywords:** coral reefs; cox1; H3; crustacea; molecular systematics; morphotypes

---

## 1. Introduction

Coral communities are highly productive ecosystems that harbor a high biodiversity and biomass of small crustaceans [1], among which decapods constitute nearly one-third of the total species [2]. Their relevance is associated not only with the high biomass they represent for the ecosystem, since Crustaceans can affect and/or modify their communities through herbivory, predation, consumption of detritus and nonorganic components and oxygenation of the sediment through burrow construction [3]. Furthermore, crustaceans form numerous, complex trophic relationships that are prone to change in response to both natural and anthropogenic stressors that can threaten the integrity of the coral reef ecosystem, including of crustacean assemblages [4].

Members of the genus *Trapezia* (Decapoda, Brachyura, Trapeziidae) are ecologically important crustaceans that form obligate symbioses with hermatypic corals of the most widely distributed and abundant genus of the Tropical Eastern Pacific (TEP): *Pocillopora* [5]. *Trapezia* crabs live their entire lives, sheltered among the coral branches, providing to the whole colony a defense system against conspecifics and against predators that threaten the coral, such as the corallivorous crown-of-thorns sea star (*Acanthaster* sp.) [6]. The crabs not only defend the coral, but they remove detritus and promote branch elongation [7]. In exchange, the crabs benefit from the coral-provided shelter, as well as nourishment in the form of the detritus that accumulates in the coral branch mucus [8]. In terms of within-colony *Trapezia* distribution, there are typically size-based hierarchies in which mature couples dominate and are located among the central branches. Smaller-sized individuals (new recruits, juveniles and even adults) mainly live at the base of the coral colony [8], promoting a hierarchy in the use of resources according to the size of the organisms. Due the complex ecological role of the carbs, the *Trapezia–Pocillopora* relationship is now considered as mutualism rather than an obligate commensalism [9,10].

The distribution of *Trapezia* predominantly mirrors that of their coral hosts [11]. Despite the TEP region having been historically characterized as a region with suboptimal conditions for coral growth/development, there are important coral communities present. However, they do face seasonal upwelling, internal waves, high turbidity, eutrophic conditions, high interannual variation in seawater quality and frequent and intense ENSO events [12,13], all of which can negatively affect corals, decreasing their cover and therefore the available habitat and food supply for the crabs. Nevertheless, the TEP harbors well-developed, shallow coral communities, and the northern TEP is considered especially biodiverse [14,15].

To date four *Trapezia* species have been recorded in the TEP: *Trapezia digitalis* and *Trapezia ferruginea*, which are widely distributed across the Indian and Pacific oceans and the endemic *T. corallina* and *T. formosa* [16]. Nevertheless, due to the almost imperceptible morphologic differences among species, some taxonomists have considered there to be only two [16]: *T. corallina* (synonymized with *T. digitalis*) and *T. formosa* (synonymized with both *T. bidentata* and *T. ferruginea*). Therefore, despite the existence of historical records of *Trapezia* within pocilloporid corals, their taxonomic classification remains unclear. Given (1) their importance to coral and coral reef health, (2) the inability to distinguish between putatively different species and (3) the need to understand whether certain crab species demonstrate specificity to (or a preference for) certain coral species, we sought to use morphometric and molecular taxonomic tools to better understand the ecology of the crab–coral symbiosis in the TEP.

## 2. Materials and Methods

### 2.1. Study Site

*Trapezia* spp. crabs were sampled from *Pocillopora* spp. colonies located in the Islas Marietas National Park (IMNP); a coral community located in the Central Mexican Pacific (CMP) region (Figure 1). This Natural Protected Area (NPA) harbors one of the most important coral communities of the Northeastern Tropical Pacific, one composed mainly of branching corals of the genus *Pocillopora* and massive and submassive corals of the genera *Porites* and *Pavona*. These three genera constitute most the hard coral cover [17] from 0 m to 20 m, as is typical elsewhere in the TEP [18]. The NPA is an oceanographic transition area where three oceanic currents converge: the Costa Rican Coastal Current, the California Current and the water mass of the Gulf of California [19]. The surface ocean temperature ranges from 23 to 30 °C during the year, with minimum and maximum values in March and September, respectively. Local conditions, such as upwelling and internal waves, cause the appearance of both daily fluctuations of up to 5 °C [20] and of a thermocline at depths of up to 20 m, resulting in highly variable conditions of sea temperature and dissolved oxygen [19,21,22]. This region is influenced by seasonal hurricanes, storms and natural stressors, and also by ENSO events that are associated with abnormal increases (El Niño) or decreases (La Niña) in surface temperature for several weeks or even months. Such temperature anomalies can affect the corals detrimentally, as can other environmental factors, such as altered pH and nutrient levels [23].

### 2.2. Sampling and Taxonomic Identification

*Trapezia* crabs were collected at three sites within IMNP (Figure 1). Each crab was obtained from a different *Pocillopora* colony at a depth of 5–7 m. Crabs were collected carefully (using metallic tweezers to extract them from the live coral colony) and immediately placed individually in 50-mL plastic tubes filled with seawater. Following collection, the crabs were preserved in 96% ethanol and transported to the laboratory. Crabs were then observed using a stereoscope (Stemi 508-Zeiss®, Oberkochen, Germany) and classified to species level following the taxonomic criteria described by Castro [16,24]; carapace shape and size, color, cheliped size and propodus color were all considered. All crabs were photo-documented using a Canon Elph PowerShot 180 camera (Canon Inc., Tokyo,

Japan), and, prior to long-term storage, a sample of muscle tissue was dissected from the cephalothorax and preserved in 96% ethanol for DNA extraction. Carapace and cheliped width and length were measured individually using Image J software (Open Source, Developed at the National Institutes of Health and the Laboratory for Optical and Computational Instrumentation, University of Wisconsin, Madison, WI, USA); values are expressed per species and sex as mean mm ± standard deviation.

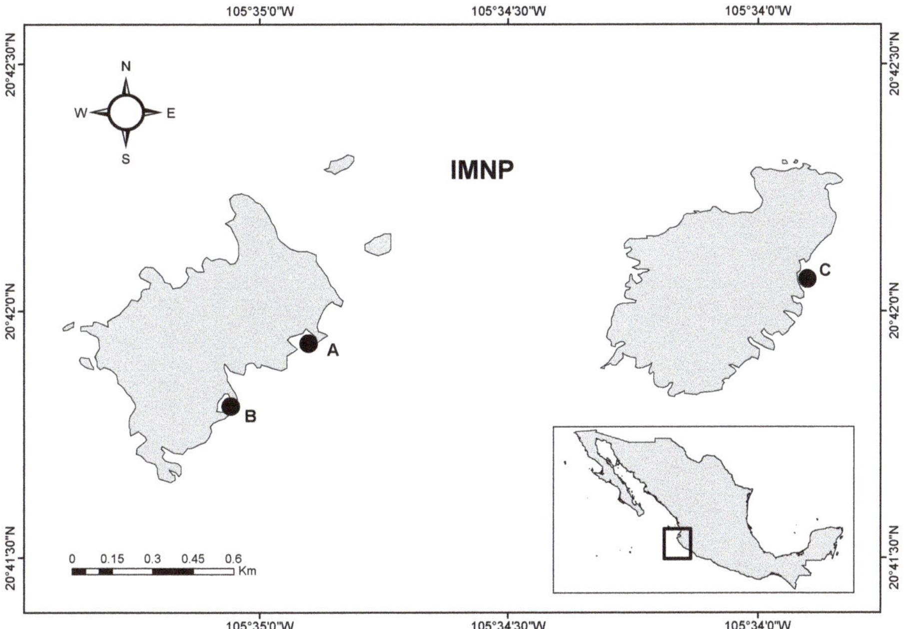


**Figure 1.** Study area in the Islas Marietas National Park (IMNP), Nayarit, Mexico, showing the location of sampling sites. All sampling sites have a high cover of *Pocillopora* sp. (15–33%). (**A**) Zona de Restauración (20.698860° N, 105.580997° W); (**B**) Cueva del Muerto (20.697389° N, 105.582806° W) at Isla Larga; (**C**) Plataforma Pavonas (20.700908° N, 105.565304° W) at Isla Redonda.

### 2.3. DNA Extraction, Amplification and Sequencing

Genomic DNA from the muscle tissue was extracted using the Wizard® Genomic DNA Purification kit (Promega, Madison, WI, USA) according to the manufacturer's protocol. Partial sequences of the nuclear gene histone H3 (H3) and the mitochondrial gene cytochrome c oxidase subunit 1 (*cox*1) were PCR amplified as follows: H3 gene fragments (347 bp) were amplified using the primers H3af 5′-ATGGCTCGTACCAAGCAGACVGC-3′ and H3ar 5′-ATATCCTTRGGCATRATRGTGAC-3′ of Lai et al. [25]. *Cox*1 gene fragments (600 bp) were amplified using the primers LCOI490 (5′-GGTCAACAAATCATAAAGAYATYGG-3′) and HCOI21908 (5′-TAAACTTCAGGGTGACCAAAR AAYCA-3′) from Folmer et al. [26]. PCR mix comprised 7.23 μL nuclease-free of $H_2O$, 0.75 μL of $MgCl_2$, 0.66 μL of dNTPs, 2.5 μL of 10× buffer, 0.13 μL of each primer, 0.10 μL of Taq polymerase (Promega) and 1.2 μL of DNA0. After an initial denaturation step at 94 °C for 5 min, 40 cycles, comprising 94 °C for 1 min, 62 °C for 1 min (H3) or 48.4 °C for 1 min (*cox*1) and 72 °C for 1 min, were carried out, with a final extension step of 5 min at 72 °C.

PCR products were visualized on 2%-TAE (Tris-acetate–EDTA) agarose gels. The final products were purified with the Wizard SV gel and PCR clean-up system (both from Promega) and sent to Macrogen, Inc. (Seoul, Korea). The forward and reverse sequences obtained were manually edited using Geneious Prime 2019.2.3 software in order to obtain a consensus sequence for each gene.

To confirm the species identity, the consensus sequences were queried against the National Center for Biotechnology Information database (i.e., GenBank) via BLAST. Each gene sequence was submitted to NCBI with the following accession numbers: (H3) MT720697, MT720698, MT720699, MT720700, MT720701, MT720702, MT720703; (cox1) MN852247, MN852248, MN852249, MN852250, MN852251, MN852252, MN852253, MN852254.

Some additional sequences from members of the family Trapeziidae were obtained from GenBank and aligned with the sequences generated herein using Clustal W within Mega-X [27]. It is important to emphasize that, to date, there are no *T. corallina* and *T. formosa* H3 and *cox*1 sequences available and that for *T. bidentata* no published H3 sequences exist. Molecular data sets (H3 and *cox*1) were analyzed with maximum likelihood (ML) and Bayesian information (BI) methods.-ML analyses were performed using Mega-X software with 1000 bootstraps. TN93+G+I (Tamura–Nei model+gamma distributed rate of substitution+estimated proportion of invariant sites) was the best-fitting nucleotide substitution model for *cox*1, with T92+G (Tamura 3-parameter+gamma distributed rate of substitution) the ML best-fit for H3. Bayesian analyses were conducted with MrBayes v.3.1.2 [28] and appropriate DNA substitution models were determined separately for the H3 and *cox*1 datasets. The Bayesian analysis was conducted by computing 10,000 Markov chain Monte Carlo generations with 7 H3 and 8 *cox*1 sequences. Using Mega-X software and *cox*1 marker, pairwise genetic distances were calculated among the *T. bidentata* and *T. formosa* sequences from the present work, as well as *T. bidentata* sequences from GenBank. Finally, the species *Kempina mikado* (Crustacea: Stomatopoda) was used as the outgroup.

## 3. Results

### 3.1. Morphologic Identification

A total of 26 crabs were sexed and identified taxonomically based on coloration patterns, the shape of the cephalothorax and the size and shape of the chelipeds. Taxonomic characteristics based on Castro [16,24] determined that the crabs belonged to three *Trapezia* species, and a mix of males, females and juveniles were obtained and described as follows (Figure 2): The crabs identified as *T. formosa* (3 females, 4 males) were all characterized as small adults, with cephalothoracic widths ranging from 4 to 7 mm (Table 1). They usually presented a reddish-orange color, with the lower margin of the propodus (chelipeds) yellowish and with chelipeds lengths 3-fold greater than widths (Table 1). Their carapaces were globose, with the anterolateral sides strongly curved (up to 45°; Figure 2A). The crabs identified as *T. bidentata* (7 females, 5 males and 1 juvenile) showed most of the morphologic characteristics of *T. formosa* with the key difference being that the specimens identified as *T. bidentata* showed a darker red coloration and presented less curved anterolateral sides of the carapace (Figure 2B); for this species the specimens examined presented higher carapace and cheliped sizes than *T. formosa* (Table 1). Finally, the crabs identified as *T. corallina* (5 males and 1 juvenile) presented a dark orange-brown color, brown reticulations on the propodus, and, in general, thick and bulky chelipeds; they were also characterized by a complete suture between the (i) second and third thoracic sternites and (ii) the ischium of the endognath of the third pair of maxillipeds; also, this species showed a rough appearance in the middle portion of the inner margin (Figure 2C). Finally, *T. corallina* presented both the largest cheliped and largest carapace of the species collected (Table 1).

**Table 1.** *Trapezia* carapace and cheliped measurements per species and sex. Values reported as means ± SD. N/A corresponds to no available data.

| Species | Carapace Width | | Carapace Length | | Cheliped Width | | Cheliped Length | |
| --- | --- | --- | --- | --- | --- | --- | --- | --- |
| | Males | Females | Males | Females | Males | Females | Males | Females |
| *Trapezia formosa* | 6.10 ± 0.99 | 5.90 ± 1.31 | 5.06 ± 0.64 | 5.19 ± 0.78 | 2.54 ± 0.50 | 2.85 ± 0.87 | 6.41 ± 1.27 | 6.09 ± 0.52 |
| *Trapezia bidentata* | 6.95 ± 0.78 | 8.93 ± 2.43 | 5.86 ± 1.76 | 6.32 ± 1.91 | 3.26 ± 0.19 | 3.15 ± 1.52 | 7.69 ± 1.83 | 6.156 ± 0.921 |
| *Trapezia corallina* | 9.03 ± 1.56 | N/A | 8.17 ± 3.01 | N/A | 4.19 ± 0.52 | N/A | 9.78 ± 2.33 | N/A |

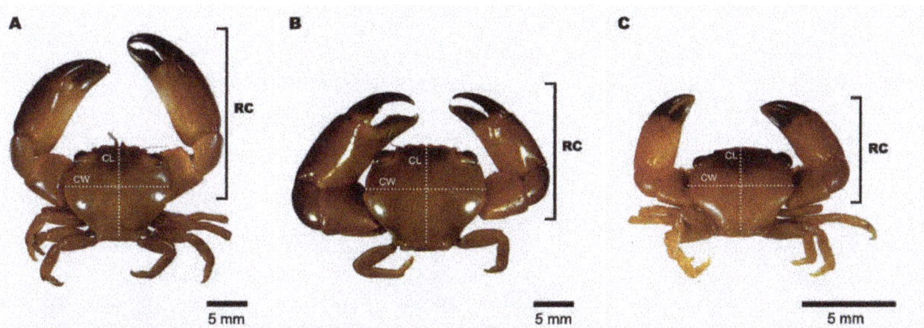

**Figure 2.** *Trapezia* crabs of the Tropical Eastern Pacific. (**A**) *Trapezia formosa*; (**B**) *T. bidentata*; (**C**) *T. corallina*. CW—carapace width; CL—carapace length; RC—right cheliped.

## 3.2. Molecular Identification

BLAST results revealed that all sequences analyzed (*n* = 15) were 99–100% similar to those of published sequences from the *Trapezia* genus on GenBank. Phylogenetic trees were built for each gene from the sequences obtained (Figures 3 and 4) within the Trapeziidae family: seven TEP H3 sequences (264 bp) and eight TEP *cox*1 sequences (582 bp). Both ML and BI methods revealed that both molecular markers clustered the species *T. bidentata* and *T. formosa* into the same group; this result was also supported by high posterior probabilities and bootstrapping (Figures 3 and 4). *T. corallina* clustered with *T. tigrina* in a different, sister group. Pairwise genetic distances were zero between *T. bidentata* and *T. formosa*. When comparing *cox*1 sequences from *T. bidentata* and *T. formosa* from the present work with those from *T. bidentata* obtained from GenBank, pairwise genetic distances were 0.001.

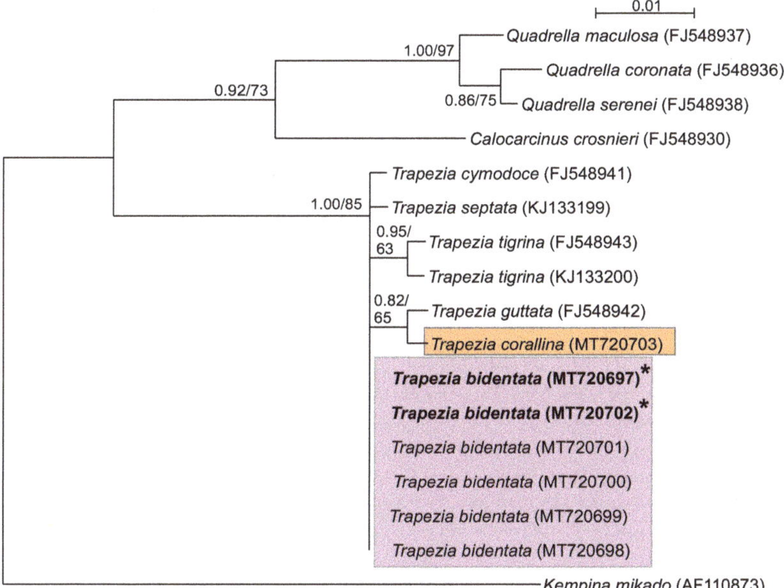


**Figure 3.** Consensus tree of the Trapeziidae family obtained using maximum likelihood and Bayesian information (BI) methods based on the H3 nDNA gene. Bootstrap values/posterior probabilities (ML/BI) are indicated under the principal node. Clade A: Quadrelliinae; B: Calocarciniinae; and C: Trapeziinae. *Kempina mikado* (Crustacea: Stomatopoda) was used as the outgroup. * = Denotes sequences of crabs taxonomically identified as *T. formosa*.

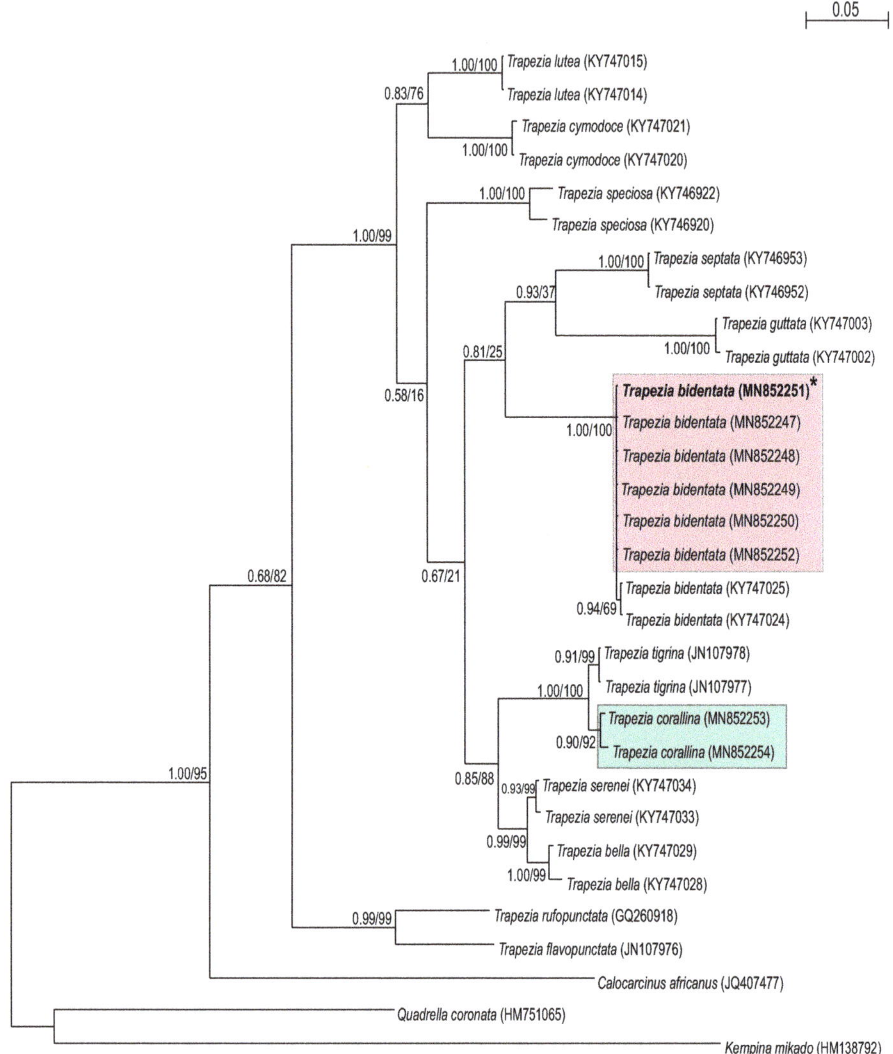

**Figure 4.** Consensus tree of the Trapeziidae family obtained using maximum likelihood and Bayesian information (BI) analyses based on the *cox*1 mtDNA gene. Bootstrap values/posterior probabilities (ML/BI) are indicated under the principal node. Clades A: Trapeziinae; B: Calocarciniinae; and C: Quadrelliinae. *Kempina mikado* (Crustacea: Stomatopoda) was used as the outgroup. * = Denotes sequences of crabs taxonomically identified as *T. formosa*.

## 4. Discussion

The use of genetic tools for investigating crustacean phylogenetics is widespread and both *cox*1 and H3 have proven suitable for resolving crab identities to species level [25,29,30]. By comparing *cox1* sequences of specimens with published available data for related taxa, we identified two clades: *T. corallina* and *T. bidentata*. The *cox1* mitochondrial gene, while overall highly conserved [31], contains a hyper-variable region that is useful for taxonomy. By performing the same procedure with the H3 gene, the same two clades were resolved: *T. corallina* and *T. bidentata.* The H3 nuclear gene evolves more slowly than mitochondrial genes [32], but nevertheless corroborated the *cox1* findings.

The delimitation of species is problematic for recently diverged lineages that are not yet morphologic distinct. Furthermore, phenotypic convergence can also lead to low morphologic divergence [33]. The integrative use of morphologic (traditional taxonomy) and genetic identification approaches employed herein could be extended to include behavioral and ecological characteristic [34,35] so as to serve as an even more holistic identification method. Using traditional taxonomic characterizations alone, *T. corallina, T. formosa* and *T. bidentata* were considered three different species. This conclusion was based on taxonomical parameters that were generally considered valid for discrimination between decapod species: the shape of the carapace, chelipeds and external coloration [24,25,36–39]. However, the molecular data support the conclusion that *T. formosa* is a morphotype of the species *T. bidentata* and thus should not be considered as a separate species.

Accurate identification of members of the *Trapezia* genus is of value not only for the future study of decapods, but also for study of their coral hosts. *Pocillopora* is considered the main reef-builder of the TEP region [40–42] and, unlike dominant corals in other parts of the Indo-Pacific, are markedly resilient to environmental change [43,44]. Given that adult healthy colonies seem always to be associated with at least one *Trapezia* sp. crab, it is possible that this high acclimatization capacity is linked to crab presence [40,45]. It is nevertheless unclear what implications future climate change-driven impacts will have on this mutualistic symbiosis and a more rigorous assessment of ecological implications of crab–coral mutualism in the TEP would be of value. In this case additional molecular markers and other traits (e.g., behavior or physiology of the crabs) could be incorporated into an investigation in order to gain new insights, including into the role of commensals in promoting resilience of the TEP's coral communities.

**Author Contributions:** H.M.C.-F., A.P.R.-T., A.L.C.-M. and E.B.-G. carried out the research. H.M.C.-F. and E.B.-G. analyzed the data. A.P.R.-T., E.B.-G. and A.L.C.-M. contributed reagents, laboratory equipment and analytical tools. H.M.C.-F., A.P.R.-T. and E.B.-G. wrote the manuscript. All authors have read and agreed to the published version of the manuscript.

**Funding:** HMCF was funded by a doctoral fellowship from the Consejo Nacional de Ciencia y Tecnologia (CONACYT-No. 632506). The present work was supported by the National Geographic Society (NGS-55349R-19 to APRT) and the project "Programa Integral de Fortalecimiento Institucional" of the Universidad de Guadalajara (P/PIFI-2010-14MSU0010Z-10 to ACM).

**Acknowledgments:** The authors thank Ismael Huerta for image editing and Peter Castro for advice on crab identification. A special thanks to Vladimir Pérez and Jeimy Santiago for their help in field and laboratory work. In addition, we thank the authorities of the Islas Marietas National Park (CONANP) for their assistance with field operations. Finally, we thank to Anderson Mayfield and Rupert Ormond for the English proofreading of the manuscript.

**Conflicts of Interest:** The authors declare no conflict of interest.

## References

1. Richter, C.; Wunsch, M.; Rasheed, M.; Koetter, I.; Badran, M.I. Endoscopic exploration of Red Sea coral reefs reveals dense populations of cavity dwelling sponges. *Nature* **2001**, *413*, 726–730. [CrossRef] [PubMed]
2. Stella, J.S.; Pratchett, M.; Hutchings, P.; Jones, G. Coral-associated invertebrates: Diversity, ecological importance and vulnerability to disturbance. *Oceanogr. Mar. Biol.* **2011**, *49*, 43–104.
3. Gagnon, J.M.; Beaudin, L.; Silverberg, N.; Mauviel, A. Mesocosm and in situ observations of the burrowing shrimp *Calocaris templemani* (Decapoda: Thalassinidea) and its bioturbation activities in soft sediments of the Laurentian Trough. *Mar. Biol.* **2013**, *160*, 2687–2697. [CrossRef]
4. Rice, M.M.; Ezzat, L.; Burkepile, D.E. Corallivory in the Anthropocene: Interactive effects of anthropogenic stressors and corallivory on coral reefs. *Front. Mar. Sci.* **2018**, *5*, 525. [CrossRef]
5. Cortés, J. Marine biodiversity baseline for Área de Conservación Guanacaste, Costa Rica: Published records. *ZooKeys* **2017**, *652*, 129. [CrossRef] [PubMed]
6. Weber, J.N.; Woodhead, P.M. Ecological studies of the coral predator *Acanthaster planci* in the South Pacific. *Mar. Biol.* **1970**, *6*, 12–17. [CrossRef]

7. Garth, J.S. The Crustacea Decapoda (Brachyura and Anomura) of Eniwetok Atoll, Marshall Islands, with special reference to the obligate commensals of branching corals. *Micronesica* **1964**, *1*, 137–144.

8. Stimson, J. Stimulation of fat-body production in the polyps of the coral *Pocillopora damicornis* by the presence of mutualistic crabs of the genus Trapezia. *Mar. Biol.* **1990**, *106*, 211–218. [CrossRef]

9. Glynn, P.W. Fine-scale interspecific interactions on coral reefs: Functional roles of small and cryptic metazoans. *Smithson. Contr. Mar. Sci.* **2013**, *39*, 229–248.

10. Patton, W.K. Community structure among the animals inhabiting the coral *Pocillopora damicornis* at Heron Island Australia. In *Symbiosis in the Sea*, 1st ed.; Vernberg, W., Ed.; Univ. South Carolina Press: Columbia, SC, USA, 1974; pp. 219–243.

11. Garth, J.S. Decapod crustaceans inhabiting reef-building corals of Ceylon and the Maldive Islands. *J. Mar. Biol. Assoc. India.* **1974**, *15*, 195–212.

12. Nyenzi, B.; Lefale, P.F. El Nino southern oscillation (ENSO) and global warming. *Adv. Geosci.* **2006**, *6*, 95–101. [CrossRef]

13. Glynn, P.W.; Alvarado, J.J.; Banks, S.; Cortés, J.; Feingold, J.S.; Jiménez, C.; Maragos, J.E.; Martínez, P.; Maté, J.L.; Moanga, D.A.; et al. Eastern Pacific Coral Reef Provinces, Coral Community Structure and Composition: An Overview. In *Coral Reefs of the Eastern Tropical Pacific*, 1st ed.; Glynn, P., Manzello, D., Enochs, I., Eds.; Springer: Dordrecht, The Netherlands, 2017; Volume 8, pp. 107–176. [CrossRef]

14. Richmond, R.H. Energetics, competency, and long-distance dispersal of planula larvae of the coral *Pocillopora damicornis*. *Mar. Biol.* **1987**, *93*, 527–533. [CrossRef]

15. Toth, L.T.; Aronson, R.B.; Vollmer, S.V.; Hobbs, J.W.; Urrego, D.H.; Cheng, H.; Enochs, I.C.; Combosch, D.J.; van Woesik, R.; Macintyre, I.G. ENSO drove 2500-year collapse of eastern Pacific coral reefs. *Science* **2012**, *337*, 81–84. [CrossRef] [PubMed]

16. Castro, P. Eastern Pacific species of *Trapezia* (Crustacea, Brachyura: Trapeziidae), sibling species symbiotic with reef corals. *Bull. Mar. Sci.* **1996**, *58*, 531–554.

17. Reyes-Bonilla, H.; Calderón Aguilera, L.E.; Cruz-Piñón, G.; Medina-Rosas, P.; López-Perez, R.A.; Herrero-Perezrul, M.D.; Leyte-Morales, G.E.; Cupul-Magaña, A.L.; Carriquiry-Beltran, J.D. *Atlas de Corales Pétreos (Anthozoa: Scleractinia) del Pacífico Mexicano*, 1st ed.; Sociedad Mexicana de Arrecifes Coralinos AC, CICESE, CONABIO, CONACYT, DBM/UABCS, CUC/UdeG, Umar: Mexico City, México, 2005; p. 129.

18. CONANP (Comisión Nacional de Áreas Naturales Protegidas). *Programa de Conservación Y Manejo*; Parque Nacional Islas Marietas: Mexico City, México, 2007.

19. Wyrtki, K. Surface currents of the eastern tropical Pacific Ocean. *Bull. Inter. Am. Trop. Tuna Commn.* **1965**, *9*, 268–305.

20. Roden, G.I.; Groves, G.W. Recent oceanographic investigations in the Gulf of California. *J. Mar. Res.* **1959**, *18*, 10–35.

21. Griffiths, R.C. Physical, chemical and biological oceanography at the entrance to the Gulf of California, spring of 1960. *US Fish. Wild. Serv. Spec. Sci. Rep. Fish.* **1968**, *573*, 1–47.

22. Fiedler, P.C. Seasonal climatologies and variability of eastern tropical Pacific surface waters. *NOAA Tech. Rep.* **1992**, *109*, 1–65.

23. Wang, C.; Fiedler, P.C. ENSO variability and the eastern tropical Pacific: A review. *Prog. Oceanogr.* **2006**, *69*, 239–266. [CrossRef]

24. Castro, P.; Ng, P.K.; Ahyong, S.T. Phylogeny and systematics of the Trapeziidae Miers, 1886 (Crustacea: Brachyura), with the description of a new family. *Zootaxa* **2004**, *643*, 1–70. [CrossRef]

25. Lai, J.C.; Ahyong, S.T.; Jeng, M.S.; Ng, P.K. Are coral-dwelling crabs monophyletic? A phylogeny of the Trapezioidea (Crustacea: Decapoda: Brachyura). *Invertebr. Syst.* **2009**, *23*, 402–408. [CrossRef]

26. Folmer, O.; Black, M.; Hoeh, W.; Lutz, R.; Vrijenhoek, R. DNA primers for amplification of mitochondrial cytochrome c oxidase subunit I from diverse metazoan invertebrates. *Mol. Mar. Biol. Biotech.* **1994**, *3*, 294–299.

27. Kumar, S.; Stecher, G.; Li, M.; Knyaz, C.; Tamura, K. MEGA X: Molecular evolutionary genetics analysis across computing platforms. *Mol. Biol. Evol.* **2018**, *35*, 1547–1549. [CrossRef] [PubMed]

28. Ronquist, F.; Huelsenbeck, J.P. MrBayes 3: Bayesian phylogenetic inference under mixed models. *Bioinformatics* **2003**, *19*, 1572–1574. [CrossRef]

29. Moritz, C. Defining 'evolutionarily significant units' for conservation. *Trends. Ecol. Evol.* **1994**, *9*, 373–375. [CrossRef]

30. da Silva, J.M.; Creer, S.; Dos Santos, A.; Costa, A.C.; Cunha, M.R.; Costa, F.O.; Carvalho, G.R. Systematic and evolutionary insights derived from mtDNA COI barcode diversity in the Decapoda (Crustacea: Malacostraca). *PLoS ONE* **2011**, *6*, e19449. [CrossRef]

31. Fransen, C.H.J.M.; De Grave, S. Evolution and Radiation of Shrimp-Like Decapods: An Overview. In *Decapod Crustacean Phylogenetics (Crustacean Issues Book 18)*, 1st ed.; Martin, J.W., Crandall, K.A., Felder, D.L., Eds.; Taylor & Francis/CRC Press: Boca Raton, FL, USA, 2016; pp. 245–259.

32. Moriyama, E.N.; Powell, J.R. Synonymous substitution rates in *Drosophila*: Mitochondrial versus nuclear genes. *J. Mol. Evol.* **1997**, *45*, 378–391. [CrossRef] [PubMed]

33. Niemiller, M.L.; Near, T.J.; Fitzpatrick, B.M. Delimiting species using multilocus data: Diagnosing cryptic diversity in the Southern cavefish, *Typhlichthys subterraneus* (Teleostei: Amblyopsidae). *Evolution* **2012**, *66*, 846–866. [CrossRef] [PubMed]

34. Wenzel, J.W. Behavioral homology and phylogeny. *Annu. Rev. Ecol. Evol. Syst.* **1992**, *23*, 361–381. [CrossRef]

35. Miller, J.S.; Wenzel, J.W. Ecological characters and phylogeny. *Annu. Rev. Entomol.* **1995**, *40*, 389–415. [CrossRef]

36. Castro, P. Notes on symbiotic decapod crustaceans from Gorgona Island, Colombia, with a revision of the eastern Pacific species of *Trapezia* (Brachyura, Xanthidae), symbionts of scleractinian corals. *An. Inst. Inv. Mar. Punta Betín* **1982**, *12*, 9–17. [CrossRef]

37. Castro, P. Shallow-water Trapeziidae and Tetraliidae (Crustacea: Brachyura) of the Philippines (Panglao 2004 Expedition), New Guinea, and Vanuatu (Santo 2006 Expedition). *Raffles Bull. Zool.* **2009**, *20*, 271–281.

38. Castro, P. Systematic status and geographic distribution of *Trapezia formosa* Smith, 1869 (Crustacea, Brachyura, Trapeziidae), a symbiont of reef corals. *Zoosystema* **1998**, *2*, 177–181.

39. Cabezas, P.; Macpherson, E.; Machordom, A. Taxonomic revision of the genus *Paramunida* Baba, 1988 (Crustacea: Decapoda: Galatheidae): A morphological and molecular approach. *Zootaxa* **2010**, *1*, 1–60. [CrossRef]

40. Glynn, P.W. Some physical and biological determinants of coral community structure in the eastern Pacific. *Ecol. Monogr.* **1976**, *4*, 431–456. [CrossRef]

41. Jiménez, C.; Cortés, J. Growth of seven species of scleractinian corals in an upwelling environment of the eastern Pacific (Golfo de Papagayo, Costa Rica). *Bull. Mar. Sci.* **2003**, *1*, 187–198.

42. Reyes-Bonilla, H. Coral reefs of the Pacific coast of Mexico. In *Latin American Coral Reefs*, 1st ed.; Cortés, J., Ed.; Elsevier Science: Amsterdam, The Netherlands; London, UK, 2003; pp. 331–349. [CrossRef]

43. Rodríguez-Troncoso, A.P.; Carpizo-Ituarte, E.; Cupul-Magaña, A.L. Physiological response to high temperature in the Tropical Eastern Pacific coral *Pocillopora verrucosa*. *Mar. Ecol.* **2016**, *37*, 1168–1176. [CrossRef]

44. Martínez-Castillo, V.; Rodríguez-Troncoso, A.P.; Santiago-Valentín, J.D.; Cupul-Magaña, A.L. The influence of urban pressures on coral physiology on marginal coral reefs of the Mexican Pacific. *Coral Reef.* **2020**, *39*, 625–637. [CrossRef]

45. Knudsen, J.W. *Trapezia* and *Tetralia* (Decapoda, Brachyura, Xanthidae) as obligate ectoparasites of pocilloporid and acroporid corals. *Pac. Sci.* **1967**, *21*, 51–57.

*Article*

# The Status of the Coral Reefs of the Jaffna Peninsula (Northern Sri Lanka), with 36 Coral Species New to Sri Lanka Confirmed by DNA Bar-Coding

Ashani Arulananthan [1,*], Venura Herath [2], Sivashanthini Kuganathan [3], Anura Upasanta [4] and Akila Harishchandra [5]

[1] Postgraduate Institute of Agriculture, University of Peradeniya, Kandy 20000, Sri Lanka
[2] Department of Agricultural Biology, University of Peradeniya, Peradeniya 20000, Sri Lanka; venura@agri.pdn.ac.lk
[3] Department of Fisheries Science, University of Jaffna, Thirunelvely 40000, Sri Lanka; sivashanthini@univ.jfn.ac.lk
[4] Faculty of Fisheries and Ocean Sciences, Ocean University of SL, Tangalle 81000, Sri Lanka; kumaraW@ocu.ac.lk
[5] School of Marine Sciences, University of Maine, Orono, ME 04469, USA; kanahera.harishchandra@maine.edu
[*] Correspondence: ashaarul1904@gmail.com

**Abstract:** Sri Lanka, an island nation located off the southeast coast of the Indian sub-continent, has an unappreciated diversity of corals and other reef organisms. In particular, knowledge of the status of coral reefs in its northern region has been limited due to 30 years of civil war. From March 2017 to August 2018, we carried out baseline surveys at selected sites on the northern coastline of the Jaffna Peninsula and around the four largest islands in Palk Bay. The mean percentage cover of live coral was $49 \pm 7.25\%$ along the northern coast and $27 \pm 5.3\%$ on the islands. Bleaching events and intense fishing activities have most likely resulted in the occurrence of dead corals at most sites (coral mortality index > 0.33). However, all sites were characterised by high values of diversity (H' $\geq$ 2.3) and evenness (E $\geq$ 0.8). The diversity index increased significantly with increasing coral cover on the northern coast but showed the opposite trend on the island sites. One hundred and thirteen species of scleractinian corals, representing 16 families and 39 genera, were recorded, as well as seven soft coral genera. Thirty-six of the scleractinian coral species were identified for the first time on the island of Sri Lanka. DNA barcoding using the mitochondrial *cytochrome oxidase subunit I* gene (*COI*) was employed to secure genetic confirmation of a few difficult-to-distinguish new records: *Acropora aspera*, *Acropora digitifera*, *Acropora gemmifera*, *Montipora flabellata*, and *Echinopora gemmacea*.

**Keywords:** Jaffna Peninsula; coral mortality index; DNA barcoding; phylogeny; biodiversity; conservation

**Citation:** Arulananthan, A.; Herath, V.; Kuganathan, S.; Upasanta, A.; Harishchandra, A. The Status of the Coral Reefs of the Jaffna Peninsula (Northern Sri Lanka), with 36 Coral Species New to Sri Lanka Confirmed by DNA Bar-Coding. *Oceans* **2021**, 2, 509–529. https://doi.org/10.3390/oceans2030029

Academic Editor: Rupert Ormond

Received: 12 March 2021
Accepted: 16 July 2021
Published: 26 July 2021

**Publisher's Note:** MDPI stays neutral with regard to jurisdictional claims in published maps and institutional affiliations.

## 1. Introduction

Corals are foundational species that appeared 425 million years ago and are responsible for creating the structural complexity and high productivity of coral reef ecosystems [1]. They have radiated into more than 1500 species and nearly 900 scleractinian corals. Almost all reef corals are hermatypic, that is, they contain in their tissues photosynthetic algae of the family Symbiodiniaceae [1–3], which, living in a symbiotic relationship with corals, are ultimately responsible for the high biomass and productivity of reef habitats [1,4]. The high species diversity of coral reefs has led to their designation as oceanic "rain forests" [5]. However, unprecedented declines in live coral cover and phase shifts in coral reef ecosystems have arisen from the impacts of anthropogenic activities. The most recent widespread degradation of coral reefs, involving coral bleaching and the consequent death and loss of corals, is mostly due to climate change and ocean acidification [6,7]. Averting the effects of climate change will be a considerable challenge if we are to secure the ecological, economic, and social values of coral reefs in the marine biome.

Considering the global nature of these ongoing threats to coral reefs, we investigated the most neglected coral reef area on the Island of Sri Lanka (SL), which is located 40 km away from the southeast coast of the Indian mainland and has 1338 km of open coastline and 2791 km located within coastal lagoons [8]. Fringing-type coral reefs occupy about 2% of the coast and cover 475.70 sq km of Sri Lankan territorial waters [9], with more extensive and well-developed fringing reefs being present on the north-western and northern coasts than in the southern and eastern regions [10,11]. The first detailed records of these coral reefs date from the 19th century and were provided by Ridley (1883) and Ortmann (1889) [12,13]. Subsequently, further studies of Sri Lankan reefs were published through the 20th century [14–19], but most of these concerned the western, south-western, southern, and north-western parts of the country [20,21]. In contrast, the northern region of SL has, in the recent past, received very little attention, due to the decades of civil war. As a result, little is known about the distribution and diversity of corals in that area, yet the region includes the largest coral reef ecosystem in SL, located near the Gulf of Mannar and known as "Bar Reef". It is recognised as a "High Regional Priority Area" by the International Union for the Conservation of Nature (IUCN)/National Oceanic and Atmospheric Administration (NOAA) USA "World Heritage Biodiversity" Project [22,23]. Further north lie Palk Strait and Palk Bay, which are highly productive shallow water areas containing coral reefs, seagrass beds, coastal lagoons, estuaries, mangroves, and salt marshes [22–27].

At the northernmost end of Sri Lanka is the Jaffna Peninsula, with around 293 km of coastline. Extensive fringing coral reefs are present along the northern coast of the peninsula and around the adjacent islands [28], many of which were previously unsurveyed. However, during and following the decades of civil war (which ended in 2009), these reefs were exposed to destructive fishing practices, such as the use of explosives, as well as over-fishing of many species, including ecologically important herbivorous fish and invertebrates, the loss of which has often been associated with a phase shift from coral dominated to algal dominated reefs [24–26]. Studying the ecology and biogeography of northern Sri Lanka represents a considerable challenge, since access presents problems and little information is available about the marine topography and substratum [29].

The present study aimed to report on the current status of reefs around the Jaffna Peninsula and to extend our knowledge of the understudied benthic community, including the diversity of coral species. To do so, we combined molecular tools and traditional morphometric approaches [30–32] and used DNA barcoding, which is known to have revolutionised the understanding of the evolution and systematics of scleractinian corals and resulted in an extensive taxonomic revision of the order [33–37]. Our surveys revealed that the coral reefs of the Jaffna Peninsula have a relatively high percentage of live coral cover and a greater diversity of scleractinian corals than other parts of the country.

## 2. Materials and Methods

### 2.1. Study Area

The Jaffna Peninsula (JP) and its islands surround an almost enclosed body of shallow water, located on the southern side of the 25–50 km wide Palk Strait. They form a reef and island complex (known as "Sethusamudram") that lies at the south-eastern tip of the Indian mainland, partially enclosed off the Bay of Bengal (BOB) (Figure 1). The islands themselves are often referred to as the "Coral Islands Archipelago" and are believed to have been formed as a result of sea level rise since the late Holocene [8,24,38]. On the SL mainland, fossilised limestone rocks are present in the near-shore area, extending up to 50 m seawards, along the northern coast between Point Pedro and Keerimalai, while sandy to muddy seabed extends westwards from Keerimalai, as well as around the islands of the JP. In contrast, the coastline along the south-eastern shore extending from Point Pedro towards the SL coastline is composed of sandy beaches.

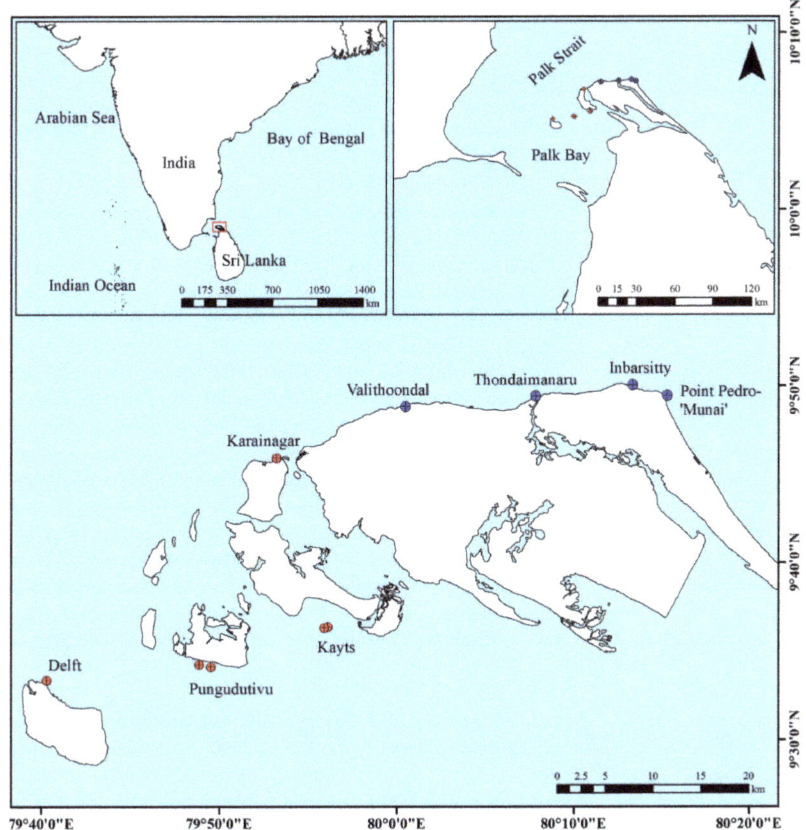

**Figure 1.** Maps showing (**above left**) the location of the Jaffna Peninsula (within the red rectangle) in relation to Sri Lanka and southern India, (**above right**) the relation of the Jaffna Peninsula to the Palk Strait and Palk Bay, and (**below**) the locations of the study sites (blue circles) on the northern side and the island sites (red circles) of the Jaffna Peninsula and its adjacent islands.

Long stretches of the reef, called "paar" by local fishermen and made up mostly of fringing reef, extend along the northern coast and around the islands at depths between 1 and 8 m [24,27]. Typically, there is a reef flat, 15–20 m wide and 1 to 2 m deep, extending seawards from the shore, with a substrate mainly of dead coral rocks and coral gravel [24]. Beyond this, the reef front (including reef crest) and reef slope region are dominated by a mix of live coral and dead coral to a depth of nearly 6 m. The maximum depth of the reef base is no more than 8 m [24], so no surveys were undertaken beyond this depth.

Study Sites

Eight reef areas were surveyed between March 2017 and August 2018. These can be divided into two groups, those on the island reefs located in the Palk Bay and those along the northern coast of the Palk Strait (Figure 1). Within these eight reef areas, ten actual sites were surveyed: the Delft reef front (Del), the Karainagar reef front (Knr), the Kayts outer reef slope (Kt1), the Kayts outer reef flat (Kt2), the Pungudutivu outer reef slope (Pu1), the Pungudutivu outer reef flat (Pu2), and the reef fronts at Point Pedro (Munai or Ptm), Inbarsitty (Inb), Thondaimanaru (Tho), and Valithoondal (Val) (Figure 1). These were either shallow, outer reef flat sites at 1 to 3 m depth, or reef front sites at 3 to 6 m depth. Visual inspections of the reef flats on the north coast revealed that all but the most protected sites supported very low, mostly nil, live coral cover; similarly, inner reef flats and reef lagoons at the island sites were mostly composed of seagrass beds and broken coral rubbles. Given this, quantitative surveys were not carried out at these locations. In addition, the Delft, which is located in Palk Bay, is known as the island of dead corals, most pieces of which are believed to be of Miocene origin. In contrast, the islands of Pungudutivu and Kayts are well-known locally to have well-developed coral reefs extending to a depth of about 6 m.

*2.2. Surveys and Sampling*

Surveys were conducted using the standard Reef Check survey method (see http://reefcheck.org/ecoaction/Monitoring_Instruction.hp, accessed on 30 August 2018) [39]. A 100 m transect was divided into four 20 m long segments, each separated by a 5 m interval from the next one, so that the four segments could be considered as replicates. On each segment, substrate cover was recorded beneath the transect line at 0.5 m intervals, so that 40 data points were recorded per segment (160 per 100 m transect). The transect lines were laid down parallel to the shore, along the long axis of the reef. The work was undertaken using SCUBA at 2 to 6 m depth, and snorkel at 1 to 2 m depth. The substrate was recorded using the following categories (and codes): hard coral (HC), soft coral (SC), dead coral with macroalgae or nutrient indicating algae (NIA), recently killed coral (RKC), sponge (SP), coral rubble (RB), hard substrate of dead coral more than one year old covered by turf algae or encrusting coralline algae (RC), sand (SD), silt (SI), and others (including sea anemones, tunicates, gorgonians, or non-living substrate) (OT) [39]. Digital photography and videography were employed (using a GoPro Digital Hero 5 Black, GoPro Inc., (San Mateo, CA, USA) to provide a visual record of the reef habitats and coral assemblages along the transects. In addition, images of live coral specimens were recorded with a dedicated underwater camera (Canon PowerShot A620, Canon, USA).

*2.3. Coral Reef Cover Analysis*

The percentage cover of the substrates at the different study sites was compiled and processed using the Reef Check Substrate data entry sheet in the Microsoft Excel application [39]. To investigate differences in the benthic community between the study sites, Principal Component Analysis and Bray–Curtis Cluster Analysis were carried out using PAST version 3.22 (Øyvind Hammer, Oslo, Norway) [40]. The mean percentage covers of substrate categories were analysed using IBM SPSS Statistics (Version 27.0, IBM, Armonk, NY, USA). Mann–Whitney $U$-tests were used to test for differences in the substrate between the Palk Bay and Palk Strait reefs. Differences in proportions of main benthic categories (HC, SC, RC, NIA, RB) amongst sites were tested using a Kruskal–Wallis test. To assess the status of the reefs, a coral mortality index (CMI) was calculated by dividing the percentage values of dead coral cover by the combined value of live and dead coral cover. If the CMI was greater than 0.33, the reef was considered likely to have been impacted by natural or anthropogenic stressors [41]. Sites were also classified based on the amount of live coral cover as excellent (75–100% live coral), good (50–74.9% live coral), fair (25–49.9% live coral), or poor (0–24.9% live coral) [41].

*2.4. Diversity of Coral Species*

Underwater photographs were taken during the surveys at each site and were used to aid morphological taxonomic identification of coral species. In addition, as far as practicable, corals were identified in the field to genus level based on morphological characters following the Indo-Pacific Coral Finder Tool Kit [42]. Subsequently, traditional diagnostic features such as corallite wall structure, the presence of paliform lobes, verrucae, valleys, and columella, and other colony characters (as described in the Corals of the World by Veron [43]) were used for species discrimination. Species diversity, dominance, and evenness indices were calculated based on the numbers of species recorded along the transect lines [44].

*2.5. DNA Extraction, Amplification, Sequencing, and Barcoding of Coral Species*

A commercially available spin-column-based DNeasy® Blood & Tissue Kit (Qiagen, Hilden, Germany) was used for DNA extraction from coral tissues, according to the manufacturer's instructions. Small pieces of the coral samples were excised with sterile forceps, and 20–25 mg samples were placed in a sterile 1.5 mL microcentrifuge tube and subjected to extended lysis for 24 h or longer. The extracted DNA was stored at $-20\,°C$. A specific region of the 658 bp fragment of the cytochrome oxidase gene (subunit I) (COI) was amplified by polymerase chain reaction (PCR) using the universal primers LCO 1490 and HCO 2198 [45]. The PCR reaction mix consisted of $1 \times$ PCR colourless buffer (pH 8.3) (GoTaq, Promega), 4 mM $MgCl_2$, 0.2 mM dNTPs, 0.25 µM of each primer, $0.01$ mg mL$^{-1}$ bovine serum albumin, 0.625 U Taq polymerase (GoTaq, Promega), and 50 ng µL$^{-1}$ genomic DNA in a 25 µL reaction volume. The PCR thermal regime consisted of one initial denaturation cycle for 2 min at 95 °C; five 1-minute denaturation cycles at 95 °C, annealing for 1.5 min at 45 °C and an extension for 1.5 min at 72 °C; 35 1-minute denaturation cycles at 95 °C, annealing for 1.5 min at 50 °C and an extension for 1 min at 72 °C, followed by a final extension for 5 min at 72 °C.

The PCR products were sequenced in both the forward and reverse directions by Macrogen Inc. (Seoul, Korea). All DNA sequence chromatograms were assembled and edited using Unipro (Novosibirsk, Russia), UGENE 1.29.0. The sequences were then aligned using ClustalW [46], and the aligned outputs queried to identify the species, along with the reference sequences of various coral species that were retrieved from the National Centre for Biotechnology Information (NCBI) (www.ncbi.nlm.nih.gov, accessed on 25 July 2021) [47], GenBank and Barcode of Life Data System (BOLD) databases, using the Basic Local Alignment Search Tool (BLAST).

*2.6. Ethics Statement*

Samples were collected with permits from, and in compliance with, the laws and regulations of the Sri Lanka Department of Wildlife Conservation.

## 3. Results

*3.1. Benthic Substrate, Coral Distribution, and the Status of Reefs*

The extent of the benthic categories varied significantly across the study sites (Figure 2). The mean percentage cover of benthic categories was compared between the sites on the Palk Bay islands and the sites on the northern coastal reefs in the Palk Strait. Nearly 50% of the substrate at the reef sites on the northern coast was composed of hard corals, while the remaining part was mostly composed of dead corals (RC and NIA).

In contrast, the substrate cover at the island sites was composed principally of dead coral rocks covered with turf algae, with the rest mostly consisting of hard corals, rubbles, and fleshy macroalgae. There was a significant difference in mean percentage cover of hard corals (HC) between the northern coastal reefs ($n = 16$, median = 45.31 $\pm$ 10.0 range) and the island reefs ($n = 24$, median = 25.31 $\pm$ 25.62 range) (Mann–Whitney $U$-test, $p = 0.0001$). In addition, the mean percentage cover of soft corals (SC) was significantly higher on the northern coast compared to the islands (Mann–Whitney $U$-test, $p < 0.05$). In contrast, there were no significant differences in mean percentage cover of fleshy macroalgae (NIA), of coral rocks covered with turf algae (RC), or of coral rubble (RB) between the northern coast and island reef sites (Kruskal–Wallis test, $p > 0.05$).

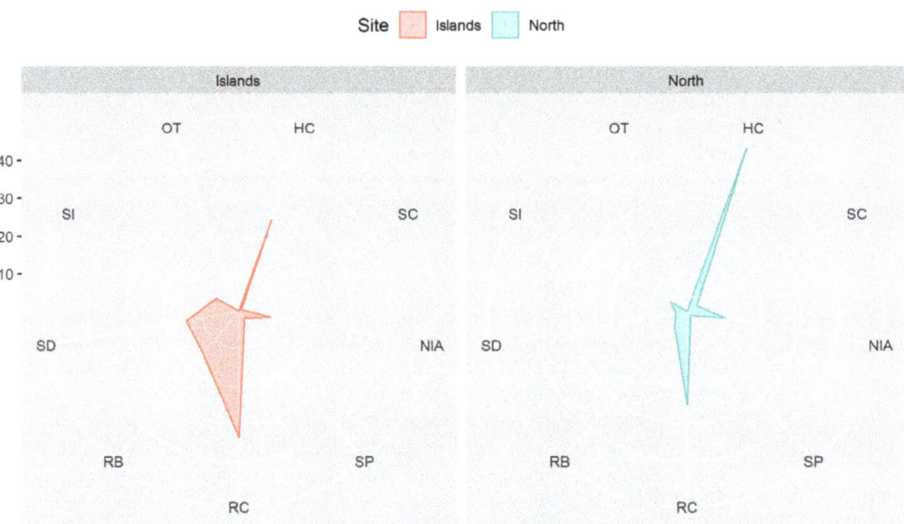


**Figure 2.** Radar plot giving an overview of benthic cover distribution on the coral reef sites on the Palk Bay islands (**left**) in comparison with those on the northern coast of the Jaffna Peninsula (**right**). (HC- hard coral, SC- soft coral, NIA- dead coral with macroalgae or nutrient indicating algae, SP- sponge, RB- coral rubble, RC- hard substrate of dead coral more than one year old covered by turf algae or encrusting coralline algae, SD- sand, SI- silt, and OT- others (including sea anemones, tunicates, gorgonians, or non-living substrate)).

Comparing individual sites, live coral cover was greatest at the Point Pedro (HC: 52% ± 7; SC: 4% ± 3), Inbarsitty (HC: 44% ± 3; SC: 3% ± 2), and Thondaimanaru (HC: 46% ± 6; SC: 2% ± 1) (±indicates SD) (Figure 3). The Valithoondal, Karainagar, Pungudutivu, and Kayts sites also had live coral cover of between 25% and 50%, corresponding to fair health status as defined here. Live coral cover was lowest (15–30%) at the five island sites Knr, Pu2, Kt1, Kt2, and Del (excluding Pu1). Punkudutviu and Kayts islands were characterised by the highest mean percentage cover of dead coral (61% ± 7 and 58.5% ± 10.5, respectively) followed by Delft reef front (46% ± 6) (Figure 3). Thus, a high live coral cover (49% ± 7.25) was observed on the northern coast and a low live coral cover (5%) on the island sites, except at Del (Figures 2 and 3). In all cases, the coral mortality index was greater than 0.33 (Figure 3), implying that all the reefs were stressed or impacted.

Figure 4 revealed two major groups of sites that cluster at a 60% similarity level. Sites Knr, Val, Ptm, Tho, Inb, Pu1, and Pu2 fall into the upper group, and Kt1 and Kt2 into the lower group, within which the Delft reef front site (Del) appears to be an outlier. There were three sub-groups in the upper group of the dendrogram, consisting of Inb, Tho, and Ptm (showing >85% homogeneity), Pu1 and Pu2 (>80% homogeneity), and Val and Knr (>70% homogeneity). Thus, all of the north coast sites plus two of the island sites (Pu1 and Pu2) fall into the upper group and the remaining island sites into the lower group, so supporting the distinction made above based on coral cover alone.

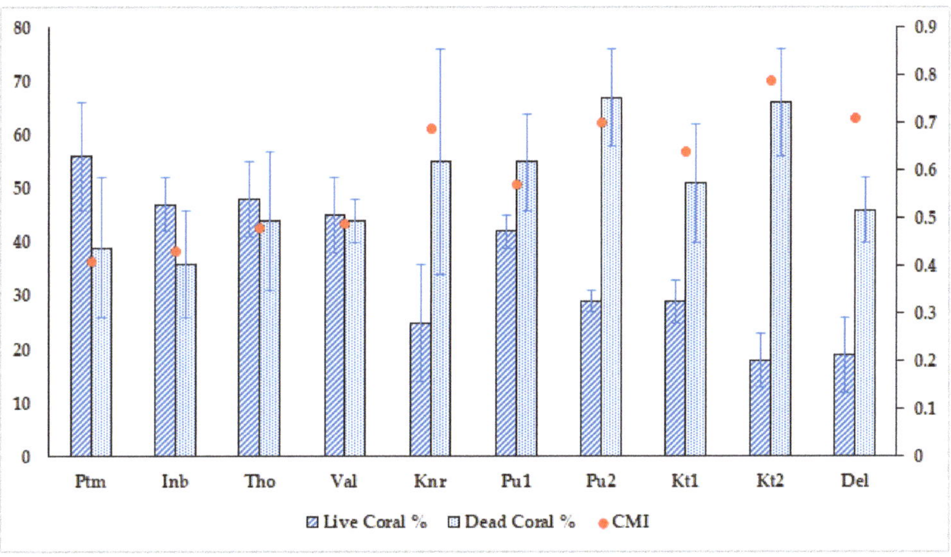

**Figure 3.** Percentages of live coral cover and dead coral cover, together with the values of the Coral Mortality Index (CMI) on coral reefs at each of the study sites.

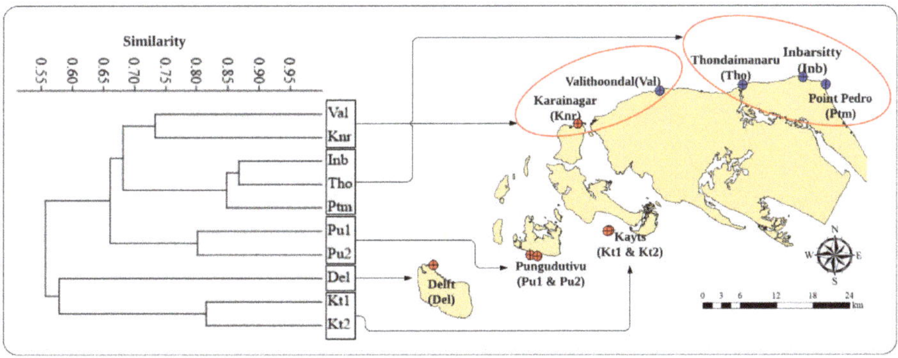

**Figure 4.** Bray–Curtis Cluster analysis based on differences in the benthic cover on the reefs of the Jaffna Peninsula, together with a map showing the locations of the sites. (Abbreviations as shown on the map.)

Turf algae-dominated dead coral rocks (RC) were observed ubiquitously at the island reefs. The greatest extent of dead corals covered with turf algae (48.5% ± 13) occurred at Kayts island and of coral rubble at Pungudutivu island (31.5% ± 7). The Delft reef front site (outlier group) in the dendrogram was distinguished from the other sites by the presence of a very low percentage of live hard coral cover (16% ± 5 SE) and a high percentage of abiotic substrates, such as sand (31% ± 5), although the highest proportion of abiotic forms were observed at Pungudutivu (mean of RB 31.5% ± 7; SD 2.5% ± 2).

Valithoondal and Karainagar reef sites had similar amounts of soft coral cover, and dead coral rocks covered with macroalgae such as *Turbinaria* sp., *Caulerpa* sp., and *Sargassum* sp. (Figure 5, Supplementary Table S1). Valithoondal reef region had the highest live coral cover (45% ± 7) of massive growth forms, including various species (Table 1). At Point Pedro, Valithoondal, and Karainagar, numbers of giant clams, sea cucumbers, tunicates, and sea anemones were present (documented as others, OT) (Supplementary Table S1).

Soft corals were uncommon but were occasionally found on transects on the northern coast. No soft corals were recorded at Pungudutivu or Kayts islands, and only a few at Karianagar island.

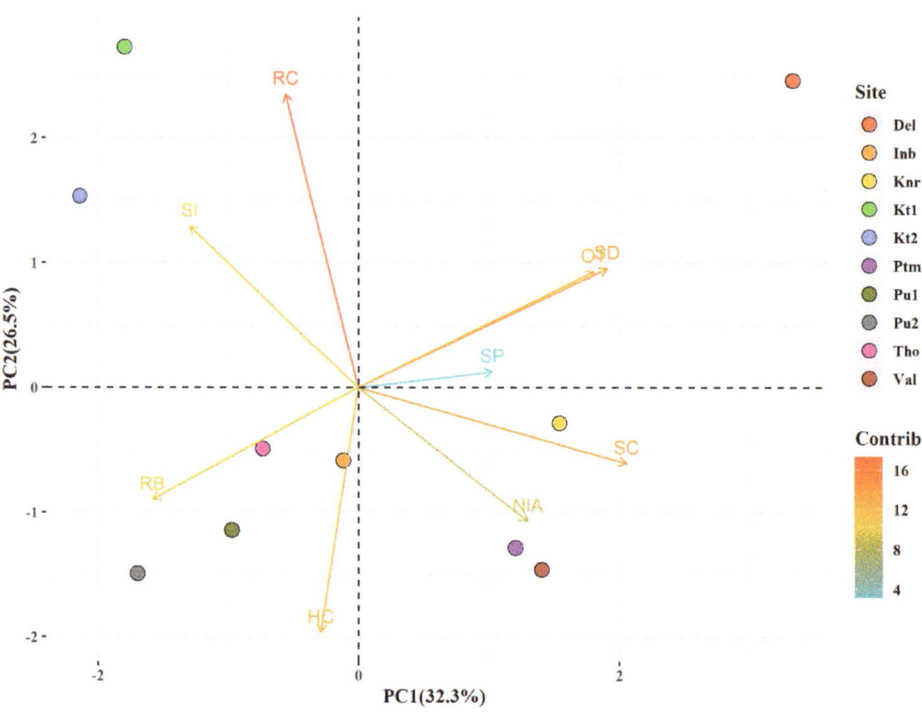


**Figure 5.** Principal component analysis (PCA) showing the pattern of differences between study sites and how they correlate with different benthic components (HC- hard coral, SC- soft coral, NIA- dead coral with macroalgae or nutrient indicating algae, SP- sponge, RB- coral rubble, RC- hard substrate of dead coral more than one year old covered by turf algae or encrusting coralline algae, SD- sand, SI- silt, and OT- others (including sea anemones, tunicates, gorgonians, or non-living substrate)).

**Table 1.** List of the coral families recorded, showing the total numbers of genera and species identified from the study area.

| No. | Family | Jaffna | |
|---|---|---|---|
| | | Genera | Species |
| 1 | Acroporidae Verrill, 1902 | 4 | 28 |
| 2 | Agariciidae Gray, 1847 | 2 | 3 |
| 3 | Dendrophylliidae Gray, 1847 | 1 | 5 |
| 4 | Diploastraeidae Chevalier & Beauvais, 1987 | 1 | 1 |
| 5 | Euphylliidae Alloiteau, 1952 | 1 | 2 |
| 6 | Fungiidae Dana, 1846 | 2 | 3 |
| 7 | Merulinidae Verrill, 1866 | 12 | 39 |
| 8 | Leptastreidae Rowlett, 2020 | 2 | 3 |
| 9 | Lobophylliidae Dai & Horng, 2009 | 5 | 8 |
| 10 | Oulastreidae Vaughan, 1919 | 1 | 1 |
| 11 | Plesiastreidae Dai & Horng, 2009 | 1 | 1 |
| 12 | Pocilloporidae Gray, 1842 | 2 | 3 |
| 13 | Poritidae Gray, 1842 | 2 | 13 |
| 14 | Psammocoridae Chevalier & Beauvais, 1987 | 1 | 1 |
| 15 | Siderastreidae Vaughan & Wells, 1943 | 1 | 2 |
| 16 | Scleractinia incertae sedis (temporary name) | 1 | 1 |
| | Total | 39 | 113 |

This distinction between reef areas was further explained by the principal component analysis (PCA) which illustrated the overall pattern of variation amongst sites and the dependence of this pattern on different benthic components (Figure 5). The analysis explains 58.8% of the total variance, of which principal component one (PC1) accounted for 32.3% of the variance and had a positive correlation (PC1 > 0) with the benthic substrates soft coral (SC), dead corals covered with macro algae (NIA), sponges (SP), sand (SD), and others (OT), while principal component two (PC2) accounted for 26.5% of the variance and showed a positive correlation (PC2 > 0) with the substrates rocks covered with turf algae (RC) and silt (SI). Accordingly, sites Pu1, Pu2, Tho, and Inb are assigned to the PC1 < 0/PC2 < 0 domain; sites Ptm, Val and Knr to the PC1 > 0/PC2 < 0 domain; sites Kt1 and Kt2 to the PC1 < 0/PC2 > 0 domain; and site Del to the PC1 > 0/PC2 > 0 (Figure 5).

The shallow areas (1 to 3 m depth) of reef front on the northern coast were dominated by macroalgae from October to April due to the Northwest Monsoon. The data were tested for any relationship between live coral cover (hard coral (HC) and soft coral (SC)) and algal cover (RC and NIA) across all sites, showing a significant negative correlation (r = −0.73, $p < 0.05$) between live coral cover and algal cover (Supplementary Figure S1).

*3.2. Diversity, Taxonomic Composition, and Distribution of Coral Species*

The taxonomic study across the eight reef areas surveyed revealed a total of 113 scleractinian coral species, from 39 genera and 16 families, together with seven soft coral genera (Table 1). Among these, 36 species of hard corals appear to be new records for Sri Lanka (Table 2), including species of the genera *Oulastrea*, *Coeloeris*, and *Siderastrea*, although all have been recorded in other parts of the Central Indian Ocean. The full species list (Table 1) includes 39 members of the family Merulinidae (44% of the total), 28 species of Acroporidae (32%), and 13 species of Poritidae (15%). The predominant genera present were *Acropora* (17 spp.), *Dipsastraea* (10), *Porites* (7), *Favites* (6), *Montipora* (6), *Goniopora* (6), *Turbinaria* (5), and *Platygyra* (5) (Supplementary Table S2).

**Table 2.** Numbers of families, genera, and species of scleractinian corals recorded in each of the eight reef areas studied around the Jaffna Peninsula, Sri Lanka.

|  | Ptm | Inb | Tho | Val | Knr | Pu1, Pu2 | Kt1, Kt2 | Del |
|---|---|---|---|---|---|---|---|---|
| Families | 12 | 07 | 12 | 08 | 09 | 06 | 08 | 10 |
| Genera | 24 | 27 | 24 | 18 | 17 | 15 | 18 | 18 |
| Species | 49 | 30 | 44 | 25 | 36 | 34 | 33 | 37 |

The numbers of genera and species recorded differed between the study sites (Table 2). On the northern coast, massive coral growth forms were predominant and the principal genera present were *Dipsastraea* (6 spp.), *Porites* (6), *Montipora* (6), *Favites* (5), *Platygyra* (4), *Goniastrea* (3), and *Symphyllia* (3). Encrusting growth forms of the genera *Symphyllia*, *Lobophyllia* and *Acanthastrea* were also more common on the northern coast. In contrast, around the island sites, tabulate and branching *Acropora*, foliose *Montipora*, massive *Porites*, and smaller *Goniastrea* predominated, with branching and tabulate *Acropora* and foliose growth forms occupying large areas. The Pungudutivu and Kayts island sites, in particular, were dominated by branching and tabulate *Acropora*, followed in substrate cover by *Porites*. The genus *Echinopora* was observed at Point Pedro, Thondaimanaru, Pungudutivu, and Kayts Islands. Only a single colony of *Pocillopora verrucosa* was documented, being at Point Pedro, while scattered *Pocillopora damicornis* were recorded on the Inabarsitty and Thondaimanru reefs. The newly recorded species *Acropora gemmifera* was found at all the study sites, but *Acropora aspera*, *Acropora digitifera*, *Montipora flabellata*, and *Echinopora gemmacea* were uncommon, with only one colony recorded at any site.

Soft coral colonies representing the genera *Capnella*, *Cladiella*, *Clavalaria*, *Lobophyton*, *Subergorgia*, and *Sarcophyton* (Supplementary Table S1) were uncommon and made only a minor contribution (2–4%) to substrate cover, but they were found on the northern coast and on the islands of Karainagar and Delft.

A comparison of the number of species present at different sites showed that the northern coastal sites had higher richness and diversity than the island sites (Tables 3 and 4). Calculation of Shannon's diversity and evenness indices suggested that all the reefs, excluding Pungudutivu, had comparable diversity and evenness values ($H' \geq 2.5$; $E \geq 0.9$), with a low dominance index ($D \leq 0.1$). The Point Pedro reef front had nearly 50 hard coral species, followed by Thondaimanaru with 44 hard coral species (Table 3), but the Shannon–Weiner diversity index at Thondaimanaru was higher ($H' = 3.020$) than that at Point Pedro ($H' = 2.927$) (Table 4).

The relationships between mean values of live coral cover and the different diversity indices (Shannon's Diversity Index—$H'$, Simpson's Dominance Index—$D$, and Evenness index—$E$) were investigated separately for each of the two parts of the study area, i.e., the island sites (Karainagar, Pungudutivu, Kayts, and Delft) and the northern coastline (Point Pedro, Inbarsitty, Thondaimanaru, Valithoondal). The relationship between these indices on the northern coast and on the island sites was further analysed with a Pearson correlation. On the northern coast, there was a moderate positive relationship between diversity and live coral cover (Pearson correlation: $r = 0.64$, $p > 0.05$). In contrast, among the island sites, there was a strong negative correlation between both diversity index (Pearson correlation: $r = -0.96$, $p < 0.05$) and evenness (Pearson correlation: $r = -0.98$, $p < 0.05$) on the one hand and live coral cover on the other. Similarly, there was a negative correlation (Pearson correlation: $r = -0.51$, $p > 0.05$) between the dominance index and mean live coral cover on the northern coast, but a strong positive correlation (Pearson correlation: $r = 0.96$, $p < 0.05$) between these variables among the island sites (Supplementary Figure S2).

**Table 3.** List of the species newly recorded in Sri Lanka during the present study, showing whether they were recorded on the north coast at the Palk Strait sites (PS sites) or on the islands in Palk Bay (PB sites), together with their previously documented occurrence (indicated by √) in other South Asia, Indian Ocean and Arabian Sea regions (LWI—Lakshadweep Islands, GOM—Gulf of Mannar, ANI—Andaman and Nicobar islands, Arabian Sea (AS) (including Red Sea), CIO—central Indian Ocean region) [43].

| No. | Family | Genus | Species | PS Sites | PB Sites | LWI | GOM | ANI | AS | CIO |
|---|---|---|---|---|---|---|---|---|---|---|
| 1 | Acroporidae Verrill, 1901 | *Acropora* Oken, 1815 | *Acropora aspera* Dana, 1846 | √ |  | √ |  | √ | √ | √ |
| 2 |  |  | *Acropora digitiferaDana,* 1846 | √ |  | √ | √ | √ | √ | √ |
| 3 |  |  | *Acropora gemmifera* Brook, 1892 | √ |  | √ |  | √ | √ | √ |
| 4 |  |  | *Acropora latistella* Brook, 1891 |  | √ |  |  | √ | √ | √ |
| 5 |  |  | *Acropora pulchra* Brook, 1891 |  | √ |  |  | √ | √ | √ |
| 6 |  | *Alveopora* Blainville, 1830 | *Alveopora allingi* Hoffmeister, 1925 | √ | √ |  |  | √ | √ | √ |
| 7 |  | *Astreopora* Blainville, 1830 | *Astreopora myriophthalma* Lamarck, 1816 | √ | √ | √ | √ | √ | √ | √ |
| 8 |  |  | *Astreopora listeri* Bernard, 1896 | √ |  | √ |  | √ | √ | √ |
| 9 |  |  | *Astreopora ocellata* Bernard, 1896 | √ |  | √ |  |  |  | √ |
| 10 |  | *Montipora* Blainville, 1830 | *Montipora flabellata* Studer,1901 | √ |  |  |  |  |  | √ |
| 11 |  |  | *Montipora informis* Bernard, 1897 | √ | √ | √ | √ | √ | √ | √ |
| 12 | Merulinidae Verrill, 1865 | *Goniastrea* Milne Edwards & Haime, 1848 | *Goniastrea minuta* Veron, 2000 |  | √ |  |  | √ |  | √ |
| 13 |  | *Coelastrea* Verrill, 1866 | *Coelastrea palauensis* Yabe & Sugiyama, 1936 |  | √ |  |  |  |  | √ |
| 14 |  | *Cyphastrea* Milne Edwards & Haime, 1848 | *Cyphastrea japonica* Yabe & Sugiyama, 1936 |  | √ |  |  |  |  | √ |
| 15 |  |  | *Cyphastrea microphthalma* Lamarck, 1816 |  | √ | √ | √ | √ | √ | √ |
| 16 |  | *Platygyra* Ehrenberg, 1834 | *Platygyra acuta* Veron, 2000 | √ |  |  |  | √ | √ | √ |
| 17 |  | *Echinopora* Lamarck, 1816 | *Echinopora gemmacea* Lamarck, 1816 | √ |  |  |  | √ | √ | √ |
| 18 |  | *Dipsastraea* Blainville, 1830 | *Dipsastraea amicorum* Milne Edwards & Haime, 1850 |  | √ |  |  |  | √ | √ |
| 19 |  |  | *Dipsastraea lizardensis* Veron, Pichon & Wijsman-Best, 1977 | √ | √ |  |  |  | √ | √ |
| 20 |  |  | *Dipsastraea rotumana* Gardiner, 1899 |  | √ |  |  | √ | √ | √ |
| 21 | Scleractinia incertae sedis | *Pachyseris* Milne Edwards & Haime, 1849 | *Pachyseris gemmae* Nemenzo, 1955 | √ |  |  |  | √ |  | √ |
| 22 | Oulastreidae Vaughan, 1919 | *Oulastrea* Milne Edwards & Haime, 1848 | *Oulastrea crispata* Lamarck, 1816 | √ |  |  |  | √ |  | √ |
| 23 | Poritidae Gray,1842 | *Porites* Link, 1807 | *Porites evermanni* Vaughan, 1907 | √ |  |  |  | √ |  | √ |
| 24 |  |  | *Porites murrayensis* Vaughan, 1918 | √ | √ |  |  | √ |  | √ |
| 25 |  |  | *Porites pukoensis* Vaughan, 1907 |  | √ |  |  |  |  | √ |
| 26 |  | *Goniopora* de Blainville, 1830 | *Goniopora lobata* Milne Edwards, 1860 | √ | √ |  |  |  | √ | √ |
| 27 |  |  | *Goniopora minor* Crossland, 1952 | √ | √ | √ | √ |  | √ | √ |
| 28 |  |  | *Goniopora somaliensis* Vaughan, 1907 |  | √ |  |  |  | √ | √ |
| 29 |  |  | *Goniopora tenuidens* Quelch, 1886 |  | √ |  |  | √ | √ | √ |
| 30 | Dendrophylliidae Gray, 1847 | *Turbinaria* Oken, 1815 | *Turbinaria frondens* Dana, 1846 | √ |  |  |  |  | √ |  |
| 31 |  |  | *Turbinaria reniformis* Bernard, 1896 | √ | √ |  |  | √ | √ |  |
| 32 |  |  | *Turbinaria stelluta* Lamarck, 1816 | √ | √ | √ |  | √ | √ |  |
| 33 | Lobophylliidae Dai & Horng, 2009 | *Acanthastrea* Milne Edwards & Haime, 1848 | *Acanthastrea ishigakiensis* Veron, 1990 | √ |  |  |  | √ | √ |  |
| 34 |  | *Micromussa* Veron, 2000 | *Micromussa amakusensis* Veron, 1990 | √ |  |  |  |  |  |  |
| 35 | Agariciidae Gray, 1847 | *Coeloseris* Vaughan, 1918 | *Coeloseris mayeri* Vaughan, 1918 |  | √ |  |  |  | √ | √ |
| 36 | Siderastreidae Vaughan & Wells, 1943 | *Siderastrea* Blainville, 1830 | *Siderastrea savignyana* Milne Edwards & Haime, 1849 | √ | √ |  | √ |  | √ | √ |

**Table 4.** The richness and values of diversity and evenness indices for hard corals at each of the study sites of the Jaffna Peninsula.

| Reef Regions | Number of Species | Richness | Shannon's Diversity Index (H′) | Simpson's Dominance Index (D) | Simpson's Diversity Index (D′) | Evenness Index (E) |
|---|---|---|---|---|---|---|
| Point Pedro | 49 | 3.429 | 2.927 | 0.066 | 0.934 | 0.921 |
| Inbarsitty | 30 | 2.921 | 2.61 | 0.084 | 0.916 | 0.900 |
| Thondaimanaru | 44 | 3.769 | 3.02 | 0.060 | 0.94 | 0.938 |
| Valithoondal | 25 | 3.470 | 2.752 | 0.069 | 0.931 | 0.971 |
| Karainagar | 35 | 2.704 | 2.575 | 0.089 | 0.911 | 0.929 |
| Pungudutivu | 32 | 2.475 | 2.323 | 0.127 | 0.873 | 0.880 |
| Kayts | 34 | 3.258 | 2.743 | 0.083 | 0.917 | 0.932 |
| Delft | 37 | 2.959 | 2.713 | 0.082 | 0.918 | 0.939 |

*3.3. DNA Barcoding of Selected Coral Species*

Although most species could be identified from morphological characteristics alone, some that were difficult to distinguish morphologically and those considered rare were subjected to DNA barcoding to identify them with greater certainty. Based on the NCBI GenBank database sequences, the best match for each of the samples studied is shown in Table 5, together with the respective accession numbers. There were seven mitochondrial *COI* gene barcode sequences recorded that represented six new records for Sri Lanka: four species from the genus *Acropora*, including two samples of *A. gemmifera*, and one species each from the genera *Montipora* (*M. flabellata*) and *Echinopora* (*E. gemmacea*) (Figure 6, Table 5). The COI barcode sequences for all the samples showed a >99% similarity to the respective sequences recorded in GenBank.

**Table 5.** List of samples sequenced in this study and the most closely related barcodes found in GenBank, together with their accession numbers, and the publisher of the sequence.

| Sample ID | Accession ID | Sequence Length | Molecular Identification (Closest Relative in GenBank) | Identity | Reference |
|---|---|---|---|---|---|
| COJP001 | MN689059 | 709 | *Acropora aspera* (KF448532.1) | 99.69% | [48] |
| COJP002 | MN689060 | 717 | *Acropora gemmifera* (MG383839.1) | 99.83% | [49] |
| COJP003 | MN689061 | 721 | *Acropora digitifera* (KR401100.1) | 99.22% | [50] |
| COJP004 | MN689062 | 711 | *Acropora gemmifera* (MG383839.1) | 99.83% | [49] |
| COJP006 | MN689067 | 715 | *Acropora hyacinthus* (LC326547.1) | 99.85% | [51] |
| COJP007 | MN689068 | 718 | *Echinopora gemmacea* (HE654561.1) | 99.84% | [52] |
| COJP009 | MN689065 | 714 | *Montipora flabellata* (HQ246603.1) | 99.51% | [53] |

**Figure 6.** Species subjected to DNA barcoding: (**a**) Acropora aspera, (**b**) Acropora digitifera, (**c**) Acropora gemmifera, (**d**) Acropora hyacinthus, (**e**) Montipora flabellata, (**f**) Echinopora gemmacea.

## 4. Discussion

### 4.1. Status of Reefs and Extent of Impacts

Located close to the southern tip of the Indian sub-continent, Sri Lanka's coral reefs are surrounded by other well-developed reef areas in both the Western Indian Ocean and the Bay of Bengal. India has three coral reef regions, one of which, Palk Bay, lies adjacent to the present study area, on the opposite side of the Palk Strait. The others, lying much further away, are the Andaman and Nicobar Islands in the Bay of Bengal and the Lakshadweep islands located between the west coast of India and the Maldives. Further to the west lie the Maldives and the Chagos Archipelagoes [54,55]. According to Arthur [56], the reef faunas of the Gulf of Mannar and Palk Bay islands are closely related to those of other Sri Lankan coral reefs. However, the differences described here in benthic composition between islands in the Palk Bay and northern coast sites in the Palk Strait suggest that, in terms of coral cover and species present, the Palk Strait reefs may also be compared to those in the above-mentioned neighbouring regions (Figure 7).

To date, it was presumed that all the coral assemblages of Sri Lanka are less well-developed and diverse than those in these other regions. Due to the limited accessibility created by the civil war, previous surveys around the Jaffna Peninsula were essentially limited to confirmation of the presence of fringing coral reefs around the islands [10,27,28]. The first quantitative report was prepared only in 2005, as part of an environmental impact assessment of the Sethusamudram Ship Channel project [27]. Even then, only rapid surveys were conducted at four reef sites (Pungudutivu, Eluvaitivu, and two locations in Kankeasanthurai harbour), while two other fringing reefs, on Analaithivu and Karathivu islands, were inspected visually to assess their condition. The study reported live coral cover on the four reefs surveyed to be between 35% and 58% [24]. In contrast, this present study found that live coral cover at the majority of sites varied between 27% and 49%, with only the Point Pedro site having more than 50% live coral cover. While this difference might relate to differences in methodology or to the present study being restricted to shallow depths, an alternative explanation is that coral cover has declined since 2005 as a result of one or more impacts.

The structure of the benthic cover at the island sites in the Palk Bay is somewhat similar to reefs in nearby regions that have been similarly subject to climate-related coral bleaching and to fishing related impacts; thus, reports of findings on the Indian reefs may provide some understanding of the impacts likely to have affected the study area [56–60]. At most sites, turf algae growing over old corals appeared to be the predominant substrate and the contribution of coral rubble to the substrate was also noticeable, suggesting that coral cover may have been higher in the past. Algal-dominated reefs are often associated with post-bleaching impacts [60,61]. A more recent study has described the extent of bleaching in the region, with bleaching being much higher in Palk Bay (71.48%) at the depth of 2–3 m than towards the Gulf of Mannar (46.04%), where the coral communities are deeper than 6 m [60]. Quantitative studies in Lakshadweep, the Gulf of Mannar, and the Andaman and Nicobar Islands have described a reduction in live coral cover and declining reef health primarily due to bleaching events from 1998 to 2015 [56,58]. More than 70% of coral cover was bleached in the Andaman Islands, Gulf of Mannar, and Lakshadweep islands during the two global bleaching events in 1998 and 2010 [60,62–68]. The bleaching event of 1998 also caused extensive coral mortality in the southern, south-western, and north-western parts of Sri Lanka [69–71], but did not affect the coral reefs in the Pigeon Island, Northern Park, and Dutch Bay areas of eastern Sri Lanka [71]. The tsunami event of December 2004 is also known to have affected the eastern and southern coral reef regions of Sri Lanka [70,71].

On the other hand, coral reefs in the Andaman and Nicobar Islands, Lakshadweep, Gulf of Mannar, Chagos Archipelago, and Maldives have shown considerable recovery following past bleaching events [57–59,64–68]. Post-bleaching monitoring observations in Lakshadweep showed a reduction in the dead coral-algal substrate, coral recruitment, and an increase in live coral cover supporting a phase shift back from algal-dominated reefs to live coral-dominated substrates [58,61]. A similar scenario might have been experienced in the present study area by the Palk Bay islands, resulting in the observed high percentage cover of standing dead coral with turf algae and of coral rubble. In addition, another recent report on the biodiversity of northern Sri Lanka indicates a similar pattern of coral benthic composition on many of the islands in Palk Bay, including Punkudutivu [24].

However, it may be suggested that the recovery of corals in the Gulf of Mannar and Palk Bay has been severely hindered by illegal fishing by vessels. These vessels operate even in the shallowest areas, stirring up the bottom and causing sedimentation [58,61,69]; they are also involved in bottom trawling, which overturns corals and causes direct damage [24–26,72]. The area is also subject to illegal fishing practices such as blast fishing [72]. More generally, the inshore reefs, especially along the northern coast of JP, are subject to considerable fishing pressure, with 21 fish landing sites present within 48 km of coastline, and a higher number of fishing vessels operating than anywhere else in the nation [73]. These reefs have also been subject to other impacts, including increased terrestrial runoff and direct damage and sedimentation resulting from dredging and the creation of boat channels at Pungudutivu, Kayts, Point Pedro, and Valithoondal. Given the scarcity of available data, further detailed research will be required to elucidate the interacting effects of environmental conditions, seasonal variation, local anthropogenic pressure, and global climate change on the community structure of these reefs.

### 4.2. Coral Species Diversity

This study has significantly improved knowledge of the coral species present in the study area. Previously, no comprehensive coral inventory had been produced for these reefs [11,20,21,27,28], which now appear to perhaps be some of the most biodiverse in Sri Lanka.

Ridley (1883) made the first note recording 49 coral species in Ceylon (Sri Lanka), his specimens being deposited in the British Museum [12]. Subsequent studies on coral species and their distribution were mostly undertaken in the more southern parts of Sri Lanka at sites including Galle, Hikkaduwa, the Kalpitiya Peninsula, Unawantuna, Weligama, Polhena, Tangalle, the Great and Little Basses, and Trincomalee [9–11,19–21,28]. In comparison, the coral reefs of northern Sri Lanka have been under-studied, except for those in the Gulf of Mannar (in north-west Sri Lanka). There was, however, some preliminary information on aspects of the coral fauna of the Jaffna area [24,27]. Initial rapid surveys by Rajasuriya [27] noted at least 43 species of hard corals belonging to 20 genera. He commented that the coral reefs on the north coast and around Pungudutivu Island were more diverse than those at other study sites in the area. In contrast, the present study recorded a total of 113 hard coral species belonging to 40 genera from just eight reef areas, with the highest number of species being recorded at Point Pedro reef front (50 spp.), followed by Thondaimanaru (44 spp.) (Table 2).

Several species recorded in the present study have also been found at other coral reef sites in Sri Lanka [9], while many were listed in the corals of India [74–77]. Thus, in terms of coral species diversity, the reefs appear more comparable to other Indian territorial reefs (Figure 7). According to Raghuraman et al., 2012 [76], the various reef areas of India support a total of 478 species, representing 19 families and 89 genera. Considering the coral species recorded, the closest faunistic relationship of the Jaffna Peninsula reefs is to the reefs of the Andaman and Nicobar Islands [12,16]. However, fewer species (133) were reported here than in the Andaman Islands, where around 400 species of corals have been recorded. Coral species diversity was also similar to the Gulf of Mannar, further to the west, which has about 100 species [75]. Of the 36 coral species reported here as new records for Sri Lanka, most appear to be widely but patchily distributed across the central Indian Ocean and the Arabian Sea (Table 3) [75], with the largest proportions of these also recorded from the Andaman and Nicobar Islands (23 spp.), the Arabian Sea (21 spp.), and the Central Indian Ocean (30 spp.) [61,74–76].

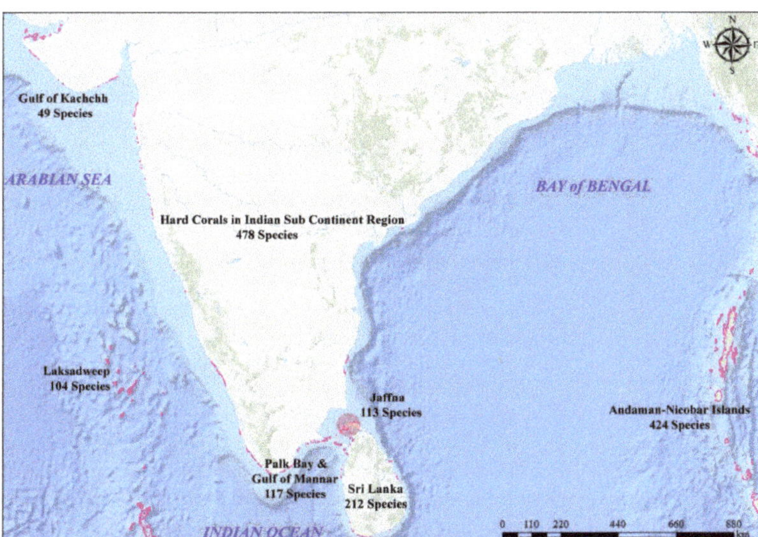


**Figure 7.** Comparison of the numbers of scleractinian corals recorded in different parts of India and Sri Lanka in relation to the Jaffna Peninsula and its islands. Pink colouring indicates the location of reef areas; the pink circle in the centre indicates the location of the present study areas at the northern end of Sri Lanka) [24,62,75,77].

Since the ability of corals to survive and recover after a bleaching event varies between species, greater species richness may endow a habitat with greater community tolerance and a greater ability to adapt to sea surface temperature changes [78–81]. At the same time, species abundance and overall coral cover will also be influenced by natural environmental factors such as wave action, salinity, exposure to light, and sediment loading [82]. In the present study, thermally tolerant encrusting and massive coral growth forms such as *Porites, Favites, Goniastrea*, and *Dipsastraea* were found to have survived to a greater extent than branching species of the genera *Acropora, Pocillopora*, and *Stylophora*, which are more susceptible to thermal events in many Indo-Pacific regions [78,79]. Nevertheless, large colonies of tabulate and branching *Acropora* and tiered plates of *Echinopora* and *Montipora* dominated the islands, while massive coral colonies dominated the northern coast. Similar patterns of coral distribution have been described in earlier studies [24], with large domes of stress-tolerant *Dipsastraea, Favites, Goniastrea, Platygyra* and *Porites* [79,80,83] occupying shallow reef flats down to 6 m depth. Notably, there were only a few small colonies of *Pocillopora* and *Stylophora* observed during the present study; this may result from their susceptibility to past bleaching events.

Of the species recorded in the present study, three are listed as "vulnerable" by the International Union for the Conservation of Nature (IUCN): *Acropora aspera, Acropora specifera*, and *Acropora hemprichii*. These were found at the Punkudutivu, Kayts, and Karainagar island sites, respectively. In addition, "threatened" species of *Montipora flabellata, Turbinaria frondens, Turbinaria stellulata*, and a rare species of *Astreopora ocellata* were also recorded during this study.

### 4.3. DNA Confirmation of Selected Coral Species

The genus *Acropora*, belonging to the family Acroporidae, is the most diverse and widespread coral genus. The species show both enormous intraspecific morphological variability and striking similarities between species within a geographical region, making species identification a challenge [84]. Consequently, for several species whose growth patterns and corallite arrangements appeared atypical, DNA barcoding was employed in parallel with morphological analysis. Notably, *A. aspera, A. digitifera*, and *A. gemmifera* were confirmed as present in Sri Lanka for the first time, although they have been widely recorded elsewhere in the Indian Ocean. The mitochondrial *COI* gene sequence of each species was found to have more than 99% similarity with the published genome sequences (Table 5). Sample COJP001 has a 99.69% similarity with the genomic region of the *A. aspera* and the morphological features also confirmed this identification (Table 5). Based on the description provided by Veron [43], *A. aspera* is a branching species with scale-like radial corallites. It is often confused with the similar species *A. millepora*; however, the corymbose clumps of colonies with branches and the two various sizes of axial corallites serve to differentiate *A. aspera* from *A. millepora*. Samples COJP002 and COJP004 were matched to samples of *A. gemmifera*, with the closest similarity being to a Malaysian sample *MG383839.1* [49]. COJP003 from the Valithoondal site was confirmed as *A. digitifera* and found to have the closest genetic relationship to a sample from the Egyptian coast of the Red Sea, *A. digitifera_KR401100.1* (Table 5) [50]. This species is widely distributed in the Indian Ocean, Red Sea, and eastern Pacific Ocean [43] (Supplementary Table S2).

The colonies of *M. flabellata*, one of the world's uncommon species, were recorded on the Point Pedro reef front. They had an encrusting form, were blue and pinkish to brown in colour, and had prominent thecal papillae covering the colony. Septa could not be observed, making species separation from *M. undata* difficult. However, sequences from this species had the closest match with reference sequences from the Hawaiian Islands, HQ246603.1 [53]. The sample COJP007, which was confirmed as being of *Echinopora gemmacea*, was found only at Point Pedro and Inbarsitty. According to [43], *E. gemmacea* is an uncommon species. This specimen has the closest kinship to *E. gemmacea HE654561.1* from Yemen (Table 5) [52].

While the *COI* gene-based sequence relates to only a narrow portion of the genome, nevertheless, as found by others, it proved invaluable in confirming species-level identification of specimens from difficult to identify genera [33,85]. Apart from any intraspecific genetic variations, corals show a marked morphological plasticity that is considered to be largely adaptive [30]; molecular genetic studies are allowing a better understanding of this plasticity [86]. More generally, the use of DNA barcoding is enabling taxonomists to overcome the pitfalls of morphology-based identification and classification. In the present study, the use of barcoding has helped validate the finding of six scleractinian coral species new to Sri Lanka.

In summary, the northern coast of Jaffna Peninsula has the highest live coral cover among the regions of Sri Lanka; however, the reefs are prone to impacts from increasing fishing pressure and environmental pollution. These new insights into the coral biodiversity, distribution, and benthic biota of the coral reefs of the Jaffna Peninsula contribute knowledge required for future conservation and management. The present study indicates that the conservation of the coral reefs in Jaffna is critical, not only because of their direct ecological and economic importance, but because they may provide a reservoir of different coral species able to withstand the effects of future climate change.

**Supplementary Materials:** The following are available online at https://www.mdpi.com/article/10.3390/oceans2030029/s1.

**Author Contributions:** Conceptualisation, A.A. and V.H.; data curation, A.A., V.H. and S.K.; funding acquisition, A.A. and V.H.; investigation, A.A. and A.U.; methodology, A.A., V.H., A.U. and A.H.; project administration, V.H.; resources, V.H.; supervision, V.H. and S.K.; validation, A.U.; visualisation, A.A., A.U. and A.H.; writing—original draft, A.A.; writing—review and editing, V.H. and S.K. All authors have read and agreed to the published version of the manuscript.

**Funding:** This research received funding from IDEAWILD and from the Research Facilitation Fund (RFF) of the Postgraduate Institute of Agriculture, University of Peradeniya.

**Institutional Review Board Statement:** Samples were collected with permits from, and in compliance with, the laws and regulations of the Sri Lanka Department of Wildlife Conservation.

**Informed Consent Statement:** Not applicable.

**Data Availability Statement:** Not applicable.

**Acknowledgments:** We acknowledge the International Coral Reef Society (ICRS) for granting a graduate fellowship to the senior author. We also thank the Department of Agricultural Biology, Faculty of Agriculture, University of Peradeniya for providing molecular laboratory facilities, the Department of Fisheries Science, Faculty of Science, University of Jaffna for their technical support We also thank the editor, Rupert Ormond, for extensive support in improving the text.

**Conflicts of Interest:** The authors declare that they have no known competing financial interest or personal relationship that could have appeared to influence the research studies presented in this paper.

# References

1. Kitahara, M.V.; Fukami, H.; Benzoni, F.; Huang, D. The New Systematics of Scleractinia: Integrating Molecular and Morphological Evidence. In *The Cnidaria, Past, Present and Future: The World of Medusa and Her Sisters*; Goffredo, S., Dubinsky, Z., Eds.; Springer International Publishing: Cham, Switzerland, 2016; pp. 41–59, ISBN 978-3-319-31305-4.
2. Harrison, P.L. Sexual Reproduction of Scleractinian Corals. In *Coral Reefs: An Ecosystem in Transition*; Dubinsky, Z., Stambler, N., Eds.; Springer: Dordrecht, The Netherlands, 2011; pp. 59–85, ISBN 978-94-007-0114-4.
3. LaJeunesse, T.C.; Parkinson, J.E.; Gabrielson, P.W.; Jeong, H.J.; Reimer, J.D.; Voolstra, C.R.; Santos, S.R. Systematic Revision of Symbiodiniaceae Highlights the Antiquity and Diversity of Coral Endosymbionts. *Curr. Biol.* **2018**, *28*, 2570–2580.e6. [CrossRef]
4. Muller-Parker, G.; D'Elia, C.F.; Cook, C.B. Interactions between Corals and Their Symbiotic Algae. In *Coral Reefs in the Anthropocene*; Birkeland, C., Ed.; Springer: Dordrecht, The Netherlands, 2015; pp. 99–116, ISBN 978-94-017-7249-5.
5. Reaka-Kudla, M.L.; Wilson, D.E.; Wilson, E.O. *Biodiversity II: Understanding and Protecting Our Biological Resources*; Joseph Henry Press: Washington, DC, USA, 1996; ISBN 978-0-309-52075-1.
6. Eakin, C.M.; Sweatman, H.P.A.; Brainard, R.E. The 2014–2017 Global-Scale Coral Bleaching Event: Insights and Impacts. *Coral Reefs* **2019**, *38*, 539–545. [CrossRef]

7.  Hughes, T.P.; Anderson, K.D.; Connolly, S.R.; Heron, S.F.; Kerry, J.T.; Lough, J.M.; Baird, A.H.; Baum, J.K.; Berumen, M.L.; Bridge, T.C.; et al. Spatial and Temporal Patterns of Mass Bleaching of Corals in the Anthropocene. *Science* **2018**, *359*, 80–83. [CrossRef]

8.  Silva, E.I.L.; Katupotha, J.; Amarasinghe, O.; Manthrithilake, H.; Ariyaratne, B.R. *Lagoons of Sri Lanka: From the Origins to the Present*; International Water Management Institute: Anand, India, 2013; ISBN 978-92-9090-777-0.

9.  Rajasuriya, A. Provisional Checklist of Corals in SL. In *The National Red List 2012 of SL Conservation Status of the Fauna and Flora*; Weerakoon, D.K., Wijesundara, S., Eds.; Ministry of Environment: Colombo, Sri Lanka, 2012; pp. 411–430, ISBN 978-955-0033-55-3.

10. Rajasuriya, A.; White, A.T. Coral Reefs of Sri Lanka: Review of Their Extent, Condition, and Management Status. *Coast. Manag.* **1995**, *23*, 77–90. [CrossRef]

11. Rajasuriya, A. Coral reefs of Sri Lanka: Current status and resource Management. In *Regional Workshop on the Conservation and Sustainable Management of Coral Reefs*; National Aquatic Resources Research and Development Agency: Colombo, Sri Lanka, 1997.

12. Ridley, S.O. XXXIV.—The Coral-Fauna of Ceylon, with Descriptions of New Species. *Ann. Mag. Nat. Hist.* **1883**, *11*, 250–262. [CrossRef]

13. Ortmann, A. Beobachtungen an Steinkorallen von der Südküste Ceylons [Observations on the Stone Corals of the South Coast of Sri Lanka]. *Zool. J.* **1889**, *4*, 493–590.

14. Bourne, G.C. *Report on the Solitary Corals Collected by Professor Herdman at Ceylon in 1902*; Rep. Govt. Ceylon Pearl Oyster Fish: Gulf Mannar, India, 1905; Volume 29, pp. 187–242.

15. Pillai, C.S.G. Stony Corals of the Seas around India. In *Proceedings of the First International Symposium on Corals and Coral Reefs*; Marine Biological Association of India: Mandapam, India, 1969.

16. Mergner, H.; Scheer, G. The Physiographic Zonation and the Ecological Conditions of Some South Indian and Ceylon Reefs. In *Proceedings of the 2nd International Coral Reef Symposium*; Great Barrier Reef Committee: Brisbane, Australia, 1974; Volume 2, pp. 3–30.

17. De Silva, M.W.R.N. Status of coral reefs of Sri Lanka, Singapore and Malaysia. In *Coral Reef Newsletter*; International Union for the Conservation of Nature: Grand, Switzerland, 1981; Volume 3, pp. 34–37.

18. Scheer, G. The Distribution of Reef-Corals in the Indian Ocean with a Historical Review of Its Investigation. *Deep-Sea Res.* **1984**, *31*, 885–900. [CrossRef]

19. De Silva, M.W.R.N. Status of the coral reefs of Sri Lanka. In Proceedings of the Fifth International Coral Reef Congress, Tahiti, Polynesia, 27 May–1 June 1985; Volume 6, pp. 515–518.

20. Rajasuriya, A.; De Silva, M.W.R.N. Stony corals of fringing reefs of the Western, South-Western and Southern coasts of Sri Lanka. In Proceedings of the 6th International Coral Reef Symposium, Townsville, Australia, 8–12 August 1988; Volume 3.

21. Rajasuriya, A. Status of Coral Reefs in the Northern, Western and Southern Coastal Waters of SL. Ten Years after bleaching-facing the consequences of climate change in the Indian Ocean. In *CORDIO Status Report*; Obura, D., Tamelander, J., Linden, O., Eds.; CORDIO/Sida-SAREC: Mombasa, Kenya, 2008; pp. 11–22.

22. Kularatne, R.K.A. Suitability of the Coastal Waters of Sri Lanka for Offshore Sand Mining: A Case Study on Environmental Considerations. *J. Coast. Conserv.* **2014**, *18*, 227–247. [CrossRef]

23. Arachchige, G.M.; Jayakody, S.; Mooi, R.; Kroh, A. A Review of Previous Studies on the Sri Lankan Echinoid Fauna, with an Updated Species List. *Zootaxa* **2017**, *4231*, zootaxa.4231.2.1. [CrossRef]

24. Weerakoon, D.; Goonatilake, S.D.A.; Wijewickrama, T.; Rajasuriya, A.; Perera, N.; Kumara, T.P.; De Silva, G.; Miththapala, S.; Mallawatantri, A. *Conservation and Sustainable Use of Biodiversity in the Islands and Lagoons of Northern Sri Lanka*; IUCN, International Union for Conservation of Nature: Grand, Switzerland, 2020; ISBN 978-2-8317-2088-3.

25. Dodangodage, P.K. Illegal Fishing by Indian Trawlers Violating the Maritime Boundary of Sri Lanka and Its Impact on Livelihood and the Indo-Sri Lanka Relations. In Proceedings of the 10th International Research Conference of KDU, Ratmalana, Sri Lanka, 3–4 August 2017.

26. Wijesundara, S.; Amunugama, D. Bottom Trawling in Palk Bay Area: Human and Environmental Implications. In Proceedings of the 10th International Research Conference of KDU, Ratmalana, Sri Lanka, 3–4 August 2017.

27. Rajasuriya, A. Coral reefs in the Palk Strait and Palk Bay in 2005. *J. Natl. Aquat. Resour. Res. Dev. Agency* **2007**, *38*, 77–86.

28. Rajasuriya, A. Status report on the reef conditions of coral reefs in Sri Lanka. In *Coral Reef Degradation in the Indian Ocean Status Report*; Lindén, O., Souter, D., Wilhelmsson, D., Obura, D., Eds.; CORDIO, Dept. of Biology and Environmental Science, University of Kalmar: Kalmar, Sweden, 2002.

29. Sachithananthan, K.; Perera, W.K.T. Topography and substratum of the Jaffna Lagoon. *Bull. Fish. Res. Stn. Ceylon* **1970**, *21*, 75–85.

30. Todd, P.A. Morphological Plasticity in Scleractinian Corals. *Biol. Rev.* **2008**, *83*, 315–337. [CrossRef] [PubMed]

31. Souter, P. Hidden Genetic Diversity in a Key Model Species of Coral. *Mar. Biol.* **2010**, *157*, 875–885. [CrossRef]

32. Veron, J.E.N. Coral Taxonomy and Evolution. In *Coral Reefs: An Ecosystem in Transition*; Dubinsky, Z., Stambler, N., Eds.; Springer: Dordrecht, The Netherlands, 2011; pp. 37–45, ISBN 978-94-007-0114-4.

33. Hebert, P.D.N.; Cywinska, A.; Ball, S.L.; de Waard, J.R. Biological Identifications through DNA Barcodes. *Proc. R. Soc. Lond. Ser. B Biol. Sci.* **2003**, *270*, 313–321. [CrossRef] [PubMed]

34. Kitahara, M.V.; Cairns, S.D.; Stolarski, J.; Blair, D.; Miller, D.J.A. Comprehensive Phylogenetic Analysis of the Scleractinia (Cnidaria, Anthozoa) Based on Mitochondrial CO1 Sequence Data. *PLoS ONE* **2010**, *5*, e11490. [CrossRef] [PubMed]

35. Benzoni, F.; Stefani, F.; Pichon, M.; Galli, P. The name game: Morpho-molecular species boundaries in the genus *Psammocora* (Cnidaria, Scleractinia). *Zool. J. Linn. Soc.* **2010**, *160*, 421–456. [CrossRef]

36. Benzoni, F.; Arrigoni, R.; Waheed, Z.; Stefani, F.; Hoeksema, B. Phylogenetic Relationships and Revision of the Genus Blastomussa (Cnidaria: Anthozoa: Scleractinia) with Description of a New Species. *Raffles Bull. Zool.* **2014**, *62*, 358–378.
37. Gittenberger, A.; Reijnen, B.T.; Hoeksema, B.W. A Molecularly Based Phylogeny Reconstruction of Mushroom Corals (Scleractinia: Fungiidae) with Taxonomic Consequences and Evolutionary Implications for Life History Traits. *Contrib. Zool.* **2011**, *80*, 107–132. [CrossRef]
38. Cooray, P.G. *An Introduction to the Geology of Sri Lanka (Ceylon)*; National Museums of Sri Lanka Publication: Colombo, Sri Lanka, 1984.
39. Hodgson, G.; Hill, J.; Kiene, W.; Maun, L.; Mihaly, J.; Liebeler, J.; Shuman, C.; Torres, R. *Reef Check Instruction Manual: A Guide to Reef Check Coral Reef Monitoring*; Reef Check Foundation: Pacific Palisades, CA, USA, 2006; ISBN 0-9723051-1-4.
40. Hammer, O.; Harper, D.; Ryan, P. Paleontological Statistics Software Package for Education and Data Analsis. *Palaeontol. Electron.* **2001**, *4*, 9–18.
41. Gomez, E.D.; Aliño, P.M.; Yap, H.T.; Licuanan, W.Y. A review of the status of Philippine reefs. *Mar. Pollut. Bull.* **1994**, *29*, 62–68. [CrossRef]
42. Kelley, R. *Indo Pacific Coral Finder*, 3rd ed.; BYO Guides: Townsville, Australia, 2016; ISBN 978-0-646-52326-2.
43. Veron, J.E.N. *Corals of the World: Vol 1–3*; Australian Institute of Marine Science: Townsville, MC, Australia, 2000; ISBN 978-0-642-32236-4.
44. Thukral, A.K.; Bhardwaj, R.; Kumar, V.; Sharma, A. Corrigendum to "New Indices Regarding the Dominance and Diversity of Communities, Derived from Sample Variance and Standard Deviation" [Heliyon 5 (10) (October 2019) E02606]. *Heliyon* **2019**, *5*, e03017. [CrossRef]
45. Folmer, O.; Black, M.; Hoeh, W.; Lutz, R.; Vrijenhoek, R. DNA primers for amplification of mitochondrial cytochrome c oxidase subunit I from diverse metazoan invertebrates. *Mol. Mar. Biol. Biotechnol.* **1994**, *3*, 294–299. [PubMed]
46. Okonechnikov, K.; Golosova, O.; Fursov, M. UGENE team Unipro UGENE: A unified bioinformatics toolkit. *Bioinformatics* **2012**, *28*, 1166–1167. [CrossRef] [PubMed]
47. Bold Systems V4. Available online: https://v4.boldsystems.org/ (accessed on 10 April 2020).
48. Chan, C.-L.; Chen, C.A. Multiplex Next Generation Sequencing of Scleractinian Mitochondrial Genomes. Sequence Submitted (25-JUL-2013) Biodiversity Research Center, Academia Sinica, 128 Academia Road Section 2, Nangang District, Taipei 115, Taiwan. National Center for Biotechnology Information. *Acropora aspera* mitochondrion, complete genome. Available online: https://www.ncbi.nlm.nih.gov/nuccore/KF448532.1 (accessed on 12 July 2021).
49. Robert, R.; Rodrigues, K.F.; Waheed, Z.; Kumar, S.V. Extensive Sharing of Mitochondrial *COI* and *CYB* Haplotypes among Reef-Building Staghorn Corals (*Acropora* Spp.) in Sabah, North Borneo. *Mitochondrial DNA Part A* **2019**, *30*, 16–23. [CrossRef] [PubMed]
50. Kamel, M.A.M.; Ahmed, M.I.; Madkour, F.F.; Hanafy, M.H. Comparative Molecular Ecology Studies on some Scleractinia (Cnidaria, Anthozoa), in the Arabian Gulf and the Egyptian Coast of the Red Sea. Sequence Submitted (01-MAY-2015) Marine Science Department, Faculty of Science, Port Said University, 23 December, Port Said 42526, Egypt. National Center for Biotechnology Information. *Acropora digitifera* isolate ADS *cytochrome oxidase subunit 1* (*COI*) gene, partial cds; mitochondrial. Available online: https://www.ncbi.nlm.nih.gov/nuccore/KR401100.1 (accessed on 12 July 2021).
51. Wijayanti, D.P.; Indrayanti, E.; Nuryadi, H.; Aini, S.N.; Rintiantoto, S.A. Genetic Diversity of Two Coral Species, Pocillopora damicornis and Acropora hyacinthus from Wakatobi waters, Indonesia. Submitted (04-OCT-2017) Contact:Diah Permata Wijayanti Diponegoro, University, Marine Science Department; Prof. Soedarto, SH. Semarang, Central Java 50275, Indonesia URL:http://www.fpik.undip.ac.id. National Center for Biotechnology Information. *Acropora hyacinthus* Mitochondrial *H_Ah17* Gene for Cytochrome Oxidase Subunit I, Partial cds. Available online: https://www.ncbi.nlm.nih.gov/nuccore/LC326547.1 (accessed on 12 July 2021).
52. Arrigoni, R.; Stefani, F.; Pichon, M.; Galli, P.; Benzoni, F. Molecular Phylogeny of the Robust Clade (Faviidae, Mussidae, Merulinidae, and Pectiniidae): An Indian Ocean Perspective. *Mol. Phylogenet. Evol.* **2012**, *65*, 183–193. [CrossRef]
53. Forsman, Z.H.; Concepcion, G.T.; Haverkort, R.D.; Shaw, R.W.; Maragos, J.E.; Toonen, R.J. Ecomorph or Endangered Coral? DNA and Microstructure Reveal Hawaiian Species Complexes: Montipora dilatata/flabellata/turgescens & M. patula/verrilli. *PLoS ONE* **2010**, *5*, e15021. [CrossRef]
54. Venkataraman, K.; Satyanarayan, C.; Alfred, J.R.B.; Wolstenholme, J. *Hand Book on Hard Corals of India*; Director, Zoological Survey of India: Kolkata, India, 2003; ISBN 81-8171-20-7.
55. Bahuguna, A.; Chaudhury, R.N.; Bhattji, N.; Navalgund, R.R. Spatial Inventory and Ecological Status of Coral Reefs of the Central Indian Ocean Using Resourcesat-1. *Indian J. Geo-Mar. Sci.* **2013**, *42*, 684–696.
56. Arthur, R. Coral Bleaching and Mortality in Three Indian Reef Regions during an El Niño Southern Oscillation Event. *Curr. Sci.* **2000**, *79*, 1723–1729.
57. Rajasuriya, A.; Zahir, H.; Muley, E.V.; Subramanian, B.R.; Venkataraman, K.; Wafar, S.M.; Hannan Khan, M.; Whittingham, E. Status of Coral Reefs in South Asia: Bangladesh, India, Maldives and Sri Lanka. In *Status of Coral Reefs of the World*; Wilkinson, C., Ed.; Australian Institute of Marine Science: Townsville, Australia, 2000; pp. 95–115.
58. Patterson, E.; Gilbert, M.; Raj, K.D.; Thinesh, T.; Patterson, J.; Tamelander, J. Coral Reefs of Gulf of Mannar, India -Signs of Resilience. In Proceedings of the 12th International Coral Reef Symposium, Cairns, Australia, 9–13 July 2012.
59. Arthur, R. Patterns of Benthic recovery in Lakshadweep atolls. In *Ten Years after Bleaching–Facing the Consequences of Climate Change in the Indian Ocean CORDIO Status Report*; Obura, D.O., Tamelander, J., Linden, O., Eds.; Coastal Oceans Research and Development in the Indian Ocean/Sida–SAREC: Mombasa, Kenya, 2008; pp. 39–44.

60. Krishnan, P.; Purvaja, R.; Sreeraj, C.R.; Raghuraman, R.; Robin, R.S.; Abhilash, K.R.; Mahendra, R.S.; Anand, A.; Gopi, M.; Mohanty, P.C.; et al. Differential Bleaching Patterns in Corals of Palk Bay and the Gulf of Mannar. *Curr. Sci.* **2018**, *114*, 679. [CrossRef]

61. Manikandan, B.; Ravindran, J. Differential Response of Coral Communities to *Caulerpa Spp.* Bloom in the Reefs of Indian Ocean. *Environ. Sci. Pollut. Res.* **2017**, *24*, 3912–3922. [CrossRef] [PubMed]

62. Rajan, R.; Satyanarayan, C.; Raghunathan, C.; Koya, S.S.; Ravindran, J.; Manikandan, B.; Venkataraman, K. Status and Review of Health of Indian Coral Reefs. *J. Aquat. Biol. Fish.* **2015**, *3*, 1–14.

63. Marimuthu, N.; Yogesh Kumar, J.S.; Raghunathan, C.; Vinithkumar, N.V.; Kirubagaran, R.; Sivakumar, K.; Venkataraman, K. North-South Gradient of Incidence, Distribution and Variations of Coral Reef Communities in the Andaman and Nicobar Islands, India. *J. Coast. Conserv.* **2017**, *21*, 289–301. [CrossRef]

64. Machendiranathan, M.; Ranith, R.; Senthilnathan, L.; Saravanakumar, A.; Thangaradjou, T. Resilience of Coral Recruits in Gulf of Mannar Marine Biosphere Reserve (GOMMBR), India. *Reg. Stud. Mar. Sci.* **2020**, *34*, 101055. [CrossRef]

65. Sukumaran, S.; George, R.M.; Kasinathan, C. Community Structure and Spatial Patterns of Hard Coral Biodiversity in Kilakarai Group of Islands in Gulf of Mannar, India. *J. Mar. Biol. Assoc. India* **2008**, *50*, 79–86.

66. Krishnan, P.; Dam Roy, S.; George, G.; Srivastava, R.C.; Anand, A.; Murugesan, S.; Kaliyamoorthy, M.; Vikas, N.; Soundararajan, R. Elevated sea surface temperature during May 2010 induces mass bleaching of corals in the Andaman. *Curr. Sci.* **2011**, *100*, 111.

67. Sheppard, C.; Harris, A.; Sheppard, A. Archipelago-Wide Coral Recovery Patterns since 1998 in the Chagos Archipelago, Central Indian Ocean. *Mar. Ecol. Prog. Ser.* **2008**, *362*, 109–117. [CrossRef]

68. Smith, L.D.; Gilmour, J.P.; Heyward, A.J. Resilience of Coral Communities on an Isolated System of Reefs Following Catastrophic Mass-Bleaching. *Coral Reefs* **2008**, *27*, 197–205. [CrossRef]

69. Rajasuriya, A.; Karunaratne, C. Post-bleaching status of the coral reefs of Sri Lanka. Coral reef degradation in the Indian Ocean. In *Coral Reef Degradation in the Indian Ocean: Status Report 2000*; Souter, D., Obura, D., Linden, O., Eds.; CORDIO, SAREC Marine Science Program, Stockholm University: Stockholm, Sweden, 2000; pp. 54–63.

70. Rajasuriya, A. Status of coral reefs in Sri Lanka in the aftermath of the 1998 coral bleaching event and 2004 tsunami. In *Coral Reef Degradation in the Indian Ocean: Status Report 2005*; Souter, D., Linden, O., Eds.; CORDIO, Department of Biology and Environmental Science, University of Kalmar: Småland, Sweden, 2005; pp. 83–96.

71. Rajasuriya, A.; Perera, N.; Fernando, M. Status of coral reefs in Trincomalee, Sri Lanka. In *Coral Reef Degradation in the Indian Ocean: Status Report 2005*; Souter, D., Lindèn, O., Eds.; CORDIO, Department of Biology and Environmental Science, University of Kalmar: Småland, Sweden, 2005; pp. 97–103.

72. Kularatne, R.K.A. Unregulated and Illegal Fishing by Foreign Fishing Boats in Sri Lankan Waters with Special Reference to Bottom Trawling in Northern Sri Lanka: A Critical Analysis of the Sri Lankan Legislation. *Ocean Coast. Manag.* **2020**, *185*, 105012. [CrossRef]

73. Survey Department, Statistics Unit, Ministry of Fisheries and Aquatic Resources Development. 2018. Available online: https://www.fisheriesdept.gov.lk/web/images/pdf/Fisheries_Statistics_2018.pdf (accessed on 19 May 2019).

74. Hoeksema, B.W.; Cairns, S. World List of Scleractinia. 2021. Available online: http://www.marinespecies.org/scleractinia (accessed on 18 April 2021).

75. Venkataraman, K.; Rajan, R.; Satyanarayana, C.H.; Raghunathan, C.; Venkatraman, C. *Marine Ecosystems and Marine Protected Areas of India*; Director, Zoological Survey of India: Kolkata, India, 2012; ISBN 978-81-8171-312-4.

76. Raghuraman, R.; Sreeraj, C.R.; Raghunathan, C.; Venkataraman, K. Scleractinian coral diversity in Andaman Nicobar Island in comparison with other Indian reefs. In *International Day for Biological Diversity Marine Biodiversity 2012*; Uttar Pradesh State Biodiversity Board: Uttar Pradesh, India, 2012; pp. 75–92.

77. Venkataraman, K.; Rajan, R. Status of Coral Reefs in Palk Bay. *Rec. Zool. Surv. India* **2013**, *113*, 1–11.

78. Loya, Y.; Sakai, K.; Yamazato, K.; Nakano, Y.; Sambali, H.; van Woesik, R. Coral Bleaching: The Winners and the Losers. *Ecol. Lett.* **2001**, *4*, 122–131. [CrossRef]

79. Woesik, R.; van Sakai, K.; Ganase, A.; Loya, Y. Revisiting the winners and the losers a decade after coral bleaching. *Mar. Ecol. Prog. Ser.* **2011**, *434*, 67–76. [CrossRef]

80. Wilson, S.K.; Robinson, J.P.W.; Chong-Seng, K.; Robinson, J.; Graham, N.A.J. Boom and bust of keystone structure on coral reefs. *Coral Reefs* **2019**, *38*, 625–635. [CrossRef]

81. Lough, J.M. Small change, big difference: Sea surface temperature distributions for tropical coral reef ecosystems, 1950–2011. *J. Geophys Res. Ocean.* **2012**, *117*. [CrossRef]

82. Polónia, A.R.M.; Cleary, D.F.R.; de Voogd, N.J.; Renema, W.; Hoeksema, B.W.; Martins, A.; Gomes, N.C.M. Habitat and Water Quality Variables as Predictors of Community Composition in an Indonesian Coral Reef: A Multi-Taxon Study in the Spermonde Archipelago. *Sci. Total Environ.* **2015**, *537*, 139–151. [CrossRef] [PubMed]

83. Gunarathna, A.B.A.K.; Rajasuriya, A. Extent and status of coral reefs in northern coastal waters of Sri Lanka. In Proceedings of the 25th Anniversary Scientific Conference of NARA on Tropical Aquatic Research towards Sustainable Development, Colombo, Sri Lanka, 15–16 February 2007; p. 41.

84. Veron, J.E.N.; Stafford-Smith, M.; DeVantier, L.; Turak, E. Overview of distribution patterns of zooxanthellate Scleractinia. *Front. Mar. Sci.* **2015**, *1*. [CrossRef]

85. Evans, N.; Paulay, G. DNA Barcoding Methods for Invertebrates. In *DNA Barcodes: Methods and Protocols*; Kress, W.J., Erickson, D.L., Eds.; Methods in Molecular Biology; Humana Press: Totowa, NJ, USA, 2012; pp. 47–77, ISBN 978-1-61779-591-6.
86. Duncan, E.J.; Gluckman, P.D.; Dearden, P.K. Epigenetics, plasticity, and evolution: How do we link epigenetic change to phenotype? *J. Exp. Zool. B Mol. Dev. Evol.* **2014**, *322*, 208–220. [CrossRef]

*Article*

# Cold-Water Coral Reefs in the Langenuen Fjord, Southwestern Norway—A Window into Future Environmental Change


**Katriina Juva [1,*], Tina Kutti [2], Melissa Chierici [3], Wolf-Christian Dullo [1] and Sascha Flögel [1]**

[1]    GEOMAR Helmholtz Centre for Ocean Research Kiel, Wischhofstraße 1-3, 24148 Kiel, Germany;
       cdullo@geomar.de (W.-C.D.); sfloegel@geomar.de (S.F.)
[2]    Institute of Marine Research in Norway (IMR), Nordnesgaten 50, 5005 Bergen, Norway; tina.kutti@hi.no
[3]    Institute of Marine Research in Norway, Fram Centre, 9007 Tromsø, Norway; melissa.chierici@hi.no
*    Correspondence: katriina.juva@iki.fi



**Abstract:** Ocean warming and acidification pose serious threats to cold-water corals (CWCs) and the surrounding habitat. Yet, little is known about the role of natural short-term and seasonal environmental variability, which could be pivotal to determine the resilience of CWCs in a changing environment. Here, we provide continuous observational data of the hydrodynamic regime (recorded using two benthic landers) and point measurements of the carbonate and nutrient systems from five *Lophelia pertusa* reefs in the Langenuen Fjord, southwestern Norway, from 2016 to 2017. In this fjord setting, we found that over a tidal (<24 h) cycle during winter storms, the variability of measured parameters at CWC depths was comparable to the intra-annual variability, demonstrating that single point measurements are not sufficient for documenting (and monitoring) the biogeochemical conditions at CWC sites. Due to seasonal and diurnal forcing, parts of the reefs experienced temperatures up to 4 °C warmer (i.e., >12 °C) than the mean conditions and high $C_T$ concentrations of 20 $\mu$mol kg$^{-1}$ over the suggested threshold for healthy CWC reefs (i.e., >2170 $\mu$mol kg$^{-1}$). Combined with hindcast measurements, our findings indicate that these shallow fjord reefs may act as an early hotspot for ocean warming and acidification. We predict that corals in Langenuen will face seasonally high temperatures (>18 °C) and hypoxic and corrosive conditions within this century. Therefore, these fjord coral communities could forewarn us of the coming consequences of climate change on CWC diversity and function.

**Keywords:** ocean acidification; ocean warming; carbonate chemistry dynamics; biogeochemical dynamics; in situ monitoring; natural variability of environmental conditions; *Lophelia pertusa*



**Citation:** Juva, K.; Kutti, T.; Chierici, M.; Dullo, W.-C.; Flögel, S. Cold-Water Coral Reefs in the Langenuen Fjord, Southwestern Norway—A Window into Future Environmental Change. *Oceans* **2021**, *2*, 583–610. https://doi.org/10.3390/oceans2030033

Academic Editor: Luis Somoza

Received: 30 October 2020
Accepted: 10 August 2021
Published: 25 August 2021

**Publisher's Note:** MDPI stays neutral with regard to jurisdictional claims in published maps and institutional affiliations.

## 1. Introduction

Cold-water coral (CWC) reefs form complex habitats in the deep sea, supporting rich associated fauna [1–4]. They play a major role in the circulation of organic carbon in the deep sea [5,6]. The reefs are recognized as vulnerable marine ecosystems (VMEs) by the United Nations [7] and as threatened and declining habitats by the OSPAR commission [8]. Warming and acidifying waters (i.e., decreasing pH and aragonite saturation state, $\Omega_{Ar}$) are predicted to pose imminent and serious threats to CWC reefs within the next decades [9,10]. Hence, there has been an intensified focus to assess the range of biogeochemical and physical conditions tolerated by CWCs in situ [11–16] and in laboratory conditions [17–21]. Finally, a focus is on determining the baseline for optimal conditions for coral growth and reef development [22]. However, there is very little data on the small-scale temporal and spatial variability of biogeochemical and physical conditions in CWC settings, which could be pivotal in understanding their resilience to climate change, as observed for tropical coral reefs [23].

In the North Atlantic, CWC reefs are most commonly build by *Lophelia pertusa* (syn. *Desmophyllum pertusum* [24]), with around a third of all known *L. pertusa* occurrences being from Norwegian waters [25]. *L. pertusa* reefs are generally found on and around elevated bathymetric structures such as mounds, offshore banks, seamounts, and sills [26–30], where relatively strong bottom flow associated with those structures enhances the flux of food particles [22,31–33] and facilitates sediment removal from CWCs or the seafloor to provide suitable settlement substrates for coral larvae [25,34]. While local hydrodynamics are important in regulating the distribution of CWCs, water column characteristics such as temperature, nutrient availability, carbonate chemistry (mainly aragonite saturation, $\Omega_{Ar}$), and oxygen levels [2,12,15,35] act as additional and strong drivers in regulating coral growth. The latter two are vital in maintaining calcification and aerobic metabolism in corals, while low-temperature regimes decelerate food decay (enhancing food availability) and reduce metabolic energy demands [36].

Rising global ocean temperatures have been observed throughout the water column since the 1960s [37]. In coastal Norway, rapid warming of the Norwegian coastal water (NCW) and the North Atlantic water (NAW, $S_A < 35$ g kg$^{-1}$) in mid-layer waters has been reported since the 1980s [38]. Ocean warming is expected to affect both the fitness and the geographical distribution of many marine species in the coming decades. CWCs are usually found in waters with temperatures <12 °C [13,29,39], but occasionally they thrive in warmer waters of ~15 °C [15,40]. Even though CWCs can tolerate substantial fluctuations in temperature [41,42], a rise of only 2 °C for a few hours will significantly increase its energy demands [36,41], which could deplete energy stores if they are not met by increased food availability and/or uptake rates [35]. Ocean warming can also indirectly affect CWCs by changing primary production patterns [43] and reducing oxygen availability [44]. In southwestern Norway, the reduction in oxygen concentration has been ~0.5 mL L$^{-1}$ per decade over the past 40 years in the fjord basin waters [45].

Long-term decreasing trends of pH and aragonite saturation have been observed in the open ocean around the globe [46,47] as well as in coastal waters off Norway. Currently, the aragonite saturation horizon (ASH, $\Omega_{Ar} < 1$) of the Norwegian Sea is at ~2000 m depth. At these depths, an annual decrease of 0.003 in $\Omega_{Ar}$ and 0.008 in pH has been observed between 2007 and 2017 [48]. The observed decrease inside the fjords, where $\Omega_{Ar}$ is currently >1 across the water column, is larger. A two-fold annual decrease of 0.007 in $\Omega_{Ar}$ and of 0.02 in pH compared to offshore and 2000 m depth has been observed at bottom water depths of ~670 m in Korsfjorden south of Bergen between 2007 and 2017 [48]. Calcifying marine organisms such as CWCs will be directly impacted by the shoaling of the ASH. Below the ASH depth, mineral dissolution occurs, dissolving the exposed dead coral framework that constitutes a major component of the reef, and waters are less favorable for skeletal growth [29,49,50]. The skeleton of *L. pertusa* is based on aragonite, the most labile form of calcium carbonate. Hence, the reefs are considered particularly susceptible to a decreasing $\Omega_{Ar}$. If current trends continue at unchanged rates, the ASH is estimated to reach the surface by the year 2070 $\pm$ 10 in southwestern Norwegian fjords [51], while deeper waters (>600 m) off Norway will become corrosive already within the next 20 years [48]. Most of the Norwegian CWCs are observed in between these depths, so they are likely to experience corrosive conditions in the next 20–50 years.

The seasonal changes in carbonate chemistry in high latitude surface waters are driven by temperature, salinity, air-sea $CO_2$ exchange, formation and decay of organic matter, and advection and vertical mixing with deeper, carbon-rich coastal water [51]. In western Norwegian fjords, the spring phytoplankton bloom yields maximum pH and reduced dissolved inorganic carbon ($C_T$) values. Seasonal $\Omega_{Ar}$ variations align with those of surface temperature, both peaking in late summer. This seasonal forcing also affects the carbonate chemistry at the depths of living CWCs (i.e., 80–250 m). Generally, the $C_T$ and carbon dioxide ($CO_2$) concentrations increase with depth. This is caused by enhanced respiration and remineralization of organic matter, which also yield to decreased pH [48,52–55]. Apart from Findlay et al. (2014) [53], who provided a high-resolution nutrient and carbonate chemistry data over a tidal cycle showing the influence of local upwelling for Scottish and Irish CWC reefs, very little data exist on diurnal variability and not least the seasonal dynamics of carbonate chemistry for deep waters. At present, carbonate chemistry and nutrient data are primarily available from global data sets [56,57]. These are useful for making large-scale habitat predictions [13,58], but they lack the fine-scale spatial and temporal coverage that is required to understand the baseline of physicochemical dynamics around CWC reefs that plays a key role in determining the resilience of CWC reefs to future climate change.

This study investigated the natural variability of biogeochemical and physical conditions at five CWC reefs located within a narrow fjord in western Norway. This was achieved by collating water column measurements of carbonate chemistry and inorganic nutrient parameters for multiple stations and timepoints over 1.5 years and combining those with high temporal resolution data from two benthic observatories deployed at one of the reefs. Comparisons of environmental data from reefs growing on vertical walls with those growing on the fjord floor were used to elucidate the relative importance of topography-flow interaction and biogeochemical conditions for CWC occurrences.

## 2. Materials and Methods

### 2.1. Study Site

The Langenuen Fjord in southwestern Norway, south of Bergen (Figure 1), is a 35 km long north-south trending water passage connecting the Korsfjorden with the Hardangerfjord. The system opens up to the Atlantic through Korsfjorden at its northern edge, through Selbjørnsfjorden in the central parts, and through Bømlafjorden at its southern end. Langenuen is ~300 m deep. The deepest point measures 568 m and is located in the northern portion between the islands of Huftarøy and Reksteren. The average width of the fjord is 2.9 km in the northern section and 1.7 km in the southern section. The fjord acts as an ocean water inlet to the Hardangerfjord [59]. It hosts several *L. pertusa*-dominated CWC reefs on the eastern side of the fjord, at depths between 80 and 240 m [60,61]. This study concentrates on five of them in two different settings: corals covering an elevated topographic feature on the fjord bottom here called "bank reef" and corals living on steep walls or cliffs, "wall reefs". The two bank reefs are Nakken (NK, 59.830° N, 5.550° E) and the northernmost site Bekkjarvik (BV, 59.983° N, 5.276° E). The southernmost site Huglhammaren near the main sill of the Hardangerfjord (HH, 59.815° N, 5.590° E), Straumsneset (SN, 59.941° N, 5.467° E), and Hornaneset (HN, 59.890° N, 5.539° E) are wall reefs on the near-vertical fjord wall. Sites NK and SN are previously described by Fosså et al. (2015) [60].

### 2.2. Water Column Measurements

Water samples were taken at several depths using a 12-bottle Niskin rosette (Ro) system. For water column hydrography profiling, an instrument measuring in situ conductivity, temperature, and pressure (CTD) attached to the water sampling system was used. A total of 12 CTD and 52 CTD/Ro (CTD/rosette water sampling) casts were carried out across the CWC sites (HH, NK, HN, SN, BK). Measurements were taken during six cruises between February 2016 and August 2017, onboard RV *Håkon Mosby*, RV *Kristine Bonnevie*, RV *H. Brattström*, and MS *Periphylla* (Table 1). In October 2016 and January 2017, water sampling was performed over a tidal cycle at HH (26-h) and NK (16-h), respectively. On other sites and times, water sampling was performed a maximum of once per site (Table 1). Data collected in October 2016 and August 2017 have been reported earlier in Skjelvan et al. (2016) [62] and Jones et al. (2018) [48], respectively.

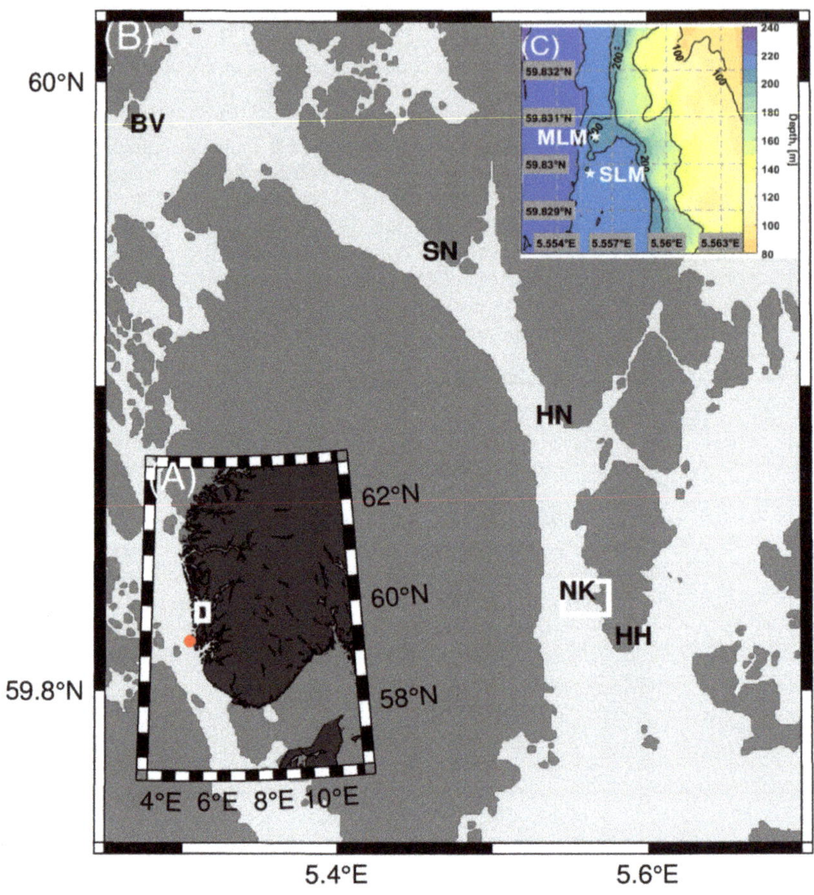



**Figure 1.** The map of (**A**) Langenuen on coastal Norway (white polygon) and Utsira coastal station (red dot), (**B**) CWC sites in Langenuen Fjord: bank reef Bekkjarvik (BV), live corals at depths 200–210, wall reef Straumsneset (SN), live corals at depths 80–220 m, wall reef Hornaneset (HN), live corals at depths 220–240 m, bank reef Nakken (NK), live corals at depths 190–220 m (white polygon) and wall reef Huglhammaren (HH), live corals at depths 80–220 m and (**C**) the Nakken reef area with lander deployment locations ~125 m apart from each other shown with stars. MLM = lander deployed on the CWC bank, SLM = lander deployed south of the CWC bank.

**Table 1.** Summary of cruises, visited locations, and measurements between February 2016 and August 2017. Abbreviations; HM = Håkon Mosby, KB = Kristine Bonnevie, NK = Nakken, HH = Huglhammaren, HN = Hornaneset, SN = Straumsneset, BK = Bekkjarvik. C = carbonate chemistry ($A_T$ and $C_T$) samples, N = nutrient samples, I = isotope samples.

| Cruise | Num of Stations | Date | Sites and Taken Water Samples | Remark |
|---|---|---|---|---|
| HM2016603 | 9 | 16–17 February2016 | $NK^C$ | CTD with oxygen |
| HM2016625 | 9 | 9–13 May 2016 | $NK^{C,N,I}$, $HH^{C,N,I}$, $HN^{C,N,I}$, $SN^{C,N,I}$ | Lander deployments |
| 2016982 | 18 | 20–21 October 2016 | $NK^C$, $HH^C$, $HN^C$, $SN^C$, $BK^C$ | CTD with fluorescence and turbidity |
| KB2017601 | 14 | 2–4 January 2017 | $NK^{C,N,I}$, $HH^{C,N,I}$, $HN^{C,N}$, $SN^{C,N}$, $BK^{C,N}$ | CTD with oxygen and fluorescence |

**Table 1.** *Cont.*

| Cruise | Num of Stations | Date | Sites and Taken Water Samples | Remark |
|---|---|---|---|---|
| KB2017609 | 11 | 17–24 April 2017 | NK$^{C,N,I}$, HH$^{C,N}$, HN$^{C,N}$, SN$^{C,N}$ | Lander recoveries, CTD with oxygen and fluorescence |
| 2017955 | 3 | 15–17 August 2017 | HH$^{C}$, HN$^{C}$, SN$^{C}$ | |

Processing of the raw CTD data was performed using the software package SBE Data Processing[63]. The wall reefs (HH, HN, and SN) were investigated above, below, and at the respective depths of living *L. pertusa* corals. At the bank reefs (NK and BV), samples were taken at the surface, at intermediate depths (at 80 or 100 m and at 150 m), and at ~5 m above the reef.

### 2.3. Benthic Landers

Two seafloor monitoring lander systems were deployed at NK bank reef to study the topographical effect of the bank on the flow between May 2016 and April 2017. One of the landers (SLM) was deployed south of the CWC bank at 63.608° N 9.383° E at 211 m water depth, while the other one (MLM) was deployed on the CWC bank at 63.609°N 9.382° E at 196 m water depth, ~125 m apart from each other (Figure 1C). Both landers were equipped with an Acoustic Doppler Current Profiler (ADCP, Teledyne RD Instruments) to measure water column flow speed and direction at 6-min intervals in between 3 to 30 m above the sea floor (with SLM) and from 5 m above the sea floor upward to the surface (with MLM). The landers were also equipped with a CTD system (SeaBird Electronics) combined with sensors to monitor dissolved oxygen, turbidity, fluorescence, and pH. Sampling intervals were set to 10 or 30 min. Processing of the raw CTD data was performed using the software package SBE Data Processing [63] and the raw ADCP data with the RDADCP Matlab package [64]. The accuracy and sensitivity of these instruments are shown in Table 2.

**Table 2.** Instrumentation details and metadata for landers deployed at Nakken bank reef area. Abbreviations: WD: water depth.

| | SLM | MLM |
|---|---|---|
| **Deployment information** | | |
| Deployment location | Off CWC bank 63.608° N 9.383° E | On CWC bank 63.609° N 9.382° E |
| Deployment depth (m) | 211 | 196 |
| **Instrumentation (sampling frequency)** | | |
| ADCP (6 min) | RDI Workhorse sentinel 600 kHz | RDI Workhorse sentinel 300 kHz |
| velocity accuracy and resolution | 0.30% | 0.50% |
| CTD (30 min) | SBE16+ | SBE16+ |
| accuracy/resolution | $\pm0.005/0.0001$ °C, $\pm0.5/0.05$ mS m$^{-1}$, $\pm 0.1/0.002\%$ WD$^{-1}$ | |
| Fluorometer and turbidity (30 min) | FLNTU(RT)D | FLNTU(RT)D |
| range/sensitivity | 0.1–50 µg L$^{-1}$/0.01 µg L$^{-1}$, 0.01–25 NTU/0.01 NTU | |
| Oxygen (10 min) | Optode | Optode |
| accuracy/resolution | <2 µM/<0.1 µM | <2 µM/<0.1 µM |
| Additional sensors | SBE 27 pH | SBE 27 pH, OIS-camera |
| pH accuracy/range | $\pm0.1$ pH/0–14 pH | $\pm0.1$ pH/0–14 pH |

Unfortunately, the lander on the NK CWC bank (MLM) tilted due to high flow velocities in mid-August. This resulted in a separation of the bottom weight and the buoyancy compartment that includes all sensors. Thus, the sensor head was released and floated upwards to ~100 m water depth after 3.5 months deployment time in mid-August, where it continued to measure all parameters. The SLM remained at deployment depth until its recovery in April 2017. ADCP, CTD, oxygen, and SLM turbidity sensors provided reliable time-series data. pH and fluorescence sensors malfunctioned. The conductivity sensor had an instrumental drift that amounted to $-0.5$ g kg$^{-1}$ over the deployment period. This drift has been taken into account when the hydrographic variables ($\Theta$, $S_A$, $\sigma_\Theta$) were calculated.

### 2.4. Analysis of Carbonate Chemistry and Inorganic Nutrients

Seawater was collected from the CTD/Ro system for the determination of dissolved inorganic carbon ($C_T$) and total alkalinity ($A_T$) into borosilicate glass bottles (250 mL) with ground glass stoppers. The samples ($n$ = 190) were preserved with 0.05 mL of saturated solution of mercuric chloride (HgCl$_2$) and stored refrigerated and dark until post-cruise analysis. Samples ($n$ = 93) for the determination of nutrients and ammonia were collected directly after the samples had been taken for the $A_T$ and $C_T$. For nutrient analysis, 20 mL of water was collected in polyethylene scintillation vials, preserved with 0.2 mL chloroform, and kept refrigerated (4 °C) until analysis occurred within a few weeks after sample collection. For ammonium analysis, waters were collected in the same type of vials and kept frozen ($-20$ °C) without preservatives until analysis.

$A_T$ and $C_T$ were measured at the IMR's CO$_2$ Laboratory following standard procedures [48,62,65]. The remaining carbonate system parameters (pH in total scale pH$_T$, partial CO$_2$ pressure pCO$_2$, aragonite saturation $\Omega_{Ar}$ and calcite saturation $\Omega_{Ca}$) were calculated from $A_T$ and $C_T$ with the program CO2SYS [66] using the thermodynamic constants from Mehrbach et al. (1973) [67] and KSO4 from Dickson (1990) [68], and refitted by Dickson and Millero (1987) [69] on the total scale. The concentrations of silicate and phosphate were set to zero during the calculations, and the errors introduced by this simplification were negligible compared to uncertainties from other sources.

Inorganic nutrient analyses were carried out at IMR's Chemical Laboratory using an Alpkem Flow Solution IV autoanalyzer (RFA methodology) for the colorimetric determination of inorganic nutrients: nitrite (NO$_2^-$), nitrate (NO$_3^-$) [70,71], phosphate (PO$_4^{3-}$) [72] and silicate (SiO$_4^4$) [72]. Nitrate concentrations were calculated by subtracting the nitrite from combined nitrate + nitrite concentration. Ammonium (NH$_4^+$) was measured fluorometrically using an excitation/emission filter combination with an Alpkem Flow Solution IV autoanalyzer [73,74].

### 2.5. Analysis of Stable Isotopes

For the carbon ($\delta^{13}$C) and oxygen ($\delta^{18}$O) stable isotope analysis, water samples ($n$ = 37) were collected in 100 mL glass vials and treated with HgCl$_2$ to prohibit biological activity. The samples were stored in a cool, dark place until measurements were carried out at the isotope laboratory at the Friedrich-Alexander University Erlangen-Nürnberg. Samples were analyzed through a standard procedure with an isotope-ratio-mass spectrometer (Gasbench 2; Thermo Fisher Scientific, Bremen, Germany), corrected to instrumental drift, and normalized to the VPDB (Vienna Pee Dee Belemnite) or to the VSMOW/SLAP (Vienna Standard Mean Ocean Water/Standard Light Antarctic Precipitation) scale, respectively [75]. An acid treatment on a Gasbench was used for carbonate stable isotope analyses. The precision of the control sample was better than 0.1‰ (1 sigma) for $\delta^{13}$C and better than 0.05‰ ($\pm1$ sigma) for $\delta^{18}$O. Oxygen samples were not corrected for the isotope salt effect as this effect has been reported to be neglectable for seawater consisting mainly of NaCl [76,77].

### 2.6. Data Analysis of Hydrography and Flow

Data visualization and CTD and ADCP data conversions were performed with Matlab R2018a. For subsequent data analysis, the raw CTD and ADCP data were converted. CTD variables were converted according to TEOS-10 standard with GSW Oceanographic Toolbox to absolute salinity ($S_A$), conservative temperature ($\Theta$), potential density ($\rho_\Theta$) and potential density anomaly ($\sigma_\Theta$), i.e., sigma-theta values [78,79]. The water column stratification was estimated from CTD casts with Brunt–Väisälä, or buoyancy, frequency

$N^2 = g^2 \rho_\Theta{}^{-1} \Delta\rho_\Theta \Delta z^{-1}$ [79], where g is the gravitational acceleration and z is the depth. The flow measurements were corrected for the local magnetic declination based on International Geomagnetic Reference Field, IGRF-11 model data [80]. To estimate the mean flow direction, the horizontal velocity components (eastward, $u_e$, and northward, $u_n$) were rotated using variance ellipses from the jlab data analysis toolbox [81]. Accordingly, the mean direction velocity components ($u_m$) are in the direction of the most energetic fluctuations, while components perpendicular are interpreted as cross-flow ($u_c$). The vertical velocity component, $u_w$, was not altered. The tidal frequencies, $\omega$, and their amplitudes, a, were analyzed with the harmonic analysis toolbox T_Tide [82]. The tidal signals were analyzed by using bottom pressure and horizontal velocity fields at 3 (SLM) and 5 (MLM) meters above the sea floor. Only signals with a signal-to-noise ratio >2 were considered to be significant.

The health statuses of the five included reefs were estimated using the known spatial extent of the reefs combined with data on density and vertical height of living coral colonies [15,22]. The flow state based on topography-flow interaction at the Nakken bank reef was determined following methods described in [22].

### 3. Results

*3.1. Cold-Water Coral Occurrences*

The shallowest reefs in the study area are HH and SN, where framework-building scleractinian corals *L. pertusa* and *Madrepora oculata* are observed at depths of 100–200 m and 80–220 m, respectively. At Straumsneset, the densest coral cover is recorded between 130 and 180 m depth. The maximum continuous *L. pertusa* cover (length < 6.5 m, width < 3.2 m, colony height < 1.2 m) is larger than maximum continuous *M. oculata* cover (length < 2.5 m, width < 0.2 m, colony height < 0.5 m, with 0.2–0.3 m dead coral skeleton). The deepest *L. pertusa* colonies are found at HN, where they live at depths between 220 and 240 m. At NK and BV bank reefs, scattered *L. pertusa* patches live at 190–220 m and 200–210 m depths, respectively. At Nakken, patches have heights of 1–2 m and diameters < 4 m. The bank is ~200 m wide. All sites have rich communities of associated megafauna, including the bivalve *Acesta excavata*, sponges *Geodia* sp. and *Mycale lingua*, and octocorals *Paragorgea arborea*, *Primnoa resedaeformis*, *Paramuricea placomus*, and *Anthothela grandiflora*. Based on coral coverage, spatial extent, and the proportion of living to dead corals, Langenuen CWC reefs are all in health category II [22], with deeper bank reefs NK and BV and wall reef HN having more coral rubble and dead coral framework than shallow wall reefs HH and SN.

*3.2. Hydrography*

3.2.1. Water Column Hydrography

Waters in Langenuen consisted of seasonally forced surface waters down to ~100 m. The Norwegian coastal water (NCW) and the North Atlantic water (NAW, $S_A$ > 35 g kg$^{-1}$) were present beneath the surface waters (Figure 2). The temperature at water depths above the CWC reefs (<80 m) ranged between 5.3 and 16.3 °C with the cool temperatures (<8 °C) recorded in winter (January 2017 and February 2016) and spring (April 2017 and May 2016) and the warm (>9 °C) temperatures observed in autumn (August 2017 and October 2016). Waters with temperatures >12 °C reached the depths of the uppermost corals (>80 m) at the three southernmost sites (HH, NK, and HN) in October 2016. The thermocline depth varied from 30 m in April 2017 to 100 m in October 2016. Beneath 80 m, the annual temperature range was between 7.5 and 12.2 °C, with a mean temperature of ~8 °C at CWCs living depths. Opposite to the surface waters, the cool bottom water temperatures (<7.8 °C, at depths >200 m) were measured in May 2016, August 2017, and October 2016, while the warm bottom temperatures (>8 °C, at depths >200 m) were recorded in January 2017, February 2016, and April 2017 (Figure 2, Table 3).

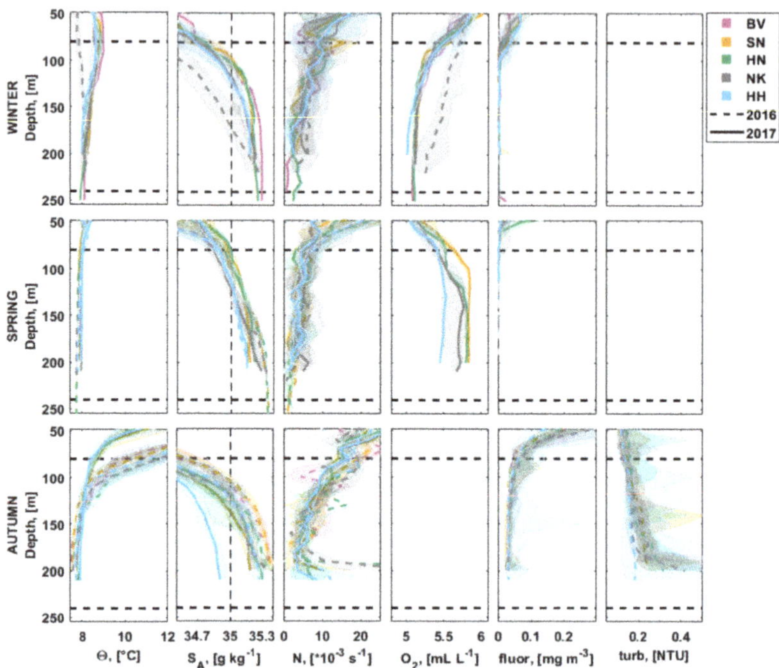



**Figure 2.** Depth profiles (y-axes) of hydrographic water properties at depths 50–250 m of the Langenuen Fjord: (first row) winter (February 2016 and January 2017), (second row) spring (April 2017 and May 2016), and (third row) autumn (August 2017 and October 2016), for (first column) temperature, (second column) salinity with 35 g kg$^{-1}$ marked, (third column) stratification, (fourth column) oxygen, (fifth column) fluorescence, and (sixth column) turbidity. The shaded area covers the range (min, max) of the variable for the station (HH—blue, NK—gray, HN—green, SN—orange, BV—purple) over the cruises and the solid (for year 2017 cruises) or the dashed (for year 2016 cruises) line shows the median value. The dashed horizontal lines represent the definition for CWC living depths (80–240 m).

**Table 3.** The range (min–max) of environmental conditions at the surface (10–80 m), at CWC walls at 80–220 m (sites HH and SN), and at CWC banks at 190–220 m (sites NK and BV) for winter (January 2017 and February 2016), spring (April 2017 and May 2016) and late summer and autumn (August 2017 and October 2016). Measured values include temperature, salinity, sigma-theta, buoyancy, oxygen, total alkalinity, dissolved organic carbon, nitrate, phosphate, silicate, stable isotopes, monthly mean speed, monthly maximum speed, southward direction, and monthly mean flow direction. Also shown are calculated values for in situ pH$_T$ (total scale), pCO$_2$, and saturation states of aragonite and calcite. These parameters were calculated using CO2SYS [66]. For flow measurements, columns show periods of December 2016–February 2017, March 2017–April 2017 and May 2016, June 2016–August 2016/September 2016–November 2016. * Flow data only from 180 to 200 m depth.

| | Surface (10–80 m) | | | Wall | | | Bank | | |
|---|---|---|---|---|---|---|---|---|---|
| | Winter | Spring | Autumn | Winter | Spring | Autumn | Winter | Spring | Autumn |
| Temperature, $\Theta$ (°C) | 5.31–8.96 | 6.66–10.63 | 8.30–16.2 | 7.92–8.82 | 7.72–8.08 | 7.51–12.2 | 7.88–8.35 | 7.72–8.01 | 7.53–7.74 |
| Salinity, S$_A$ (g kg$^{-1}$) | 29.33–34.89 | 29.69–35.00 | 26.58–34.78 | 34.45–35.23 | 34.86–35.33 | 33.93–35.39 | 35.03–35.28 | 35.19–35.34 | 35.2–35.34 |
| Sigma-theta, $\sigma_\Theta$ (kg m$^{-3}$) | 23.81–26.88 | 22.69–27.15 | 19.33–26.47 | 26.62–27.3 | 27.03–27.44 | 25.66–27.48 | 27.17–27.35 | 27.29–27.45 | 27.35–27.5 |

**Table 3.** *Cont.*

| | Surface (10–80 m) | | | Wall | | | Bank | | |
|---|---|---|---|---|---|---|---|---|---|
| | Winter | Spring | Autumn | Winter | Spring | Autumn | Winter | Spring | Autumn |
| Buoyancy, N ($10^{-3}$ $s^{-1}$) | <54.1 | <55.5 | <79.2 | <12.76 | <9.83 | <32.18 | <9.24 | <8.17 | <35.55 |
| Oxygen, $O_2$ (mL $L^{-1}$) | 5.3–6.43 | 5.04–7.11 | | 4.96–5.77 | 5.41–5.81 | | 5.07–5.56 | 5.51–5.75 | |
| $A_T$, ($\mu$mol $kg^{-1}$) | 2127–2308 | 2201–2312 | 1948–2286 | 2318–2330 | 2310–2324 | 2300–2328 | 2317–2343 | 2319–2328 | 2329–2337 |
| $C_T$, ($\mu$mol $kg^{-1}$) | 1992–2132 | 2016–2149 | 1795–2286 | 2136–2167 | 2140–2183 | 2159–2187 | 2135–2192 | 2160–2175 | 2175–2176 |
| $pH_T$, (–) | 7.979–8.084 | 7.989–8.15 | 8.003–8.078 | 7.999–8.063 | 7.951–8.053 | 7.944–7.985 | 7.96–8.081 | 7.968–8.014 | 8.001–8.018 |
| $pCO_2$, ($\mu$atm) | 359.7–443.6 | 293.2–452.4 | 346.3–440.1 | 374.1–438.7 | 386.5–497.7 | 456.2–500.8 | 356.4–485.6 | 423.6–476.6 | 421.5–505.8 |
| $\Omega_{Ar}$, (–) | 1.41–2.06 | 1.56–2.03 | 1.74–2.28 | 1.71–1.98 | 1.53–1.91 | 1.50–1.68 | 1.56–2.01 | 1.59–1.74 | 1.68–1.76 |
| $\Omega_{Ca}$, (–) | 2.24–3.24 | 2.48–3.23 | 2.75–3.56 | 2.70–3.12 | 2.41–3.00 | 2.36–2.64 | 2.46–3.16 | 2.50–2.74 | 2.65–2.76 |
| Nitrate, $NO_3^-$ ($\mu$mol $L^{-1}$) | 5.75–6.82 | 1.54–10.08 | | 8.64–11.61 | 9.92–10.83 | | 10.49–11.37 | 10.4–11.03 | |
| Phosphate, $PO_4^{3-}$ ($\mu$mol $L^{-1}$) | 0.33–0.52 | 0.22–0.80 | | 0.67–0.96 | 0.68–0.89 | | 0.76–0.89 | 0.69–0.77 | |
| Silicate, $SiO_4^{4-}$ ($\mu$mol $L^{-1}$) | 2.02–3.42 | 1.60–5.52 | | 4.37–6.92 | 4.64–6.00 | | 5.15–6.00 | 4.86–5.46 | |
| $\delta^{13}C$, (‰) | | | | 0.27–0.52 | 0.32–0.48 | | 0.32–0.43 | 0.43–0.52 | |
| $\delta^{18}O$, (‰) | | | | 0.21–0.42 | 0.28–0.45 | | 0.36–0.41 | 0.36–0.44 | |
| Flow mean (cm $s^{-1}$) | | 77.8–205.0 | 22.5–63.0/– | 16.0–21.9 * | 10.7–79.8 | 11.6–27.2/ 16.0–24.9 * | 14.6–21.9 | 18.2–23.1 | 12.7–17.8/ 15.5–24.1 |
| Flow max (cm $s^{-1}$) | | 247–529 | 61.9–316/– | 58.1–66.0* | 36.1–309 | 28.7–215/ 57.8–67.1 * | 47.8–63.1 | 35.4–57.6 | 38.2–54.6/ 47.9–65.0 |
| Flow southward (%) | | 78–100 | 0–100 | 30–71 * | 46–89 | 27–84/ 52–65* | 30–77 | 68–97 | 33–74/ 54–67 |
| Flow dir (°) | | 6–176 | 0–174 | 159–165 * | 158–180 | 158–180/ 158–165 * | 153–162 | 153–161 | 153–160/ 152–162 |

Above 80 m, salinity varied from <30 g $kg^{-1}$ at the surface (<20 m) to close to 35 g $kg^{-1}$ at depths ~80 m. The halocline depth varied from 20 to 60 m (not shown) except in February 2016, when the water column was well mixed (Figure 2). The 35 g $kg^{-1}$ isohaline depth (indicating North Atlantic water) varied from 220 m at the HH reef located south of the Langenuen in August 2017 to 75 m at the SN in April 2017. At the depths where CWCs occur, the salinity varied from <34.5 g $kg^{-1}$ at wall settings at depths 80–110 in February 2016 and in autumn to >35.3 g $kg^{-1}$ at depths >180 m in May 2016 and at depths >140 m in August 2017 (Figure 2, Table 3). In general, the northern sites were saltier than the southern sites indicating the NAW route from Korsfjord and fresh water output from Hardangerfjord. The surface layer (<80 m) was well stratified (N > 7 × $10^3$ $s^{-1}$) throughout the year. The strongest stratification occurred in May and August 2017 and in October 2016. Beneath 80 m, stratification varied from 0.2 to 20 × $10^3$ $s^{-1}$ (Figure 2) without a clear seasonal or latitudinal signal. The waters at the CWC living depths were less dense at shallower wall reefs (HH, SN) than deeper bank and wall settings (NK, BV, HN) (Table 3). Sigma-theta values >27.35 kg $m^{-3}$ were measured at all sites in May 2016, August 2017, and October 2016 (Table 3). Across the water column and seasons, dissolved oxygen ranged from 4.96 to 7.11 mL $L^{-1}$. In winter (January 2017 and February 2016), oxygen decreased with depth, and in April 2017, oxygen increased from 50 to 100 m and was well mixed beneath 100 m (Figure 2). At CWC living depths, oxygen ranged from 4.96 to 5.81 mL $L^{-1}$. Fluorescence reached values >1 mg $m^{-3}$ in the uppermost 20 m in April 2017 and in October 2016 (not shown). Beneath 50 m, the fluorescence values were <0.05 mg $m^{-3}$ (Figure 2). In October 2016, turbidity reached values >1 NTU at the surface and at the bottom. The mean turbidity through the water column was ~0.2 NTU (Figure 2).

### 3.2.2. Bottom Water Hydrography

The annual bottom temperature at Nakken bank reef ranged from 7.4 to 8.3 °C, with a mean temperature of 7.8 °C. This included two distinct phases: cool temperatures (<7.8 °C) observed from May to mid-November and warm temperatures (>7.8 °C) measured from late November to April (Figure 3), indicating heat dissipation from the surface during autumn and water column mixing in spring. March was the warmest month. September and October were the coolest months (Figures 3 and 4). Bottom water salinity varied between 35.00 and 35.34 g kg$^{-1}$ with the lowest salinities (<35.2 g kg$^{-1}$) observed in autumn and winter, and the highest salinities observed in late spring, indicating the water source variability between North Atlantic water (S$_A$ > 35 g kg$^{-1}$)-dominated and Norwegian coastal water-dominated waters. The salinities measured by MLM were ~0.05 g kg$^{-1}$ higher than those measured by SLM (Figure 3) during the time both landers were at the sea bed, whereas measured temperatures were similar. Relatively high sigma-theta (density) values of >27.35 kg m$^{-3}$ were measured from May until October (Figure 3). Bottom water oxygen values varied from 6.58 to 6.68 mL L$^{-1}$, with the highest values in October and lowest values in January (Figure 4) following temperature variations. Turbidity values reached up to 1.5 NTU between July and August. Turbidity values dropped in October to around 0.15 NTU. They remained low (<0.15 NTU) with few turbid events (>0.2 NTU) until the end of the deployment (Figure 4).

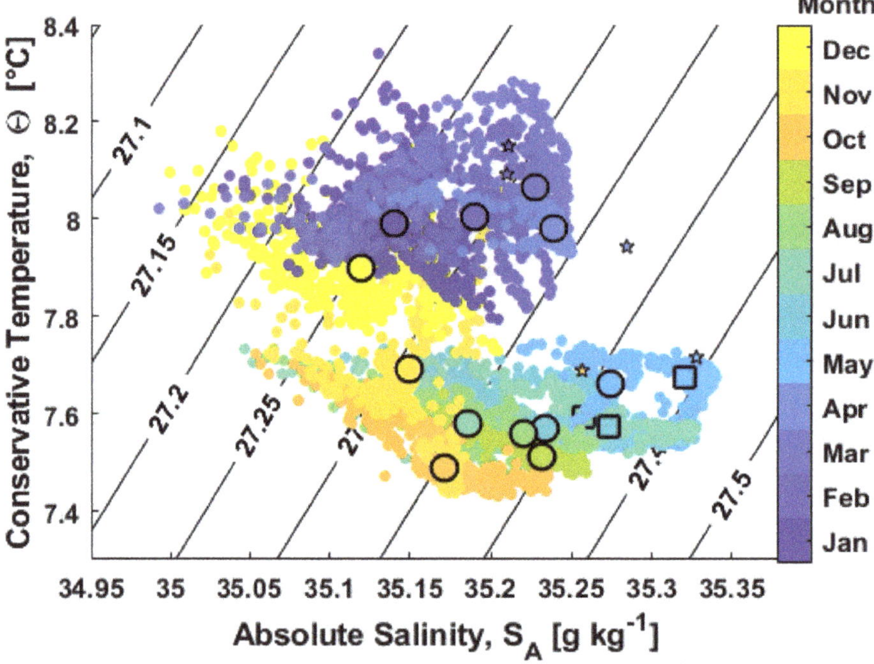


**Figure 3.** Bottom water Θ, S$_A$-values at Nakken bank reef. Months are color coded. The larger circles (SLM) and squares (MLM) mark the monthly median values, and stars show the cruise CTD cast median values for 205–215 m depth.

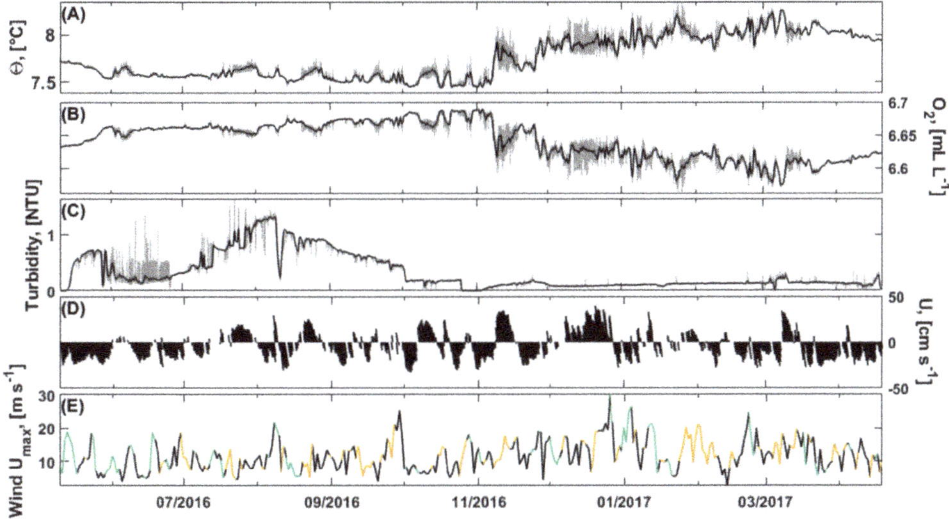

**Figure 4.** Nakken bottom water (~210 m) characteristics for (**A**) temperature, (**B**) oxygen, (**C**) turbidity, (**D**) daily mean horizontal flow velocity (up northward and down southward flow) at Nakken, and (**E**) daily maximum wind speed from Utsira coastal station based on data from MET Norway. The prevailing wind directions are marked with green (northerly from NW (315°) to NE (45°)) and yellow (southerly from SE (135°) to SW (225°)). The thicker line (in three first rows) shows the 1-d running mean.

### 3.3. Carbonate Chemistry in the Water Column

Concentrations of dissolved inorganic carbon ($C_T$) and total alkalinity ($A_T$) increased with depth at all sites and times. Below 50 m (Figure 5), the variability between the sites was largest in winter (February 2016 and January 2017) and smallest in autumn (August 2016 and October 2017). Variability was higher at 200 m than at intermediate depths of 100–150 m. Relatively low surface values ($A_T$ < 2100 µmol kg$^{-1}$ and $C_T$ < 2000 µmol kg$^{-1}$) occurred in autumn (August 2017 and October 2016) (Table 3). Both the lowest ($C_T$ = 2135 µmol kg$^{-1}$) and the highest ($C_T$ = 2192 µmol kg$^{-1}$) $C_T$ values at CWC living depths were measured at the NK bank reef at 200 m depth during the 16 h sampling period in January 2017. The highest $A_T$ value (2343 µmol kg$^{-1}$) was also measured during the same period and place. The lowest $A_T$ of 2300 µmol kg$^{-1}$ was measured at HH wall reef at 200 m depth in August 2017. In general, the northernmost sites (BV and SN) had higher $A_T$ in autumn and higher $C_T$ in winter and autumn at 200 m depth than sites farther north, but short-term variability recorded over tidal cycle at a single location (HH or NK) was larger than this between sites difference.

$pH_T$ ranged from 7.94 to 8.08 throughout the water column over the study period (Figure 5, Table 3). Low values (<7.98) were measured at depths >140 m at BV bank reef and at depths >100 m at NK bank reef in winter (January 2017 and February 2016), at ~80 m depth at wall reefs HH and SN in May 2016, at ~200 m depth at HH and NK in April 2017 and at depths >100 m at all sites except BV in autumn with a minimum at 150 m depth. High values of >8.03 were measured at the surface (<40 m) at all sites and times (Table 3). Below 80 m, high values (>8.03) were measured at HH at ~100 m and at NK at ~ 200 m depth in winter (January 2017 and February 2016) and in April 2017. The $pCO_2$ ranged from 293 to 452 µatm above 80 m. Beneath 80 m, it ranged from 356 µatm at ~200 m at NK in January 2017 to 506 µatm at ~200 m at HH in October 2016 (Figure 5, Table 3). Above 80 m, $\Omega_{Ar}$ ranged from 1.41 at NK in January 2017 to 2.28 at NK in August 2017, and below 80 m, it ranged from 1.50 at ~200 m at HH in October 2016 to 2.00 at ~200 m at NK in January 2017. $\Omega_{Ar}$ and $\Omega_{Ca}$ had maxima at the surface in autumn

(August 2017 and October 2016) ($\Omega_{Ar} > 2.1$ and $\Omega_{Ca} > 3.3$). The low ($\Omega_{Ar} < 1.59$) aragonite saturation levels were measured at 150–200 m depths at wall reefs (HH, SN, HN) in autumn (August 2017 and October 2016), at ~200 m at HH in April 2017, and at depths >140 m at BV in January 2017 (Figure 5, Table 3).

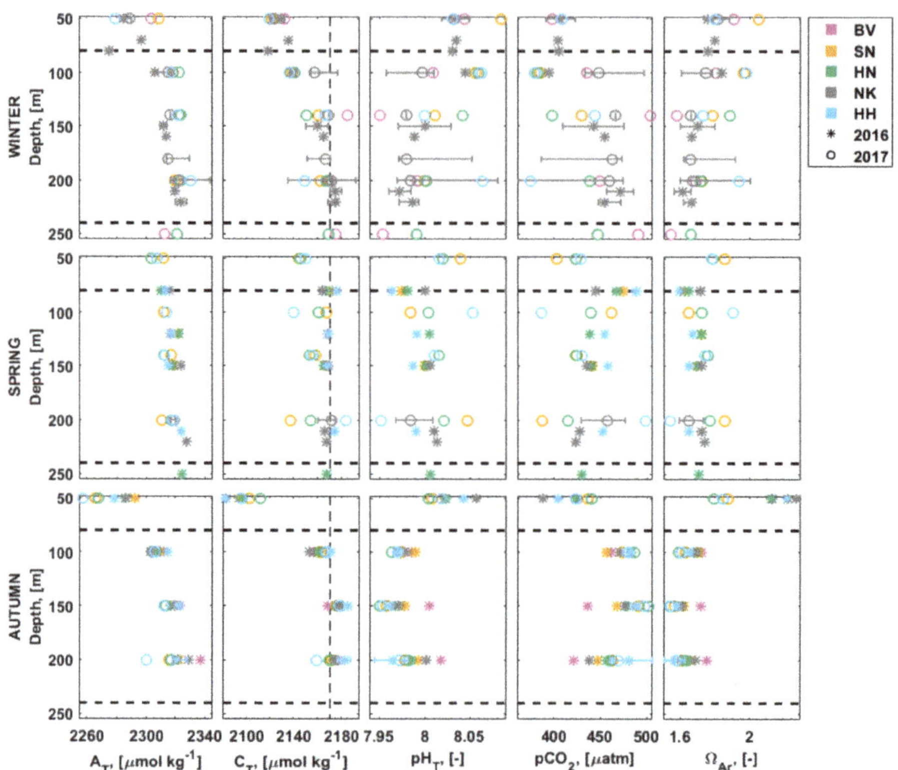


**Figure 5.** Depth profiles (y-axes) of carbonate chemistry at depths 50–250 m of the Langenuen Fjord: (first row) winter (February 2016 and January 2017), (second row) spring (April 2017 and May 2016), and (third row) autumn (August 2017 and October 2016), for (first column) total alkalinity, $A_T$, (second column) total inorganic carbon, $C_T$, with 2070 $\mu$mol kg$^{-1}$ marked, (third column) $pH_T$, (fourth column) partial $CO_2$ pressure, $pCO_2$, and (fifth column) aragonite saturation, $\Omega_{Ar}$. The errorbars show the range of the variable (min, max) for cruises with multiple CTD cast samples at one station, and the marker shows the median for the station (HH—blue, NK—gray, HN—green, SN—orange, BV—purple) over the cruises. The 2016 measured values are shown with asterisks (*) and 2017 with circles (o). The dashed line represents the definition for depths of the living CWCs (80–240 m).

Carbonate chemistry sampling was performed over several hours at two stations at two different times: At wall reef HH (corals at depths 80–220) in October 2016 and at bank reef NK (corals at depths 190–220) in January 2017. During a 26 h sampling period at 200 m depth at the HH wall reef in October 2016, $\Theta$ and $S_A$ were relatively stable with ranges of $\Delta\Theta = 0.14\,°C$ and $\Delta S_A = 0.11$ g kg$^{-1}$ (Figure 6). Over the sampling period, the carbonate system parameters changed by: $\Delta A_T = 10$ $\mu$mol kg$^{-1}$, $\Delta C_T = 10$ $\mu$mol kg$^{-1}$, $\Delta pH_T = 0.032$, $\Delta pCO_2 = 45$ $\mu$atm, $\Delta\Omega_{Ar} = 0.12$, and $\Delta\Omega_{Ca} = 0.19$. This is comparable to the seasonal variability between winter and summer from single measurements at HH at this depth $\Delta A_T = 11$ $\mu$mol kg$^{-1}$, $\Delta C_T = 34$ $\mu$mol kg$^{-1}$, $\Delta pH_T = 0.112$, $\Delta pCO_2 = 123.6$ $\mu$atm and $\Delta\Omega_{Ar} = 0.4$ (Figure 6). The flow direction at nearby Nakken reef was southward with speeds between 4 and 43 cm s$^{-1}$.

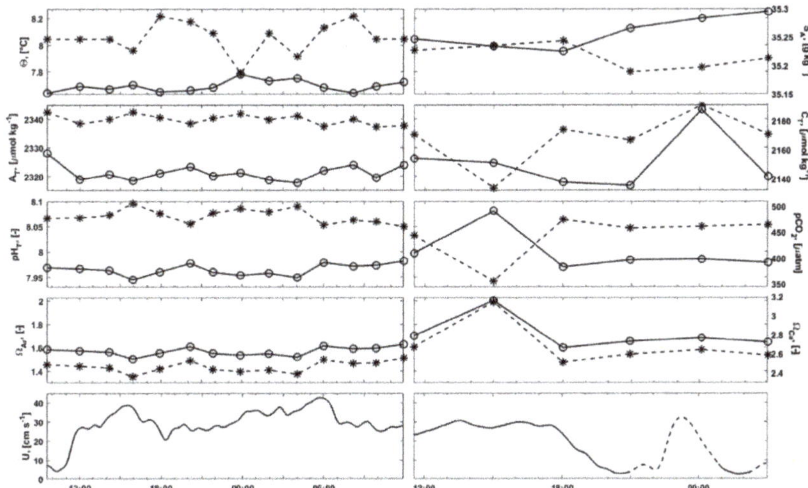

**Figure 6.** Short time series (x-axes) of carbonate system fluctuations on 20–21 October 2016 at Huglhammaren wall reef (left panels) and on 3–4 January 2017 at Nakken bank reef (right panels) at 200 m depth for (first row) temperature, $\Theta$ (o) and salinity, $S_A$ (*), (second row) total alkalinity, $A_T$ (o) and total inorganic carbon, $C_T$ (*), (third row) $pH_T$ (o) and partial $CO_2$ pressure, $pCO_2$ (*), (fourth row) aragonite saturation, $\Omega_{Ar}$ (o) and calcite saturation, $\Omega_{Ca}$ (*), and (fifth row) 1 h running mean of the horizontal flow at 200 m depth from SLM lander at NK bank reef, U with southward (solid line), and northward (dashed line) flow.

Short-term temporal variability of carbonate parameters at Nakken bank reef in January 2017 was larger than the short-term variability at Huglhammaren in October 2016. During this sampling period of 16 h, $\Theta$ and $S_A$ at 200 m varied from warm ($\Theta$ >8 °C) and less saline ($S_A$ < 35.23 g kg$^{-1}$) conditions to cool and saltier conditions and back. The overall ranges were $\Delta\Theta$ = 0.32 °C and $\Delta S_A$ = 0.05 g kg$^{-1}$ (Figure 6). The carbonate chemistry followed the changes in flow direction with high $C_T$ and low $A_T$ values observed with northward flow (=warm and fresh phase). The highest $A_T$ value (2343 µmol kg$^{-1}$) was also measured during northward flow conditions. Over the sampling period, the carbonate system parameters showed large ranges of: $\Delta A_T$ = 26 µmol kg$^{-1}$, $\Delta C_T$ = 57 µmol kg$^{-1}$, $\Delta pH_T$ = 0.11, $\Delta pCO_2$ = 118 µatm, $\Delta\Omega_{Ar}$ = 0.41, and $\Delta\Omega_{Ca}$ = 0.64. This is larger than the seasonal variability between spring and summer from single measurements at NK at this depth $\Delta A_T$ = 10 µmol kg$^{-1}$, $\Delta C_T$ = 41 µmol kg$^{-1}$, $\Delta pH_T$ = 0.04, $\Delta pCO_2$ = 47.6 µatm, and $\Delta\Omega_{Ar}$ = 0.14 (Figure 6).

### 3.4. Inorganic Nutrients in the Water Column

Within the uppermost 80 m, nitrite ($NO_2^-$) and ammonium ($NH_4^+$) concentrations ranged from 0.064 to 0.175 µmol L$^{-1}$ and from 0.075 to 1.04 µmol L$^{-1}$, respectively, with the highest concentrations measured in May 2016 (not shown). Below 80 m, their concentrations were too low (<0.06 µmol L$^{-1}$) for reliable measurements.

Nitrate ($NO_3^-$), phosphate ($PO_4^{3-}$), and silicate ($SiO_4^{4-}$) concentrations all increased with depth in January 2017. In spring (May 2016 and April 2017), the concentrations were mixed beneath 80 m (Table 3, Figure 7). In January 2017 and April 2017, nutrient concentrations were higher at the southernmost (HH and NK) sites at depths >100 m than at sites farther north. In January 2017, the surface (<40 m) nutrient concentrations (<6.65, <0.51, and <3.26 µmol L$^{-1}$) were higher than during spring (May 2016 and April 2017) (<5.13, <0.4, and <2.68 µmol L$^{-1}$) (Table 3). Beneath 80 m, both the highest and the lowest nutrient concentrations were measured in January 2017. The highest concentrations were measured at ~200 m depth at HH and the lowest at ~100 m depth at BV, HH, and NK. At CWC living depths, nutrient concentrations at the shallow wall (HH and SN) and deeper bank (NK and BV) reefs were on average for $NO_3^-$ ~10.34 and ~10.80 µmol $L^{-1}$, for $PO_4^{3-}$ ~0.78 and ~0.77 µmol $L^{-1}$, and for $SiO_4^{4-}$ ~5.48 and ~5.41 µmol $L^{-1}$, respectively (Table 3).

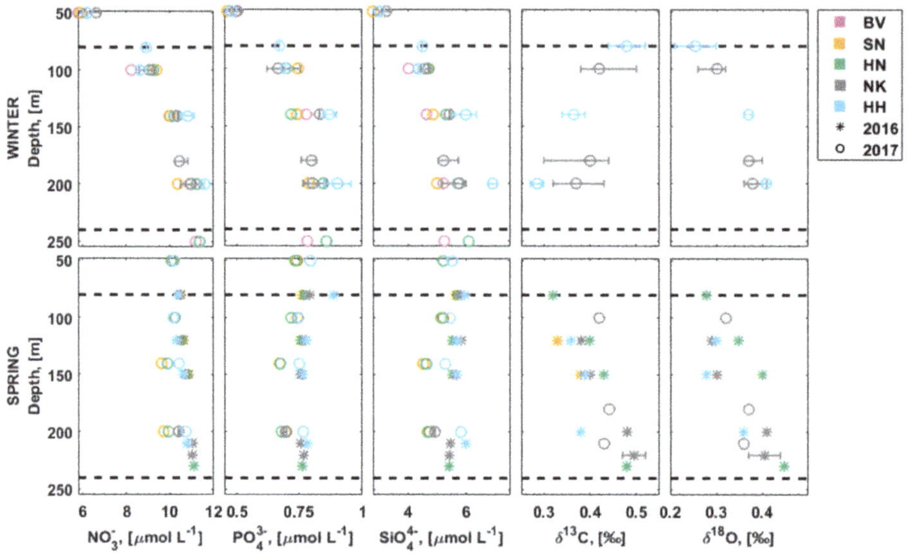

**Figure 7.** Depth profiles (y-axes) of biogeochemical water properties at depths 50–250 m of the Langenuen Fjord: (first row) January 2017, (second row) April 2017 and May 2016, for (first column) nitrate, $NO_3^-$, (second column) phosphate, $PO_4^{3-}$, (third column) silicate, $SiO_4^{4-}$, (fourth column) carbon isotope, $\delta^{13}C$, (fifth column) oxygen isotope, $\delta^{18}O$. The errorbars show the range of the variable (min, max) for cruises with multiple CTD cast samples from one site, and the marker shows the median for the site (HH—blue, NK—gray, HN—green, SN—orange, BV—purple) over the cruises. The 2016 measured values are shown with asterisks (*) and 2017 with circles (o). The dashed horizontal lines represent the definition for depths of the living CWCs (80–240 m).

At Nakken, CTD data and nutrient samples were collected from 100 to 200 m depth every three hours for 16 h between January 3–4, 2017 (Figure 8). The hydrographic range at 200 m is presented in Section 3.3. At 100 m depth, the variation patterns in $\Theta$ and $S_A$ differed from 200 m. They varied from cool ($\Theta < 8.7\ °C$) and saltier ($S_A > 34.9$ g $kg^{-1}$) conditions to warm and less saline and back. The overall ranges were $\Delta\Theta = 0.14\ °C$ and $\Delta S_A = 0.10$ g $kg^{-1}$ (Figure 8). The range of concentrations at 200 m and 100 m depth over the sampling period reached $\Delta NO_3^-$ 0.88 and 0.73 µmol $L^{-1}$, for $\Delta PO_4^{3-}$ 0.09 and 0.13 µmol $L^{-1}$, and for $\Delta SiO_4^{4-}$ 0.84 and 0.32 µmol $L^{-1}$, respectively. As for the carbonate system parameters, the change from low nutrient concentrations and low-temperature conditions to higher nutrient concentrations and warmer temperatures at 200 m depth coincided with flow velocity shift from strong southward velocity to low northward velocity (Figures 6 and 8).

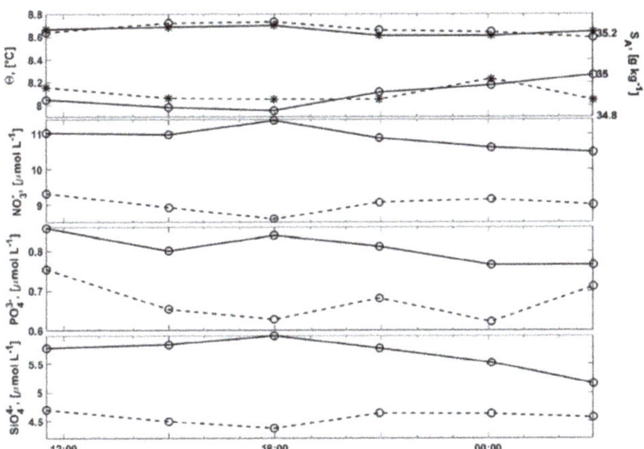

**Figure 8.** Short time series (x-axes) of inorganic nutrient fluctuations on 3–4 January 2017 at Nakken bank reef at 200 m (solid line) and 100 m (dashed line) depth for (first row) temperature, $\Theta$ (o) and salinity, $S_A$ (*), (second row) nitrate, $NO_3^-$, (third row) phosphate, $PO_4^{3-}$, and (fourth row) silicate, $SiO_4^{4-}$.

### 3.5. Stable Isotopes

The distribution of $\delta^{13}C$ within the water column varied between the different sampling times (Figure 7). In January 2017, $\delta^{13}C$ decreased with depth. In April 2017, the water column was well mixed with respect to $\delta^{13}C$, and in May 2016, $\delta^{13}C$ increased with depth. The $\delta^{13}C$ value was generally higher at bank reefs (0.41‰ ± 0.06‰, depths 190–220 m) than at shallower wall reefs (0.37‰ ± 0.07‰, depths 80–220 m) (Table 3). $\delta^{18}O$ increased with depth and had similar ranges for all sampling times. Therefore, $\delta^{18}O$ was higher at bank reefs (0.39‰ ± 0.03‰) than at shallower wall reefs (0.34‰ ± 0.06‰) (Table 3). In May 2016, when samples were taken from most of the sites, $\delta^{18}O$ had the largest range at ~150 m depth of $\Delta\delta^{18}O = 0.12$ ‰ (Figure 7). With a relatively low sample size ($n = 37$), these results are preliminary and only provide us the first look at the distribution of stable isotopes in Langenuen. Thus, these results have to be confirmed by future research.

### 3.6. Current Measurements

#### 3.6.1. Flow Regime

The flow below 180 m was strong (mean 18 cm s$^{-1}$, max 60 cm s$^{-1}$) with a semidiurnal tidal component (Figures 4 and 9, Table 3). The monthly mean flow speed ($U_{mean}$) varied between 12 and 25 cm s$^{-1}$. Low mean flow speeds ($U_{mean} < 15$ cm s$^{-1}$) were measured in June and July, and high mean flow speeds ($U_{mean} > 20$ cm s$^{-1}$) were measured in March and in late autumn in between October and December. During the remaining months, $U_{mean}$ varied between 16 and 18 cm s$^{-1}$. At depths above 180 m (recorded with MLM), flow speeds varied more. From May to August, $U_{mean}$ was < 15 cm s$^{-1}$ in between 115 and 180 m depth. $U_{mean}$ reached maxima at 90 m and at <50 mm depth with 27 and >45 cm s$^{-1}$, respectively (Figure 9A).

The maximum monthly flow speeds ($U_{max}$) below 180 m depth varied between 30 and 70 cm s$^{-1}$. $U_{max}$ was <45 cm s$^{-1}$ in May and June and ~60 cm s$^{-1}$ between October and December. During the other months, $U_{max}$ varied between 50 and 55 cm s$^{-1}$. Between 125 and 180 m water depth, $U_{max}$ was <50 cm s$^{-1}$. At depths of <117 m, $U_{max}$ was >100 cm s$^{-1}$ except in August, with peak speeds (>200 cm s$^{-1}$) measured around 60 and 100 m depths (Figure 9B).

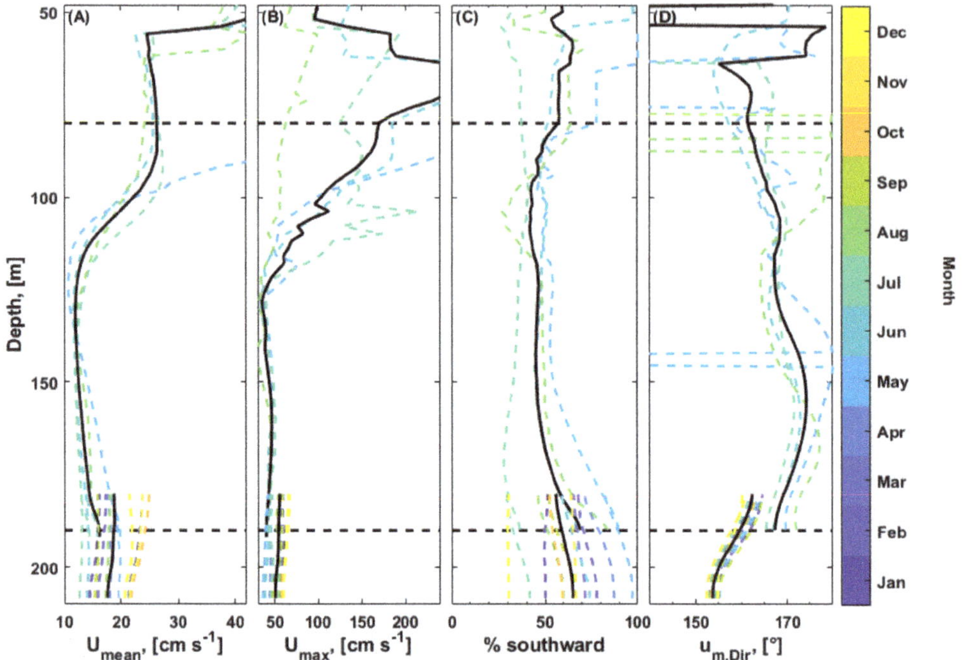

**Figure 9.** Monthly water column values for (**A**) mean flow speed, (**B**) maximum flow speed, (**C**) percentage of southward flow (90–270°), and (**D**) direction of most energetic fluctuations at SLM (depths 180–210 m) and MLM (depths 50–190 m) lander positions. The thick black line marks the mean of the recorded period. The dashed horizontal lines represent the definition for surface waters (<80 m) and the depths of the living CWCs at Nakken (>190 m). The values were calculated from a 1 h running median to avoid outliers.

The current direction was determined by the orientation of the fjord and was dominantly southward (between 90° and 270°) beneath 180 m, varying in between 80 and 180 m, and southward (May and June) or northward (July and August) near the surface (Figure 9C, Table 3). In July and December, the flow direction was dominantly northward (Figure 9C). The direction of the most energetic flow, $u_m$, was dominantly southward across the water column (150°–163°, at depths >180–208 m and 160°–170° at depths <180 m). The northward flow was measured at ~150 m depth in May and depths <100 m in August (Figure 9D).

The flow regime at NK bank reef was mostly subcritical, but it supported hydraulic jumps with flow velocities >$U_{mean}$. These are observed as rapid changes in hydrographic parameters. For example, in the winter months, temperature, salinity, and turbidity fluctuated within 12 h $\Delta\Theta = 0.4\ °C$, $\Delta S_A = 0.15\ g\ kg^{-1}$, and $\Delta$turbidity >0.2 NTU. Following the categorization of Juva et al. (2020) [22], the flow state is thus partially subcritical ($PB_{sc}$).

### 3.6.2. Tides

Tidal analysis of the pressure data explained 87.2% and 95.6% of the pressure fluctuations with 28 and 22 significant constituents at SLM and MLM, respectively. At both lander sites, the semidiurnal ($M_2$) signal generated the largest amplitude of 0.34 dbar. The other significant constituents were diurnal, semidiurnal, and higher frequency constituents (Table 4).

**Table 4.** Tidal analysis for bottom pressure and flow record based on the harmonic analysis toolbox T_Tide [82]. Shown are the harmonics with amplitudes >0.02 dbar and >1.5 cm s$^{-1}$ for pressure and flow, respectively, in three categories: diurnal, semidiurnal, and short period constituents. Explained variance through the tidal model in percent is given next to the parameter. Abbreviations: p: pressure, $u_e$ $u_n$: bottom horizontal flow.

| Site (Record in Days) | % | Most Significant Tidal Constituents (and Their Amplitudes) | | |
| --- | --- | --- | --- | --- |
| | | Diurnal | Semidiurnal | Higher |
| NK$_{SLM}$ (345 days) | | | | |
| p (dbar) | 87.2 | O$_1$ (0.02), K$_1$ (0.03) | N$_2$ (0.06), M$_2$ (0.34), S$_2$ (0.13), K$_2$ (0.04) | M$_6$ (0.02) |
| $u_e$ $u_n$ (cm s$^{-1}$) | 7.2 | – | N$_2$ (1.7), M$_2$ (5.4), MKS$_2$ (2.0) | M$_6$ (2.2), 2MS$_6$ (1.6) |
| NK$_{MLM}$ (87 days) | | | | |
| p (dbar) | 95.6 | O$_1$ (0.02), K$_1$ (0.03) | N$_2$ (0.06), M$_2$ (0.34), S$_2$ (0.10) | M$_6$ (0.03) |
| $u_e$ $u_n$ (cm s$^{-1}$) | 14.1 | O$_1$ (1.8) | N$_2$ (2.1), M$_2$ (5.7), S$_2$ (1.8) | 2MN$_6$ (1.7), M$_6$ (3.0), 2MS$_6$ (2.0) |

The analyses of the tidal constituents from the near bottom horizontal velocity records revealed a different picture with more significant higher frequency harmonics than in pressure. With 14 and 12 significant tidal constituents, tidal analysis explained 7.2% and 14.1% of the horizontal bottom velocity fluctuations at SLM and MLM, respectively. The M$_2$ generated amplitudes of ~5 cm s$^{-1}$ with flow direction to NNW (330°–341°), the opposite of the mean flow direction (Table 4).

## 4. Discussion

In this study, we investigated the dynamics of the flow and environmental (hydrography, carbonate chemistry, and inorganic nutrients) conditions around three *L. pertusa*-dominated wall reefs and two bank reefs in a narrow fjord on diurnal, seasonal, and annual time scales. Our results suggest that both hydrodynamics and hydro-biogeochemistry regulate the distribution and health of the reefs in Langenuen and that the fjord is likely to be in a state of change that has been ongoing since the 1980s.

### 4.1. Flow Dynamics

All five CWC reefs are healthy, but their limited spatial extent suggests that areas with suitable living conditions for CWCs are narrow both vertically and horizontally. CWC occurrences, both on the walls and banks coincide with the most dynamical parts of the fjord.

The flow regime in Langenuen is controlled by the Norwegian coastal current and modified by bathymetry, wind conditions, and tidal forcing [83–86]. In the upper 100 m, the flow is strong with peak speeds of >100 cm s$^{-1}$. At *L. pertusa* living depths (100–220 m), water flow is predominantly southward with mean speeds of <20 cm s$^{-1}$ and peak speeds of ~60 cm s$^{-1}$ (Figure 9). This is comparable to other CWC sites in the NE Atlantic [27,31,87]. The flow down to ~120 m is driven by southerly winds and Norwegian coastal water entering the fjord above the sill depth from the south [88,89]. Flow in deeper water layers is driven by seasonal density differences outside the fjord system. For example, the flow is reversed to northward-dominated at 120–200 m depth in December and in January (Figure 9C), when prevailing northerly winds (Figure 4D) [90,91] create coastal upwelling that pushes warm and deep North Atlantic water over the sill and into the deep fjord basin [92]. Intrusions of high salinity and relatively warm North Atlantic water renew the basin water in the fjords [89,93] and maintain aerobic conditions at the bottom [94]. At Nakken, the tidal flow was small compared to the residual flow (~5 vs. ~20 cm s$^{-1}$).

This is common in these coastal waters [95,96]. Similar amplitudes have been recorded in Hardangerfjord, the main fjord to which Langenuen is a tributary [93].

In the Langenuen Fjord system, the distribution of CWC reefs, both on walls and banks, seems to be governed by small-scale hydrodynamics. At the wall reefs of HH and SN, *L. pertusa* corals live just beneath the seasonal thermocline and the fastest flow layer (>100 m). The upper vertical limit of the corals could be set by this fast flow layer as in strong currents, the polyps could bend backward, reducing the feeding surface [97], and prey could escape from the polyps [98], restricting energy uptake in the coral. The 100 m depth also coincides with the strongly stratified layer in late summer and autumn ($N > 20 \times 10^{-3}$ s$^{-1}$, Figure 2), which could reduce the zooplankton migration to CWC living depths [99,100]. Below 100 m, flow is slower, and the framework of the CWCs further reduces flow speeds due to friction [28] toward the efficient prey capture speeds of the *L. pertusa* [34,98]. Moreover, the CWC framework acts as a natural sediment trap allowing the corals to capture and use the enhanced particle delivery. At these depths, thriving colonies of *L. pertusa* and *M. oculata* occur on the vertical walls. The lower vertical distribution limit of the CWC growth on vertical walls is created by the lack of hard bottom substrate when moving beneath 200 m depth at Straumsneset and Huglhammaren and below 240 m depth at Hornaneset.

At the bank reefs, the topography-flow interaction is suggested to directly influence the health and growth of CWCs [22]. At the Nakken bank reef, the flow supports periodic hydraulic jumps at high flow speeds. This will create a link between the surface and the deep reefs. During these turbulent events, resuspended organic particles that have settled to the sea bed would locally elevate food supply to the reefs compared to the surrounding deep sea-bed. These processes could be particularly important during periods of food limitation [101]. However, during the periods of periodic hydraulic jumps also mineral particles will be resuspended and settled. It is assumed that patchy reefs with similar flow regimes would have high vertical growth rates to prevent burial [22]. Maier et al. (2020) [101] estimated the linear growth at NK to be 13 mm year$^{-1}$ for new polyps. This is 30% to 100% higher than the linear growth measured for other Norwegian reefs located farther north [102,103]. These areas occur likely in a more dynamic flow state preventing the settling of particles and keeping the reef clear from heavy sedimentation. This would reduce the need for large vertical growth while increasing the need for horizontal growth and bridging between polyps, enforcing the coral skeleton to withstand high physical forcing without breaking [22]. At the Nakken bank reef, turbulent conditions created by hydraulic jumps are only present in winter, i.e., from October to December ($U_{mean}$ >20 cm s$^{-1}$, and $U_{max}$ >50 cm s$^{-1}$, Figure 9). It is plausible that the emergence and growth of new polyps, taking place at this particular reef from December to March [101], is initiated by the increased sedimentation caused by the periodic hydraulic jumps. Growth rates 60% higher than those measured at NK have been documented for *L. pertusa* in the Gulf of Mexico, where corals form similar patches as observed in NK [104].

*4.2. Hydrographic and Biogeochemical Conditions*

Seasonal signals in hydrography and biogeochemistry in the Langenuen Fjord are strong, similar to the general area [92]. All CWC reefs in Langenuen occur in well-oxygenated ($O_2$ > 4.9 mL L$^{-1}$) and inorganic carbon-rich ($C_T$ > 2135 µmol kg$^{-1}$) waters. The hydrographic ($\Theta$ and $S_A$), alkalinity ($A_T$), and inorganic nutrient ($NO_3^-$, $PO_4^{3-}$, and $SiO_4^{4-}$) ranges are within previously reported values from NE Atlantic CWC sites (Table 5) [11,13,29,39,53].

Due to surface warming in spring and summer, the seasonal isopycnal reaches ~100 m. The uppermost CWCs on the wall reef setting (at 80–100 m, HH) experience temperatures >12 °C in late summer before the water is mixed in late autumn. This is ~4 °C warmer than the mean temperature at these shallow reefs and likely enhances the metabolism of *L. pertusa* and increases their energetic demand [36,41]. Together with strong flow speeds (limiting prey capture rates in *L. pertusa*), seasonal high temperatures could restrict the upper limit of the CWCs on the fjord wall. Beneath the seasonal warming layer of 100 m, annual ranges of hydrographic variables were small ($\Delta\Theta$ = 1.92 °C, $\Delta S_A$ = 0.96 g kg$^{-1}$) particularly when compared to the largest measured temperature fluctuation at any CWC site, namely the ~9 °C (5.8–15.2 °C) temperature fluctuation within a day registered in the Cape Lookout area, NW Atlantic [105]. Bottom temperatures decrease in late winter due to dissipation from the surface and remain low throughout summer due to the pronounced water column stratification established by the spring freshwater flood [106]. Due to relatively warm winter temperatures in the deeper layer and the freshwater influence, reefs occur in less dense waters than suggested for healthy CWC sites in NE Atlantic (i.e., $\sigma_\Theta$ = 27.35–27.65 kg m$^{-3}$, [12]).

At Langenuen, the dissolved inorganic carbon was at times high, and a maximum value of 2192 µmol kg$^{-1}$ was observed. This is higher compared to what is previously linked to thriving NE Atlantic CWC occurrences ($C_T$ < 2170 µmol kg$^{-1}$, [15]). On the other hand, total alkalinity (2300–2343 µmol kg$^{-1}$) was similar to other NE Atlantic CWC sites (2287–2377 µmol kg$^{-1}$, Table 5). For comparison, the typical carbonate chemistry properties in Atlantic core water in the Norwegian Sea is about 2160 µmol kg$^{-1}$ and 2310 µmol kg$^{-1}$ for $C_T$ and $A_T$, respectively [107]. Regional studies of carbonate chemistry surrounding CWC ecosystems in the Gulf of Mexico [14], Gulf of Cadiz [15], Mauretania [15], Mediterranean [10,15], and Marmara Sea [10] have shown that $C_T$ > 2170 µmol kg$^{-1}$ is common for CWC sites outside the NE Atlantic (Table 5) while significantly higher ($A_T$ > 2500 µmol kg$^{-1}$) alkalinity levels are measured only at the Mediterranean and the Marmara Sea CWC sites. Together with regional hydrographic ($\Theta$, $S_A$) conditions, the relatively high $A_T$ result that the aragonite saturation is also relatively high ($\Omega_{Ar}$ > 2.5) in the Mediterranean CWC sites compared to other basins, where values <2.0 are common (Table 5). Consequently, this study emphasizes the importance of estimating both $C_T$ and $A_T$ since it is the relationship between them that determines the $CaCO_3$ saturation. These regional ranges are obtained from single time point measurements that do not include the variability on the diurnal level.

**Table 5.** The range (published min–max) of environmental conditions at different regions, specifically at the depth and site of known *Lophelia pertusa, Madreora oculata,* or *Desmophyllum dianthus* occurrences, comparing data from this study with data available in the literature. Abbreviations: GoC: Gulf of Cadiz, Ma: Mauretania, Me: Mediterranean, GoM: Gulf of Mexico, CF: Chilean Fjords, MS: Marmara Sea, TS: Tassman Seamount, C-Refs.: References for carbon system data, N-Refs.: References for nutrient data, nd: no data reported, * Only mean salinity reported in [108].

| Location | This Study | NE Atlantic | GoC | Ma | Me | GoM | CF | MS | TS |
|---|---|---|---|---|---|---|---|---|---|
| Depth, (m) | 80–220 | 100–1000 | 606–1322 | 451–568 | 200–850 | 307–620 | 20–200 | 932 | 1050 |
| Temperature, (°C) | 7.4–12.2 | 5.9–10.6 | 9.2–11.0 | 9.7–11.7 | 12.8–14.0 | nd | 10.6–12.5 | 14.5 | 4.59 |
| Salinity, (-) | 33.93–35.39 | 35.0–35.7 | 35.6–36.0 | 35.2–35.4 | 38.6–38.8 | 35.05 * | 31.7–33.0 | 38.8 | 34.4 |
| $A_T$, (µmol kg$^{-1}$) | 2300–2343 | 2287–2377 | 2332–2342 | 2314–2375 | 2577–2742 | 2259–2391 | 2136–2235 | 2610 | 2315 |
| $C_T$, (µmol kg$^{-1}$) | 2135–2192 | 2088–2186 | 2180–2200 | 2183–2240 | 2307–2333 | 2135–2231 | 2025–2188 | 2470 | 2218 |
| $\Omega_{Ar}$, (–) | 1.50–2.01 | 1.35–3.03 | 1.43–1.83 | 1.31–1.58 | 2.59–4.06 | 1.19–1.69 | 0.78–1.60 | 1.46 | 1.02 |
| C-refs. | | [15,53] | [15] | [15] | [10,15,108] | [108] | [10,109] | [10] | [10] |
| $PO_4^{3-}$, (µmol L$^{-1}$) | 0.67–0.96 | 0.4–1.6 | | | 0.20–0.41 | | | | |
| $NO_3^-$, (µmol L$^{-1}$) | 8.64–11.61 | 4.1–23.4 | | | nd | | | | |
| $NH_4^+$, (µmol L$^{-1}$) | <0.6 | 0.5–1.6 | | | 0–0.29 | | | | |
| $SiO_4^{4-}$, (µmol L$^{-1}$) | 4.37–6.92 | 2.1–46.6 | | | nd | | | | |
| N.-Refs. | | [11,53] | | | [108] | | | | |

The short-term variability in carbonate chemistry in Langenuen is large, and the highest values are linked to the tidal cycle. Similarly, high $C_T$ concentrations have been observed in dynamical NE Atlantic CWC mounds in the southern Rockall Bank [53]. There, values up to 2186 µmol kg$^{-1}$ were measured over a tidal cycle, indicating that *L. pertusa* is resilient to $C_T$ levels exceeding 2170 µmol kg$^{-1}$, at least over short time scales. At Rockall Bank, the tidal range of $C_T$ was 58 µmol kg$^{-1}$. This is similar to short-term (within 16-h) variability measured at NK reef ($\Delta C_T \approx 57$ µmol kg$^{-1}$) in January 2017 and to the annual ranges of $C_T$ at CWC living depths at other Langenuen reefs ($\Delta C_T \approx 51$ µmol kg$^{-1}$ at HH, and $\Delta C_T \approx 39$ µmol kg$^{-1}$ at SN). The short-term changes in carbonate chemistry at Langenuen reefs cannot be explained by salinity and temperature changes alone. For example, the pCO$_2$ change is about 4.2% per 1 °C, implying that the temperature effect on pCO$_2$ could explain ~5 µatm of the 124 µatm difference recorded at Nakken reef during 16 h in January 2017. This large variability is most likely caused by rapid change in the dominant water mass from Norwegian coastal water to North Atlantic water. NAW has high pCO$_2$, thus large $C_T$, but also relatively high $A_T$. Therefore, pH$_T$ and $\Omega_{Ar}$ are generally higher in NAW than in waters containing freshwater (such as NCW), and large changes in the whole carbonate chemistry are possible during short time periods. It is also plausible that the presence of North Atlantic water with higher $\Omega_{Ar}$ compared to Norwegian coastal water is buffering ocean acidification in Langenuen, and strengthening of the coastal stratification outside the fjord system caused by warming could reduce its presence at the Langenuen coral reefs in the future.

Observed differences between the five reefs were not consistent between the different time periods or seasons sampled. This variability is at least partly a consequence of the short-term variability that is not captured with single time point measurements or replicates taken close to each other in time. The observed variability over the 16 h sampling period in January 2017 was larger both between the sites and at NK compared to other months. The high fluctuation in winter could be caused by a combination of the storm event and dynamical winter conditions (i.e., weaker stratification, stronger flow, water source variation between the north (Korsfjorden) and south (~Utsira) more than during other seasons), see Figures 2, 4, 6 and 9. This kind of extreme event is likely to occur each year, and based on meteorological reports [90,91], it is likely that even more energetic events occurred in winter 2016–2017. Since the tidal component causes large variability in the carbonate chemistry, the full seasonal and diurnal variability of the carbonate and nutrient chemistry should be measured in further ocean acidification studies [48].

### 4.3. Long-Term Changes

The measured time period (from February 2016 to August 2017) is not long enough to fully capture the interannual variability nor the long-term trends of environmental conditions occurring in the Langenuen Fjord. However, because of the water exchange between Langenuen and the open ocean above sill depth, the mean hydrographic conditions at Langenuen CWC depths can be correlated to the fixed coastal station Utsira south of Langenuen (Figures 1A and 10) [110]. Sætre et al. (2003) [38] reported that both temperature and salinity decreased between 1950 and 1989 along the Norwegian coast. At Utsira, the winter temperature (JFM) decreased from 7.6 to 7.0 °C and salinity from 35 to 34.8 [38] at 150 m depth during this period. These trends are affected by basin-wide phenomena such as "great salinity anomalies" (GSAs) [111–113] and the North Atlantic oscillation (NAO) [114]. GSAs are low-salinity/low-temperature events that are formed by cold winters, freshwater outflow, strong northerly winds, and a large sea ice extent in the northwestern Atlantic [112]. A positive NAO index is associated with mild winters, an increase in westerly winds, higher winter precipitation over Scandinavia, and a deeper Norwegian coastal current. GSA salinity minima are reported at Utsira coastal station in 1977–1978, 1987–1989, and 1994–1996 [38]. High temperatures and salinities in between these periods are caused by an increase in the Atlantic inflow combined with the atmospheric conditions associated with periods of positive NAO and a general rise in temperature over the past decades [115]. The decadal forcing acts together with climate change to determine the past, the present, and the future deep-water hydrography within the region and at Langenuen. The whole water column had warmed after the GSA in the 1980s when only salinity returned to previous levels (Figure 10), indicating ocean warming. At Utsira, the increase has been 0.45 and 0.25 °C per decade at 50 and 200 m depths, respectively, since 1975 (Figure 10). If we assume a similar warming trend in bottom temperature at Langenuen, the waters at CWC living depths were ~1 °C cooler in the 1970s compared to the study period 2016–2017. Furthermore, if these warming rates continue at Langenuen, temperatures could seasonally increase to >18 °C at 50 m depth and >10 °C at 200 m depth (Figure 10). The summer thermocline with temperatures >12 °C would reach depths >100 m by the year 2100. The warming within the past 40 years has likely already increased the energy demand of CWCs compared to that in the 1970s, with possible effects on coral energy reserves and reproductive output if these are not met by increased food uptake rates [41].

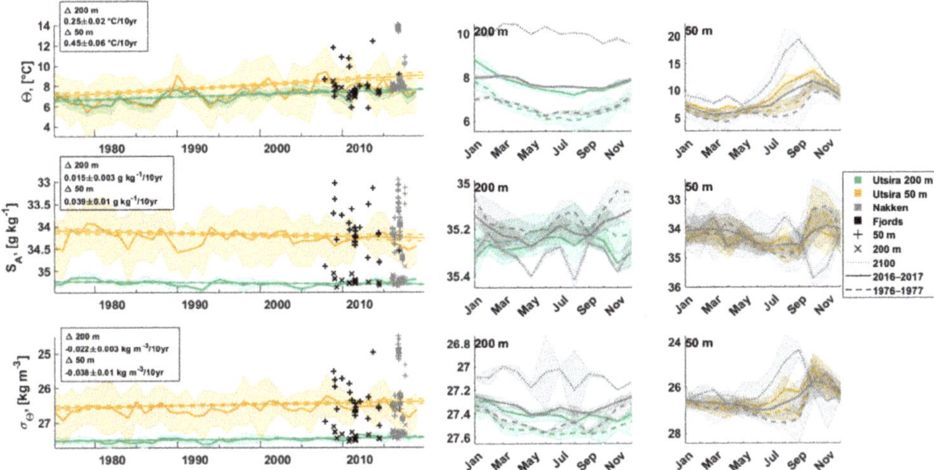

**Figure 10.** Temperature (first row), salinity (second row) and density (third row) at Utsira coastal station (color) ©Havforskningsinstituttet, Korsfjorden-Langenuen-Bømlafjorden area (black) ©Havforskningsinstituttet, and at Langenuen/Nakken (gray). The first column shows Utsira annual mean ± 1σ and linear fit with 95% confidence for 1975–2020 at 50 m (orange line) and at 200 m (green line) depths and individual measurements inside the fjord system at 50 m (plus signs) and 200 m (cross signs) depths. The second column shows the annual cycle at 200 m depth, and the third column the annual cycle at 50 m depth. In the second and third columns, the dashed line shows the average year ± 1σ for Utsira for years 1976–1977, the solid line shows the average year ± 1σ for Utsira 2016–2017, and measurements from Nakken and dotted gray line is the linear estimation at Nakken for the year 2100.

The warming of coastal waters caused by climate change has likely led to a decreased frequency of intrusions of dense, oxygen-rich North Atlantic water with relatively high $\Omega_{Ar}$ levels and has caused a general oxygen decline in the basin waters of some western Norwegian fjords with shallow sills (i.e., <100 m) [45,116]. Furthermore, warmer water contains less oxygen. In the Masfjorden, north of Bergen, this multidecadal decline corresponds to a loss of 2.0 mL L$^{-1}$ over 42 years and an associated 1 °C rise in temperature [45]. A similar decline in oxygen has been reported in Byfjorden, off the city of Bergen, with an accompanying shift in benthic communities toward domination of opportunistic benthic species [116]. Given the observed warming in Utsira, oxygen concentrations have likely decreased in Langenuen as well and are likely to decrease even further in the future. Hebbeln et al. (2020) [35] suggested that *L. pertusa* populations are highly sensitive to low oxygen conditions of 40%–50% lower than the ambient oxygen values. From observed conditions, a decline of 40% from present values would set the lower oxygen limit in Langenuen to ~2.5 mL $L^{-1}$ compared to the global limit of <1.5 mL $L^{-1}$ [36,117,118]. If a decline of 0.5 mL $L^{-1}$ per decade continued linearly in Langenuen, corals would be exposed to hypoxic conditions by the end of the century and to conditions <2.5 mL $L^{-1}$ by the year 2070.

Besides warming waters and oxygen decline, salinity in the surface layer has been decreasing along the Norwegian coast [38]. At Utsira, the decline at 50 m depth has been up to $-0.05$ g $kg^{-1}$ per decade between May and September since 1975 (Figure 10), but salinity has increased slightly during other months and at other depths. This long-term decrease in salinity at the surface is partly caused by increased precipitation and retreat/melting of glaciers, substantially increasing freshwater run-off. Førland and Hanssen-Bauer (2000) [119] demonstrated that precipitation in northern Norway has increased by 1.7% per decade during 1961–1990. Hanssen-Bauer et al. (2003) [120] projected an annual increase of 15% in precipitation from 1980–1999 to 2030–2049 in western Norway. During the last decade, the glaciers of Norway have continuously retreated ([121] and references therein). This could further increase the surface stratification and reduce the surface-bottom interaction during the melt water period.

Driven by warming, the water column has become less dense within the past 40 years in Utsira at a rate of 0.037 kg $m^{-3}$ per decade at 50 m and ~0.02 kg $m^{-3}$ per decade at 200 m depth. If density has changed similarly in Langenuen, waters at CWC living depths would have been within the suggested sigma-theta range for thriving CWC sites in the NE Atlantic (>27.35 kg $m^{-3}$) [12] during the 1970s but are lower than that now. Such changes may be critical because the depth zonation of CWCs on the vertical walls may be related to physical boundary conditions at specific depths that act to concentrate food particles [122], and corals cannot physically move to adjust to that.

Freshwater from rivers, rain, and melting glaciers have been observed to decrease $A_T$ and $\Omega_{Ar}$ substantially and play a large role in increasing ocean acidification in surface waters [123,124]. These waters could reach the coral depths during autumn and winter mixing even though the summer stratification is likely getting stronger. Therefore, there is a weak positive indication that the influence of fresh coastal water into the deeper water masses of the fjord may locally accelerate ocean acidification at CWC living depths.

Our hindcast suggests that ocean warming and acidification may already today affect the functioning of Langenuen CWCs. As oceanic uptake of atmospheric $CO_2$ continues [115], warming and ocean acidification continue in the western Norwegian fjords. We, therefore, stipulate that these shallow fjord reefs could serve as windows into the future and forewarn about the likely effects of ocean acidification and warming on CWC reef biodiversity and productivity. Studies from tropical coral sites suggest that the rate of calcification is lower under recent conditions (~400 ppm) compared to pre-industrial (280 ppm) times [125,126], and enhanced reef growth has been observed in situ in lagoons where $pCO_2$ levels have been manipulated to pre-industrial levels. McGrath et al. (2012) [127] showed that $C_T$ has increased in subsurface waters in the NE Atlantic between 1991 and 2010. This was concomitant with a decrease in $\Omega_{Ar}$ and a shoaling of the ASH. Within this century, ~70% of the known CWC habitats are predicted to be in corrosive waters [49]. The ocean acidification monitoring program of Norwegian waters [48] has shown an annual decrease in $\Omega_{Ar}$ and pH of 0.007 and 0.02 in Korsfjorden at depths of ~670 m between 2007 and 2017, respectively. The decrease in $\Omega_{Ar}$ and pH has been reported to be greater in coastal areas than in deeper waters (~2000 m) offshore [48]. If the carbonate system changed at similar rates at ~200 m depth in Langenuen, CWCs would be exposed to corrosive conditions ($\Omega_{Ar} < 1$) by 2090. Given their shallower location in a narrow fjord, it is likely that the rate of change is even larger, and $\Omega_{Ar} < 1$ conditions would occur even sooner [51]. The living coral can withstand some corrosive conditions with pH up-regulation [10,128], and in the Mediterranean, $\Omega_{Ar} < 0.92$ was found to be the calcifying limit for *L. pertusa* [129]. However, the exposed dead coral skeleton, which frequently forms the largest portion of the CWC reef, starts to dissolve in corrosive conditions [130].

## 5. Conclusions

The distribution of cold-water corals in the narrow Langenuen Fjord in southwestern Norway (health category II) is limited by physical environmental boundaries and the hydrodynamical setting driven by seasonal and short-term forcings. Summer thermocline with temperatures of over 12 °C reached the uppermost populations on the vertical wall reefs (~80 m). This is >4 °C warmer than the mean temperature (7.8 °C) at the coral living depths and may be over their tolerance limit, thus limiting the vertical extent of coral growth. Variability of chemical parameters during a recorded winter storm ($\Delta A_T = 26$ µmol kg$^{-1}$, $\Delta C_T = 57$ µmol kg$^{-1}$) was comparable to the measured annual variability ($\Delta A_T = 42$ µmol kg$^{-1}$, $\Delta C_T = 57$ µmol kg$^{-1}$). This short-term variability was driven by rapid changes in flow conditions and water masses. For all reef settings, but in particular the bank reefs, our findings highlight the importance of sampling at different phases within the tidal cycle when collecting samples for environmental monitoring programs.

Norwegian coastal waters outside Langenuen have warmed since the 1970s on average at a rate of 0.25 °C per decade at depths >100 m and up to 0.44 °C per decade at 50 m depth. Waters have warmed more during summer and autumn compared to the remaining seasons. This warming has reduced water mass density and its oxygen concentration and strengthened summer and autumn stratification, changing the physical boundaries toward "non-optimal" for coral growth in Langenuen. The depth zonation of CWCs on the vertical walls may be related to physical boundary conditions at specific depths that act to concentrate food particles, and corals cannot physically move to adjust to that. Clearly, this aspect needs more attention and should be tackled in future research.

If warming continues at a similar rate in the future, the summer thermocline with temperatures >12 °C would reach depths >100 m, and corals would be exposed to hypoxic and corrosive events by 2100. In conclusion, more frequent short-term stress events and the gradual change of mean conditions may restrict the depth zonation of corals and change the benthic species composition in Langenuen, as well as other CWC sites, dramatically.

**Author Contributions:** Conceptualization, S.F., T.K., M.C. and K.J.; software, K.J.; validation, K.J., T.K., M.C. and S.F.; formal analysis, K.J.; investigation, T.K., M.C., K.J. and S.F.; resources, T.K. and S.F.; data curation, K.J., T.K. and S.F.; writing—original draft preparation, K.J.; writing—review and editing, K.J., T.K., M.C., W.-C.D. and S.F.; visualization, K.J.; project administration, T.K. and S.F. All authors have read and agreed to the published version of the manuscript.

**Funding:** This research is part of FATE—Fate of cold-water coral reefs—identifying drivers of ecosystem change project funded by the Research Council of Norway (project no. 244604/E40). We are grateful for additional financial support from the Osk. Huttunen foundation doctorate research grant and the Norwegian Environment Agency—Ocean Acidification Monitoring Program.

**Institutional Review Board Statement:** Not applicable.

**Informed Consent Statement:** Not applicable.

**Data Availability Statement:** The data presented in this study are available on request from the corresponding author.

**Acknowledgments:** We would like to thank captains, crew, ROV pilots, chief scientists, and scientific teams during cruises on RV Håkon Mosby and on RV Kristine Bonnevie (2016–2017) as well as GEOMAR technicians Thorben Berghäuser and Asmus Petersen for their help with the benthic landers. We are also grateful for the valuable feedback provided by the three reviewers and final comments by the academic editors.

**Conflicts of Interest:** The authors declare no conflict of interest.

## References


1. Henry, L.-A.; Roberts, J.M. Biodiversity and ecological composition of macrobenthos on cold-water coral mounds and adjacent off-mound habitat in the bathyal Porcupine Seabight, NE Atlantic. *Deep Sea Res. Part I Oceanogr. Res. Pap.* **2007**, *54*, 654–672. [CrossRef]
2. Roberts, J.M.; Henry, L.-A.; Long, D.; Hartley, J.P. Cold-water coral reef frameworks, megafaunal communities and evidence for coral carbonate mounds on the Hatton Bank, north east Atlantic. *Facies* **2008**, *54*, 297–316. [CrossRef]
3. Freiwald, A.; Fosså, J.H.; Grehan, A.; Koslow, T.; Roberts, J.M. Cold-Water Coral Reefs. *Environment* **2004**, *22*, 615–625.
4. Rueda, J.L.; Urra, J.; Aguilar, R.; Angeletti, L.; Bo, M.; García-Ruiz, C.; González-Duarte, M.M.; López, E.; Madurell, T.; Maldonado, M.; et al. Water Coral Associated Fauna in the Mediterranean Sea and Adjacent Areas. In *Mediterranean Cold-Water Corals: Past, Present and Future*; Orejas, C., Jiménez, C., Eds.; Springer: Cham, Switzerland, 2019; pp. 295–333. [CrossRef]
5. Cathalot, C.; Van Oevelen, D.; Cox, T.J.S.; Kutti, T.; Lavaleye, M.; Duineveld, G.; Meysman, F.J.R. Cold-water coral reefs and adjacent sponge grounds: Hotspots of benthic respiration and organic carbon cycling in the deep sea. *Front. Mar. Sci.* **2015**, *2*, 37. [CrossRef]
6. Rovelli, L.; Attard, K.M.; Bryant, L.D.; Flögel, S.; Stahl, H.; Roberts, J.M.; Linke, P.; Glud, R.N. Benthic $O_2$ uptake of two cold-water coral communities estimated with the non-invasive eddy correlation technique. *Mar. Ecol. Prog. Ser.* **2015**, *525*, 97–104. [CrossRef]
7. FAO. *Report of the Technical Consultation on International Guidelines for the Management of Deep-Sea Fisheries in the High Seas*; FAO: Rome, Italy, 2009.
8. OSPAR Commission. *Case Reports for the OSPAR List of Threatened and/or Declining Species and Habitats*; Biodiversity Series; OSPAR Commission: London, UK, 2008.
9. Hoegh-Guldberg, O.; Poloczanska, E.S.; Skirving, W.; Dove, S. Coral Reef Ecosystems under Climate Change and Ocean Acidification. *Front. Mar. Sci.* **2017**, *4*, 158. [CrossRef]
10. McCulloch, M.; Trotter, J.; Montagna, P.; Falter, J.; Dunbar, R.; Freiwald, A.; Försterra, G.; López Correa, M.; Maier, C.; Rüggeberg, A.; et al. Resilience of cold-water scleractinian corals to ocean acidification: Boron isotopic systematics of pH and saturation state up-regulation. *Geochim. Cosmochim. Acta* **2012**, *87*, 21–34. [CrossRef]
11. Davies, A.J.; Wisshak, M.; Orr, J.C.; Murray Roberts, J. Predicting suitable habitat for the cold-water coral *Lophelia pertusa* (Scleractinia). *Deep Sea Res. Part I Oceanogr. Res. Pap.* **2008**, *55*, 1048–1062. [CrossRef]
12. Dullo, W.C.; Flögel, S.; Rüggeberg, A. Cold-water coral growth in relation to the hydrography of the Celtic and Nordic European continental margin. *Mar. Ecol. Prog. Ser.* **2008**, *371*, 165–176. [CrossRef]
13. Davies, A.J.; Guinotte, J.M. Global Habitat Suitability for Framework-Forming Cold-Water Corals. *PLoS ONE* **2011**, *6*, e18483. [CrossRef]
14. Lunden, J.J.; Georgian, S.E.; Cordes, E.E. Aragonite saturation states at cold-water coral reefs structured by *Lophelia pertusa* in the northern Gulf of Mexico. *Limnol. Oceanogr.* **2013**, *58*, 354–362. [CrossRef]
15. Flögel, S.; Dullo, W.-C.; Pfannkuche, O.; Kiriakoulakis, K.; Rüggeberg, A. Geochemical and physical constraints for the occurrence of living cold-water corals. *Deep Sea Res. Part II Top. Stud. Oceanogr.* **2014**, *99*, 19–26. [CrossRef]
16. Castellan, G.; Angeletti, L.; Taviani, M.; Montagna, P. The Yellow Coral *Dendrophyllia cornigera* in a Warming Ocean. *Front. Mar. Sci.* **2019**, *6*, 692. [CrossRef]
17. Form, A.U.; Riebesell, U. Acclimation to ocean acidification during long-term $CO_2$ exposure in the cold-water coral *Lophelia pertusa*. *Glob. Chang. Biol.* **2012**, *18*, 843–853. [CrossRef]
18. Maier, C.; Schubert, A.; Berzunza Sànchez, M.M.; Weinbauer, M.G.; Watremez, P.; Gattuso, J.-P. End of the Century $p\,CO_2$ Levels Do Not Impact Calcification in Mediterranean Cold-Water Corals. *PLoS ONE* **2013**, *8*, e62655. [CrossRef] [PubMed]
19. Hennige, S.J.; Wicks, L.C.; Kamenos, N.A.; Bakker, D.C.E.; Findlay, H.S.; Dumousseaud, C.; Roberts, J.M. Short-term metabolic and growth responses of the cold-water coral *Lophelia pertusa* to ocean acidification. *Deep Sea Res. Part II Top. Stud. Oceanogr.* **2014**, *99*, 27–35. [CrossRef]
20. Büscher, J.V.; Form, A.U.; Riebesell, U. Interactive Effects of Ocean Acidification and Warming on Growth, Fitness and Survival of the Cold-Water Coral *Lophelia pertusa* under Different Food Availabilities. *Front. Mar. Sci.* **2017**, *4*, 101. [CrossRef]
21. Maier, C.; Weinbauer, M.G.; Gattuso, J.-P. Fate of Mediterranean Scleractinian Cold-Water Corals as a Result of Global Climate Change. A Synthesis. In *Mediterranean Cold-Water Corals: Past, Present and Future*; Orejas, C., Jiménez, C., Eds.; Springer: Cham, Switzerland, 2019; pp. 517–529. [CrossRef]
22. Juva, K.; Flögel, S.; Karstensen, J.; Linke, P.; Dullo, W.-C. Tidal Dynamics Control on Cold-Water Coral Growth: A High-Resolution Multivariable Study on Eastern Atlantic Cold-Water Coral Sites. *Front. Mar. Sci.* **2020**, *7*, 132. [CrossRef]
23. Carilli, J.; Donner, S.D.; Hartmann, A.C. Historical Temperature Variability Affects Coral Response to Heat Stress. *PLoS ONE* **2012**, *7*, e34418. [CrossRef] [PubMed]
24. Addamo, A.M.; Vertino, A.; Stolarski, J.; García-Jiménez, R.; Taviani, M.; Machordom, A. Merging scleractinian genera: The overwhelming genetic similarity between solitary *Desmophyllum* and colonial *Lophelia*. *BMC Evol. Biol.* **2016**, *16*, 108. [CrossRef]
25. Järnegren, J.; Kutti, T. *Lophelia pertusa in Norwegian Waters. What Have We Learned since 2008?* NINA Report 1028; Norwegian Institute for Nature Research: Trondheim, Norway, 2014.
26. Wilson, J.B. 'Patch' development of the deep-water coral *Lophelia Pertusa* (L.) on Rockall Bank. *J. Mar. Biol. Assoc. U. K.* **1979**, *59*, 165–177. [CrossRef]

27. Mortensen, P.B.; Hovland, T.; Fosså, J.H.; Furevik, D.M. Distribution, abundance and size of *Lophelia pertusa* coral reefs in mid-Norway in relation to seabed characteristics. *J. Mar. Biol. Assoc. U. K.* **2001**, *81*, 581–597. [CrossRef]
28. Roberts, J.M.; Wheeler, A.J.; Freiwald, A.; Cairns, S.D. *Cold-Water Corals: The Biology and Geology of Deep-Sea Coral Habitats*; Cambridge University Press: Cambridge, UK, 2009. [CrossRef]
29. Roberts, J.M.; Wheeler, A.J.; Freiwald, A. Reefs of the Deep: The Biology and Geology of Cold-Water Coral Ecosystems. *Science* **2006**, *312*, 543–547. [CrossRef]
30. Rüggeberg, A.; Flögel, S.; Dullo, W.-C.; Hissmann, K.; Freiwald, A. Water mass characteristics and sill dynamics in a subpolar cold-water coral reef setting at Stjernsund, northern Norway. *Mar. Geol.* **2011**, *282*, 5–12. [CrossRef]
31. Thiem, Ø.; Ravagnan, E.; Fosså, J.H.; Berntsen, J. Food supply mechanisms for cold-water corals along a continental shelf edge. *J. Mar. Syst.* **2006**, *60*, 207–219. [CrossRef]
32. Davies, A.J.; Duineveld, G.C.A.; Lavaleye, M.S.S.; Bergman, M.J.N.; Van Haren, H. Downwelling and deep-water bottom currents as food supply mechanisms to the cold-water coral *Lophelia pertusa* (Scleractinia) at the Mingulay Reef Complex. *Limnol. Oceanogr.* **2009**, *54*, 620–629. [CrossRef]
33. Cyr, F.; Van Haren, H.; Mienis, F.; Duineveld, G.; Bourgault, D. On the influence of cold-water coral mound size on flow hydrodynamics, and vice versa. *Geophys. Res. Lett.* **2016**, *43*, 775–783. [CrossRef]
34. Orejas, C.; Gori, A.; Rad-Menéndez, C.; Last, K.S.; Davies, A.J.; Beveridge, C.M.; Sadd, D.; Kiriakoulakis, K.; Witte, U.; Roberts, J.M. The effect of flow speed and food size on the capture efficiency and feeding behaviour of the cold-water coral *Lophelia pertusa*. *J. Exp. Mar. Biol. Ecol.* **2016**, *481*, 34–40. [CrossRef]
35. Hebbeln, D.; Wienberg, C.; Dullo, W.-C.; Freiwald, A.; Mienis, F.; Orejas, C.; Titschack, J. Cold-water coral reefs thriving under hypoxia. *Coral Reefs* **2020**, *39*, 853–859. [CrossRef]
36. Dodds, L.A.; Roberts, J.M.; Taylor, A.C.; Marubini, F. Metabolic tolerance of the cold-water coral *Lophelia pertusa* (Scleractinia) to temperature and dissolved oxygen change. *J. Exp. Mar. Biol. Ecol.* **2007**, *349*, 205–214. [CrossRef]
37. Domingues, C.M.; Church, J.A.; White, N.J.; Gleckler, P.J.; Wijffels, S.E.; Barker, P.M.; Dunn, J.R. Improved estimates of upper-ocean warming and multi-decadal sea-level rise. *Nature* **2008**, *453*, 1090–1093. [CrossRef]
38. Sætre, R.; Aure, J.; Danielssen, D.S. Long-Term Hydrographic Variability Patterns off the Norwegian Coast and in the Skagerrak. *ICES Mar. Sci. Symp.* **2003**, *219*, 150–159.
39. Roberts, S.; Hirshfield, M. Cold-Water Corals: Out of Sight—No Longer Out of Mind. *Front. Ecol. Environ.* **2004**, *2*, 123–130. [CrossRef]
40. Roder, C.; Berumen, M.L.; Bouwmeester, J.; Papathanassiou, E.; Al-Suwailem, A.; Voolstra, C.R. First biological measurements of deep-sea corals from the Red Sea. *Sci. Rep.* **2013**, *3*, 2802. [CrossRef]
41. Dorey, N.; Gjelsvik, Ø.; Kutti, T.; Büscher, J.V. Broad Thermal Tolerance in the Cold-Water Coral *Lophelia pertusa* From Arctic and Boreal Reefs. *Front. Physiol.* **2020**, *10*, 1636. [CrossRef] [PubMed]
42. Mienis, F.; Duineveld, G.C.A.; Davies, A.J.; Ross, S.W.; Seim, H.; Bane, J.; van Weering, T.C.E. The influence of near-bed hydrodynamic conditions on cold-water corals in the Viosca Knoll area, Gulf of Mexico. *Deep Sea Res. Part I Oceanogr. Res. Pap.* **2012**, *60*, 32–45. [CrossRef]
43. Henson, S.; Cole, H.; Beaulieu, C.; Yool, A. The impact of global warming on seasonality of ocean primary production. *Biogeosciences* **2013**, *10*, 4357–4369. [CrossRef]
44. Breitburg, D.; Levin, L.A.; Oschlies, A.; Grégoire, M.; Chavez, F.P.; Conley, D.J.; Garçon, V.; Gilbert, D.; Gutiérrez, D.; Isensee, K.; et al. Declining oxygen in the global ocean and coastal waters. *Science* **2018**, *359*, eaam7240. [CrossRef] [PubMed]
45. Aksnes, D.L.; Aure, J.; Johansen, P.-O.; Johnsen, G.H.; Vea Salvanes, A.G. Multi-decadal warming of Atlantic water and associated decline of dissolved oxygen in a deep fjord. *Estuar. Coast. Shelf Sci.* **2019**, *228*, 106392. [CrossRef]
46. Lauvset, S.K.; Gruber, N.; Landschützer, P.; Olsen, A.; Tjiputra, J. Trends and drivers in global surface ocean pH over the past 3 decades. *Biogeosciences* **2015**, *12*, 1285–1298. [CrossRef]
47. Bates, N.R.; Astor, Y.M.; Church, M.J.; Currie, K.; Dore, J.E.; González-Dávila, M.; Lorenzoni, L.; Muller-Karger, F.; Olafsson, J.; Santana-Casiano, J.M. A Time-Series View of Changing Ocean Chemistry Due to Ocean Uptake of Anthropogenic $CO_2$ and Ocean Acidification. *Oceanography* **2014**, *27*, 126–141. [CrossRef]
48. Jones, E.; Chierici, M.; Skjelvan, I.; Norli, M.; Børsheim, K.Y.; Lødemel, H.H.; Kutti, T.; Sørensen, K.; King, A.L.; Jackson, K.; et al. *Monitoring Ocean Acidification in Norwegian Seas in 2017*; Rapport, Miljødirektoratet, M-XXX I 2018; Norwegian Environment Agency: Trondheim, Norway, 2018.
49. Guinotte, J.M.; Orr, J.; Cairns, S.; Freiwald, A.; Morgan, L.; George, R. Will Human-Induced Changes in Seawater Chemistry Alter the Distribution of Deep-Sea Scleractinian Corals? *Front. Ecol. Environ.* **2006**, *4*, 141–146. [CrossRef]
50. Turley, C.M.; Roberts, J.M.; Guinotte, J.M. Corals in deep-water: Will the unseen hand of ocean acidification destroy cold-water ecosystems? *Coral Reefs* **2007**, *26*, 445–448. [CrossRef]
51. Omar, A.M.; Skjelvan, I.; Erga, S.R.; Olsen, A. Aragonite saturation states and pH in western Norwegian fjords: Seasonal cycles and controlling factors, 2005–2009. *Ocean Sci.* **2016**, *12*, 937–951. [CrossRef]
52. Findlay, H.S.; Artioli, Y.; Moreno Navas, J.; Hennige, S.J.; Wicks, L.C.; Huvenne, V.A.I.; Woodward, E.M.S.; Roberts, J.M. Tidal downwelling and implications for the carbon biogeochemistry of cold-water corals in relation to future ocean acidification and warming. *Glob. Chang. Biol.* **2013**, *19*, 2708–2719. [CrossRef]

53. Findlay, H.S.; Hennige, S.J.; Wicks, L.C.; Navas, J.M.; Woodward, E.M.S.; Roberts, J.M. Fine-scale nutrient and carbonate system dynamics around cold-water coral reefs in the northeast Atlantic. *Sci. Rep.* **2014**, *4*, 3671. [CrossRef] [PubMed]

54. Findlay, H.S.; Tyrrell, T.; Bellerby, R.G.J.; Merico, A.; Skjelvan, I. Carbon and nutrient mixed layer dynamics in the Norwegian Sea. *Biogeosciences* **2008**, *5*, 1395–1410. [CrossRef]

55. Kitidis, V.; Hardman-Mountford, N.J.; Litt, E.; Brown, I.; Cummings, D.; Hartman, S.; Hydes, D.; Fishwick, J.R.; Harris, C.; Martinez-Vicente, V.; et al. Seasonal dynamics of the carbonate system in the Western English Channel. *Cont. Shelf Res.* **2012**, *42*, 30–40. [CrossRef]

56. Garcia, H.E.; Locarnini, R.A.; Boyer, T.P.; Antonov, J.I.; Baranova, O.K.; Zweng, M.M.; Reagan, J.R.; Johnson, D.R. *World Ocean Atlas 2013, Volume 4 : Dissolved Inorganic Nutrients (Phosphate, Nitrate, Silicate)*; National Centers for Environmental Information: Silver Spring, MD, USA, 2013; Volume 4.

57. Lauvset, S.K.; Key, R.M.; Olsen, A.; van Heuven, S.; Velo, A.; Lin, X.; Schirnick, C.; Kozyr, A.; Tanhua, T.; Hoppema, M.; et al. A new global interior ocean mapped climatology: The $1° \times 1°$ GLODAP version. *Earth Syst. Sci. Data* **2016**, *8*, 325–340. [CrossRef]

58. Burgos, J.M.; Buhl-Mortensen, L.; Buhl-Mortensen, P.; Ólafsdóttir, S.H.; Steingrund, P.; Ragnarsson, S.; Skagseth, Ø. Predicting the Distribution of Indicator Taxa of Vulnerable Marine Ecosystems in the Arctic and Sub-arctic Waters of the Nordic Seas. *Front. Mar. Sci.* **2020**, *7*, 131. [CrossRef]

59. Ervik, A.; Agnalt, A.-L.; Asplin, L.; Aure, J.; Bekkvik, C.; Døskeland, I.; Hageberg, A.A.; Hansen, T.; Karlsen, Ø.; Oppedal, F.; et al. *AkvaVis-Dynamisk GIS-Verktøy for Lokalisering Av Oppdrettsanlegg for Nye Oppdrettsarter Miljøkrav for Nye Oppdrettsarter Og Laks*; Havforskningsinstituttet: Bergen, Norway, 2008.

60. Fosså, J.H.; Kutti, T.; Buhl-Mortensen, P.; Skjoldal, R.H. *Rapport Fra Havforskningen: Vurdering Av Norske Korallrev*; Rapport fra Havforskningen, Nr-8-2015; Institute of Marine Research: Bergen, Norway, 2015.

61. Tambs-Lyche, H. Zoogeographical and Faunistic Studies on West Norwegian Marine Animals. In *Årbok for Universitetet i Bergen*; Matematisk-Naturvitenskaplig Serie 7:3-241958; University of Bergen: Bergen, Norway, 2019.

62. Skjelvan, I.; Chierici, M.; Sørensen, K.; Jackson, K.; Kutti, T.; Lødemel, H.; King, A.; Reggiani, E.; Norli, M.; Bellerby, R.; et al. *Havforsuring i Vestlandsfjorder Og $CO_2$—Variabilitet i Lofoten*; Miljødirektoratet-642; Norwegian Environment Agency: Trondheim, Norway, 2016.

63. Sae-Bird Scientific. *Software Manual Seasoft V2: SBE Data Processing*; Sea-Bird Scientific: Bellevue, WA, USA, 2017; p. 177.

64. Pawlowicz, R. RDADCP, mar10 v. 0. Available online: https://www.eoas.ubc.ca/~{}rich/#RDADCP (accessed on 23 February 2017).

65. Dickson, A.G.; Sabine, C.L.; Christian, J.R. *Guide to Best Practices for Ocean $CO_2$ Measurements*; PICES Spec.; North Pacific Marine Science Organization: Patricia Bay, BC, Canada, 2007. [CrossRef]

66. Pierrot, D.; Lewis, E.; Wallace, D.W.R. *MS Excel Program Developed for $CO_2$ System Calculations*; ORNL Environmental Sciences Division: Oak Ridge, TN, USA, 2011. [CrossRef]

67. Mehrbach, C.; Culberson, C.H.; Hawley, J.E.; Pytkowicx, R.M. Measurement of the apparent dissociation constants of carbonic acid in seawater at atmospheric pressure. *Limnol. Oceanogr.* **1973**, *18*, 897–907. [CrossRef]

68. Dickson, A.G. Standard Potential of the Reaction: $AgCl(s) + 1/2H_2(g) = Ag(s) + HCl(Aq)$, and the Standard Acidity Constant of the Ion $HSO_4-$ in Synthetic Sea Water from 273.15 to 318.15 K. *J. Chem. Thermodyn.* **1990**, *22*, 113–127. [CrossRef]

69. Dickson, A.G.; Millero, F.J. A comparison of the equilibrium constants for the dissociation of carbonic acid in seawater media. *Deep Sea Res. Part A Oceanogr. Res. Pap.* **1987**, *34*, 1733–1743. [CrossRef]

70. Bendschneider, K.; Robinson, R.J. A New Spectrophotometric Method for the Determination of Nitrite in Sea Water. *J. Mar. Res.* 1952. Available online: https://digital.lib.washington.edu/researchworks/bitstream/handle/1773/15938/52-1.pdf?sequence=1&isAllowed=y (accessed on 1 August 2021).

71. RFA Methodology. *Nitrate + Nitrite Nitrogen. A303-S170 Revision 6–89*; Alpkem: College Station, TX, USA, 1986.

72. Grasshoff, K. *On the Automatic Determination of Phosphate, Silicate and Fluorine in Sea Water*; ICES Hydrogr. Comm. Rep. 129; International Council for the Exploration of the Sea: Copenhagen, Denmark, 1965.

73. Kérouel, R.; Aminot, A. Fluorometric determination of ammonia in sea and estuarine waters by direct segmented flow analysis. *Mar. Chem.* **1997**, *57*, 265–275. [CrossRef]

74. Holmes, R.M.; Aminot, A.; Kérouel, R.; Hooker, B.A.; Peterson, B.J. A simple and precise method for measuring ammonium in marine and freshwater ecosystems. *Can. J. Fish. Aquat. Sci.* **1999**, *56*, 1801–1808. [CrossRef]

75. Brand, W.A.; Coplen, T.B.; Vogl, J.; Rosner, M.; Prohaska, T. Assessment of international reference materials for isotope-ratio analysis (IUPAC Technical Report). *Pure Appl. Chem.* **2014**, *86*, 425–467. [CrossRef]

76. Horita, J.; Wesolowski, D.J.; Cole, D.R.; Wesolowski, D.J. The Activity-Composition Relationship of Oxygen and Hydrogen Isotopes in Aqueous Salt Solutions: II. Vapor-Liquid Water Equilibration of Mixed Salt Solutions from 50 to 100 °C and Geochemical Implications. *Geochim. Cosmochim. Acta* **1993**, *57*, 4703–4711. [CrossRef]

77. Bourg, C.; Stievenard, M.; Jouzel, J. Hydrogen and oxygen isotopic composition of aqueous salt solutions by gas–water equilibration method. *Chem. Geol.* **2001**, *173*, 331–337. [CrossRef]

78. McDougall, T.J.; Barker, P.M. *Getting Started with TEOS-10 and the Gibbs Seawater (GSW) Oceanographic Toolbox*; WG127; SCOR: Paris, France; IAPSO: Copenhagen, Denmark, 2011; 28p.

79. IOC; SCOR; IAPSO. *The International Thermodynamic Equation of Seawater—2010: Calculation and Use of Thermodynamic Properties*; Manuals and Guides No. 56; UNESCO: Paris, France, 2010.

80. Finlay, C.C.; Maus, S.; Beggan, C.D.; Bondar, T.N.; Chambodut, A.; Chernova, T.A.; Chulliat, A.; Golovkov, V.P.; Hamilton, B.; Hamoudi, M.; et al. International Geomagnetic Reference Field: The eleventh generation. *Geophys. J. Int.* **2010**, *183*, 1216–1230. [CrossRef]

81. Lilly, J.M. jLab: A Data Analysis Package for Matlab, v. 1.6.5. Available online: http://www.jmlilly.net/jmlsoft.html (accessed on 20 December 2017).

82. Pawlowicz, R.; Beardsley, B.; Lentz, S. Classical tidal harmonic analysis including error estimates in MATLAB using TDE. *Comput. Geosci.* **2002**, *28*, 929–937. [CrossRef]

83. Stigebrandt, A. Hydrodynamics and Circulation of Fjords. In *Encyclopedia of Earth Sciences Series*; Springer Nature: Cham, Switzerland, 2012; p. 344. [CrossRef]

84. Brattegard, T.; Høisaster, T.; Sjetun, K.; Fenchel, T.; Uiblein, F. Norwegian fjords: From natural history to ecosystem ecology and beyond. *Mar. Biol. Res.* **2011**, *7*, 421–424. [CrossRef]

85. Klinck, J.M.; O'Brien, J.J.; Svendsen, H. A Simple Model of Fjord and Coastal Circulation Interaction. *J. Phys. Oceanogr.* **1981**, *11*, 1612–1626. [CrossRef]

86. Asplin, L.; Salvanes, A.G.V.; Kristoffersen, J.B. Nonlocal wind-driven fjord-coast advection and its potential effect on plankton and fish recruitment. *Fish. Oceanogr.* **1999**, *8*, 255–263. [CrossRef]

87. White, M.; Dorschel, B. The importance of the permanent thermocline to the cold water coral carbonate mound distribution in the NE Atlantic. *Earth Planet. Sci. Lett.* **2010**, *296*, 395–402. [CrossRef]

88. Sætre, R.; Aure, J.; Ljøen, R. Wind effects on the lateral extension of the Norwegian Coastal Water. *Cont. Shelf Res.* **1988**, *8*, 239–253. [CrossRef]

89. Erga, S.R. Ecological studies on the phytoplankton of Boknafjorden, western Norway. The effect of water exchange processes and environmental factors on temporal and vertical variability of biomass. *Sarsia* **1989**, *74*, 161–176. [CrossRef]

90. Olsen, A.; Graner, M.; Hygen, H.O.; Mamen, J. *MET-Info*; Ekstremværrapport: Hendelse: Urd 26. Desember 2016: Nr. 18/2017; Meteorologisk Institutt: Bergen, Norway, 2017; 19p.

91. *MET-Info*; Ekstremværrapport: Hendelse: Vidar 12. Januar 2017: Nr. 14/2017; Meteorologisk Institutt: Bergen, Norway, 2017; 12p.

92. Bakke, J.L.W.; Sands, N.J. Hydrographical studies of Korsfjorden, western Norway, in the period 1972–1977. *Sarsia* **1977**, *63*, 7–16. [CrossRef]

93. Asplin, L.; Johnsen, I.A.; Sandvik, A.D.; Albretsen, J.; Sundfjord, V.; Aure, J.; Boxaspen, K.K. Dispersion of salmon lice in the Hardangerfjord. *Mar. Biol. Res.* **2014**, *10*, 216–225. [CrossRef]

94. Matthews, J.B.L.; Sands, N.J. Ecological studies on the deep-water pelagic community of Korsfjorden, western Norway. The topography of the area and its hydrography in 1968–1972, with a summary of the sampling programmes. *Sarsia* **1973**, *52*, 29–52. [CrossRef]

95. Gjevik, B.; Straume, T. Model simulations of the M2and the K1tide in the Nordic Seas and the Arctic Ocean. *Tellus A* **1989**, *41A*, 73–96. [CrossRef]

96. Haugan, P.M.; Evensen, G.; Johannessen, J.A.; Johannessen, O.M.; Pettersson, L.H. Modeled and observed mesoscale circulation and wave-current refraction during the 1988 Norwegian Continental Shelf Experiment. *J. Geophys. Res.* **1991**, *96*, 10487–10506. [CrossRef]

97. Fabricius, K.E.; Genin, A.; Benayahu, Y. Flow-dependent herbivory and growth in zooxanthellae-free soft corals. *Limnol. Oceanogr.* **1995**, *40*, 1290–1301. [CrossRef]

98. Purser, A.; Larsson, A.I.; Thomsen, L.; van Oevelen, D. The influence of flow velocity and food concentration on *Lophelia pertusa* (Scleractinia) zooplankton capture rates. *J. Exp. Mar. Biol. Ecol.* **2010**, *395*, 55–62. [CrossRef]

99. Van Engeland, T.; Godø, O.R.; Johnsen, E.; Duineveld, G.C.A.; van Oevelen, D. Cabled ocean observatory data reveal food supply mechanisms to a cold-water coral reef. *Prog. Oceanogr.* **2019**, *172*, 51–64. [CrossRef]

100. Jónasdóttir, S.H.; Visser, A.W.; Richardson, K.; Heath, M.R. Seasonal copepod lipid pump promotes carbon sequestration in the deep North Atlantic. *Proc. Natl. Acad. Sci. USA* **2015**, *112*, 12122–12126. [CrossRef]

101. Maier, S.R.; Bannister, R.J.; van Oevelen, D.; Kutti, T. Seasonal Controls on the Diet, Metabolic Activity, Tissue Reserves and Growth of the Cold-Water Coral *Lophelia pertusa*. *Coral Reefs* **2020**, *39*, 173–187. [CrossRef]

102. Mortensen, P.B.; Rapp, H.T. Oxygen and Carbon Isotope Ratios Related to Growth Line Patterns in Skeletons of *Lophelia pertusa* (L.) (Anthozoa, Scleractinia): Implications for Determination of Linear Extension Rates. *Sarsia* **1998**, *83*, 433–446. [CrossRef]

103. Büscher, J.V.; Wisshak, M.; Form, A.U.; Titschack, J.; Nachtigall, K.; Riebesell, U. In Situ Growth and Bioerosion Rates of *Lophelia pertusa* in a Norwegian Fjord and Open Shelf Cold-Water Coral Habitat. *PeerJ* **2019**, *2019*, e7586. [CrossRef] [PubMed]

104. Brooke, S.; Young, C.M. In Situ Measurement of Survival and Growth of *Lophelia pertusa* in the Northern Gulf of Mexico. *Mar. Ecol. Prog. Ser.* **2009**, *397*, 153–161. [CrossRef]

105. Mienis, F.; Duineveld, G.C.A.; Davies, A.J.; Lavaleye, M.M.S.; Ross, S.W.; Seim, H.; Bane, J.; Van Haren, H.; Bergman, M.J.N.; De Haas, H.; et al. Cold-Water Coral Growth under Extreme Environmental Conditions, the Cape Lookout Area, NW Atlantic. *Biogeosciences* **2014**, *11*, 2543–2560. [CrossRef]

106. Husa, V.; Steen, H.; Sjøtun, K. Historical Changes in Macroalgal Communities in Hardangerfjord (Norway). *Mar. Biol. Res.* **2014**, *10*, 226–240. [CrossRef]

107. Jones, E.; Chierici, M.; Skjelvan, I.; Norli, M.; Frigstad, H.; Børsheim, K.Y.; Lødemel, H.H.; Kutti, T.; King, A.L.; Sørensen, K.; et al. *Monitoring Ocean Acidification in Norwegian Seas in 2019*; Norwegian Environment Agency: Begen, Norway, 2020.

108. Maier, C.; Watremez, P.; Taviani, M.; Weinbauer, M.G.; Gattuso, J.P. Calcification Rates and the Effect of Ocean Acidification on Mediterranean Cold-Water Corals. *Proc. R. Soc. B Biol. Sci.* **2012**, *279*, 1716–1723. [CrossRef]

109. Jantzen, C.; Häussermann, V.; Försterra, G.; Laudien, J.; Ardelan, M.; Maier, S.; Richter, C. Occurrence of a Cold-Water Coral along Natural pH Gradients (Patagonia, Chile). *Mar. Biol.* **2013**, *160*, 2597–2607. [CrossRef]

110. Aure, J.; Østensen, Ø. Hydrografiske Normaler Og Langtidsvariasjoner i Norske Kystfarvann. *Fisk. Og Havet* **1993**, *6*, 1–75.

111. Belkin, I.M. Propagation of the "Great Salinity Anomaly" of the 1990s around the Northern North Atlantic. *Geophys. Res. Lett.* **2004**, *31*, L08306. [CrossRef]

112. Belkin, I.M.; Levitus, S.; Antonov, J.; Malmberg, S.A. "Great Salinity Anomalies" in the North Atlantic. *Prog. Oceanogr.* **1998**, *41*, 1–68. [CrossRef]

113. Dickson, R.R.; Meincke, J.; Malmberg, S.A.; Lee, A.J. The "Great Salinity Anomaly" in the Northern North Atlantic 1968. *Prog. Oceanogr.* **1988**, *20*, 103–151. [CrossRef]

114. Hurrell, J.W. Decadal Trends in the North Atlantic Oscillation: Regional Temperatures and Precipitation. *Science* **1995**, *269*, 676–679. [CrossRef]

115. IPCC. *IPCC Special Report on the Ocean and Cryosphere in a Changing Climate*; PoOrtner, H.-O., Roberts, D.C., Masson-Delmotte, V., Zhai, P., Tignor, M., Poloczanska, E., Mintenbeck, K., Alegría, A., Nicolai, M., Okem, A., et al., Eds.; Intergovernmental Panel on Climate Change: Geneva, Switzerland, 2019.

116. Johansen, P.O.; Isaksen, T.E.; Bye-Ingebrigtsen, E.; Haave, M.; Dahlgren, T.G.; Kvalø, S.E.; Greenacre, M.; Durand, D.; Rapp, H.T. Temporal Changes in Benthic Macrofauna on the West Coast of Norway Resulting from Human Activities. *Mar. Pollut. Bull.* **2018**, *128*, 483–495. [CrossRef]

117. Lunden, J.J.; McNicholl, C.G.; Sears, C.R.; Morrison, C.L.; Cordes, E.E. Acute Survivorship of the Deep-Sea Coral Lophelia Pertusa from the Gulf of Mexico under Acidification, Warming, and Deoxygenation. *Front. Mar. Sci.* **2014**, *1*, 78. [CrossRef]

118. Hanz, U.; Wienberg, C.; Hebbeln, D.; Duineveld, G.; Lavaleye, M.; Juva, K.; Dullo, W.C.; Freiwald, A.; Tamborrino, L.; Reichart, G.J.; et al. Environmental Factors Influencing Benthic Communities in the Oxygen Minimum Zones on the Angolan and Namibian Margins. *Biogeosciences* **2019**, *16*, 4337–4356. [CrossRef]

119. Førland, E.J.; Hanssen-Bauer, I. Increased Precipitation in the Norwegian Arctic: True or False? *Clim. Chang.* **2000**, *46*, 485–509. [CrossRef]

120. Hanssen-Bauer, I.; Förland, E.J.; Haugen, J.E.; Tveito, O.E. Temperature and Precipitation Scenarios for Norway: Comparison of Results from Dynamical and Empirical Downscaling. *Clim. Res.* **2003**, *25*, 15–27. [CrossRef]

121. Paul, F.; Bolch, T. Glacier Changes Since the Little Ice Age. In *Geomorphology of Proglacial Systems*; Heckmann, T., Morche, D., Eds.; Springer: Cham, Switzerland, 2019; pp. 23–42. [CrossRef]

122. Huvenne, V.A.I.; Tyler, P.A.; Masson, D.G.; Fisher, E.H.; Hauton, C.; Hühnerbach, V.; Bas, T.P.; Wolff, G.A. A Picture on the Wall: Innovative Mapping Reveals Cold-Water Coral Refuge in Submarine Canyon. *PLoS ONE* **2011**, *6*, e28755. [CrossRef]

123. Fransson, A.; Chierici, M.; Nomura, D.; Granskog, M.A.; Kristiansen, S.; Martma, T.; Nehrke, G. Effect of Glacial Drainage Water on the $CO_2$ System and Ocean Acidification State in an Arctic Tidewater-Glacier Fjord during Two Contrasting Years. *J. Geophys. Res. Ocean.* **2015**, *120*, 2413–2429. [CrossRef]

124. Chierici, M.; Fransson, A. Calcium Carbonate Saturation in the Surface Water of the Arctic Ocean: Undersaturation in Freshwater Influenced Shelves. *Biogeosciences* **2009**, *6*, 2421–2431. [CrossRef]

125. Albright, R.; Caldeira, L.; Hosfelt, J.; Kwiatkowski, L.; Maclaren, J.K.; Mason, B.M.; Nebuchina, Y.; Ninokawa, A.; Pongratz, J.; Ricke, K.L.; et al. Reversal of Ocean Acidification Enhances Net Coral Reef Calcification. *Nature* **2016**, *531*, 362–365. [CrossRef]

126. Leclercq, N.; Gattuso, J.P.; Jaubert, J. $CO_2$ Partial Pressure Controls the Calcification Rate of a Coral Community. *Glob. Chang. Biol.* **2000**, *6*, 329–334. [CrossRef]

127. McGrath, T.; Kivimäe, C.; Tanhua, T.; Cave, R.R.; McGovern, E. Inorganic Carbon and pH Levels in the Rockall Trough 1991–2010. *Deep Res. Part I Oceanogr. Res. Pap.* **2012**, *68*, 79–91. [CrossRef]

128. Wall, M.; Ragazzola, F.; Foster, L.C.; Form, A.; Schmidt, D.N. pH Up-Regulation as a Potential Mechanism for the Cold-Water Coral Lophelia Pertusa to Sustain Growth in Aragonite Undersaturated Conditions. *Biogeosciences* **2015**, *12*, 6869–6880. [CrossRef]

129. Maier, C.; Popp, P.; Sollfrank, N.; Weinbauer, M.G.; Wild, C.; Gattuso, J.P. Effects of Elevated $p$ $CO_2$ and Feeding on Net Calcification and Energy Budget of the Mediterranean Cold-Water Coral Madrepora Oculata. *J. Exp. Biol.* **2016**, *219*, 3208–3217. [CrossRef] [PubMed]

130. Hennige, S.J.; Wicks, L.C.; Kamenos, N.A.; Perna, G.; Findlay, H.S.; Roberts, J.M. Hidden Impacts of Ocean Acidification to Live and Dead Coral Framework. *Proc. R. Soc. B Biol. Sci.* **2015**, *282*, 20150990. [CrossRef] [PubMed]

*Article*

# Reef Structural Complexity Influences Fish Community Metrics on a Remote Oceanic Island: Serranilla Island, Seaflower Biosphere Reserve, Colombia

Diana Castaño [1], Diana Morales-de-Anda [2,*], Julián Prato [1,3], Amílcar Leví Cupul-Magaña [2], Johanna Paola Echeverry [4] and Adriana Santos-Martínez [1,3]

[1] Universidad Nacional de Colombia—Sede Caribe, Archipiélago de San Andrés, Providencia y Santa Catalina, San Andrés Isla 880008, Colombia; dcastano@unal.edu.co (D.C.); jprato@unal.edu.co (J.P.); asantosma@unal.edu.co (A.S.-M.)

[2] Laboratorio de Ecología Marina, Centro de Investigaciones Costeras, Centro Universitario de la Costa, Universidad de Guadalajara, Puerto Vallarta 48280, Mexico; amilcar.cupul@gmail.com

[3] Corporation Center of Excellence in Marine Science-CEMarin, Bogotá 111311, Colombia

[4] Dirección General Marítima, Subdirección de Desarrollo Marítimo, Infraestructura de Datos Espaciales, Marítima, Fluvial y Costera, Bogotá 111321, Colombia; jpecheverryh@unal.edu.co

[*] Correspondence: dianamorales9009@gmail.com; Tel.: +52-322-213-7717

**Citation:** Castaño, D.; Morales-de-Anda, D.; Prato, J.; Cupul-Magaña, A.L.; Echeverry, J.P.; Santos-Martínez, A. Reef Structural Complexity Influences Fish Community Metrics on a Remote Oceanic Island: Serranilla Island, Seaflower Biosphere Reserve, Colombia. *Oceans* 2021, 2, 611–623. https://doi.org/10.3390/oceans2030034

Academic Editor: Rupert Ormond

Received: 30 October 2020
Accepted: 18 August 2021
Published: 3 September 2021

**Publisher's Note:** MDPI stays neutral with regard to jurisdictional claims in published maps and institutional affiliations.

**Abstract:** Serranilla is a protected island of the Seaflower Biosphere Reserve, far from dense human population. These characteristics could help sustain structurally complex coral reefs, often associated with higher biodiversity, abundance, and biomass of reef-associated organisms, including reef fish. However, the multiple threats present in Serranilla, including intense illegal fishing, can impact coral ecosystems generally and also specific key groups, such as the parrotfish, in particular. During the "Seaflower Research Expedition 2017", we assessed how structural habitat complexity influences reef fish assemblages. In addition, we explored differences in parrotfish species (family: Scaridae) between Serranilla and San Andrés, the most populated island in the Archipelago. On Serranilla, we found that habitat structure, rugosity, and coral cover accounted for up to 66% of variation in reef fish diversity, abundance, and biomass, with values being higher on more complex reefs. Parrotfish species differed between the islands, with larger species supporting higher biomasses at Serranilla, by comparison with San Andrés; however, the abundance, biomass, and lengths of parrotfish species were low in both areas compared with those reported from other protected Caribbean reefs. Our study indicates that despite the evident relationship between structurally complex habitats and reef fish, other threats in Serranilla could be affecting parrotfish populations, such as illegal fishing, a widespread activity in the area.

**Keywords:** structural complexity; coral reef; Caribbean; biodiversity; overfishing; parrotfish; Seaflower Biosphere Reserve

## 1. Introduction

Coral reefs are widely recognized as one of the most diverse marine ecosystems, providing essential habitats for numerous species; although coral reefs cover only 0.02% of the ocean's surface, they are believed to harbor nearly 30% of all known marine species [1–4]. Reef ecosystems provide multiple ecosystem services [5–7], including coastal protection [8,9] and the provision of food and livelihoods through fishing activities [10]. However, in recent decades, coral reef ecosystems worldwide have been facing degradation due to multiple environmental and human stressors [11,12]. Currently, ~60% of the world's coral reefs are considered to be threatened [13]. In the Caribbean, by 2003, at least 80% of the coral cover registered in the 1970s was lost [14,15], while in the Atlantic, 75% of coral reefs are considered at risk [12]. This loss of coral cover has often produced a decline in the reef's

101

structural complexity, affecting associated organisms including reef fish [16]. The replacement of structurally complex coral species represents a more silent problem [17]; branching coral species have often been replaced by "weedy" species of coral, which display less rugosity and tridimensional structure, so that even if coral cover recovers, the system may possess lower structural complexity [18]. On Caribbean reefs, a considerable loss of species that provide high structural complexity, such as *Acropora palmata*, *A. cervicornis*, and *Montastrea cavernosa*, was evident from the 1970s, aggravated by mass coral mortality as the result of reef erosion caused by the sea urchin *Diadema antillarum* through an increasing prevalence of coral disease and by coral bleaching events [17,19]. Such declines in structural complexity on coral reefs can have negative consequences, including loss of ecological function, biodiversity, and ecosystem services [2,7,20]. However, it is essential to identify, beyond structural complexity, other features of reef ecosystems which are involved in the maintenance of essential functions and processes [21].

Herbivory is one of the key processes in coral reefs that can prevent the overgrowth of corals by algae through food web control over turf and macroalgae [11,22,23]. Herbivorous fishes are considered essential to reef ecosystems' health, especially after the radical decline in the Caribbean of other large herbivores, including sea turtles and manatees [24], and previously dense populations of *D. antillarum* [25]. Among herbivorous fish, parrotfishes (Family: Scaridae) are considered important in promoting the recovery of coral reefs after disturbances (i.e., bleaching events or storms) and in preventing phase shifts from coral to algal-dominated communities [16,26–28]. Notably, in the Caribbean, some Marine Protected Areas (MPAs) have been found to possess high parrotfish biomass, low macroalgae cover, and high coral cover, with abundant coral recruits [28,29]. However, parrotfish are high among the fisheries' target species, crucial for most Caribbean economies [23]. Fishing has been found to deplete many coral reef fish species, promoting reef degradation [30–32].

The Seaflower Biosphere Reserve (Seaflower BR) is one of the largest protected areas in the Caribbean, covering 180,000 km$^2$ and incorporating a large percentage of Colombia's coral reefs [33]. Although the Seaflower BR is an extensive MPA and experiences oceanic conditions that might be expected to support pristine coral reefs with high structural complexity and biodiversity, the archipelago islands nevertheless experience illegal fishing, including overfishing, together with other anthropogenic pressures [23]. Of the nine islands in the archipelago, Serranilla, northwest of the Seaflower BR, is furthest from the mainland and it was here that recent research efforts were focused during the "Seaflower Research Expedition 2017". We designed our study to evaluate the status of fish assemblages in relation to reef habitat structural complexity, paying particular attention to comparing the parrotfish species of Serranilla from that found at San Andrés.

## 2. Materials and Methods

### 2.1. Study Site

Serranilla island (15°50′ N, 079°50′ W), officially named Isla Cayos de Serranilla, is located at the northwest end of the Seaflower Biosphere Reserve (Figure 1). Seaflower BR, with an area of 180,000 km$^2$, was declared as a Biosphere Reserve by UNESCO in 2000; within this larger area, the Colombian Government designated 65,000 km$^2$ as Marine Protected Area in 2005 [33]. In fact, Serranilla is not a single island, but a reef complex formed on a submarine mountain, on which emerged cays are scattered. The principal cay is Beacon Cay, located in the south east of the complex (Figure 1c). The underwater seascape consists of a carbonate platform with reef habitats such as reef crests, coral patches, sand flats, hardground areas, and also seagrass beds [34]. This island is located a long way (>400 km) from the most populated island of the archipelago, San Andrés, and lies only 320 km south of Jamaica (Figure 1a,b; Figure S1). San Andrés, in contrast, has a terrestrial area of 27 km$^2$, with a maximum length of 12.6 km and maximum width of 3 km. On the west side of San Andrés there is a submarine hard bottom platform extending seawards for 200–500 m, and a mainly continuous rocky shore, but no barrier reef. On the east side of the island, the platform extends seawards for more than 2 km with some soft sand bottom areas, small coral patches, mangroves, and sandy beaches on the coast, and, at the east edge of the northern part, a discontinuous barrier reefs of about 9 km in length [33].

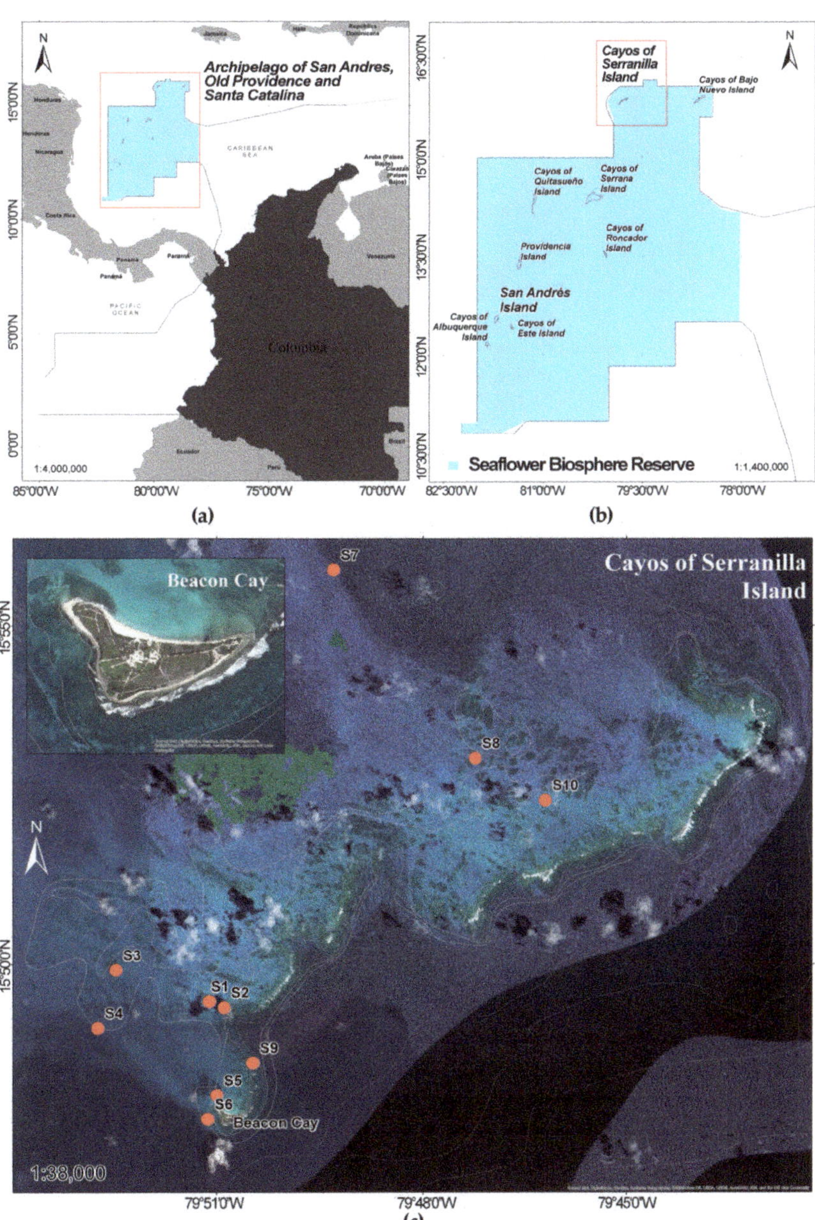



**Figure 1.** Maps showing the position of the study area: (**a**) the location of the Seaflower BR in the south west of the Caribbean, (**b**) map of the Seaflower BR showing the San Andres archipelago, including San Andrés, Old Providence, and Serranilla island, and (**c**) the locations of the survey sites at Serranilla (red points: S1–S10) together with (inserted) an aerial image of Beacon Cay, the largest cay on Serranilla (top left).

*2.2. Data Collection*

The surveys were conducted at ten sites at Serranilla during the "Seaflower Research Expedition 2017". Ten sites were selected to capture the heterogeneity of Serranilla (Table S1). At each site, habitat structure (benthic components and rugosity) and fish assemblage structure and composition were assessed using underwater visual census (UVC) techniques conducted by a paired-diver team along 50 m × 2 m belt transects (n per site = 2 or 3). On each transect, the fish species, abundance, and visual estimated size per individual (interval length) were recorded by the first diver, following established sampling methodologies [35–37]. To highlight the specific relationships of the key fish family of parrotfish (Scaridae) with habitat, the abundance of parrotfish according to the size was also calculated; we distinguished large species (>50.1 cm total length) and small species (<50 cm total length) based on available values from FishBase [38]. Valid species names were verified, according to Fricke et al. [39].

Habitat structure was evaluated as a combination of the percentage cover of the main benthic components and of reef rugosity. To characterize the coral community, we selected the percentage cover of five morpho-functional benthic groups: branching corals, brain corals, submassive corals, spherical corals, and algae. The variables were recorded with a video camera that was held 0.4 m above the seabed while being moved along each of the 50 transects. From each such video transect, 40 frames were selected, and 50 points in each frame were sampled to estimate the percentage of the benthic groups. Rugosity was estimated with the rugosity index (RI), obtained using the chain-link method [40]. Within each belt transect, the second diver laid down a 10 m chain over the seabed closely following the substrate contour and then measured the linear distance occupied by the chain. This was done three times within each transect. RI was then calculated as RI = 1 − d/L, where d is the horizontal distance covered by the chain, and L is the chain's true (i.e., stretched) length. This method is quick, objective, and highly replicable [36].

Fish data for San Andrés were obtained during the same sampling period as the Serranilla surveys (September 2017). Data were collected from transects at three sites on the west side (Luna Verde, Wild Life and Bajo Bonito) and one on the east side (Bahía Honda) of the island. Fish species, abundance, and estimated fish size in San Andrés were recorded following the same UVC described above for Serranilla. To compare parrotfish, we selected 12 transects for each location (Serranilla and San Andrés), all placed parallel to the coast, with comparable values of RI (0.19–0.49).

*2.3. Analysis*

We used two approaches to analyze the reef fish assemblages at Serranilla and San Andrés. The first approach aimed to characterize the relationship of fish assemblages to habitat structure. The second focused on the parrotfish assemblage, evaluating differences between Serranilla and San Andrés.

2.3.1. Relationship of Reef Fish Assemblages to Habitat Structure

For each transect, we estimated diversity, abundance, and biomass. Biomass (g m$^{-2}$) was calculated using the allometric weight equation:

$$W = a\,L^{b} \tag{1}$$

where L is the weighted mean of the estimated length for each species per transect, and a and b are the constants of length–weight obtained from FishBase [38].

We evaluated diversity using Hill's effective number of species of order 0, 1, and 2 [41], where $^{0}D$ is species richness, $^{1}D$ considers all species and their abundance, $^{2}D$ reflects the most abundant species, and $^{2/1}D$ reflects their evenness. To explore the relationships of reef fish assemblages to habitat, we performed multiple linear regressions of fish assemblage measures against the main explanatory variables that define habitat structure (benthic cover and rugosity). The preferred models were selected based on the lowest values of Akaike's

Information Criterion (AICc). The models' variables included the average parameters of the models with <2 $\Delta$AICc [42]. The *p*-value and adjusted $R^2$ of the model with all the explanatory variables and the selected variables were reported. For each model, the validity of the linear models was examined with the normality of the residuals [43].

### 2.3.2. Differences in Parrotfish Assemblages between Islands (Serranilla and San Andrés)

We evaluated variation in the metrics ($^0$D, biomass, and abundance) and composition (abundance and biomass) of the Scaridae species present between the two islands, Serranilla and San Andrés. We also examined the mean sizes observed for each species at Serranilla and contrasted these with the available information about common and maximum size obtained from FishBase [38].

Two ANOVAs based on permutations (10,000) were performed to test the differences in biomass and abundance using Euclidean matrices from fourth-root transformed data. We used a model with one factor:

$$Y_{ij} = \mu + Island_i + e_{ij} \tag{2}$$

where $\mu$ is the general mean; $Island_i$ is the factor with two levels (Serranilla and San Andrés) and 12 replicates per site, and $e_{ij}$ is the associated error.

Two permutational multivariate analyses of variance (PERMANOVA) were performed to assess parrotfish assemblage composition differences in biomass and abundance. We used the previous ANOVA model and verified the homogeneity of dispersion with the PERMDISP test and non-metric multidimensional scaling (NMDS) [44]. Similarity percentage analysis (SIMPER) were used to detect the parrotfish responsible for the dissimilarities between the islands. We plotted in the NMDS the small and large parrotfish abundance and biomass to visualize the differences in composition. For PERMANOVA, PERMDISP, SIMPER, and NMDS, we used a Bray–Curtis similarity matrix with a fourth root transformed data. Statistical analyses were performed in R and Primer v6.1 PERMANOVA+ [45,46], and plots for SIMPER results generated in SigmaPlot v11 software.

## 3. Results

A total of 8137 individual fish belonging to 68 species and 22 families were recorded across the sites on Serranilla. The families represented by the most species were: Scaridae (12 spp), Pomacentridae (9 spp), and Labridae (7 spp). Three wrasses (Labridae) were the most abundant species, representing 50% of the total fish abundance: *Halichoeres bivittatus*, *H. garnoti*, and *Thalassoma bifasciatum*. However, because of the larger sizes of individuals, the highest mean biomasses were represented by *Haemulon album*, *Melichthys niger*, and *Mulloidichthys martinicus*, which together contributed >30% of the total biomass (Table S2).

### 3.1. Relationship of Reef Fish Assemblages to Habitat Structure

The results of the multiple linear regressions supported the interpretation that at Serranilla overall fish abundance, biomass, and diversity were related to habitat structure (Table 1). Up to 66% of the variation in measures of reef fish assemblage structure was significantly explained, except for variation in biomass and evenness. Among all the measured explanatory variables, rugosity, and coral cover of branching and brain corals were most often significant. Of the diversity indices, species richness ($^0$D) was the measure with the greatest percentage of variation described (59%), in that case by only one variable (rugosity). In contrast, 66% of variation in the abundance of small parrotfish species was explained by the combination of all six measures of habitat structure (Table 1).

**Table 1.** Habitat structure drivers of reef fish diversity and abundance at Serranilla; regressions of fish assemblage measures against reef habitat measures. The results shown within each pair of lines are for (upper line) linear models using all potential explanatory variables and (lower line) for the best fit model selected using AICc criteria. *p*-values are shown for the adjusted $R^2$ estimates as well as for the entire model.

| Indices | Num. of Variables | *p*-Values of Variables in the Model | | | | | | Total Model | |
|---|---|---|---|---|---|---|---|---|---|
| | | Rugo | Bran | Sphe | Brain | Subm | Algae | $R^2_{adj}$ | *p*-Value |
| Biomass | All | 0.46 | 0.81 | 0.87 | 0.41 | 0.71 | 0.16 | 0.364 | 0.066 |
| | 5 | 0.54 | 0.34 | 0.37 | | 0.87 | 0.22 | 0.378 | 0.427 |
| Abundance | All | 0.001 | 0.94 | 0.49 | 0.59 | 0.09 | 0.51 | **0.558** | **0.01** |
| | 3 | 0.14 | | 0.56 | | 0.41 | | **0.385** | **0.016** |
| $^0$D | All | **0.03** | 0.9 | 0.81 | 0.7 | 0.94 | 0.71 | **0.463** | **0.028** |
| | 1 | **<0.001** | | | | | | **0.592** | **<0.001** |
| $^1$D | All | 0.3 | 0.83 | 0.71 | 0.98 | 0.52 | 0.14 | **0.389** | **0.034** |
| | 3 | 0.23 | | | | 0.39 | 0.06 | **0.502** | **0.003** |
| $^2$D | All | 0.36 | 0.56 | 0.54 | 0.92 | 0.51 | 0.07 | 0.357 | 0.07 |
| | 3 | **0.02** | | | | 0.73 | **0.01** | **0.459** | **0.006** |
| $^{2/1}$D | All | 0.22 | 0.23 | 0.24 | 0.49 | 0.64 | 0.12 | 0.237 | 0.156 |
| | 4 | 0.16 | 0.6 | | | 0.48 | 0.26 | 0.261 | 0.083 |
| Small parrotfish abundance | All | 0.28 | **<0.01** | 0.12 | **<0.01** | 0.082 | 0.83 | **0.661** | **0.002** |
| | 4 | 0.53 | **<0.01** | | **<0.01** | 0.49 | | **0.408** | **0.006** |
| Large parrotfish abundance | All | 0.02 | 0.15 | 0.11 | 0.77 | 0.23 | 0.61 | 0.376 | 0.06 |
| | 2 | **<0.01** | | 0.22 | | | | **0.408** | **0.006** |

Abbreviations. Rugo: rugosity; Bran: branching coral; Subm: submassive coral; Sphe: spherical coral; Brain: brain coral.

### 3.2. Parrotfish Differences between Islands (Serranilla and San Andrés)

We recorded 11 parrotfish species present within one or both of the two islands, 11 on Serranilla and six on San Andrés (*Scarus coelestinus*, *S. coeruleus*, *S. guacamaia*, *S. iseri*, *S. taeniopterus*, *S. vetula*, *Sparisoma atomarium*, *S. aurofrenatum*, *S. chrysopterum*, *S. rubripinne*, *S. viride*). *S. viride* was the species with both the highest abundance and biomass on Serranilla, and on San Andrés, *S. taeniopterus* and *S. aurofrenatum* had the highest abundance and biomass, respectively (Table S3). Comparison of metrics between Serranilla and San Andrés showed that there was no statistical difference in overall abundance (Pseudo-*F* = 2.137, *p* = 0.146) or species richness per transect ($^0$D; Pseudo-*F* = 2.057, *p* = 0.105) between the two island areas, but there was a statistically significant difference in parrotfish biomass per transect (Pseudo-*F* = 4.637; *p* = 0.040).

Nevertheless, parrotfish assemblage composition did appear significantly different between the two islands when examined both in terms of abundance (Pseudo-*F* = 16.031, *p* < 0.001) and in terms of biomass (Pseudo-*F* = 12.170, *p* < 0.001) with NMDS plots; these showed, in relation to both measures, a strong spatial segregation between sites on the two islands and a notably higher dispersion among the Serranilla sites (Figure 2a,b). As shown by the sizes of the bubbles, at Serranilla, most of the parrotfish contributing to abundance and biomass were larger species, whereas at San Andrés these were mainly small-size parrotfish (Figure 2c–f).

The SIMPER analysis showed that just three species were responsible for >50% of island dissimilarity, two of which had higher abundance and biomass in San Andrés (*S. taeniopterus* and *S. aurofrenatum*), and a third which had a considerably higher abundance in Serranilla (*S. viride*) (Figure 3; Table S4). Overall, the mean parrotfish size of each species was considerably smaller than the common and maximum lengths reported in FishBase (Figure S2).

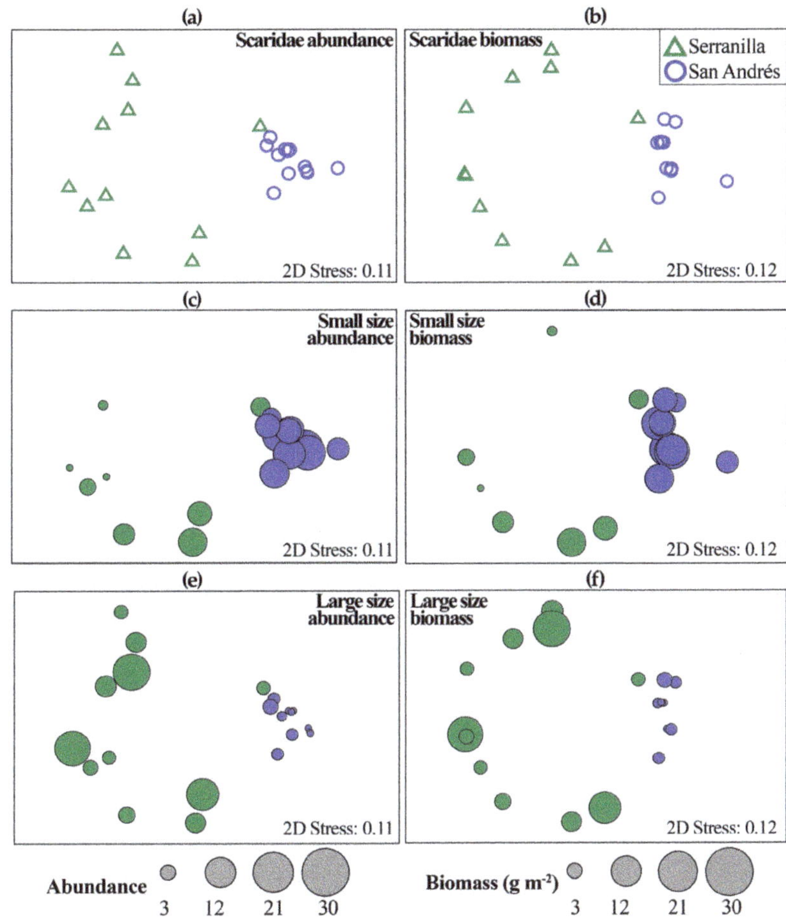



**Figure 2.** Non-metric multidimensional scaling (NMDS) plots of parrotfish abundance and biomass at sites at Serranilla and at San Andrés. Plots of abundance are shown for (**a**) all parrotfish, (**c**) small-size parrotfish (<50 cm), and (**e**) large-size parrotfish (>50.1 cm). Corresponding plots of biomass are shown for (**b**) all parrotfish, (**d**) small-size parrotfish, and (**f**) large-size parrotfish. Bubble sizes indicate the abundance or biomass values for each site.

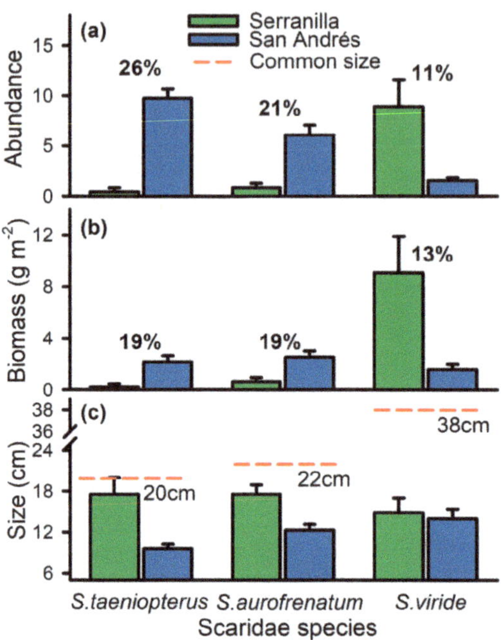



**Figure 3.** Comparison of mean (**a**) abundance, (**b**) biomass, and (**c**) size at Serranilla and at San Andrés of the three Scaridae species that together contribute at least 50% to the accumulated dissimilarity between these two study areas, as estimated by the SIMPER analysis. The contributions of the individual species are shown adjacent to the bars in (**b**). The error bars are ±SE. The red lines and written values in the lower part indicate the common size of each species, as reported from FishBase [38].

## 4. Discussion

Our study demonstrates that variation in reef fish diversity, biomass, and abundance in Serranilla can be partly explained by two benthic components, coral cover and rugosity, that contribute to habitat complexity in coral reefs [18,19]. Nevertheless, the mean values of parrotfish biomass and length at Serranilla were low, which could reflect the effect of anthropogenic pressure in the area, mainly through illegal fishing [30–32].

Structural complexity in coral reefs has been evaluated for its ecological relevance, enhancing abundance, biomass, and biodiversity [47,48]. Multiple studies have found that the complexity of healthy coral reefs favors coastal protection, reducing the effects of wave action and other environmental phenomena and maintaining shoreline equilibrium [9,49,50]. A relationship of fish assemblage with the structural complexity, such as we observed at Serranilla in the Seaflower BR, has also been observed at other locations in the Atlantic [51,52] and in other regions of the world [53,54].

Reef habitat structures provide microhabitats and refuge areas that benefit multiple organisms, including reef fish [55,56], with different rugosity levels favoring multiple fish groups [52]. Loss of habitat complexity can negatively affect multiple species [47,57], including reef fish [58], leading to species losses [11] and localized extinctions [59], including species of commercial interest such as parrotfish. Rogers et al. [52] predicted that a loss of complexity could cause a more than three-fold reduction in fishery productivity. Therefore, preserving complexity in coral reefs is crucial, particularly since structural complexity is currently in rapid decline in the Caribbean [17], with one of the largest protected areas in the Caribbean, the Seaflower BR, remaining highly vulnerable to anthropogenic pressure [33].

Coral reef degradation may also result from the loss of critical groups, such as the parrotfish. Their loss, through increased in macroalgae, decreased coral cover, and subsequently a diminished habitat complexity, may lead to a decline in associated organisms [22,60]. At Serranilla it was found that transects closer to Beacon Cay showed lower rugosity and diversity of reef fish. Although greater habitat complexity may have helped maintain higher abundance and biomass on Serranilla's reefs compared with those on San Andrés, direct anthropogenic pressure may still limit any positive effect, and lead to a lower abundance and diversity of reef fish, including parrotfish [22,46,47,52].

Anthropogenic pressure is among the most recurrent threats to coastal ecosystems, since human populations reside nearby and depend on coastal areas for commerce and survival [61,62]. Studies have found a relationship between reef fish assemblages and the distance from dense human populations [63], and found that sites closer to human populations tend to have lower abundance, biomass, and richness of reef fish than areas that are more remote, with an exponential increase in reef fish biomass as the distance from the significant human population increases [64,65].

Considering that Serranilla is far from any significant human population, we expected to find higher biomasses of reef fish, particularly of key coral reef groups such as the parrotfish. Beacon Cay, Serranilla, has only around 12 permanent inhabitants (equivalent to 9.2 inhabitants per 0.1 km$^2$), while San Andrés has more than 65,000 (equivalent to 241 inhabitants per 0.1 km$^2$ [33]. Additionally, more than 730,000 tourists visit San Andrés each year [65]. However, our data showed that despite the high reef rugosity, most parrotfish at Serranilla had low biomass and abundance values, similar to those found at San Andrés.

Differences in the size and abundance of some parrotfish species between Serranilla and San Andrés seem likely due to a lower fishing pressure at Serranilla. Nevertheless, fishing pressure may have modified the parrotfish assemblages at Serranilla. On San Andrés, fishing is one of the most important commercial activities, and includes spearfishing targeting multiple parrotfish species, such as *S. coeruleus*, *S. coelestinus*, and *S. guacamaia* [66], that were rarely observed during this study. While permitted fishing activity is mainly artisanal, illegal vessels have been reported carrying tons of illegally caught parrotfish [67]. In addition, Bruckner [34] reported illegal fishing boats at Serranilla island during the "Global Reef Expedition" in 2011, and such boats were also observed during the present "Seaflower Research Expedition" in 2017. Fishers have in recent years expressed concern over the absence of large parrotfish in the lagoon as a result of overfishing [68].

Herbivorous fish biomass has been used as one of the indicators for evaluating reef health. In the report on Mesoamerican reefs by the Healthy Reefs Initiative, they recorded 270 kg ha$^{-1}$ of herbivorous fish on the healthier reefs compared with 83 kg ha$^{-1}$ on reefs in critical conditions [69]. In particular, a relationship has been reported between fishing pressure and parrotfish biomass and size, making parrotfish abundance a useful biological indicator of fishing pressure [63]. On a small island in the Caribbean, exploited reefs, with areas open to artisanal fisheries, were found to have a parrotfish biomass of around 320 kg ha$^{-1}$, with smaller parrotfish contributing most of this biomass (~200 kg ha$^{-1}$).

We found that Serranilla BR had larger parrotfish than San Andrés, though the biomass in Serranilla BR was nevertheless lower (143 kg·ha$^{-1}$) than those values reported from other Caribbean MPAs with adequate protection and low local human populations, in which parrotfish biomass values can exceed 500 kg·ha$^{-1}$ [70]. In contrast, in areas with high local human population densities and inadequate protection, parrotfish biomass can be rapidly reduced to 50 kg·ha$^{-1}$ or less [22,29]. However, other factors may also influence herbivorous fish abundance; for example, on the Alacranes reef, an MPA in the southern part of the Gulf of Mexico, where human activities are restricted, and fishing pressure is low, the values of parrotfish biomass were nevertheless relatively low (~58–128 kg h$^{-1}$), although the parrotfish present had larger sizes than those observed on Caribbean reefs [71].

As with many other species, small individuals have become a target of fisheries due to the loss of larger individuals and the decline of other commercial fish species more generally, leading to decreased sizes [67]. Parrotfish size is likely related to reef resilience [23,57], since larger individuals can remove more algae from the reef and have more effect on benthic communities than smaller ones [24,29]. In the present study, we found that 90% of the parrotfish species at Serranilla had smaller mean sizes than the typical size reported in FishBase [38]. Notably, the observed mean size of *S. viride* and *S. chrysopterum* were both below the size at first maturity (L50), as recorded in the Colombian Caribbean [72], suggesting a higher fishing pressure at Serranilla than expected.

### 5. Conclusions

This study on Serranilla island, which is part of the Seaflower BR, has demonstrated that the overall reef fish assemblage is spatially variable with habitat structure explaining up to 66% of this variation [53,57]. In addition, focusing on the parrotfish and their sizes has allowed us to compare sizes and biomasses of fish on Serranilla with those on San Andrés. In combination, the results presented here, together with the intense illegal fishing activities observed in the reserve and other drivers not considered in this study, highlight the need to evaluate other factors that may influence reef fish biodiversity beyond structural complexity [26,73]. Meanwhile biological and ecological information about the current status of most of the key groups in the Seaflower BR remains scarce [67]. Maintaining healthy reef ecosystems is becoming critical, especially in areas where human well-being depends on coral reef ecosystem services, such as food production [73], so that the need for effective management and active surveillance at the Serranilla MPA is urgent.

**Supplementary Materials:** The following are available online at https://www.mdpi.com/article/10.3390/oceans2030034/s1, Figure S1: Serranilla Island and distances between islands (San Andres 415 km, Providencia 309 km, Jamaica 305 km), Figure S2: Mean observed fish sizes during census at Serranilla Island (SI) compared with the common and maximum (Max) reported in FishBase, Table S1: Site summary of the metrics of fish assemblages in Serranilla Island with average value per metric ± SD, Table S2: Summary of fish assemblage families and species in Serranilla Island. Values of abundance and biomass with the total sum, frequency, and average ± SD, Table S3: Summary of the abundance and biomass of the Scaridae species in Serranilla and San Andrés with the average values ± SD, Table S4: SIMPER analysis summary. Scaridae average contribution and cumulative contribution percentage in biomass and abundance to the dissimilarity between SI and SAI. Registered and common length of those species.

**Author Contributions:** Conceptualization, D.C., D.M.-d.-A., J.P., A.L.C.-M. and A.S.-M.; Data curation, D.C., J.P. and J.P.E.; Formal analysis, D.C., D.M.-d.-A. and J.P.; Funding acquisition, A.S.-M., D.C. and J.P.; Investigation, D.C., D.M.-d.-A., J.P., A.L.C.-M. and A.S.-M.; Methodology, D.C., D.M.-d.-A., J.P., A.L.C.-M., J.P.E. and A.S.-M.; Supervision, A.L.C.-M. and A.S.-M.; Validation, A.L.C.-M. and A.S.-M.; Visualization, D.C., D.M.-d.-A., J.P. and J.P.E.; Writing—original draft, D.C., D.M.-d.-A., J.P. and A.L.C.-M.; Writing—review and editing, D.C., D.M.-d.-A., J.P., A.L.C.-M., J.P.E. and A.S.-M. All authors have read and agreed to the published version of the manuscript.

**Funding:** This research was developed thanks to the funding of the Universidad Nacional de Colombia Sede Caribe on the frame of Seaflower Research Expedition 2017, supported by multiple institutions, including essential contributions from Colombia BIO–Colciencias project and Colombian Ocean Commission (CCO), Project management contract No. FP4484-398-2015, Armada Nacional de Colombia, Corporación para el Desarrollo Sostenible del departamento (Coralina), Gobernación de San Andrés, Providencia y Santa Catalina, and Universidad Nacional de Colombia Sede Caribe. The project under which this research was conducted is "Valoración de servicios ecosistémicos de los arrecifes de coral en los alrededores de la Isla Cayo Serranilla, Reserva de Biosfera Seaflower, Caribe Colombiano," developed and supported by Universidad Nacional de Colombia Sede Caribe. Additional funding included Colciencias Ph.D. scholarship (Conv. 757) and CEMarin (Call 14, 2018) that funded the maintenance of the CEMarin young researcher Julián Prato and the project "Relationships between coral reef complexity and ecosystem services at Caribbean oceanic islands, Seaflower Biosphere Reserve, Colombia". Universidad Nacional de Colombia Sede Caribe funded fish census at San Andrés to Diana Castaño MSc. thesis project "Estructura y Función de peces herbívoros en zonas arrecifales de San Andrés una isla oceánica en el Caribe".

**Data Availability Statement:** The datasets generated during the current study are available within the article and its supplementary materials. Additional data are available from the authors on reasonable request.

**Acknowledgments:** We express our special thanks to the Universidad Nacional de Colombia Sede Caribe for supporting our participation in the Seaflower Research Expedition 2017 "Cayos de Serranilla Island." We thank Armada Nacional de Colombia, Corporación para el Desarrollo Sostenible del Archipiélago de San Andrés, Providencia y Santa Catalina (Coralina), Nacor Bolaños, the crew of the ARC 20 de Julio vessel and Captain and fisherman Casimiro Newball for their support during sailing and sampling at the expedition. We also thank personnel from the Colombian Ocean Commission (CCO) specially to VALM Juan Manuel Soltau, CF Hurtado, CN. Guerra and Juliana Sintura for their important work to make the expedition possible. Finally, we thank Venus Avendaño and Alicia del Mar PC for their multiple support and Omar Abril Howard, Brigitte Gavio, and Violeta Posada for contributing to a successful field working team. We also thank two anonymous referees and the Special Issue editor for their great assistance in improving the manuscript.

**Conflicts of Interest:** The authors declare no conflict of interest. The funders had no role in the design of the study; in the collection, analyses, or interpretation of data; in the writing of the manuscript, or in the decision to publish the results.

# References


1. Moritz, C.; Vii, J.; Lee Long, W.; Tamelander, J.; Thomassin, A.; Planes, S. Status and Trends of Coral Reefs of the Pacific. In *Global Coral Reef Monitoring Network*; Technical Report; United Nations Environment Programme: Washington, DC, USA, 2018; pp. 47–50.
2. Komyakova, V.; Munday, P.L.; Jones, G.P. Relative Importance of Coral Cover, Habitat Complexity and Diversity in Determining the Structure of Reef Fish Communities. *PLoS ONE* **2013**, *8*, e83178. [CrossRef] [PubMed]
3. Knowlton, N.; Brainard, R.E.; Fisher, R.; Moews, M.; Plaisance, L.; Caley, M.J. Coral reef biodiversity. In *Life in the World's Oceans: Diversity Distribution and Abundance*, 1st ed.; McIntyre; Blackwell Publishing Ltd.: West Sussex, UK, 2010; pp. 65–78.
4. Plaisance, L.; Caley, M.J.; Brainard, R.E.; Knowlton, N. The diversity of coral reefs: What are we missing? *PLoS ONE* **2011**, *6*, e25026. [CrossRef] [PubMed]
5. Woodhead, A.J.; Hicks, C.C.; Norström, A.V.; Williams, G.J.; Graham, N.A.J. Coral reef ecosystem services in the Anthropocene. *Funct. Ecol.* **2019**, *33*, 1023–1034. [CrossRef]
6. Costanza, R.; de Groot, R.; Sutton, P.; van der Ploeg, S.; Anderson, S.J.; Kubiszewski, I.; Farber, S.; Turner, R.K. Changes in the Global Value of Ecosystem Services. *Glob. Environ. Chang.* **2014**, *26*, 152–158. [CrossRef]
7. Waite, R.; Burke, L.; Gray, E. *Coastal Capital: Ecosystem Valuation for Decision Making in the Caribbean*; World Resources Institute: Washington, DC, USA, 2014; pp. 1–78.
8. Elliff, C.I.; Silva, I.R. Coral Reefs as the First Line of Defense: Shoreline Protection in Face of Climate Change. *Mar. Environ. Res.* **2017**, *127*, 148–154. [CrossRef]
9. Reguero, B.G.; Beck, M.W.; Agostini, V.N.; Kramer, P.; Hancock, B. Coral Reefs for Coastal Protection: A New Methodological Approach and Engineering Case Study in Grenada. *J. Environ. Manag.* **2018**, *210*, 146–161. [CrossRef]
10. Samonte-Tan, G.P.B.; White, A.T.; Tercero, M.A.; Diviva, J.; Tabara, E.; Caballes, C. Economic Valuation of Coastal and Marine Resources: Bohol Marine Triangle, Philippines. *Coast. Manag.* **2007**, *35*, 319–338. [CrossRef]
11. Hughes, T.P.; Baird, A.H.; Bellwood, D.R.; Card, M.; Connolly, S.R.; Folke, C.; Grosberg, R.; Hoegh-Guldberg, O.; Jackson, J.B.C.; Kleypas, J.; et al. Climate Change, Human Impacts, and the Resilience of Coral Reefs. *Science* **2003**, *301*, 929–933. [CrossRef]
12. Burke, L.; Spalding, M.D.; Perry, A. *Reefs at Risk Revisited*; World Resources Institute: Washington, DC, USA, 2011; p. 114.
13. Pandolfi, J.M.; Connolly, S.R.; Marshall, D.J.; Cohen, A.L. Projecting Coral Reef Futures under Global Warming and Ocean Acidification. *Science* **2011**, *333*, 418–422. [CrossRef]
14. Gardner, T.A.; Côté, I.M.; Gill, J.A.; Grant, A.; Watkinson, A.R. Long-Term Region-Wide Declines in Caribbean Corals. *Science* **2003**, *301*, 958–960. [CrossRef]
15. Burke, L.; Maidens, J. *Reefs at Risk in the Caribbean*; World Resources Institute: Washington, DC, USA, 2004; p. 80.
16. Mumby, P.J.; Wolff, N.H.; Bozec, Y.M.; Chollett, I.; Halloran, P. Operationalizing the Resilience of Coral Reefs in an Era of Climate Change. *Conserv. Lett.* **2014**, *7*, 176–187. [CrossRef]
17. Álvarez-Filip, L.; Dulvy, N.K.; Gill, J.A.; Côté, I.M.; Watkinson, A.R. Flattening of Caribbean Coral Reefs: Region-Wide Declines in Architectural Complexity. *Proc. Royal Soc. B* **2009**, *276*, 3019–3025. [CrossRef] [PubMed]
18. Álvarez-Filip, L.; Carricart-Ganivet, J.P.; Horta-Puga, G.; Iglesias-Prieto, R. Shifts in Coral-Assemblage Composition Do Not Ensure Persistence of Reef Functionality. *Sci. Rep.* **2013**, *3*, 1–15. [CrossRef] [PubMed]
19. Zea, S.; Geister, J.; Garzón-Ferreira, J.; Díaz, J.M. Biotic Changes in the Reef Complex of San Andres Island (Southeastern Caribbean Sea, Columbia) Occuring over Three Decades. *Atoll. Res. Bull.* **1998**, *456*, 1–30. [CrossRef]
20. Guannel, G.; Arkema, K.; Ruggiero, P.; Verutes, G. The Power of Three: Coral Reefs, Seagrasses and Mangroves Protect Coastal Regions and Increase Their Resilience. *PLoS ONE* **2016**, *11*, e0158094. [CrossRef] [PubMed]

21. Pendleton, L.H.; Hoegh-Guldberg, O.; Langdon, C.; Comte, A. Multiple Stressors and Ecological Complexity Require a New Approach to Coral Reef Research. *Front. Mar. Sci.* **2016**, *3*, 36. [CrossRef]

22. Mumby, P.J. The impact of exploiting grazers (Scaridae) on the dynamics of Caribbean coral reefs. *Ecol. Appl.* **2006**, *16*, 747–769. [CrossRef]

23. Jackson, J.; Donovan, M.; Cramer, K.; Lam, V. *Status and Trends of Caribbean Coral Reefs*; Global Coral Reef Monitoring Network: Washington, DC, USA, 2014; p. 304.

24. Jackson, J.B.C. Reefs since Columbus. *Coral Reefs* **1997**, *16*, S23–S32. [CrossRef]

25. Carpenter, R.C. Mass Mortality of Diadema Antillarum II. Effects on Population Densities and Grazing Intensity of Parrotfishes and Surgeonfishes. *Mar. Biol.* **1990**, *104*, 79–86. [CrossRef]

26. Burkepile, D.E.; Hay, M.E. Herbivore Species Richness and Feeding Complementarity Affect Community Structure and Function on a Coral Reef. *Proc. Natl. Acad. Sci. USA* **2008**, *105*, 16201–16206. [CrossRef]

27. Adam, T.C.; Schmitt, R.J.; Holbrook, S.J.; Brooks, A.J.; Edmunds, P.J.; Carpenter, R.C.; Bernardi, G. Herbivory, Connectivity, and Ecosystem Resilience: Response of a Coral Reef to a Large-Scale Perturbation. *PLoS ONE* **2011**, *6*, e23717. [CrossRef] [PubMed]

28. Mumby, P.J.; Dahlgren, C.P.; Harborne, A.R.; Kappel, C.V.; Micheli, F.; Brumbaugh, D.R.; Holmes, K.E.; Mendes, J.M.; Broad, K.; Sanchirico, J.N.; et al. Fishing, Trophic Cascades, and the Process of Grazing on Coral Reefs. *Science* **2006**, *311*, 98–101. [CrossRef]

29. Bonaldo, R.; Hoey, A.; Bellwood, D. The Ecosystem Roles of Parrotfishes on Tropical Reefs. *Oceanogr. Mar. Biol. Annu. Rev.* **2014**, *52*, 81–132.

30. Bellwood, D.R.; Hughes, T.P.; Folke, C.; Nyströ, M. Confronting the Coral Reef Crisis. *Nature* **2004**, *429*, 827–833. [CrossRef]

31. Hawkins, J.P.; Roberts, C.M. Effects of Artisanal Fishing on Caribbean Coral Reefs. *Conserv. Biol.* **2004**, *18*, 215–226. [CrossRef]

32. Bozec, Y.M.; O'Farrell, S.; Bruggemann, J.H.; Luckhurst, B.E.; Mumby, P.J. Tradeoffs between Fisheries Harvest and the Resilience of Coral Reefs. *Proc. Natl. Acad. Sci. USA* **2016**, *113*, 4536–4541. [CrossRef] [PubMed]

33. Gómez-López, D.I.; Segura-Quintero, C.; Sierra-Correa, P.C.; Garay-Tinoco, J. *Atlas de La Reserva de Biósfera Seaflower. Archipiélago de San Andrés, Providencia y Santa Catalina*; Serie de Publicaciones Especiales; INVEMAR: Bogotá, Colombia, 2012; p. 180.

34. Bruckner, A. *Global Reef Expedition: San Andrés Archipelago, Colombia*; Field Report; Khaled bin Sultan Living Oceans Foundation: Landover, MD, USA, 2012; p. 52.

35. Samoilys, M.A.; Carlos, G. Determining Methods of Underwater Visual Census for Estimating the Abundance of Coral Reef Fishes. *Environ. Biol. Fishes* **2000**, *57*, 289–304. [CrossRef]

36. World Wild Fund (WWF). Mejores prácticas de pesca en arrecifes coralinos. In *Guía para la Colecta de Información que Apoye el Manejo de Pesquerías Basado en Ecosistemas*; WWF: Cancún, México; San José, Costa Rica, 2006; p. 81.

37. Caldwell, Z.R.; Zgliczynski, B.J.; Williams, G.J.; Sandin, S.A. Reef Fish Survey Techniques: Assessing the Potential for Standardizing Methodologies. *PLoS ONE* **2016**, *11*, e0153066. [CrossRef]

38. Froese, R.; Pauly, D. Fish Base. World Wide Web Electronic Publication. Available online: www.fishbase.org (accessed on 12 December 2019).

39. Fricke, R.; Eschmeyer, W.N.; Fong, J.D. Species by Family/Subfamily. Eschmeyer's Catalog of Fishes. Available online: www.researcharchive.calacademy.org/research/ichthyology/catalog/SpeciesByFamily.asp (accessed on 21 December 2019).

40. Risk, M.J. Fish Diversity on a Coral Reef in the Virgin Islands. *Atoll. Res. Bull.* **1972**, *153*, 1–6. [CrossRef]

41. Jost, L. Entropy and Diversity. *Oikos* **2006**, *113*, 363–375. [CrossRef]

42. Burnham, K.P.; Anderson, D.R. Multimodel inference: Understanding AIC and BIC in model selection. *Sociol. Methods Res.* **2004**, *33*, 261–304. [CrossRef]

43. Zar, J.H. *Biostatistical Analysis*, 5th ed.; Pearson Prentice-Hall: New Jersey, NJ, USA, 2010; p. 944.

44. Anderson, M.; Gorley, R.; Clarke, K.P. *For PRIMER: Guide to Software and Statistical Methods*; PRIMER-E: Plymouth, UK, 2008.

45. Clarke, K.R.; Gorley, R.N. *Primer v6 Permanova+*; PRIMER-E: Plymouth, UK, 2006.

46. Team, R. *A Language and Environment for Statistical Computing*; R Foundation for Statistical Computing: Vienna, Austria, 2018.

47. Harborne, A.R.; Mumby, P.J.; Ferrari, R. The Effectiveness of Different Meso-Scale Rugosity Metrics for Predicting Intra-Habitat Variation in Coral-Reef Fish Assemblages. *Environ. Biol. Fishes* **2012**, *94*, 431–442. [CrossRef]

48. Richardson, L.E.; Graham, N.A.J.; Pratchett, M.S.; Hoey, A.S. Structural Complexity Mediates Functional Structure of Reef Fish Assemblages among Coral Habitats. *Environ. Biol. Fishes* **2017**, *100*, 193–207. [CrossRef]

49. Van Zanten, B.T.; van Beukering, P.J.H.; Wagtendonk, A.J. Coastal Protection by Coral Reefs: A Framework for Spatial Assessment and Economic Valuation. *Ocean Coast. Manag.* **2014**, *96*, 94–103. [CrossRef]

50. Monismith, S.G.; Rogers, J.S.; Koweek, D.; Dunbar, R.B. Frictional Wave Dissipation on a Remarkably Rough Reef. *Geophys. Res. Lett.* **2015**, *42*, 4063–4071. [CrossRef]

51. Walker, B.K.; Jordan, L.K.B.; Spieler, R.E. Relationship of Reef Fish Assemblages and Topographic Complexity on Southeastern Florida Coral Reef Habitats. *J. Coast. Res.* **2009**, *53*, 39–48. [CrossRef]

52. Rogers, A.; Blanchard, J.L.; Mumby, P.J. Vulnerability of Coral Reef Fisheries to a Loss of Structural Complexity. *Curr. Biol.* **2014**, *24*, 1000–1005. [CrossRef] [PubMed]

53. Friedlander, A.M.; Parrish, J.D. Habitat Characteristics Affecting Fish Assemblages on a Hawaiian Coral Reef. *J. Exp. Mar. Biol. Ecol.* **1998**, *224*, 1–30. [CrossRef]

54. González-Rivero, M.; Harborne, A.R.; Herrera-Reveles, A.; Bozec, Y.M.; Rogers, A.; Friedman, A.; Ganase, A.; Hoegh-Guldberg, O. Linking Fishes to Multiple Metrics of Coral Reef Structural Complexity Using Three-Dimensional Technology. *Sci. Rep.* **2017**, *7*, 1–15.
55. Kuffner, I.B.; Brock, J.C.; Grober-Dunsmore, R.; Bonito, V.E.; Hickey, T.D.; Wright, C.W. Relationships between Reef Fish Communities and Remotely Sensed Rugosity Measurements in Biscayne National Park, Florida, USA. *Environ. Biol. Fishes* **2007**, *78*, 71–82. [CrossRef]
56. Graham, N.A.J.; Nash, K.L. The Importance of Structural Complexity in Coral Reef Ecosystems. *Coral Reefs* **2013**, *32*, 315–326. [CrossRef]
57. Plass-Johnson, J.G.; Ferse, S.C.A.; Jompa, J.; Wild, C.; Teichberg, M. Fish Herbivory as Key Ecological Function in a Heavily Degraded Coral Reef System. *Limnol. Oceanogr.* **2015**, *60*, 1382–1391. [CrossRef]
58. Paddack, M.J.; Reynolds, J.D.; Aguilar, C.; Appeldoorn, R.S.; Beets, J.; Burkett, E.W.; Chittaro, P.M.; Clarke, K.; Esteves, R.; Fonseca, A.C.; et al. Dynamic Fragility of Oceanic Coral Reef Ecosystems. *Proc. Natl. Acad. Sci. USA* **2006**, *103*, 8425–8429.
59. Pratchett, M.S.; Munday, P.; Wilson, S.K.; Graham, N.A.; Cinner, J.E.; Bellwood, D.R.; Jones, G.P.; Polunin, N.V.; McClanahan, T.R. Effects of Climate-Induced Coral Bleaching on Coral-Reef Fishes. Ecological and economic consequences. *Oceanogr. Mar. Bio Annu. Rev.* **2008**, *46*, 251–296.
60. Jackson, J.B.C.; Kirby, M.X.; Berger, W.H.; Bjorndal, K.A.; Botsford, L.W.; Bourque, B.J.; Bradbury, R.H.; Cooke, R.; Erlandson, J.; Estes, J.A. Historical Overfishing and the Recent Collapse of Coastal Ecosystems. *Science* **2001**, *293*, 629–637. [CrossRef] [PubMed]
61. Pandolfi, J.M.; Bradbury, R.H.; Sala, E.; Hughes, T.P.; Bjorndal, K.A.; Cooke, R.G.; McArdle, D.; McClenachan, L.; Newman, M.J.H.; Paredes, G.; et al. Global Trajectories of the Long-Term Decline of Coral Reef Ecosystems. *Science* **2003**, *301*, 955–958. [CrossRef] [PubMed]
62. Vallès, H.; Gill, D.; Oxenford, H.A. Parrotfish Size as a Useful Indicator of Fishing Effects in a Small Caribbean Island. *Coral Reefs* **2015**, *34*, 789–801. [CrossRef]
63. Cinner, J.E.; Graham, N.A.J.; Huchery, C.; Macneil, M.A. Global Effects of Local Human Population Density and Distance to Markets on the Condition of Coral Reef Fisheries. *Conserv. Biol.* **2013**, *27*, 453–458. [CrossRef]
64. Brewer, T.D.; Cinner, J.E.; Fisher, R.; Green, A.; Wilson, S.K. Market Access, Population Density, and Socioeconomic Development Explain Diversity and Functional Group Biomass of Coral Reef Fish Assemblages. *Glob. Environ. Chang.* **2012**, *22*, 399–406. [CrossRef]
65. Prato, J.A.; Newball, R. *Aproximación a La Valoración Económica Ambiental Del Departamento Archipiélago de San Andrés, Providencia y Santa Catalina*; Comisión Colombiana del Océano: Bogotá, Colombia, 2016.
66. Chasqui, V.; Polanco, L.A.; Acero, A.; Mejía-Falla, P.A.; Navia, A.; Zapata, L.A.; Caldas, J.P. *Libro Rojo de Peces Marinos de Colombia*; Instituto de Investigaciones Marinas y Costeras Invemar, Ministerio de Ambiente y Desarrollo Sostenible: Santa Marta, Colombia, 2017; p. 552.
67. Friedlander, A.; Nowlis, J.S.; Sanchez, J.A.; Appeldoorn, R.; Usseglio, P.; Mccormick, C.; Bejarano, S.; Mitchell-Chui, A. Designing Effective Marine Protected Areas in Seaflower Biosphere Reserve, Colombia, Based on Biological and Sociological Information. *Conserv. Biol.* **2003**, *17*, 1769–1784. [CrossRef]
68. McField, M.; Kramer, P.; Giró Petersen, A.; Soto, M.; Drysdale, I.; Craig, N.; Rueda-Flores, M. Mesoamerican Reef Report Card, Healthy Reefs Initiative. Available online: www.healthyreef.org (accessed on 28 June 2021).
69. Bruggemann, J.H.; van Rooij, J.M.; Videler, J.J.; Breeman, A.M. Dynamics and Limitations of Herbivore Populations on a Caribbean Coral Reef. Ph.D. Thesis, University of Groningen, Groningen, The Netherlands, 1995.
70. Hernández-Landa, R.C.; Aguilar-Perera, A. Structure and composition of surgeonfish (Acanthuridae) and parrotfish (Labridae: Scarinae) assemblages in the south of the Parque Nacional Arrecife Alacranes, southern Gulf of Mexico. *Mar. Biol.* **2019**, *49*, 647–662. [CrossRef]
71. Jaimes Rodríguez, L.I. Algunos Aspectos Biológico-Pesqueros de Las Principales Especies Ícticas Capturadas En El Sector de San Bernardo, Parque Nacional Natural Corales Del Rosario y de San Bernardo, Caribe Colombiano. Bachelor's Thesis, Universidad de Bogotá Jorge Tadeo Lozano, Bogotá, Colombia, 2011.
72. McClanahan, T.; Karnauskas, M. Relationships between Benthic Cover, Current Strength, Herbivory, and a Fisheries Closure in Glovers Reef Atoll, Belize. *Coral Reefs* **2011**, *30*, 9–19. [CrossRef]
73. Balzan, M.V.; Caruana, J.; Zammit, A. Assessing the Capacity and Flow of Ecosystem Services in Multifunctional Landscapes: Evidence of a Rural-Urban Gradient in a Mediterranean Small Island State. *Land Use Policy* **2018**, *75*, 711–725. [CrossRef]

*Article*

# Reef Fish Associations with Natural and Artificial Structures in the Florida Keys

Kara Noonan [1,*], Thomas Fair [2], Kristiaan Matthee [2], Kelsey Sox [2], Kylie Smith [2] and Michael Childress [2]

[1] Department of Ecology and Evolutionary Biology, Rice University, Houston, TX 77005, USA
[2] Department of Biological Sciences, Clemson University, Clemson, SC 29634, USA; tfair@g.clemson.edu (T.F.); kmatth6@g.clemson.edu (K.M.); kmsox@g.clemson.edu (K.S.); kylie4@g.clemson.edu (K.S.); mchildr@clemson.edu (M.C.)
[*] Correspondence: krn4@rice.edu; Tel.: +1-630-863-2230

**Abstract:** Throughout the Caribbean, coral reefs are transitioning from rugose, coral-dominated communities to flat, soft coral-dominated habitats, triggering declines in biodiversity. To help mitigate these losses, artificial structures have been used to re-create substrate complexity and support reef inhabitants. This study used natural and artificial structures to investigate the factors influencing the use of habitat by reef fish. During 2018 and 2019, divers added artificial structures and monitored the fish assemblages associating with both the artificial structures and naturally occurring corals. Overall, there were more fish on natural structures than on artificial structures. While structure shape did not influence fish use, there was a non-significant trend for increased use of larger structures. Fish observations did not differ across a gradient of shallow, complex reefs to deeper, flatter reefs; however, analyses of feeding guilds revealed clearer patterns: herbivores and omnivores were positively associated with low rugosity reefs where macroalgal abundance was higher, whereas invertivores preferred more rugose reefs. These results suggest that as reefs lose structural complexity, fish communities may become dominated by herbivores and omnivores. It also appears that the addition of artificial structures of the type used here may not mitigate the effects of structure loss on reef fish assemblages.

**Keywords:** reef ecology; reef fish; structure associations; artificial structures

**Citation:** Noonan, K.; Fair, T.; Matthee, K.; Sox, K.; Smith, K.; Childress, M. Reef Fish Associations with Natural and Artificial Structures in the Florida Keys. *Oceans* **2021**, 2, 634–647. https://doi.org/10.3390/oceans2030036

Academic Editor: Rupert Ormond

Received: 29 October 2020
Accepted: 18 August 2021
Published: 8 September 2021

**Publisher's Note:** MDPI stays neutral with regard to jurisdictional claims in published maps and institutional affiliations.

## 1. Introduction

Coral reefs are amongst the most diverse and complex ecosystems in the ocean, despite occupying less than 1% of the ocean floor [1,2]. The combination of soft and hard structures creates an architecturally complex marine habitat that is heavily utilized by numerous organisms and is considered to provide hotspots of diversity and endemism [3–6]. There are both structural and biotic components that make coral reefs multifaceted environments, including rugosity [7–13], algae [8,9,14], hard coral morphology [15,16], and emergent limestone ledges [17,18]; all of these have been described to increase reef fish species diversity.

Caribbean coral reefs have been experiencing severe degradation due to continual disturbances including, but not limited to, disease, sedimentation, and eutrophication, which, together, are eliminating the complex landscapes [19,20]. In many areas, Caribbean coral reef degradation far surpasses that of Indo-Pacific coral reefs, so the Caribbean has become the focal area for studies analyzing the response of reef fishes to this rapid loss of reef structural complexity. Long-term studies following species-specific responses to coral decline found that 43 out of the 72 fish species censused had experienced declines greater than 50% [21–29]. Other studies have found similar trends, with estimated density losses of 2.7–6.0% per year [30], and with predictive models estimating continued losses for particular functional feeding guilds such as invertivores in the years to come [31].



*Oceans* **2021**, 2, 634–647. https://doi.org/10.3390/oceans2030036          https://www.mdpi.com/journal/oceans

To combat the loss of complex coral structures, artificial reef structures (ARs) have been used to increase the physical complexity and substrate available to support reef fish communities. There have been many studies that indicated a positive effect of ARs, but these have identified that particular characteristics are necessary for this mitigation strategy to be effective [32–39]. The overall height of ARs has a significant impact on their effectiveness [36,40,41], while the size, surface area available, and complexity of ARs appear to influence the diversity of reef fish across an entire reef [40–42]. In fact, some artificial reefs have been observed to contain species assemblages that are more diverse than those of natural reefs, leading to the conclusion that this method of intervention can be successful [36,37,43–45].

Previous studies have assessed reef fish community responses to coral decline [6,24,27,29] and the use of ARs [32,33,36,37,39], and predicted how reef fish communities may be structured in the future [31], but there is little research investigating how reef fish are utilizing the structures that remain in the Caribbean. The goal of the present study was to investigate how reef fishes utilize both natural and artificial structures, identify structural characteristics that may influence their use, and assess whether reef location and topographic complexity influence the use of structure by different functional feeding guilds. Based on previous literature, we investigated the hypotheses that reef fishes would utilize biotically complex natural structures more often than non-biotic artificial structures, that height and surface area would be the most effective characteristics driving use, and that reef fishes' use of structures would be evident on reefs with higher rugosity but would differ between different feeding guilds. By identifying heavily utilized structures and their associated traits, we can better predict the reef fish community response to structural declines and assess whether artificial structures can mitigate further losses.

## 2. Materials and Methods

### 2.1. Site Selection and Substrate Survey

Field surveys were conducted in the summers of 2018 and 2019 across 8 reef sites, which varied in their distance from shore (1.62 to 8.86 km) and depth (3.0 to 8.1 m), in the middle of Florida Keys National Marine Sanctuary (Figure 1). Depths were calculated during the mid-tide transition using a depth finder on a boat when it was positioned over the middle of the reef. Each reef area was surveyed using a permanent 50 m transect that ran parallel to the primary axis of the reef, and 4 30 m transects that were laid perpendicular to and crossing the permanent transect at distances of 10, 20, 30, and 40 m, creating a 50 × 30 m grid. The substrate cover of the study area was recorded using digital photographs of 50 × 50 cm portions of the substrate, starting with 2 pictures on each side of the permanent 50 m transect at 0 m, 2 pictures again at 10 m, at 20 m, at 30 m, at 40 m, and again at 50 m (i.e., a total of 12 images per reef = 96 images for all reefs). Percent substrate cover by the major substrates (hard corals, soft corals, sponges, fleshy algae, turf algae, calcareous algae, sand) was estimated using 25 randomly selected points per photograph, with the aid of Coral Point Count with the Excel extension [46]. Rugosity was measured 3 times on each of the 4 30 m perpendicular transects using a chain and tape method. A 2 m chain was placed along the side of the 30 m transect and the beginning and end chain measurements were recorded and subtracted from each other. If a reef was flat, the chain would extend a full 2 m on the transect tape, giving an overall measurement of low rugosity. If a reef was more complex, the chain would not extend the full 2 m. The rugosity measurements for each site (3 per 30 m transect × 4 per 30 m transects = 12 rugosity measurements) were then averaged to obtain an overall rugosity across the site.

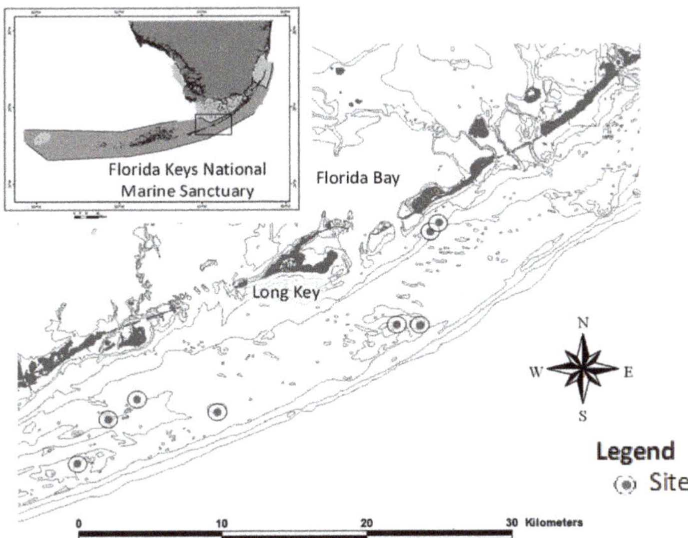


**Figure 1.** Map to show the location of the eight study sites in the middle of Florida Keys, USA, surveyed during the summers of 2018 and 2019. These eight sites are located inside the Florida Keys National Marine Sanctuary.

## 2.2. Video Transect Surveys as an Assessment of Reef Fish Abundance

Some reefs are known to have more abundant and diverse reef fish communities, and we suspected that this may affect how structures are used across a reef tract. We therefore began by establishing a baseline of abundance across the 8 sites. To evaluate the overall abundance and diversity of reef fish species within the functional feeding guilds present at the 8 sites, 8 videos (~3–4 min) were captured along each of the perpendicular 30 m transects (2 videos per transect) between the hours of 900 and 1500. A diver swam the length of the transect while holding a PVC camera frame with 2 forward-facing GoPro cameras attached at heights of 30 cm and 100 cm above the substrate, so as to capture both benthic and mid-water fish species. Each video was analyzed to identify both the fish species present and any behavioral interactions with the substrate including, but not limited to, using substrate for shelter or as a food resource. Fish species were classified into one of four functional feeding guilds (Table S1): herbivores, omnivores, invertivores, and piscivores [25,29,47]. Analysis of fish feeding guilds has been found to be adequate for comparing reef fish community structures and can distinguish functionally diverse communities from seemingly diverse communities that have functional redundancy [48–50]. Although fish of the family Haemulidae are omnivores, they were separated into their own functional feeding guild—invertivores—because of their nocturnal feeding activities and daytime use of the structure for rest.

## 2.3. Natural and Artificial Structures

On each of the 8 reef sites used for this study, 1 soft coral sea rod, and 2 healthy, 2 diseased, and 2 recently dead hard coral colonies were selected as the natural structures for further monitoring (Figure 2). Thus, a total of 8 soft coral sea rods, 16 healthy, 16 diseased, and 16 dead hard coral colonies were tracked across our 8 reef sites. The hard coral colonies included five common boulder coral species (*Colpophyllia natans, Montastraea cavernosa, Orbicella faveolata, Porites astreoides,* and *Siderastrea siderea*), whereas the soft coral was a species from the family Plexauridae. "Sea rod", "dead", "diseased", and "healthy" were used as terms to refer to these different types of natural structure. In 2018, each of these corals was tagged, photographed, and analyzed using ImageJ software to estimate the percent of live, diseased, and dead coral tissue. Estimated percent cover of dead, diseased, or live tissue was converted into surface area using the surface area formula of a half-dome ($2\,\pi r^2$), with the radius estimated as half the mean of the height, length, and width of the coral. In 2019, the resulting 56 corals were re-photographed and re-analyzed to evaluate any changes in tissue cover and surface area. Of the 16 diseased hard corals infected with Stony Coral Tissue Loss Disease (SCTLD) in 2018, 15 had survived and were classified as "healed coral colonies" for the purpose of the 2019 survey.

**Figure 2.** Reef fishes' use of structures was monitored on natural and artificial structures. Natural structures (**top**) included sea rods, and dead, actively diseased, and healthy boulder corals. Artificial structures (**bottom**) included a control and three structures that were built to mimic historically abundant natural structures including soft coral, boulder coral, and branching coral (elkhorn coral, *Acropora palmata*).

Separately from the survey of the existing corals, at each reef site, 4 artificial reef structures (ARs) were deployed, 1 of each of 4 different types, variously created out of PVC piping, concrete, rope, tomato cages, and Vexar mesh (plastic coated wire grid) (Figure 2). The 4 structures were designed to imitate different types of natural habitats. The first type consisted of just 1 single length of PVC pipe standing up from a concrete block; this was a structure that lacked physical complexity. The other artificial structures represented a soft coral (created with PVC arms and frayed rope), a boulder coral (created with Vexar mesh around a tomato cage), and a branching elkhorn coral (created with arms wrapped in Vexar extending out from the center) (Figure 2). These different types of artificial structure were referred to as "control", "soft", "boulder", and "elkhorn" respectively. These 4 artificial structures were designed to assess which characteristic (holes, surface area, edge space), if any, influenced use by the reef fish community. These artificial structures were similar in height (100 cm) and diameter (20–100 cm) to the natural boulder hard corals (height 25–200 cm and diameter 30–150 cm) and natural sea rod soft corals (height 30–125 cm and diameter 30–100 cm) present on the reef. In 2018, the artificial structures were deployed, left to acclimate for 2 weeks, surveyed, then removed from the site. In 2019, the artificial structures were redeployed for 2 weeks and surveyed again before being removed. The 2-week deployment and removing schedule was set to minimize the amount of biotic build up on the artificial structures as well as to avoid the potential for reef damage from breakage during the hurricane season.

*2.4. Reef Fish Observations on Individual Structures*

Reef fishes' use of each individual structure was estimated using time-lapsed video-photography. A single, anchored GoPro was placed facing the structure and set to take a picture every minute for 60 min, the maximum time possible, given the power of the batteries. The camera was mounted 50 cm above the substrate and 150 cm away from the center of the structure. Every photograph after the first 5 min except the last 5 min was analyzed for fish species and for the relation of the fish to the structure (near or far). Only the photos captured after the first 5 min and before the final 5 min were analyzed to avoid any diver influence on reef fishes' use of the structure. Juvenile and adult fishes were treated the same for this study. Only those fish near the camera and directly above, below, beside, or in front of the structure were considered to be associated with it. These observed fish were then assigned to the 4 functional feeding guilds referred to above (Table S1). Since observations could be influenced by multiple images of the same individual, we considered the individual structure surveys as a measure of coral use rather than an estimate of fish abundance.

Over the course of 2 summers, we visited the 8 reefs once per year, recording 64 artificial structures and 112 natural structures so as to obtain 10,496 photos, resulting in 29,279 fish observations. After eliminating those fish considered not to be directly in the vicinity of the structure, our dataset consisted of 18,881 counts of 109 species of fish from 31 families, divided among the 4 functional feeding guilds as follows: 6047 individual herbivores from 23 species, 6225 individual omnivores from 48 species, 5725 individual invertivores from 12 species, and 884 individual piscivores from 26 species (Table S1).

*2.5. Statistical Analyses*

We performed a mixed-model ANOVA for all reef fishes and each functional group of fish against structure type as the main effect, with site as a random factor. Since all models indicated a significant effect of site on reef fish counts, we performed a principal component analysis to create orthogonal component scores that characterized the sites by their physical distance from shore, depth, rugosity, algal, soft coral, hard coral, and sponge substrate covers. The three most significant component scores from this analysis accounted for 72.8% of the variation in physical traits (PC 1 = 35.1%, PC 2 = 27.0%, PC 3 = 10.7%) and were included in the ANCOVA as covariates. We then analyzed fishes' use of the structures, using a nested analysis of covariance with structure state (artificial or natural) and structure type (control, soft, boulder, elkhorn, sea rod, diseased, dead, or healthy) nested within structure state as the fixed effects, and fish abundance from the video transect surveys and the three component scores as covariates. The first principal component score was the only covariate positively correlated with either of the main effects, so we included an interaction term of PC1 and structure state in our ANCOVA model. We used Tukey's post-hoc comparisons to estimate differences within structure states.

Observations at each structure (n = 177) were natural log-transformed to meet the assumptions of normality of the residuals and homogeneity of the variances. ANCOVAs were performed separately on the log-transformed counts for all reef fishes together and on the log-transformed counts of each fish feeding guild considered separately (herbivores, omnivores, invertivores, and piscivores). We also conducted ANCOVAs on the five most abundant fish species: bicolor damselfish (*Stegastes partitus*), striped parrotfish (*Scarus iseri*), white grunts (*Haemulon plumierii*), blue-striped grunts (*Haemulon sciurus*), and schoolmaster snapper (*Lutjanus apodus*). Linear regressions were used to relate fish counts to measures of hard coral height and to percent cover of live and diseased tissues, as estimated using JMP Pro 14.1.0 software.

## 3. Results

The reef fish abundances (video transect survey) on the reef sites varied significantly by reef. The eight reef sites differed in rugosity (0.046 to 0.410), hard coral cover (0.0 to 21.4%), soft coral cover (0.0 to 27.7%), sponge cover (1.0 to 15.5%), fleshy algae cover (0.0 to 86.4%), calcareous algae cover (0.0 to 13.8%), turf algae cover (8.4 to 63.4%) and sand cover (0.0 to 42.7%). The differences in abundance (video transect survey) per site accounted for 76.9% of the variance in the use of natural structures (structure survey) and 78% of the variance in the use of artificial structures (structure survey) (Table S2). There was no significant effect of year on the use of the structures by the reef fish on either the natural (F = −1.4376, df = 1.279, $p$ = 0.1517) or artificial structures (F = −1.5778, df = 1.127, $p$ = 0.1171). The total number of reef fishes using the natural structures was significantly more than the total number of reef fishes using the artificial structures (F = 5.8812, df = 1.166, $p$ = 0.0164) (Figure 3). There was, however, no statistically significant preference for a single type of structure, although diseased colonies had the most individuals associated with them and the soft artificial structures had the least. There were significantly more omnivore (F = 5.6278, df = 1.155, $p$ = 0.0189) and invertivore individuals (F = 6.5732, df = 1.123, $p$ = 0.0116) preferring natural structures over artificial structures; they preferred natural sea rod structures the most, although not significantly so. Herbivore use of the structures was not significantly influenced by the overall state of the structure (artificial versus natural) (F = 1.7337, df = 1.162, $p$ = 0.1898), but utilized particular structure types significantly more than others, using diseased coral heads significantly more than any other structure type, and used natural sea rod structures the least. There was no significant effect of structure state or type on the number of piscivorous fishes (F = 0.2082, df = 1.115, $p$ = 0.6491).

The data for the fish species that were most prevalent in the functional feeding guilds were analyzed to determine the reef characteristics that appeared to influence their use of habitats. Among the herbivores, the presence of bicolor damselfish (*Stegastes partitus*) was strongly correlated with low-complexity reefs (offshore) (F = 11.3251, df = 1.92, $p$ = 0.0011). Striped parrotfish (*Scarus iseri*) were observed significantly more around natural and artificial structures when their abundance on the reef (video transect survey), as observed in the video transect surveys, was high (F = 8.2186, df = 1.84, $p$ = 0.0052). This was also the pattern observed with the two most frequent invertivore species: blue-striped grunts (*Haemulon sciurus*) (F = 6.8165, df = 1.57, $p$ = 0.0115) and white grunts (*Haemulon plumierii*) (F = 14.0461, df = 1.85, $p$ = 0.0003). The most common piscivore was the schoolmaster snapper (*Lutjanus apodus*), which was significantly more numerous at natural structures (F = 6.3776, df = 1.27, $p$ = 0.0117) than at artificial structures. The omnivores were mostly represented by bluehead wrasses (*Thalassoma bifasciatum*), which were also more numerous at natural structures than at artificial structures (F = 16.5063, df = 1.129, $p$ < 0.0001) and at the low-complexity offshore sites (F = 4.6595, df = 1.129, $p$ = 0.0327).

The principal component analysis identified three axes which best summarized reef character and substrate composition (Table 1). The first component score (PC1) accounted for 35.1% of the variation and loaded positively with both distance from the shore and depth, but negatively with physical rugosity. The axis characterized reef structural differences between nearshore and offshore reefs in the middle Keys [51–53]. The second component score (PC2) accounted for 27.0% of the variation and loaded positively with calcareous algae/sand cover and negatively with fleshy/turf algae cover. The third component score (PC3) accounted for 10.7% of the variance and loaded positively with sponge/hard coral cover and negatively with soft coral cover. The three component scores were used to evaluate whether fishes' use of structure varied with differences in the physical structure, algal substrates, or hard/soft coral substrate.

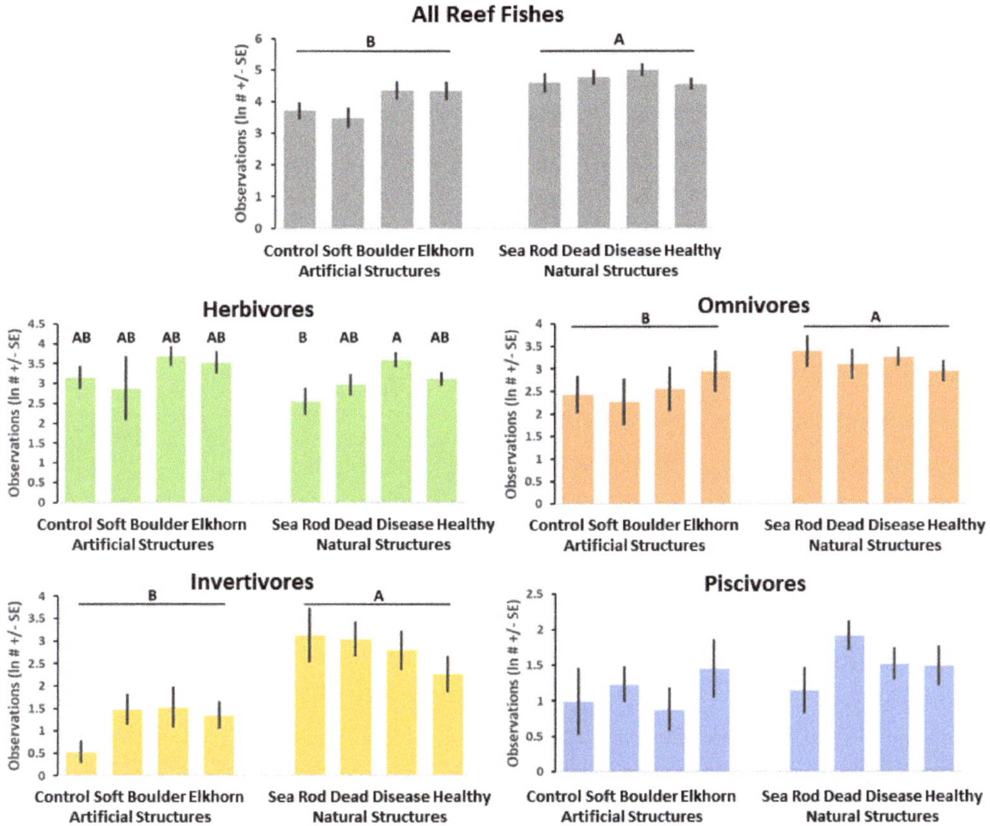

**Figure 3.** Log n (ln) mean per structure of the counts for all reef fishes combined (gray) and for each functional group—herbivores (green), omnivores (orange), predators (blue), and invertivores (yellow)—obtained during the time-lapse photography observations on the different natural and artificial corals. Error bars are ±1 SEM. Differences in the letter located over each bar indicate when values were shown to be statistically different from one another by Tukey's post hoc comparisons.

**Table 1.** Principal component score correlations for reef substrate measures. Bold indicates loadings with $p < 0.05$.

| Measures | PC 1 Topo | PC 2 Algal Cover | PC 3 Coral Cover |
|---|---|---|---|
| Rugosity | **−0.8573** | −0.0397 | −0.1667 |
| Distance to shore | **0.9207** | 0.1501 | 0.0139 |
| Depth | **0.7603** | 0.2546 | 0.0807 |
| % Sand | 0.2249 | **0.8930** | −0.1249 |
| % Fleshy algae | **0.7219** | **−0.5776** | −0.0785 |
| % Calcareous | −0.2716 | **0.7073** | −0.0971 |
| % Turf algae | −0.0191 | **−0.8884** | 0.2747 |
| % Hard coral | **−0.4116** | 0.2856 | **0.3895** |
| % Soft coral | **−0.3962** | −0.3304 | **−0.7455** |
| % Sponge | **−0.6138** | −0.0196 | **0.4706** |

We included four covariates in the ANCOVA model for fish use on individual structures including fish abundance (as estimated by video transect surveys), PC1 topography, PC2 algal cover, and PC3 coral cover (Figure 4). The fully fitted ANCOVA for all reef fishes combined was significant (F = 3.04, df = 12,164, $p$ = 0.0007) with an adjusted $r^2$ = 0.122 (Table 2). Four factors significantly contributed to the model: reef fish abundance on the video transects (F = 5.62, df = 1, $p$ = 0.0189, estimate = 0.531), structure state (whether artificial or natural) (F = 10.24, df = 1, $p$ = 0.0015, estimate (artificial) = $-0.3601$), topography (PC1) (F = 7.182, df = 1, $p$ = 0.0058, estimate = 0.2793), and the structure type by rugosity interaction (F = 6.495, df = 1, $p$ = 0.0117, estimate = 0.2427).

**Table 2.** Nested analysis of covariance with fish abundance (from video transect surveys), topography, algal cover, and coral cover as covariates and structure type (control, soft, barrel, elkhorn, sea rod, dead coral, diseased coral, and healthy coral) nested within structure state (natural versus artificial). Bolded values indicate significant values.

| Variable | Source | Df | F Ratio | $p$ | adj $r^2$ |
|---|---|---|---|---|---|
| **All Fish** | All Fish # | 1 | 5.620 | **0.0189** | 0.122 |
| | Topography (PC1) | 1 | 7.182 | **0.0058** | |
| | Algae (PC2) | 1 | 0.163 | 0.6867 | |
| | Coral (PC3) | 1 | 1.071 | 0.3021 | |
| | Structure Type | 1 | 6.495 | **0.0117** | |
| | Structure State | 6 | 10.24 | 0.0015 | |
| **Herbivores** | Herbivore # | 1 | 8.412 | **0.0042** | 0.1844 |
| | Topography (PC1) | 1 | 17.25 | **0.0001** | |
| | Algae (PC2) | 1 | 1.536 | 0.2169 | |
| | Coral (PC3) | 1 | 0.909 | 0.3417 | |
| | Structure Type | 1 | 1.733 | 0.1898 | |
| | Structure State | 6 | 2.523 | **0.0232** | |
| **Omnivores** | Omnivore # | 1 | 0.964 | 0.3275 | 0.1909 |
| | Topography (PC1) | 1 | 28.691 | **0.0001** | |
| | Algae (PC2) | 1 | 0.020 | 0.8876 | |
| | Coral (PC3) | 1 | 0.470 | 0.4938 | |
| | Structure Type | 1 | 5.627 | **0.0189** | |
| | Structure State | 6 | 0.463 | 0.8344 | |
| **Invertivores** | Invertivore # | 1 | 8.169 | **0.0050** | 0.3227 |
| | Topography (PC1) | 1 | 11.78 | **0.0008** | |
| | Algae (PC2) | 1 | 0.224 | 0.6381 | |
| | Coral (PC3) | 1 | 0.627 | 0.4300 | |
| | Structure Type | 1 | 6.573 | **0.0116** | |
| | Structure State | 6 | 0.968 | 0.4499 | |
| **Predators** | Predator # | 1 | 5.882 | **0.0170** | 0.1106 |
| | Topography (PC1) | 1 | 1.231 | 0.2697 | |
| | Algae (PC2) | 1 | 0.957 | 0.3300 | |
| | Coral (PC3) | 1 | 0.255 | 0.6143 | |
| | Structure Type | 1 | 0.208 | 0.6491 | |
| | Structure State | 6 | 0.743 | 0.6163 | |

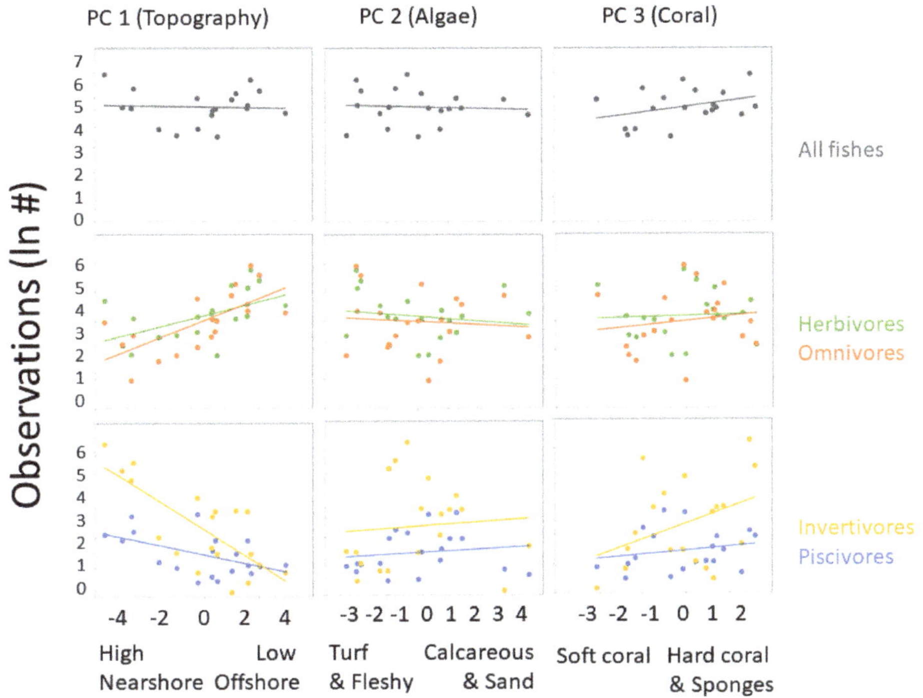


**Figure 4.** Linear regressions of mean numbers of fish observed (log-transformed) for the structure surveys within a reef site versus the three component scores that best describe the reef characteristics. The three component scores, from left to right, are topography (PC1), algae (PC2), and coral cover (PC3). The specific loadings of the reef measures for each component score are in Table 1. Graphs include all reef fish combined (gray), herbivores (green), omnivores (orange), invertivores (yellow), and predators (blue).

Nested ANCOVAs for each separate feeding guild of fish revealed differences in the use of the structures among fish guilds. For herbivores, the fully fitted ANCOVA was significant (F = 4.55, df = 11,162, $p < 0.0001$) with an adjusted $r^2$ = 0.184 (Table 2). Three factors significantly contributed to the model: herbivore abundance in the video transect surveys (F = 8.41, df = 1, $p$ = 0.0042, estimate = 0.5265), topography (F = 17.25, df = 1, $p < 0.0001$, estimate = 0.2217), and the type of natural structure (F = 2.52, df = 1, $p$ = 0.0232) (Table 2). Herbivore use of the structures demonstrated a strong negative relationship with reef rugosity (or a positive relationship with deeper, offshore reefs) but no relationship with algal cover (Figure 4). For the omnivores, there were two factors that significantly contributed to the model of structure use: depth/rugosity (F = 28.69, df = 1, $p < 0.0001$, estimate = 0.3757) and structure state (F = 5.62, df = 1, $p$ = 0.0189, estimate (artificial) = −0.3195) (Table 2). Omnivores demonstrated a strong negative relationship with reef rugosity (or a positive relationship with deeper, offshore reefs) (Figure 4). For invertivores, three factors significantly contributed to the model of structure use: invertivore abundance observed in the video transect surveys (F = 8.16, df = 1, $p$ = 0.0050, estimate = 0.3742), topography (F = 11.78, df =1, $p$ = 0.0008, estimate = −0.3693), and structure state (artificial or natural) (F = 6.57, df = 1, $p$ = 0.0116, estimate (artificial) = −0.4964) (Table 2). In contrast to the findings for herbivores and omnivores, invertivores' use of the structures demonstrated a strong positive relationship with reef rugosity (or a negative relationship with deeper, offshore reefs) (Figure 4). The abundance of piscivores across the video transect surveys was the only factor that significantly contributed to the model of structure use (F = 5.88, df = 1, $p$ = 0.0170, estimate = 0.2663) (Table 2 and Figure 4).

## 4. Discussion

We found that there were no significant effects of rugosity, hard coral cover, soft coral cover, or algal cover on total reef fish and functional feeding guild abundance across our sites. This is different from previous studies that found a greater abundance and diversity of fishes on reefs that have higher hard coral cover and structural complexity [29]. There are several factors that could explain this unexpected finding. First, some studies have found that reef fish abundance generally increases with depth [54]. Second, the shallow-water nearshore habitats of the middle Florida Keys experience a great deal of disturbance, such as nutrient loading or extreme temperatures [55,56], so that there is a possibility of a delayed response of the reef fish community to decades of continuous perturbations [57]. Alternatively, it may be that our methods were not sensitive enough to detect an effect; for example, the acclimation period may not have been long enough, or the artificial structure not large enough to influence fish distribution.

We found that the natural structures were used by fish significantly more than the artificial structures, even though these were designed to mimic the size and shape of the available natural structures. Previous studies have also found that artificial reef structures (ARs) built with numerous distinct holes and crevices were unsuccessful overall [58]. Such studies have also suggested that artificial structures are used mostly at night, a possibility that we did not investigate with our artificial structures [59]. Evidence has also suggested that fish are attracted to the auditory and chemical cues of living hard corals, so the lack of cues from the artificial structures might explain the low rates of use [60,61].

Unsurprisingly, the abundance of fishes on each reef (video transect survey) accounted for more than half of the variance found in the structures' associations. There were more fish using the structures when there were more fish available on the reef. Generally, reef fish used natural structures significantly more than artificial ones, but, surprisingly, the types of natural (sea rod, dead, diseased, and healthy hard corals) or artificial (control, soft, boulder, and elkhorn) structures did not influence fishes' use. Natural structures, on average, had similar heights and total surface area to our artificial structures, which suggests that differences in observations of use were not strictly due to space competition. However, the natural structures were covered with living organisms and algae and provided more food than our artificial structures. The small amount of biota could explain why the artificial structures were used significantly less than natural structures [9,14]. Moreover, these differences in the resources available on artificial and natural structures could affect reef fish habitat specialists differently from habitat generalists (Table S2) [29]. We found that artificial structures were occupied more on low-rugosity sites than on high-rugosity sites. This supports previous observations that artificial structures have their greatest impact in environments with fewer structures [62,63].

Further characterization of structure use by different functional feeding guilds revealed other important patterns not apparent from the analysis of use by all reef fishes combined. Herbivore observations on natural and artificial structures were lower on shallow reefs with high rugosity, but they significantly preferred diseased corals over sea rods. This may be due to the surfaces on the coral heads being newly opened for turf algal colonization, a preferred foraging substrate for all parrotfish species, which commonly inhabit these reefs [51]. Alternatively, the newly available coral tissue could be providing nutritional benefits or a greater concentration of autotrophic organisms that attracted parrotfish foragers [64]. Herbivores on artificial structures were not significantly different from on natural structures, perhaps because after 2 weeks, we observed that the artificial structures had accumulated enough biofilm to be a suitable foraging substrate (Figure 3). The functionally important role that parrotfish play on Caribbean coral reefs is well understood [65,66], and our study suggests that with the flattening of reefs, the reefs of the future may see a decline in parrotfish abundance.

Piscivores rarely used either natural or artificial structures and were unrelated to any of the three component scores of reef substrate. This is unsurprising, due to previous findings that piscivore abundance is associated more with prey availability than any habitat characteristic [67,68]. In contrast, omnivores used natural structures significantly more than artificial structures and decreased structure use on shallow reefs with high rugosity, a response that has been predicted in recent models [31]. However, omnivores' preference for low-rugosity reefs contradicts previous literature that has found that all feeding groups tend to be positively associated with increased complexity [11–13,69]. We expected that rugosity and hard coral cover would be correlated [25], but found that rugosity was not a function of hard coral cover. Presently, the reefs in the Florida Keys are composed of scattered boulder corals, abundant soft corals, and limestone ledges. Additionally, with the recent outbreak of stony coral tissue loss disease, there has been a decrease in live hard coral cover but a lingering presence of dead hard coral structures [70]. With time, overall rugosity and the presence of scattered boulder corals will continue decreasing as dead coral heads begin eroding away.

Similarly, invertivores used natural structures significantly more than artificial structures, but in contrast to the herbivores and omnivores, they increased on shallow reefs with greater rugosity. This suggests that grunts may be the one feeding guild most impacted by the loss of structural complexity and the flattening of the reef. This relationship is best explained by the reef component score associated with rugosity rather than the component score associated with hard coral cover, which suggests that it is physical habitat that matters more to grunts than the health status of the coral. Invertivores often depend on finer-scale shelters for their prey species to occupy and would explain why they, and their prey, would be negatively impacted by structural loss [71]. This negative response of invertivores to reef decline has been predicted as a response to climate change according to predictive climate change models [31].

All guilds had a predominant species that did not mirror the pattern of the rest of the guild. Bicolor damselfish (*Stegastes partitus*), striped parrotfish (*Scarus iseri*), blue-striped grunts (*Haemulon sciurus*), white grunts (*Haemulon plumierii*), schoolmaster snapper (*Lutjanus apodus*), and the bluehead wrasses (*Thalassoma bifasciatum*) had a disproportionate representation within their associated functional feeding guilds, but their patterns did not parallel the overall effect observed within their guild. Their presence may be due to their being habitat generalists, equally at home in hard coral- or soft coral-dominated reefs [67,72]. The reduced structure use by the bicolor damselfish, striped parrotfish, or bluehead wrasses could be explained by the higher abundance of piscivores, although we did not observe an increase in piscivore presence, even though these habitats had an abundance of prey, which should have driven their numbers up [67,68,71,72].

Our results suggest that use of structures, both natural and artificial structures, differs among reef fish functional feeding guilds. If the low-rugosity reefs with low hard coral cover are representative of the future reefs of the middle Florida Keys, we would predict a shift in reef fish community with increasing proportions of herbivores and omnivores, and a decreasing proportion of invertivores. These results can be used as a predictive model for reef fish community responses to changes in reef composition and may be useful in the design of future marine protected areas needed to preserve feeding guilds critical to the recovery of hard corals. Future studies should examine how the relative abundance of fish functional feeding guilds changes in response to this transition from hard-coral-dominated to soft-coral-dominated reefs in the middle Florida Keys.

**Supplementary Materials:** The following are available online at https://www.mdpi.com/article/10.3390/oceans2030036/s1. Table S1: Species included in this study, their functional feeding guild, their presence (Y or N) at artificial and natural structures, and their habitat use. Artificial structures = soft coral mimic (SC), control structure (C), boulder coral mimic (BC), and elkhorn coral mimic (EC). Natural structures = sea rods (SR), dead corals (DC), infected diseased corals (IC), and healthy corals (HC). Fish are separated into generalist or specialist habitat use classifications [29]. Table S2: Percent presence of generalist versus specialist species observed for the four artificial and four natural shelter types for each functional feeding guild. Artificial structures = soft coral mimic (SC), control structure (C), boulder coral mimic (BC), and elkhorn coral mimic (EC). Natural structures = sea rods (SR), dead corals (DC), infected diseased corals (IC), and healthy corals (HC).

**Author Contributions:** Conceptualization, K.N., K.S. (Kylie Smith) and M.C.; methodology, K.N., K.S. (Kylie Smith) and M.C.; formal analysis, K.N., T.F. and M.C.; investigation, K.N.; resources, K.N., K.S. (Kylie Smith) and M.C.; data curation, K.N., T.F., K.M. and K.S. (Kelsey Sox); writing—original draft preparation, K.N., T.F. and M.C.; writing—review and editing, K.N., K.M., K.S. (Kelsey Sox) and K.S. (Kylie Smith); supervision, K.N. and M.C.; funding acquisition, K.N. All authors have read and agreed to the published version of the manuscript.

**Funding:** This research was funded by the American Museum of Natural History Lerner-Gray Fund for Marine Research, the American Society of Ichthyologists and Herpetologists Raney Award, Clemson University's Creative Inquiry Initiative, the Explorer's Club Mamont Grant, the South Carolina Space Grant Consortium Kathryn D. Sullivan Earth and Marine Science Fellowship (# 2014105), and Clemson University.

**Institutional Review Board Statement:** This study was conducted according the guidelines of the Florida Keys National Marine Sanctuary (permit numbers FKNMS-2017-032, approved on 6 January 2017, and permit number FKNMS-2018-119, approved on 10 January 2018).

**Informed Consent Statement:** Not applicable.

**Data Availability Statement:** Data is archived at Clemson University and is available upon request to Michael Childress (mchildr@clemson.edu).

**Acknowledgments:** We thank Reanna Jeanes, Morgan Gardner, Riley Garvey, Rachel Radick, and Emma Crowfoot for assistance in data collection and processing. We also thank the two anonymous referees and the special issue editor for their great assistance in improving the manuscript.

**Conflicts of Interest:** The authors declare no conflict of interest. The funders had no role in the design of the study; in the collection, analyses, or interpretation of data; in the writing of the manuscript; or in the decision to publish the results.

## References


1. Weber, M.L.; Gradwohl, J. Life in the seas. In *Weber MLaJG (ed) The Wealth of Oceans*; W.W. Norton & Company: New York, NY, USA, 1995.
2. Spalding, M.; Grenfell, A.M. New estimates of global and regional coral reef areas. *Coral Reefs* **1997**, *16*, 225–230. [CrossRef]
3. Roberts, C.M.; McClean, C.J.; Veron, J.E.; Hawkins, J.P.; Allen, G.R.; McAllister, D.E.; Mittermeier, C.G.; Schueler, F.W.; Spalding, M.; Wells, F.; et al. Marine biodiversity hotspots and conservation priorities for tropical reefs. *Science* **2002**, *295*, 1280–1284. [CrossRef]
4. Bellwood, D.R.; Hughes, T.P.; Folke, C.; Nystrom, M. Confronting the coral reef crisis. *Nature* **2004**, *429*, 827–833. [CrossRef]
5. Allen, G.R. Conservation hotspots of biodiversity and endemism for Indo-Pacific coral reef fishes. *Aquat. Conserv. Mar. Freshw. Ecosyst.* **2008**, *18*, 541–556. [CrossRef]
6. Alvarez-Filip, L.; Gill, J.A.; Dulvy, N.K.; Perry, A.L.; Watkinson, A.R.; Côté, I.M. Drivers of region-wide declines in architectural complexity on Caribbean reefs. *Coral Reefs* **2011**, *30*, 1051–1060. [CrossRef]
7. Risk, M.J. Fish Diversity on a Coral Reef in the Virgin Islands. *Atoll Res. Bull.* **1972**, *153*, 1–4. [CrossRef]
8. Luckhurst, B.E.; Luckhurst, K.L. Analysis of the influence of substrate variables onc oral reef fish communities. *Mar. Biol.* **1978**, *49*, 317–323. [CrossRef]
9. Roberts, C.M.; Ormond, R.F.G. Habitat complexity and coral reef fish diversity and abundance on Red Sea fringing reefs. *Mar. Ecol. Prog. Ser.* **1987**, *41*, 1–8. [CrossRef]
10. Ohman, M.C.; Rajasuriya, A. Relationships between habitat structure and fish communities on coral and sandstone reefs. *Environ. Biol. Fishes* **1998**, *53*, 19–31. [CrossRef]
11. Gratwicke, B.; Speight, M.R. The relationship between fish species richness, abundance and habitat complexity in a range of shallow tropical marine habitats. *J. Fish Biol.* **2005**, *66*, 650–667. [CrossRef]
12. Gonzalez-Rivero, M.; Harborne, A.R.; Herrera-Reveles, A.; Bozec, Y.M.; Rogers, A.; Friedman, A.; Ganase, A.; Hoegh-Guldberg, O. Linking fishes to multiple metrics of coral reef structural complexity using three-dimensional technology. *Sci. Rep.* **2017**, *7*, 13965. [CrossRef]
13. Richardson, L.E.; Graham, N.A.J.; Pratchett, M.S.; Hoey, A.S. Structural Complexity Mediates Functional Structure of Reef Fish Assemblages among Coral Habitats. *Environ. Biol. Fish.* **2017**, *100*, 193–207. [CrossRef]
14. Molles, M.C. Fish species Diversity on Model and Natural Reef Patches: Experimental Insular Biogeography. *Ecol. Monogr.* **1978**, *48*, 289–305. [CrossRef]
15. Carpenter, K.E. The influence of substrate structure on the local abundance and diversity of philippine reef fishes. In Proceedings of the Fourth International Coral Reef Syposium, Marine Sciences Center, University of the Philippines, Manila, Phillippines, 18–22 May 1981.
16. Bell, J.O.; Galzin, R. Influence of live coral cover on coral-reef fish communities. *Mar. Ecol. Prog. Ser.* **1984**, *15*, 265–274. [CrossRef]

17. Guidetti, P. Differences among fish assemblages associated with nearshore *Posidonia oceanica* Seagrass Beds, Rocky-algal Reefs and Unvegetated Sand Habitats in the Adriatic Sea. Estuarine. *Coast. Shelf Sci.* **2000**, *50*, 515–529. [CrossRef]
18. Khalaf, M.A.; Kochzius, M. Changes in trophic community structure of shore fishes at an industrial site in the Gulf of Aqaba, Red Sea. *Mar. Ecol. Prog. Ser.* **2002**, *239*, 287–299. [CrossRef]
19. Harborne, A.R.; Rogers, A.; Bozec, Y.M.; Mumby, P.J. Multiple Stressors and the Functioning of Coral Reefs. *Ann. Rev. Mar. Sci.* **2017**, *9*, 445–468. [CrossRef]
20. Hughes, T.P.; Barnes, M.L.; Bellwood, D.R.; Cinner, J.E.; Cumming, G.S.; Jackson, J.B.; Palumbi, S.R. Coral reefs in the Anthropocene. *Nature* **2017**, *546*, 82–90. [CrossRef]
21. Hoekstra, J.M.; Boucher, T.M.; Ricketts, T.H.; Roberts, C. Confronting a biome crisis: Global disparities of habitat loss and protection. *Ecol. Lett.* **2005**, *8*, 23–29. [CrossRef]
22. Wilson, S.K.; Graham, N.A.J.; Pratchett, M.S.; Jones, G.P.; Polunin, N.V.C. Multiple disturbances and the global degradation of coral reefs: Are reef fishes at risk or resilient? *Glob. Chang. Biol.* **2006**, *12*, 2220–2234. [CrossRef]
23. Pratchett, M.S.; Munday, P.L.; Wilson, S.K.; Graham, N.A.J.; Cinner, J.E.; Bellwood, D.R.; Jones, G.P.; Polunin, N.V.C.; McClanahan, T.R. Effects of climate-induced coral bleaching on coral-reef fishes-ecological and economic consequences. *Mar. Biol. Annu. Rev.* **2008**, *46*, 257–302.
24. Alvarez-Filip, L.; Dulvy, N.K.; Gill, J.A.; Cote, I.M.; Watkinson, A.R. Flattening of Caribbean coral reefs: Region-wide declines in architectural complexity. *Proc. Biol. Sci.* **2009**, *276*, 3019–3025. [CrossRef]
25. Alvarez-Filip, L.; Gill, J.A.; Dulvy, N.K. Complex reef architecture supports more small-bodied and longer food chains on Caribbean reefs. *Ecosphere* **2011**, *2*, 1–17. [CrossRef]
26. Pratchett, M.S.; Hoey, A.S.; Wilson, S.K.; Messmer, V.; Graham, N.A.J. Changes in biodiversity and functioning of reef fish assemblages following coral bleaching and coral loss. *Diversity* **2011**, *3*, 424–452. [CrossRef]
27. Alevizon, W.S.; Porter, J.W. Coral loss and fish guild stability on a Caribbean coral reef: 2014, 1974–2000. *Environ. Biol. Fishes* **2015**, *98*, 1035–1045. [CrossRef]
28. Pratchett, M.S.; Hoey, A.S.; Wilson, S.K. Reef degradation and the loss of critical ecosystem goods and services provided by coral reef fishes. *Curr. Opin. Environ. Sustain.* **2014**, *7*, 37–43. [CrossRef]
29. Alvarez-Filip, L.; Paddack, M.J.; Collen, B.; Robertson, D.R.; Cote, I.M. Simplification of Caribbean Reef-Fish Assemblages over Decades of Coral Reef Degradation. *PLoS ONE* **2015**, *10*, e0126004. [CrossRef]
30. Paddack, M.J.; Reynolds, J.D.; Aguilar, C.; Appeldoorn, R.S.; Beets, J.; Burkett, E.W.; Chittaro, P.M.; Clarke, K.; Esteves, R.; Fonseca, A.C.; et al. Recent region-wide declines in Caribbean reef fish abundance. *Curr. Biol.* **2009**, *19*, 590–595. [CrossRef]
31. Inagaki, K.Y.; Pennino, M.G.; Floeter, S.R.; Hay, M.E.; Longo, G.O. Trophic interactions will expand geographically but be less intense as oceans warm. *Glob. Chang. Biol.* **2020**, *26*, 6805–6812. [CrossRef]
32. Bohnsack, J.A. Are high densities of fishes at artificial reefs the result of habitat limitation or behavioral preference? *Bull. Mar. Sci.* **1989**, *44*, 631–645.
33. Bohnsack, J.A.; Harper, D.E.; McClellan, D.B.; Hulsbeck, M. Effects of reef size on colonization and assemblage structure of fishes at artificial reefs off southeastern Florida, USA. *Bull. Mar. Sci.* **1994**, *55*, 796–823.
34. Carr, M.H.; Hixon, M.A. Artificial reefs: The importance of comparisons with natural reefs. *Fisheries* **1997**, *22*, 28–33. [CrossRef]
35. Clark, S.; Edwards, A.J. An evaluation of artificial reef structures as tools for marine habitat rehabilitation in the Maldives. *Aquat. Conserv. Mar. Freshw. Ecosyst.* **1999**, *9*, 5–21. [CrossRef]
36. Rilov, G.; Benayahu, Y. Fish assemblage on natural versus vertical artificial reefs: The rehabilitation perspective. *Mar. Biol.* **2000**, *136*, 931–942. [CrossRef]
37. Abelson, A.; Shlesinger, Y. Comparison of the development of coral and fish communities on rock-aggregated artificial reefs in Eilat, Red Sea. *ICES J. Mar. Sci.* **2002**, *59*, S122–S126. [CrossRef]
38. Edwards, R.A.; Smith, S.D. Subtidal assemblages associated with a geotextile reef in south-east Queensland, Australia. *Mar. Freshw. Res.* **2005**, *56*, 133–142. [CrossRef]
39. Clynick, B.G.; Chapman, M.G.; Underwood, A.J. Fish assemblages associated with urban structures and natural reefs in Sydney, Australia. *Austral Ecol.* **2008**, *33*, 140–150. [CrossRef]
40. Bohnsack, J.A. Habitat structure and the design of artificial reefs. In *Habitat Structure*; Springer: Dordrecht, The Netherlands, 1991; pp. 41–426.
41. Rilov, G.; Benayahu, Y. Rehabilitation of coral reef-fish communities: The importance of artificial-reef relief to recruitment rates. *Bull. Mar. Sci.* **2002**, *70*, 185–197.
42. Rilov, G.; Benayahu, Y. Vertical artificial structures as an alternative habitat for coral reef fishes in disturbed environments. *Mar. Environ. Res.* **1998**, *45*, 431–451. [CrossRef]
43. Alevizon, W.S.; Gorham, J.C. Effects of artificial reef deployment on nearby resident fishes. *Bull. Mar. Sci.* **1989**, *44*, 646–661.
44. Pérez-Ruzafa, A.; Garcıa-Charton, J.A.; Barcala, E.; Marcos, C. Changes in benthic fish assemblages as a consequence of coastal works in a coastal lagoon: The Mar Menor (Spain, Western Mediterranean). *Mar. Pollut. Bull.* **2006**, *53*, 107–120. [CrossRef] [PubMed]
45. Burt, J.; Bartholomew, A.; Usseglio, P.; Bauman, A.; Sale, P.F. Are artificial reefs surrogates of natural habitats for corals and fish in Dubai, United Arab Emirates? *Coral Reefs* **2009**, *28*, 663–675. [CrossRef]

46. Kohler, K.E.; Gill, S.M. Coral Point Count with Excel extensions (CPCe): A Visual Basic program for the determination of coral and substrate coverage using random point count methodology. *Comput. Geosci.* **2006**, *32*, 1259–1269. [CrossRef]

47. Halpern, B.S.; Floeter, S.R. Functional diversity responses to changing species richness in reef fish communities. *Mar. Ecol. Prog. Ser.* **2008**, *364*, 147–156. [CrossRef]

48. Allen, C.R.; Gunderson, L.; Johnson, A.R. The use of discontinuities and functional groups to assess relative resilience in complex systems. *Ecosystems* **2005**, *8*, 958–966. [CrossRef]

49. Folke, C. Resilience: The emergence of a perspective for social-ecological systems analyses. *Glob. Environ. Chang.* **2006**, *16*, 253–267. [CrossRef]

50. Fischer, J.; Lindenmayer, D.B.; Blomberg, S.P.; Montague-Drake, R.; Felton, A.; Stein, J.A. Functional richness and relative resilience of bird communities in regions with different land use intensities. *Ecosystems* **2007**, *10*, 964–974. [CrossRef]

51. Smith, K.M.; Quirk-Royal, B.E.; Drake-Lavelle, K.; Childress, M.J. Influences of ontogenetic phase and resource availability on parrotfish foraging preferences in the Florida Keys, FL USA. *Mar. Ecol. Prog. Ser.* **2018**, *603*, 175–187. [CrossRef]

52. Smith, K.M.; Payton, T.G.; Sims, R.J.; Stroud, C.S.; Jeanes, R.C.; Hyatt, T.B.; Childress, M.J. Impacts of consecutive bleaching events on transplanted coral colonies in the Florida Keys. *Coral Reefs* **2019**, *38*, 851–861. [CrossRef]

53. Manzello, D.P.; Enochs, I.C.; Kolodziej, G.; Carlton, R. Recent decade of growth and calcification of *Orbicella faveolata* in the florida keys: An inshore-offshore comparison. *Mar. Ecol. Prog. Ser.* **2015**, *521*, 81–89. [CrossRef]

54. Thresher, R.E.; Colin, P.L. Trophic structure, diversity and abundance of fishes of the deep reef (30–300 m) at Enewetak, Marshall Islands. *Bull. Mar. Sci.* **1986**, *38*, 253–272.

55. Ginsburg, R.N.; Shinn, E.A. Preferential distribution of reefs in the Florida reef tract: The past is the key to the present. In *Proceedings of the Colloquium on Global Aspects of Coral Reefs: Health, Hazards and History*; Ginsburg, R.N., Ed.; Rosenstiel School of Marine and Atmospheric Science, University of Miami: Coral Gables, FL, USA, 1994; pp. 21–26.

56. Leichter, J.J.; Stewart, H.L.; Miller, S.L. Episodic nutrient transport to Florida coral reefs. *Limnol. Oceanogr.* **2003**, *48*, 1394–1407. [CrossRef]

57. Bellwood, D.R.; Renema, W.; Rosen, B.R. Biodiversity hotspots, evolution and coral reef biogeography. In *Biotic Evolution and Environmental Change in Southeast Asia*; Cambridge University Press: Cambridge, UK, 2012; Volume 216.

58. Nemeth, M.; Appeldoorn, R. The distribution of herbivorous coral reef fishes within fore-reef habitats: The role of depth, light and rugosity. *Caribb. J. Sci.* **2009**, *45*, 247–253. [CrossRef]

59. Harborne, A.R.; Mumby, P.J.; Ferrari, R. The effectiveness of different meso-scale rugosity metrics for predicting intra-habitat variation in coral-reef fish assemblages. *Environ. Biol. Fishes* **2012**, *94*, 431–442. [CrossRef]

60. Gordon, T.A.; Radford, A.N.; Davidson, I.K.; Barnes, K.; McCloskey, K.; Nedelec, S.L.; Simpson, S.D. Acoustic enrichment can enhance fish community development on degraded coral reef habitat. *Nat. Commun.* **2019**, *10*, 5414. [CrossRef]

61. Hu, Y.; Majoris, J.E.; Buston, P.M.; Webb, J.F. Potential roles of smell and taste in the orientation behaviour of coral-reef fish larvae: Insights from morphology. *J. Fish Biol.* **2019**, *95*, 311–323. [CrossRef]

62. Komyakova, V.; Chamberlain, D.; Jones, G.P.; Swearer, S.E. Assessing the performance of artificial reefs as substitute habitat for temperate reef fishes: Implications for reef design and placement. *Sci. Total. Environ.* **2019**, *668*, 139–152. [CrossRef]

63. Komyakova, V.; Swearer, S.E. Contrasting patterns in habitat selection and recruitment of temperate reef fishes among natural and artificial reefs. *Mar. Environ. Res.* **2019**, *143*, 71–81. [CrossRef]

64. Gutiérrez-Isaza, N.; Espinoza-Avalos, J.; León-Tejera, H.P.; González-Solís, D. Endolithic community composition of Orbicella faveolata (Scleractinia) underneath the interface between coral tissue and turf algae. *Coral Reefs* **2015**, *34*, 625–630. [CrossRef]

65. Burkepile, D.E.; Hay, M.E. Impact of herbivore identity on algal succession and coral growth on a Caribbean reef. *PLoS ONE* **2010**, *5*, e8963. [CrossRef]

66. Cardoso, S.C.; Soares, M.C.; Oxenford, H.A.; Côté, I.M. Interspecific differences in foraging behaviour and functional role of Caribbean parrotfish. *Mar. Biodivers. Rec.* **2009**, *2*, e148. [CrossRef]

67. Hixon, M.A.; Beets, J.P. Predation, prey refuges, and the structure of coral-reef fish assemblages. *Ecol. Monogr.* **1993**, *63*, 77–101. [CrossRef]

68. Stewart, B.D.; Jones, G.P. Associations between the abundance of piscivorous fishes and their prey on coral reefs: Implications for prey-fish mortality. *Mar. Biol.* **2001**, *138*, 383–397. [CrossRef]

69. Hensel, E.; Allgeier, J.E.; Layman, C.A. Effects of predator presence and habitat complexity on reef fish communities in The Bahamas. *Mar. Biol.* **2019**, *166*, 136. [CrossRef]

70. Noonan, K.R.; Childress, M.J. Association of butterflyfishes and stony coral tissue loss disease in the Florida Keys. *Coral Reefs* **2020**, *39*, 1581–1590. [CrossRef]

71. Yates, D.C.; Lonhart, S.I.; Hamilton, S.L. Effects of marine reserves on predator-prey interactions in central California kelp forests. *Mar. Ecol. Prog. Ser.* **2020**, *655*, 139–155. [CrossRef]

72. Boaden, A.E.; Kingsford, M.J. Predators drive community structure in coral reef fish assemblages. *Ecosphere* **2015**, *6*, 1–33. [CrossRef]

*Article*

# Correlation between Coral Reef Condition and the Diversity and Abundance of Fishes and Sea Urchins on an East African Coral Reef

Pia Ditzel [1,*], Sebastian König [1], Peter Musembi [2,3] and Marcell K. Peters [1]

[1] Department of Animal Ecology and Tropical Biology, Biocenter, University of Würzburg, Am Hubland, 97074 Würzburg, Germany; sebastian.c.koenig@uni-wuerzburg.de (S.K.); marcell.peters@uni-wuerzburg.de (M.K.P.)

[2] Department of Biological Sciences, Pwani University, Kilifi 80108, Kenya; pmusembi@cordioea.net

[3] CORDIO East Africa, Mombasa 80101, Kenya

[*] Correspondence: piaditzel@gmail.com

**Abstract:** Coral reefs are one of the most diverse marine ecosystems, providing numerous ecosystem services. This present study investigated the relationship between coral reef condition and the diversity and abundance of fishes, on a heavily fished East African coral reef at Gazi Bay, Kenya. Underwater visual censuses were conducted on thirty $50 \times 5$ m belt transects to assess the abundance and diversity of fishes. In parallel, a 25-m length of each of the same transects was recorded with photo-quadrats to assess coral community structure and benthic characteristics. For statistical analyses, multi-model inference based on the Akaike Information Criterion was used to evaluate the support for potential predictor variables of coral reef and fish diversity. We found that coral genus richness was negatively correlated with the abundance of macroalgae, whereas coral cover was positively correlated with both the abundance of herbivorous invertebrates (sea urchins) and with fish family richness. Similarly, fish family richness appeared mainly correlated with coral cover and invertebrate abundance, although no correlates of fish abundance could be identified. Coral and fish diversity were very low, but it appears that, contrary to some locations on the same coast, sea urchin abundance was not high enough to be having a negative influence on coral and fish assemblages. Due to increasing threats to coral reefs, it is important to understand the relationship among the components of the coral reef ecosystem on overfished reefs such as that at Gazi Bay.

**Keywords:** coral reef ecosystem; coral reef resilience; global warming; climate change; overfishing; indicator species

**Citation:** Ditzel, P.; König, S.; Musembi, P.; Peters, M.K. Correlation between Coral Reef Condition and the Diversity and Abundance of Fishes and Sea Urchins on an East African Coral Reef. *Oceans* **2022**, 3, 1–14. https://doi.org/10.3390/oceans3010001

Academic Editors: Rupert Ormond and Fleur C. van Duyl

Received: 27 April 2021
Accepted: 14 December 2021
Published: 4 January 2022

**Publisher's Note:** MDPI stays neutral with regard to jurisdictional claims in published maps and institutional affiliations.

## 1. Introduction

Coral reefs are one of the most diverse marine ecosystems, providing numerous ecosystem services to millions of inhabitants of tropical countries. More than 500 million people depend on coral reefs [1–4]. Together with functionally and ecologically linked mangrove habitats and seagrass meadows, coral reefs likely support the highest marine biodiversity in the world [5–7]. The complex reef system is mainly built by scleractinian corals which can only survive a narrow range of environmental variation, and hence are vulnerable to many kinds of disturbance [8]. Most of the world's reefs are threatened by human activities [4,9–11], which have exacerbated a background of otherwise natural impacts, such as those caused by diseases, outbreaks of invasive species, storms, or sedimentation [10,12–15]. In addition, coral bleaching now represents a major threat to coral reefs, whereby corals expel their zooxanthellae and associated pigments, due to high sea temperatures, slow water movement, or extreme UV exposure [2,16–19].

A fundamental goal of ecology is to understand how organisms interact with other species of organisms and their physical environment, providing among other things an explanation of variation in species richness [20]. The present study aimed to explore the relationship between the condition of coral reefs and the abundance and the diversity of reef fishes and sea urchins in an area known to be subject to heavy fishing pressure, but where herbivorous invertebrates, such as sea urchins, are not subject to these fishing practices [21]. In studying coral reefs, the abundances of key organisms have frequently been used as indicators of reef health. For example, a low abundance of corals or fishes, or a high abundance of algae, seagrass, or other soft-bodied organisms, have been taken to indicate the degradation of a reef area [10,15,22,23]. Because herbivores (both fish and invertebrates, such as sea urchins) play an important role in maintaining the coral–algae balance on a reef and in creating an appropriate substrate for coral recruitment [3,24–31], overfishing of herbivores can lead to a decline in their grazing capacity and cause a phase shift resulting in algal rather than coral dominance [7,26,29,32]. Corallivores have also often been used as indicators of reef health because of their usually positive relationship with live coral cover and negative correlation with macroalgae cover [33–36]. The abundance of corallivores is presumed to be a good indicator of a healthy reef, although, in cases of extreme corallivory, an abundance of predators may result in loss of corals or make the corals more susceptible to other stressors, such as thermal stress or the spread of diseases, as the coral is already debilitated [37–39]. In addition, physical or chemical variables, such as salinity or turbidity, may affect the status of a reef and are often monitored [40].

In summary, many variables can have an impact on the condition of a coral reef, but the effects may vary at different locations [40–44]. The present study was conducted both to evaluate the condition of coral reefs in Gazi Bay, Kenya, and to investigate the relationships between coral diversity, fish diversity, and other reef variables (e.g., seagrass abundance, macroalgae abundance, or sea urchin abundance) at this particular location. Additionally, we gathered information about local fishing practices.

## 2. Materials and Methods

### 2.1. Study Region and Sites

The study was conducted on coral reefs at Gazi Bay, adjacent to Gazi Village (4°26′ S, 39°30′ E), a small fisher village located on the south coast of Kenya, about 45 km south of Mombasa [45] (Figure 1). The inner (northern) part of the bay is fringed by mangroves and seagrass meadows. The open sea lies to the south, while to the south-east, extending south from Chale Island, a submarine promontory known as Chale Reef supports the coral reef community that was the subject of this study. The reefs inside this lagoon are all relatively shallow, not exceeding 5 m in depth. The area is open to fishing and no form of protected area exists in the vicinity, so there has been increasing concern that fishing pressure is affecting the reefs. Although reef fish are known to be heavily exploited by local fishers, so far as it is known, reef invertebrates such as sea urchins are not exploited locally. Mean ambient air temperature and water temperature were both approximately 28 °C in October and November 2018 when the work was undertaken.

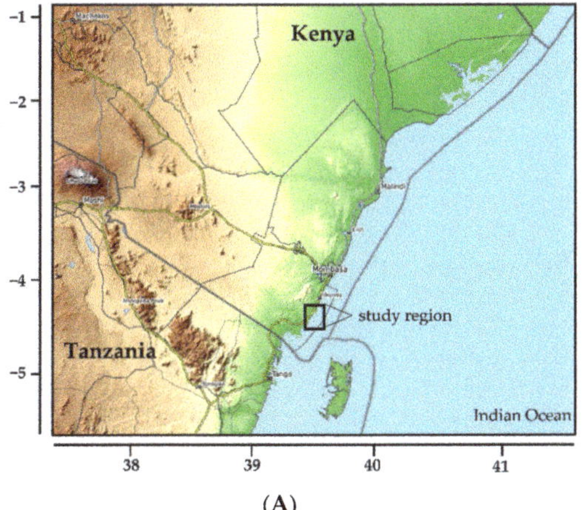 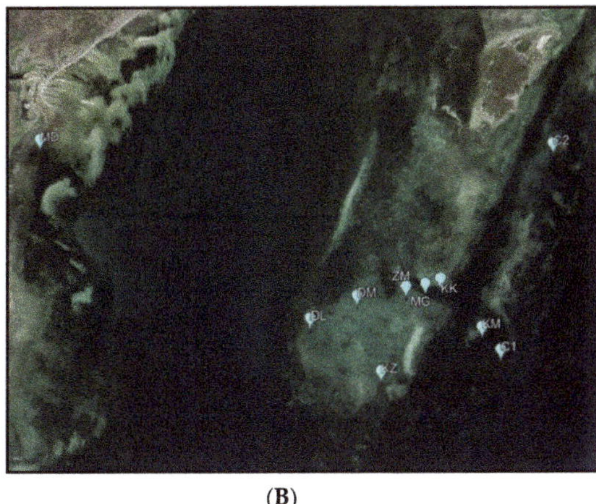

(A)          (B)

**Figure 1.** (**A**) Location of the study region on the southern Kenyan coast of East Africa; the scales indicate latitude south of the equator and longitude east of the prime meridian. Color on land indicates altitude. (**B**) Aerial image of Gazi Bay showing the location of the ten study sites: Chale 1 (C1), Chale 2 (C2), Doa Lower (DL), Doa Upper (DM), Kijamba Mkandi (KM), Kiziwa Kule (KZ), Kukuni (KK), Mikingamo (MG), Mikingamo 2 (ZM), Mwandamo (MD) (Satellite image © Google).

### 2.2. Monitoring of Coral Reef Fishes and Corals

Corals and fishes were surveyed at ten different reef sites on Chale Reef, with some transects on the seaward side of the reef and some on the landward side to provide a range of different reef conditions. Within each site, three belt transects of 5 × 50 m were placed haphazardly, but mostly parallel to the shore. Transects were located at least 150 m apart so as to achieve reasonable independence of the samples. Sampling was conducted around low tide (<70 cm) to provide standardized assessment conditions, while time of day and weather conditions varied through the study. During each survey, one person snorkeled in front, counting fishes using established underwater visual survey methodology [46], while a second person swimming behind laid out a 50 m tape along a track following the first snorkeler (this sequence was followed to avoid the fish being disturbed before they could be counted). For this study, eight fish families, which are known indicators of coral reef biodiversity, were preselected for the census: *Scaridae*, *Acanthuridae*, *Siganidae*, *Ephippidae*, *Pomacanthidae*, *Chaetodontidae*, *Lutjanidae*, and *Serranidae*, all of which are moderately speciose on Kenya's southern coast [47]. Additionally, on the first half of the transect (25-m length) the hard corals were identified to genus level and the colonies counted.

### 2.3. Benthic Analyses

After counting fish, the first half (25 m length) of each transect was investigated to assess the substrate cover of corals, turf, and macroalgae and seagrass, and of other benthic organisms such as sea urchins. Photos of the substrate were taken from 1 m above the sea floor surface, with five photos being taken at intervals of 5 m per transect (camera: Canon Powershot G7 X Mark II with the Fantasea underwater housing). Because of limitations of time and resources, only a limited number of replicate quadrats could be sampled in this way. The photos of the benthos were analyzed with the assistance of the Coral Point Count (CPC) software [48]. The program was set to generate 25 random points on each image, the substrates below each of which were then classified by the observer into 16 major benthic categories. From these values the percentage substrate covers of the following benthic categories were then calculated: "hard coral" (HC), "soft coral" (SC), "rubble" (RUB), "seagrass" (SGR), "sand" (SND), "macroalgae" (AMAC), "turf algae" (ATRF),

"*Halimeda* algae" (AHAL), "dead coral" (DC), "recently dead coral" (RDC), "bare substrate" (BS), "unidentified/others" (UNID/OT), and "obstacles/tape/shadows" (OBS/TWS), in addition to the abundance of "herbivorous invertebrates" (INV), specifically of echinoids (sea urchins).

### 2.4. Statistical Analysis

Statistical analyses were performed in the R software 3.4.4 (R Core Team, Vienna, Austria) [49] using the R packages "readxl", "gdata", "MuMIn", "mgcv", "reshape2", "tiff", and "nlme". Linear models (lm) were used to identify the key predictors of fish abundance and fish family richness and of coral cover and coral genus richness. For overall fish abundance and fish family richness (i.e., the number of fish families recorded), we tested the proportional cover of the main substrate variables, "macroalgae", "turf algae", "herbivorous invertebrates", "seagrass" and "coral", as predictor variables. For coral cover and coral genus richness (i.e., the number of coral genera recorded) we used the same substrate variables, "macroalgae", "turf algae", "herbivorous invertebrates", and "seagrass", and also "fish abundance", and "fish family richness" were tested as predictor variables. The herbivorous invertebrate data was $log_{n+1}$ transformed (as is often done with zero biased data) in order to increase linearity between predictor and response variables and to increase normality of model residuals. To evaluate the support for different combinations of predictor variables, the dredge-function of the "MuMIn" R package was used. For each model, the Akaike Information Criterion (AIC) was derived, which evaluates models based on model fit and model complexity [50]. As the sample size was relatively low compared to the number of estimated parameters, the AIC with a second-order bias correction ($AIC_C$) was used. Using $AIC_C$-based model selection, the best linear models were selected for explaining fish abundance, fish species richness, coral cover, and coral genus richness. The best model was considered to be the one with the lowest $AIC_C$ ranking [50].

## 3. Results

In total, 213 fishes from seven of the eight indicator families and 845 hard coral colonies from 22 genera were recorded. Across all sites, the *Acanthuridae* (surgeon fishes) were the dominant family in terms of individuals, followed by the *Chaetodontidae* (butterflyfishes). From the other families, relatively few individuals were recorded (Table 1).

**Table 1.** Mean number of fish families per site ± standard deviation.

| | Chale 1 | Chale 2 | Doa Lower | Doa Upper | Kijamba Mkandi | Kiziwa Kule | Kukuni | Mikingamo | Mikingamo 2 | Mwandamo |
|---|---|---|---|---|---|---|---|---|---|---|
| Acanthuridae | 4.0 ± 2.6 | 0.0 | 0.0 | 0.0 | 18 ± 4.0 | 0.7 ± 1.2 | 6.0 ± 7.9 | 2.0 ± 1.7 | 6.0 ± 5.3 | 3.0 ± 1.7 |
| Chaetodontidae | 4.3 ± 3.1 | 0.7 ± 1.2 | 0.0 | 0.3 ± 0.6 | 2.0 ± 1.7 | 1.0 ± 1.7 | 7.7 ± 7.1 | 2.0 ± 1.7 | 5.0 ± 5.0 | 1.7 ± 1.2 |
| Ephippidae | 0.0 | 0.0 | 0.0 | 0.0 | 0.0 | 0.0 | 0.0 | 0.0 | 1.0 ± 1.7 | 0.0 |
| Lutjanidae | 0.0 | 0.0 | 0.0 | 0.0 | 1.3 ± 2.3 | 0.0 | 1.0 ± 1.7 | 0.0 | 0.0 | 0.7 ± 1.2 |
| Pomacanthidae | 0.0 | 0.0 | 0.0 | 0.0 | 0.0 | 0.0 | 0.0 | 0.0 | 0.0 | 0.3 ± 0.6 |
| Scaridae | 0.0 | 0.0 | 0.0 | 0.0 | 0.0 | 0.0 | 0.3 ± 0.6 | 0.0 | 1.3 ± 1.2 | 0.0 |
| Siganidae | 0.0 | 0.0 | 0.0 | 0.0 | 0.0 | 0.0 | 0.0 | 0.0 | 0.3 ± 0.6 | 0.3 ± 0.6 |

For corals, the most abundant genera were massive *Porites* followed by *Acropora* (Table 2). The benthic analyses showed a dominance of macroalgae and turf algae at all sites (Table 2). Hard corals, sand, seagrass, and rubble also formed a large part of the benthos, whereas herbivorous invertebrates (mainly sea urchins), *Halimeda* algae and soft corals were less abundant (Table 3).

**Table 2.** Mean number of coral colonies for different genera per site ± standard deviation.

| | Chale 1 | Chale 2 | Doa Lower | Doa Upper | Kijamba Mkandi | Kiziwa Kule | Kukuni | Mikingamo | Mikingamo 2 | Mwandamo |
|---|---|---|---|---|---|---|---|---|---|---|
| Acanthastrea | 0.00 | 0.00 | 0.00 | 0.00 | 0.00 | 0.00 | 0.00 | 0.33 ± 0.58 | 0.00 | 0.67 ± 0.58 |
| Acropora | 0.00 | 4.33 ± 1.15 | 2.00 ± 1.00 | 4.00 ± 2.00 | 14.00 ± 6.56 | 3.00 ± 1.73 | 1.33 ± 1.53 | 3.33 ± 3.06 | 2.00 | 7.33 ± 7.77 |
| Astreopora | 0.00 | 1.00 ± 1.73 | 3.00 ± 1.00 | 3.67 ± 2.52 | 0.00 | 0.33 ± 0.58 | 1.00 ± 1.73 | 1.00 ± 1.00 | 1.33 ± 1.73 | 0.33 ± 0.58 |
| Diploastrea | 0.00 | 0.00 | 0.00 | 0.00 | 0.00 | 0.00 | 0.33 ± 0.58 | 0.00 | 0.00 | 0.00 |
| Echinopora | 0.33 ± 0.58 | 1.33 ± 0.58 | 1.33 ± 1.53 | 0.00 | 0.33 ± 0.58 | 0.00 | 1.67 ± 2.08 | 0.00 | 1.00 ± 1.73 | 3.33 ± 4.04 |
| Favia | 0.67 ± 1.15 | 5.33 ± 1.53 | 0.00 | 0.00 | 0.67 ± 0.58 | 0.33 ±0.58 | 0.00 | 2.00 ± 2.65 | 0.00 | 1.33 ± 1.15 |
| Favites | 0.00 | 1.33 ± 1.15 | 1.33 ± 1.15 | 0.33 ± 0.58 | 1.00 ± 1.73 | 1.67 ± 2.08 | 0.33 ± 0.58 | 2.67 ± 1.53 | 0.67 ± 0.58 | 0.67 ± 0.58 |
| Fungia | 0.00 | 0.00 | 0.00 | 0.00 | 3.00 ± 5.20 | 0.00 | 3.33 ± 2.52 | 1.00 ± 1.73 | 0.00 | 0.33 ± 0.58 |
| Galaxea | 2.33 ± 1.15 | 1.67 ± 1.53 | 0.67 ± 1.15 | 1.00 ± 1.00 | 19.33 ± 8.08 | 0.33 ± 0.58 | 2.33 ± 2.08 | 0.33 ± 0.58 | 0.67 ± 0.58 | 0.00 |
| Goniastrea | 0.00 | 0.00 | 0.00 | 0.00 | 0.00 | 0.00 | 0.00 | 0.00 | 0.00 | 0.00 |
| Goniopora | 1.67 ± 1.53 | 2.67 ± 1.15 | 1.00 ± 1.00 | 0.00 | 0.00 | 0.00 | 0.33 ± 0.58 | 4.67 ± 1.53 | 2.33 ± 3.21 | 0.67 ± 1.15 |
| Hydnophora | 0.00 | 0.00 | 0.33 ± 0.58 | 0.00 | 0.33 ± 0.58 | 0.67 ± 0.58 | 0.67 ± 0.58 | 0.33 ± 0.58 | 0.00 | 0.00 |
| Leptoria | 0.00 | 2.67 ± 1.15 | 0.00 | 0.00 | 0.00 | 0.00 | 0.00 | 0.00 | 0.33 ± 0.58 | 0.00 |
| Lobophyillia | 0.00 | 0.00 | 0.00 | 0.00 | 0.00 | 0.00 | 0.00 | 0.00 | 0.00 | 0.00 |
| Millepora | 0.00 | 0.00 | 0.00 | 0.00 | 0.00 | 0.00 | 0.00 | 0.00 | 0.00 | 0.00 |
| Montipora | 2.0 ± 1.73 | 1.33 ± 1.53 | 0.67 ± 0.58 | 0.00 | 1.00 ± 1.73 | 0.33 ± 0.58 | 0.33 ± 0.58 | 1.67 ± 1.15 | 2.00 ± 1.00 | 0.00 |
| Pavona | 0.00 | 0.00 | 0.33 ± 0.58 | 0.00 | 0.00 | 0.00 | 0.67 ± 1.15 | 1.00 ± 1.00 | 0.00 | 3.00 ± 2.00 |
| Platygyra | 0.00 | 1.33 ± 1.15 | 0.67 ± 0.58 | 0.00 | 0.67 ± 1.15 | 0.67 ± 0.58 | 1.00 ± 1.15 | 2.67 ± 0.58 | 0.00 | 2.67 ± 3.29 |
| Plesiastrea | 0.00 | 1.00 ± 1.00 | 0.67 ± 0.58 | 0.00 | 0.00 | 0.00 | 0.00 | 0.00 | 1.33 ± 0.58 | 0.00 |
| Pocillopora | 0.00 | 1.00 | 0.33 ± 0.58 | 1.33 ± 0.58 | 12.67 ± 3.21 | 1.67 ± 2.08 | 1.67 ± 1.15 | 0.67 ± 1.15 | 0.67 ± 0.58 | 7.00 ± 6.00 |
| Porites (Branching) | 0.00 | 0.00 | 0.33 ± 0.58 | 0.00 | 1.33 ± 1.15 | 0.00 | 0.00 | 0.00 | 0.67 ± 1.15 | 4.33 ± 5.13 |
| Porites (Massive) | 0.00 | 4.67 ± 0.58 | 16.67 ± 1.53 | 10.67 ± 4.73 | 0.00 | 1.33 ± 1.15 | 0.67 ± 0.58 | 17.33 ± 8.08 | 16.00 ± 9.17 | 4.33 ± 2.52 |
| Seriatopora | 0.00 | 0.00 | 0.00 | 0.33 ± 0.58 | 0.33 ± 0.58 | 0.00 | 0.00 | 0.33 ± 0.58 | 0.33 ± 0.58 | 0.00 |
| Stylophora | 0.00 | 0.00 | 0.00 | 0.67 ± 0.58 | 0.00 | 0.00 | 3.00 ± 1.73 | 4.33 ± 5.86 | 1.00 ± 1.00 | 0.00 |
| Symphilla | 0.00 | 0.00 | 0.00 | 0.00 | 0.00 | 0.00 | 0.00 | 0.00 | 0.00 | 0.00 |

**Table 3.** Percentage cover of the benthos per site ± standard deviation.

| | Chale 1 | Chale 2 | Doa Lower | Doa Upper | Kijamba Mkandi | Kiziwa Kule | Kukuni | Mikingamo 1 | Mikingamo 2 | Mwandamo |
|---|---|---|---|---|---|---|---|---|---|---|
| AHAL | 0.27 ± 0.46 | 0.00 | 0.27 ± 0.46 | 0.00 | 0.27 ± 0.46 | 0.81 ± 1.40 | 0.57 ± 0.96 | 0.00 | 1.39 ± 2.41 | 0.00 |
| AMAC | 56.15 ± 0.76 | 19.67 ± 11.41 | 2.68 ± 1.21 | 5.50 ± 1.31 | 62.51 ± 5.52 | 62.74 ± 9.48 | 41.20 ± 16.85 | 5.91 ± 3.57 | 3.74 ± 1.68 | 14.76 ± 17.35 |
| ATRF | 33.27 ± 1.89 | 51.48 ± 15.07 | 32.45 ± 8.47 | 46.41 ± 16.83 | 14.67 ± 2.46 | 18.63 ± 8.59 | 26.97 ± 20.59 | 63.63 ± 5.91 | 54.50 ± 13.95 | 28.31 ± 7.56 |
| BS | 0.00 | 0.00 | 0.00 | 0.56 ± 0.96 | 0.00 | 0.00 | 3.56 ± 3.38 | 0.28 ± 0.48 | 0.00 | 0.00 |
| HC | 0.53 ± 0.92 | 4.32 ± 1.89 | 14.22 ± 0.37 | 2.18 ± 1.72 | 10.73 ± 6.05 | 0.53 ± 0.92 | 8.33 ± 3.58 | 1.34 ± 1.66 | 10.47 ± 6.78 | 16.28 ± 8.56 |
| DC | 0.00 | 0.00 | 0.00 | 0.00 | 0.00 | 0.00 | 1.40 ± 2.42 | 0.00 | 0.00 | 0.00 |
| URC | 0.00 | 1.34 ± 1.24 | 3.77 ± 3.08 | 2.53 ± 2.28 | 1.89 ± 1.68 | 0.27 ± 0.46 | 1.88 ± 3.25 | 8.31 ± 8.37 | 2.22 ± 0.97 | 0.53 ± 0.92 |
| RDC | 0.00 | 0.00 | 0.00 | 0.00 | 0.00 | 0.00 | 3.35 ± 5.80 | 2.70 ± 4.68 | 0.00 | 0.00 |
| RUB | 0.00 | 0.54 ± 0.94 | 7.78 ± 1.71 | 3.89 ± 3.94 | 4.84 ± 4.47 | 3.49 ± 1.22 | 0.00 | 4.83 ± 5.24 | 10.46 ± 13.95 | 12.10 ± 10.67 |
| SND | 5.39 ± 5.19 | 15.94 ± 7.01 | 28.53 ± 13.15 | 16.45 ± 8.69 | 0.00 | 7.03 ± 3.27 | 4.63 ± 7.34 | 10.04 ± 9.87 | 15.04 ± 7.34 | 7.37 ± 7.03 |
| SGR | 3.06 ± 5.29 | 5.07 ± 2.31 | 8.91 ± 6.99 | 21.66 ± 20.03 | 4.27 ± 3.78 | 6.22 ± 0.88 | 9.23 ± 9.29 | 0.00 | 0.80 ± 1.39 | 16.03 ± 13.00 |
| SC | 1.33 ± 1.22 | 1.62 ± 1.41 | 1.39 ± 2.41 | 0.83 ± 1.44 | 0.81 ± 0.80 | 0.27 ± 0.46 | 0.00 | 0.27 ± 0.46 | 0.58 ± 0.50 | 1.33 ± 2.31 |
| UNID/OT | 0.00 | 0.00 | 0.00 | 0.00 | 0.00 | 0.00 | 0.29 ± 0.50 | 0.00 | 0.80 ± 1.39 | 3.29 ± 1.67 |
| OBS/TWS | 1.33 ± 0.46 | 0.80 ± 0.80 | 1.07 ± 0.46 | 2.93 ± 1.67 | 0.53 ± 0.46 | 1.33 ± 0.92 | 2.67 ± 1.22 | 1.07 ± 1.22 | 1.87 ± 1.67 | 1.07 ± 0.46 |

Based on the Akaike Information Criterion (AIC$_C$), the best supported model for coral genus richness was one in which seagrass and macroalgae both had negative effects (Table 4). Coral genus richness doubled as macroalgae cover dropped from 69 to 2% (Figure 2A). The data indicated, on average, a loss of one coral genus for every additional 20% increase in the cover of macroalgae. Coral genus richness was similarly negatively related to seagrass substrate cover (Figure 2B). In the second- and third-best models, the presence of turf algae had an additional negative effect on coral genus richness (Table 4). In the best supported model of the factors influencing coral cover, (log-transformed) herbivorous invertebrate abundance and fish family richness both showed significant positive effects (Table 4). On average, there were twice as many corals on the transects where invertebrate numbers were the highest compared to those where it was lowest (Figure 2C). Similarly, we observed a significant increase in coral cover with increasing fish species richness (Figure 2D). However, in the second-best model, it was fish abundance instead of the fish family richness that showed a positive relationship. In the third model, turf algae cover had an additional negative impact (Table 4).

**Table 4.** Results of the AIC$_C$ based model selection for the environmental variables coral genus richness (CGR) and coral cover (CA), using the predictor variables AMAC (macroalgae), ATRF (turf algae), FABUND (fish abundance), FFM (fish family richness), Lg1(INV) (log-transformed herbivorous invertebrate abundance), and SGR (seagrass). '*' $p < 0.05$; '**' $p < 0.01$; '.' $0.05 < p < 0.10$. Successive lines show the results for the best supported, and second- and third-best supported models.

| Response | AMAC | ATRF | FABUN | FFM | Lg1(INV) | SGR | AIC$_C$ | R$^2$ |
|---|---|---|---|---|---|---|---|---|
| CGR | −0.5676 ** | | | | | −0.3010 . | 146.1 | 0.3581 |
| | −0.8815 ** | −0.4119 | | | | −0.4850 * | 146.2 | 0.4139 |
| | −0.5195 ** | | | | | | 147.3 | 0.2698 |
| CA | | | 0.3333 * | 0.3908 * | 0.5389 ** | | 247.0 | 0.3573 |
| | | −0.1330 | | | 0.4886 ** | | 248.6 | 0.3211 |
| | | | | 0.3968 * | 0.6091 ** | | 249.3 | 0.3702 |

Fish family richness was best explained by a model in which coral cover had a positive impact, whereas herbivorous invertebrate abundance had a negative impact (Table 5). Fish family richness tripled along a gradient of coral cover from 2 to 69 percent (Figure 3A), whereas there were approximately twice as many fish species on sites with low invertebrate abundance as on those with high abundance (Figure 3B). In the second-best supported model, seagrass cover appeared to have an additional negative impact, whereas in the third-best model coral cover alone influenced the family richness of fish. In the best model for fish abundance (Table 5, Model B), macroalgae and coral cover both had positive effects on fish abundance. There were more than twice as many fish on the most coral rich site as on the least coral rich site (Figure 3C), while fish abundance also doubled along a gradient of macroalgae cover from 1 to 69% (Figure 3D). The second-best supported model explaining fish abundance showed a negative impact of turf algae and seagrass, whereas the third-best model contained only the positive impact of macroalgae abundance.

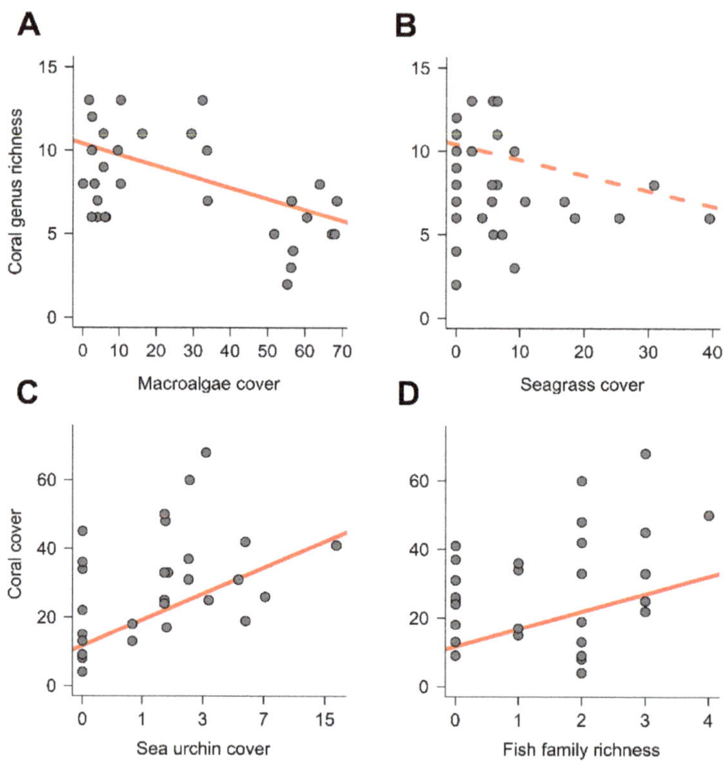

**Figure 2.** Relationship of different substrate variables to coral genus richness and coral cover at the reefs in Gazi Bay. Black lines correspond to model prediction analysis, orange lines represent the estimates derived from multiple regression analysis. Coral genus richness decreased with increasing macroalgae (**A**) and seagrass cover (**B**). Coral cover increased with increasing herbivorous invertebrate abundance (**C**) and fish family richness (**D**). Grey data points show individual values on each of the 30 transects. The dashed line indicates that parameter estimates are only marginally significant ($0.05 < p < 0.10$).

**Table 5.** Results of the $AIC_C$ based model selection for fish family richness (FFR) and fish abundance (FA) using the key predictor variables AMAC (macroalgae), ATRF (turf algae), CA (coral cover), Lg1(INV) (log-transformed invertebrate abundance), and SGR (seagrass). '*' $p < 0.05$; '.' $0.05 < p < 0.10$. Successive lines show the results for the best supported, and second- and third-best supported models.

| Response | AMAC | ATRF | CA | Lg1(INV) | SGR | $AIC_C$ | $R^2$ |
|---|---|---|---|---|---|---|---|
| FFR | | | 0.4747 * | −0.4291 * | | 98.4 | 0.21930 |
| | | | 0.4238 * | −0.4498 * | −0.2056 | 99.8 | 0.25690 |
| | | | 0.2809 | | | 100.6 | 0.07892 |
| FA | 0.3562. | | 0.3557. | | | 217.2 | 0.20680 |
| | | −0.3186 . | | | −0.3737 . | 218.8 | 0.16360 |
| | 0.2908 | | | | | 218.8 | 0.08457 |

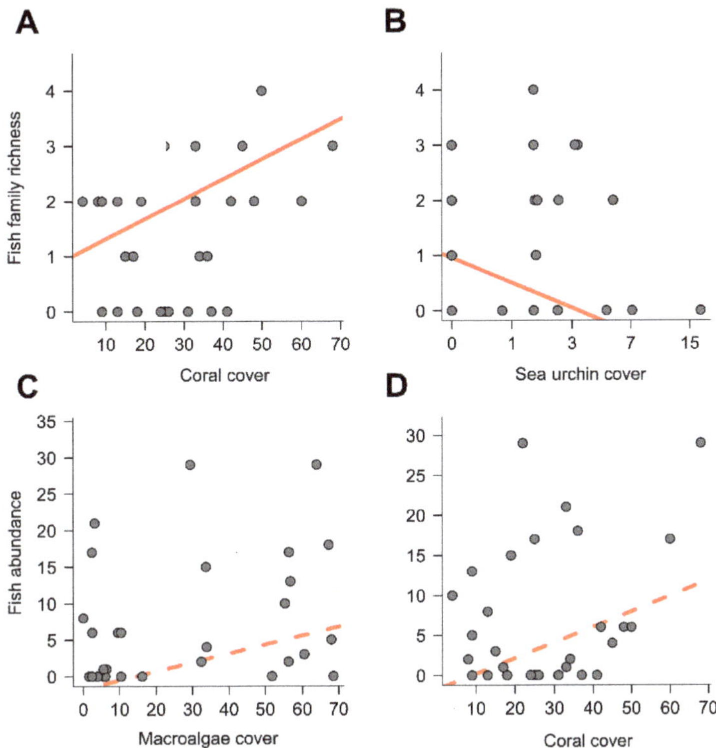

**Figure 3.** Relationship of different environmental factors to fish family richness and fish abundance. The orange lines represent the estimates derived from multiple-regression analysis. Fish family richness increases with increasing coral cover (**A**) and decreases with increasing invertebrate/sea urchin abundance (**B**). Fish abundance increases with increasing coral cover (**C**) and macroalgae cover (**D**). Grey dots show individual measures of response variables on the 30 transects.

## 4. Discussion

We found that on the reefs at Gazi Bay coral genus richness was negatively correlated with the abundance of macroalgae, whereas coral cover was positively correlated with the abundance of herbivorous invertebrates (effectively sea urchins) and with fish family richness. Fish family richness was mainly correlated with coral cover and invertebrate abundance, but no determinants of fish abundance could be convincingly identified. However, caution is required before assuming these trends apply more widely, because, for example, the abundance of fishes and the extent of live coral cover observed in Gazi Bay appeared noticeably low compared to protected reef sites on the Kenyan coast visited by the authors in the same year (e.g., in the Kisite-Mpunguti Marine National Park and the Malindi National Park).

*Acanthuridae* (surgeon fishes) were the most abundant family across the sites, followed by the *Chaetodontidae* (butterflyfishes). As specific butterflyfishes are facultative or obligate corallivores they are generally associated with high live coral cover [47,51–53], even though in some cases this correlation was found to be quite low [54,55]. We found that there are enough living corals for at least some butterflyfish species to inhabit the reefs of Gazi Bay. Nevertheless, the benthic composition was dominated by macroalgae and turf algae. This may be because of the near absence of some other herbivorous fishes, notably species of parrotfish [10,25]. Their intensive feeding promotes the settlement, growth, and survivorship of corals due to the removal of competitive macroalgae. In this way herbivores are considered important for coral reef resilience, because they keep macroalgae at low levels after any disturbance affecting corals, thus allowing the corals to recover [3,56,57]. The fact that during the study we observed no groupers and only a few parrotfishes, rabbitfishes, and snappers, is likely a consequence of overfishing, because all three families are important food-fish on the Kenyan coast.

In addition to fish, we found sea urchin abundance to be positively related to coral cover. This relationship is likely explained by the negative impact of sea urchins on algae, which otherwise compete with corals for space and light. Various field and computer studies have shown that urchin domination can help maintain coral cover by reducing competitive algae cover [32,58–62]. The study by Nozawa et al. (2020) showed that diadematid sea urchins enhance coral recruitment on Taiwanese reefs [32]. Another study in Kenya found that areas that were inaccessible to urchin grazing retained a high cover of algal turf and so were not suitable for coral recruitment, whereas in stable reef systems grazing fishes graze down the turf algae, facilitating coral recruitment [58]. The areas that were less accessible to echinoids were, in the absence of grazing fishes, dominated by large, fleshy algae which smothered corals [58]. However, the effects of sea urchins on the dynamics of phase shifts on coral reefs are density dependent, because they can play two different roles. In low abundances, reef echinoids promote coral reef resilience by controlling algal cover [56,63–65]. In high abundances, echinoids may erode large parts of both living corals and the dead coral framework (in addition to the algae growing on the latter), resulting in loss in live coral cover, reduced coral recruitment, reduced calcium carbonate accretion, and reduced associated faunal diversity [9,10,56,63–65]. Our results, however, showed a positive effect of herbivorous invertebrate/sea urchin abundances on corals, indicating that at Gazi Bay sea urchin population density is not (yet?) harmful to the coral community.

Our results also showed that, in our study area, coral genus richness was negatively related to macroalgae abundance. This is likely also because of the negative effect (just discussed) of macroalgae on hard coral cover, as overgrowth of corals by macroalgae can suppress the coral fecundity, hinder coral growth, and smother coral recruits [7,58,66,67]. Algal turfs, by comparison with macroalgae, are less successful competitors of corals [66,68]. Hopkins (2009) also found this correlation between coral and macroalgae cover, with the hard-coral cover being low when algae cover is high (and vice versa). Coral-macroalgae phase shifts have been reported from many countries, including, notably, Jamaica [69], and also East Africa [44], a phenomenon generally believed to be triggered by various impacts, such as the loss of herbivores (either urchins or fish), increased nutrient input, hurricane damage, or crown-of-thorns-starfish outbreak [63].

In our study, seagrass abundance was negatively related to coral genus richness, although this relationship was only marginally significant in the best supported model. A negative relationship between seagrass cover and coral genus richness may be largely explained by corals and seagrasses colonizing different types of substrates, but there may also be a degree of competition for space on some substrates. A previous study described seagrass colonizing the reef substratum after it was eroded by sea urchins [44]. However, other studies have shown that seagrass meadows may enhance coral reef resilience against ocean acidification by modifying the pH of the water column [68,70].

We found the cover of corals to be positively related to both fish abundance and fish family richness. However, this could be the result either of fish activity, such as grazing on algae, favoring the health of corals, or of corals providing a better habitat for fish [10,71,72]. Herbivorous fishes, which made up a large proportion in the counted fish, may clear space for coral settlement [73] by ingesting seagrass, macroalgae, or algal turfs. In particular, by reducing the amount of competitively superior macroalgae, herbivorous fishes allow corals and cementing coralline algae to survive [3,72]. Exclusion experiments such as those described by Hughes et al. [61] have repeatedly found that after excluding herbivores from parts of a reef, the coverage of macroalgae will quickly increase but will decline rapidly again when the area is made newly accessible to grazers. With increasing biomass of grazers/detritivores, macroalgal cover decreases, thereby increasing the cover of live coral [7,25,66].

Although in our study the fact that herbivorous surgeonfishes (*Acanthuridae*) were the most abundant family makes it plausible that in our location it is reef fish diversity and abundance that are influencing coral cover, a positive relationship between coral cover and fish density and diversity has also been interpreted in many studies as inferring that it is coral cover which influences fish abundance [41,42]. In particular, increasing topographic or architectural complexity, substratum diversity, and live coral cover has been associated with increasing fish abundance and diversity [25,42,56,74,75]. Because of the difficulties in coral identification, the relationship between species richness of the fish assemblage and the diversity or cover of coral is less apparent in the literature than the effects of habitat complexity [42,76]. Coral reefs can provide fish with physical refuges from predators, in addition to breeding, nursery, and feeding sites, and thus enhance fish diversity. In contrast, coral mortality has been predicted to reduce reef fish abundance and diversity and so have long-term consequences on the community [19,56]. In fact, a recent study by Darling et al. suggests that reefs without corals will no longer support diverse fish faunas but be numerically dominated by a small subset of species that prefer algal or rubble substrata [77].

In our study, the abundance of sea urchins, besides being positively related to coral cover, was negatively related to fish species diversity. As we have said, echinoids are important grazers in shallow reef ecosystems and compete for resources with herbivorous fish and other benthic organisms [58,78]. Other studies have also revealed a negative impact of sea urchin abundance on the number of fish species and their abundance [58,78,79]. McClanahan et al. (1994) found that in heavily fished Kenyan reef lagoons, sea urchins are the most abundant grazers with herbivorous fishes showing a contrasting pattern of distribution. It appears that, as the fish density is reduced by fishing, echinoid populations expand to use the newly available resources [79]. Once an urchin-dominated community is established, it can be hard for herbivorous fishes to re-establish themselves [73]; high sea urchin abundances may even suppress fish growth and recruitment and hence recovery of the fish populations [58].

Due to the correlative nature of our study, it is not possible to determine the true causal relationship between fishes and corals or fishes and sea urchins, e.g., whether low abundances of fishes cause an increase in sea urchins or if high abundances of sea urchins cause a decline in fish species populations. Nevertheless, for the persistence of herbivorous fishes, it may be important to protect sea urchin predators, such as triggerfish (*Balistidae*) [21,28,80], which prevent sea urchins from becoming dominant, because otherwise these effects are not expected to be reversible [73,81,82].

## 5. Conclusions

As the shallow reef areas in Gazi Bay had low coral cover and low fish abundance, combined with high algae cover, we consider the condition of the reef there to be far from natural [44,83]. Without a longer-term monitoring program, we cannot be certain about the causes of the poor reef condition. Nevertheless, our study demonstrates, for an East African coral reef subject to high fishing pressure, clear relationships between algae growth, live coral cover, and the abundance and diversity of the reef fish assemblages. Increased human impact on the coral reefs (e.g., by increases in fishing pressure, destruction of corals, or discharge of nutrients, the lack of which otherwise helps limit algae growth) will inevitably lead to further changes in reef condition.

**Author Contributions:** Conceptualization, P.D., P.M. and M.K.P.; methodology, P.D. and P.M.; formal analysis, P.D., S.K., M.K.P.; investigation, P.D.; resources, P.D., P.M.; writing—original draft preparation, P.D.; writing—review and editing, M.K.P., P.M., S.K.; visualization, P.D., S.K., M.K.P.; supervision, M.K.P. All authors have read and agreed to the published version of the manuscript.

**Funding:** This research received no external funding.

**Institutional Review Board Statement:** Not applicable.

**Data Availability Statement:** The data presented in this study are available in an excel table format on request from the corresponding author.

**Acknowledgments:** Thanks go to the Kenyan Marine and Fisheries Research Institute (KMFRI) for providing the first author with accommodation and their facilities. Thanks are also due to Rupert Ormond for his most helpful comments and editing as well as to the anonymous reviewers.

**Conflicts of Interest:** The authors declare no conflict of interest.

## References

1. Roberts, C.M.; McClean, C.J.; Veron, J.E.N.; Hawkins, J.P.; Allen, G.R.; McAllister, D.E.; Mittermeier, C.G.; Schueler, F.W.; Spalding, M.; Wells, F.; et al. Marine biodiversity hotspots and conservation priorities for tropical reefs. *Science* **2002**, *295*, 1280–1284. [CrossRef]
2. Veron, J.E.N. *Corals of the World*; Australian Institute of Marine Science: Townsville, Australia, 2000; Volume 1.
3. Ogden, J.C.; Lobel, P.S. The role of herbivorous fishes and urchins in coral reefs communities. *Environ. Biol. Fishes* **1978**, *3*, 49–63. [CrossRef]
4. Hughes, T.P.; Baird, A.H.; Bellwood, D.R.; Card, M.; Connolly, S.R.; Folke, C.; Grosberg, R.; Hoegh-Guldberg, O.; Jackson, J.B.C.; Kleypas, J.; et al. Climate change, human impacts, and the resilience of coral reefs. *Science* **2003**, *301*, 929–933. [CrossRef]
5. Obura, D.O.; Grimsditch, G. *Resilience Assessment of Coral Reefs—Assessment Protocol for Coral Reefs, Focusing on Coral Bleaching and Thermal Stress*; IUCN: Gland, Switzerland, 2009; p. 70.
6. Moberg, F.; Folke, C. Ecological goods and services of coral reef ecosystems. *Ecol. Econ.* **1999**, *29*, 215–233. [CrossRef]
7. Knowlton, N.; Jackson, J.B.C. Shifting baselines, local impacts, and global change on coral reefs. *PLoS Biol.* **2008**, *6*, e54. [CrossRef]
8. Maynard, J.A.; Marshall, P.A.; Parker, B.; Mcleod, E.; Ahmadia, G.; Van Hooidonk, R.; Planes, S.; Williams, G.J.; Raymundo, L.; Beeden, R.; et al. *A Guide to Assessing Coral Reef Resilience for Decision Support*; UN Environment: Nairobi, Kenya, 2017.
9. Tebbett, S.B.; Bellwood, D.R. Functional links on coral reefs: Urchins and triggerfishes, a cautionary tale. *Mar. Environ. Res.* **2018**, *141*, 255–263. [CrossRef] [PubMed]
10. Bellwood, D.R.; Hughes, T.P.; Folke, C.; Nyström, M. Confronting the coral reef crisis. *Nature* **2004**, *429*, 827–833. [CrossRef] [PubMed]
11. Ferrigno, F.; Bianchi, C.N.; Lasagna, R.; Morri, C.; Russo, G.F.; Sandulli, R. Corals in high diversity reefs resist human impact. *Ecol. Indic.* **2016**, *70*, 106–113. [CrossRef]
12. Wilkinson, C.R. Global change and coral reefs: Impacts on reefs, economies and human cultures. *Glob. Chang. Biol.* **1996**, *2*, 547–558. [CrossRef]
13. Magris, R.A.; Grech, A.; Pressey, R.L. Cumulative human impacts on coral reefs: Assessing risk and management implications for Brazilian reefs. *Diversity* **2018**, *10*, 26. [CrossRef]
14. Nepote, E.; Bianchi, C.N.; Chiantore, M.; Morri, C.; Montefalcone, M. Pattern and intensity of human impact on coral reefs depend on depth along the reef profile and on the descriptor adopted. *Estuar. Coast. Shelf Sci.* **2016**, *178*, 86–91. [CrossRef]
15. Goatley, C.H.R.; Bonaldo, R.M.; Fox, R.J.; Bellwood, D.R. Sediments and herbivory as sensitive indicators of coral reef degradation. *Ecol. Soc.* **2016**, *21*, 21. [CrossRef]
16. McClanahan, T.R.; Muthiga, N.; Mangi, S. Coral and algal changes after the 1998 coral bleaching: Interaction with reef management and herbivores on Kenyan reefs. *Coral Reefs* **2001**, *19*, 380–391. [CrossRef]
17. Safaie, A.; Silbiger, N.J.; McClanahan, T.R.; Pawlak, G.; Barshis, D.J.; Hench, J.L.; Rogers, J.S.; Williams, G.J.; Davis, K.A. High-frequency temperature variability reduces the risk of coral bleaching. *Nat. Commun.* **2018**, *9*, 1671. [CrossRef] [PubMed]
18. Douglas, A.E. Coral bleaching—How and why? *Mar. Pollut. Bull.* **2003**, *46*, 385–392. [CrossRef]
19. Rudi, E.; Iskandar, T.; Fadli, N.; Hidayati. Effects of coral bleaching on reef fish fisheries at Sabang. In Proceedings of the 12th International Coral Reef Symposium, Cairns, Australia, 9–13 July 2012.
20. Bellwood, D.R.; Hughes, T.P.; Connolly, S.R.; Tanner, J. Environmental and geometric constraints on Indo-Pacific coral reef biodiversity. *Ecol. Lett.* **2005**, *8*, 643–651. [CrossRef]
21. McClanahan, T.R.; Shafir, S.H. Causes and consequences of sea urchin abundance and diversity in Kenyan coral reef lagoons. *Oecologia* **1990**, *83*, 362–370. [CrossRef]
22. Bahartan, K.; Zibdah, M.; Ahmed, Y.; Israel, A.; Brickner, I.; Abelson, A. Macroalgae in the coral reefs of Eilat (Gulf of Aqaba, Red Sea) as possible indicator of reef degradation. *Mar. Pollut. Bull.* **2010**, *60*, 759–764. [CrossRef]
23. McClanahan, T.R.; Uku, J.N.; Machano, H. Effect of macroalgal reduction on coral-reef fish in the Watamu Marine National Park, Kenya. *Mar. Freshw. Res.* **2002**, *53*, 223–231. [CrossRef]
24. Hill, J.; Wilkinson, C. *Methods for Ecological Monitoring of Coral Reefs—A Resource for Managers*; Australian Institute of Marine Science: Townsville, Australia, 2004; p. 117.
25. Green, A.L.; Bellwood, D.R. *Monitoring Functional Groups of Herbivorous Reef Fishes as Indicators of Coral Reef Resilience—A Practical Guide for Reef Managers in the Asia Pacific Region*; IUCN working group on Climate Change and Coral Reefs: Gland, Switzerland, 2009; p. 70.

26. Humphries, A.T.; McClanahan, T.R.; McQuaid, C.D. Algal turf consumption by sea urchins and fishes is mediated by fisheries management on coral reefs in Kenya. *Coral Reefs* **2020**, *39*, 1137–1146. [CrossRef]

27. McManus, J.W.; Polsenberg, J.F. Coral-algal phase shifts on coral reefs: Ecological and environmental aspects. *Prog. Oceanogr.* **2004**, *60*, 263–279. [CrossRef]

28. Cheal, A.J.; MacNeil, M.A.; Cripps, E.; Emslie, M.J.; Jonker, M.; Schaffelke, B.; Sweatman, H. Coral-macroalgal phase shifts or reef resilience: Links with diversity and functional roles of herbivorous fishes on the Great Barrier Reef. *Coral Reefs* **2010**, *29*, 1005–1015. [CrossRef]

29. Reverter, M.; Jackson, M.; Daraghmeh, N.; von Mach, C.; Milton, N. 11-yr of coral community dynamics in reefs around Dahab (Gulf of Aqaba, ed Sea): The collapse of urchins and rise of macroalgae and cyanobacterial mats. *Coal Reefs* **2020**, *39*, 1605–1618. [CrossRef]

30. Rempel, H.S.; Bodwin, K.N.; Ruttenberg, B.I. Impacts of parrotfish predation on a major reef-building coral: Quantifying healing rates and thresholds of coral recovery. *Coral Reefs* **2020**, *39*, 1441–1452. [CrossRef]

31. Roach, N.F.; Little, M.; Arts, M.G.I.; Huckeba, J.; Haas, A.F.; George, E.E.Q.R.A.; Cobián-Güemes, A.; Naliboff, D.S.; Silveira, C.; Vermeij, M.J.A.; et al. A multiomic analysis of in situ coral-turf algal interactions. *Proc. Natl. Acad. Sci. USA* **2020**, *117*, 13588–13595. [CrossRef]

32. Nozawa, Y.; Lin, C.; Meng, P. Sea urchins (diadematids) promote coral recover via recruitment on Taiwanese reefs. *Coral Reefs* **2020**, *39*, 1199–1207. [CrossRef]

33. Leahy, S.M.; Russ, G.R.; Abesamis, R.A. Primacy of bottom-up effects on a butterflyfish assemblage. *Mar. Freshw. Res.* **2016**, *67*, 1175–1185. [CrossRef]

34. Soule, D.F.e.a. *Marine Organisms as Indicators*; Springer: New York, NY, USA, 1988.

35. Reese, E.S. Predation on corals by fishes of the family *Chaetodontidae*: Implications for conservation and management of coral reef ecosystems. *Bull. Mar. Sci.* **1981**, *31*, 594–604.

36. Chabanet, P.; Adjeroud, M.; Andrèfouet, S.; Bozec, Y.M.; Ferraris, J.; Garcia-Charton, J.; Schrimm, M. Human-induced physical disturbances and their indicators on coral reef habitats: A multi-scale approach. *Aquat. Living Resour.* **2005**, *18*, 215–230. [CrossRef]

37. Noonan, K.R.; Childress, M.J. Association of butterflyfishes and stony coral tissue loss disease in the Florida Keys. *Coral Reefs* **2020**, *39*, 1581–1590. [CrossRef]

38. Nicolet, K.J.; Chong-Seng, K.M.; Pratchett, M.S.; Willis, B.L.; Hoogenboom, M.O. Predation scars may influence host susceptibility to pathogens: Evaluating the role of corallivores as vectors of coral disease. *Sci. Rep.* **2018**, *8*, 5258. [CrossRef]

39. Rice, M.M.; Ezzat, L.; Burkepile, D.E. Corallivory in the anthropocene: Interactive effects of anthropogenic stressors and corallivory on coral reefs. *Front. Mar. Sci.* **2019**, *5*, 525. [CrossRef]

40. Hughes, T.P.; Connell, J.H. Multiple stressors on coral reefs: A long-term perspective. *Limnol. Oceanogr.* **1999**, *44*, 932–940. [CrossRef]

41. Bell, J.D.; Galzin, R. Influence of live coral on coral reef fish communities. *Mar. Ecol. Prog. Ser.* **1984**, *15*, 265–274. [CrossRef]

42. Chabanet, P.; Ralambondrainy, H.; Amanieu, M.; Faure, G.; Galzin, R. Relationships between coral reef substrata and fish. *Coral Reefs* **1997**, *19*, 93–102. [CrossRef]

43. Done, T.J. Phase shifts in coral reef communities and their ecological significance. *Hydrobiologia* **1992**, *247*, 121–132. [CrossRef]

44. McClanahan, T.R.; Glaesel, H.; Rubens, J.; Kiambo, R. The effects of traditional fisheries management on fisheries yields and the coral-reef ecosystem of southern Kenya. *Environ. Conserv.* **1997**, *24*, 105–120. [CrossRef]

45. Dahdouh-Guebas, F.; Van Pottelbergh, I.; Kairo, J.G.; Cannicci, S.; Koedam, N. Human-impacted mangroves in Gazi (Kenya): Predicting future vegetation based on retrospective remote sensing, social surveys, and tree distribution. *Mar. Ecol. Prog. Ser.* **2004**, *272*, 77–92. [CrossRef]

46. Cowburn, B.; Samoilys, M.A.; Obura, D. The current status of coral reefs and their vulnerability to climate change and multiple human sresses in the Comoros Archipelago, Western India. *MArine Pollut. Bull.* **2018**, *133*, 956–969. [CrossRef] [PubMed]

47. Lieske, E.; Myers, R. *Coral Reef Fishes*; Princeton University Press: Princeton, NJ, USA, 2002.

48. Kohler, K.E.; Gill, S.M. Coral Point Count with Excel extensions (CPCe): A Visual Basic program for the determination of coral and substrate coverage using random point count methodology. *Comput. Geosci.* **2006**, *32*, 1259–1269. [CrossRef]

49. R Core Team. *R: A Language and Environment for Statistical Computing*; R Foundation for Statistical Computing: Vienna, Austria, 2018.

50. Burnham, K.P.; Anderson, D.R. Multimodel interference—Understanding AIC and BIC in model selection. *Sociol. Methoads Res.* **2004**, *33*, 261–304. [CrossRef]

51. Brooker, R.M.; Munday, P.L.; Mcleod, I.M. Habitat preferences of a corallivorous reef fish: Predation risk versus food quality. *Coral Reefs* **2013**, *32*, 613–622. [CrossRef]

52. Smith, J.E.; Brainard, R.; Carter, A.; Grillo, S.; Edwards, C.; Harris, J.; Lewis, L.; Obura, D.; Rohwer, F.; Sala, E.; et al. Re-evaluating the health of coral reef communities: Baselines and evidence for human impacts across the central Pacific. *Proc. R. Soc. B* **2016**, *283*, 20151985. [CrossRef] [PubMed]

53. Lieske, E.; Myers, R. *Korallenriff-Führer Rotes Meer*; Franckh-Kosmos Verlags GmbH & Co. KG: Stuttgart, Germany, 2010; Volume 2.

54. Roberts, C.M.; Ormond, R. Habitat complexity and coral reef diversity and abundance on Red Sea fringing reefs. *Mar. Ecol. Prog. Ser.* **1987**, *41*, 1–8. [CrossRef]

55. Fowler, A.J. Spatial and temporal patterns of distribution and abundance of chaetodontid fishes at One Tree Reef, southern GBR. *Mar. Ecol. Prog. Ser.* **1990**, *64*, 39–53. [CrossRef]
56. Sebens, K.P. Biodiversity of coral reefs: What are we losing and why? *Am. Zool.* **1994**, *34*, 115–133. [CrossRef]
57. Burkepile, D.E.; Hay, M.E. Herbivore species richness and feeding complementarity affect community structure and function on a coral reef. *Proc. Natl. Acad. Sci. USA* **2008**, *105*, 16201–16206. [CrossRef]
58. McClanahan, T.R.; Kamukuru, A.T.; Muthiga, N.A.; Gilagabher Yebio, M.; Obura, D. Effect of sea urchin reductions to algae, coral and fish populations. *Conserv. Biol.* **1996**, *10*, 136–154. [CrossRef]
59. Sammarco, P.W. Echinoid grazing as a structuring force in coral communities: Whole reef manipulations. *J. Exp. Mar. Biol. Ecol.* **1982**, *61*, 31–35. [CrossRef]
60. Carpenter, R.C. Mass mortality of Diadema antillarum: 1. long-term effects on sea urchin population dynamics and coral reef algal communitites. *Mar. Biol.* **1990**, *104*, 67–77. [CrossRef]
61. Hughes, T.P.; Reed, D.C.; Boyle, M.J. Herbivory on coral reefs: Community structure following mass mortalities of sea urchins. *J. Exp. Mar. Biol. Ecol.* **1987**, *113*, 39–59. [CrossRef]
62. McClanahan, T.R. A coral reef ecosystem-fisheries model: Impacts of fishing intensity and catch selection on reef structure and process. *Ecol. Model.* **1995**, *80*, 1–19. [CrossRef]
63. Norström, A.V.; Nyström, M.; Lokrantz, J.; Folke, C. Alternative states on coral reefs: Beyond coral-macroalgal phase shifts. *Mar. Ecol. Prog. Ser.* **2009**, *376*, 295–2009. [CrossRef]
64. McClanahan, T.R.; Muthiga, N.A. Changes in Kenyan coral reef community structure and function due to exploitation. *Hydrobiologia* **1988**, *166*, 269–276. [CrossRef]
65. Carreiro-Silva, M.; McClanahan, T.R. Echinoid bioerosion and herbivory on Kenyan coral reefs: The role of protection from fishing. *J. Exp. Mar. Biol. Ecol.* **2001**, *262*, 133–153. [CrossRef]
66. Hughes, T.P.; Rodrigues, M.J.; Bellwood, D.R.; Ceccarelli, D.; Hoegh-Guldberg, O.; McCook, L.J.; Moltschaniwskyj, N.; Pratchett, M.S.; Steneck, R.S.; Willis, B.L. Phase shifts, herbivory, and the resilience of coral reefs to climate change. *Curr. Biol.* **2007**, *17*, 360–365. [CrossRef]
67. Littler, M.M.; Littler, D.S. Assessment of coral reefs using herbivory/nutrient assays and indicator groups of benthic primary producers: A critical synthesis, proposed protocols. and critique of management strategies. *Aquat. Conserv. Mar. Freschwater Ecosyst.* **2007**, *17*, 195–215. [CrossRef]
68. Unsworth, R.K.F.; Collier, C.J.; Henderson, G.M.; McKenzie, L.J. Tropical seagrass meadows modify seawater carbon chemistry: Implications for coral reefs impacted by ocean acidification. *Environ. Res. Lett.* **2012**, *7*, 7. [CrossRef]
69. Hughes, T.P. Catastrophes, phase shifts, and large-scale degradation of a Caribbean coral reef. *Science* **1994**, *265*, 1547–1551. [CrossRef] [PubMed]
70. Hendriks, I.E.; Olsen, Y.S.; Ramajo, L.; Basoo, L.; Steckbauer, A.; Moore, T.S.; Howard, J.; Duarte, C.M. Photosynthetic activity buffers ocean acidification in seagrass meadows. *Biogeosciences* **2014**, *11*, 333–346. [CrossRef]
71. Nanami, A. Spatial distribution and feeding substrate of butterflyfishes (family *Chaetodontidae*) on an Okinawan coral reef. *PeerJ* **2020**, *8*, e9666. [CrossRef] [PubMed]
72. Hoey, A.S.; Pratchett, M.S.; Cvitanovic, C. High macroalgal cover and low coral recruitment undermines the potential resilience of the world's southermist coral reef assemblages. *PLoS ONE* **2011**, *6*, e25824. [CrossRef] [PubMed]
73. Jennings, S.; Kaiser, M.J. The effects of fishing on marine ecosystems. *Adv. Mar. Biol.* **1998**, *34*, 203–352.
74. Luckhurst, B.; Luckhurst, K. Estimating total abundance of a large temperate reef fish using visual strip-transect. *Mar. Biol.* **1987**, *96*, 469–478.
75. Cole, A.J.; Pratchett, M.S.; Jones, G.P. Diversity and functional importance of coral-feeding fishes on tropical coral reefs. *Fish Fish.* **2008**, *9*, 286–307. [CrossRef]
76. Darling, E.S.; Graham, N.A.J.; Januchowski-Hartley, F.A.; Nash, K.L.; Pratchett, M.S.; Wilson, S.K. Relationship between structural complexity, coral traits, and reef fish assemblages. *Coral Reefs* **2017**, *36*, 561–575. [CrossRef]
77. Jones, G.P.; McCormick, M.I.; Srinivasan, M.; Eagle, J.V. Coral decline threatens fish biodiversity in marine reserves. *Proc. Natl. Acad. Sci. USA* **2004**, *101*, 8251–8253. [CrossRef] [PubMed]
78. McClanahan, T.R. Kenyan coral reef lagoon fish: Effects of fishing, substrate complexity, and sea urchins. *Coral Reefs* **1994**, *13*, 231–241. [CrossRef]
79. McClanahan, T.R.; Nugues, M.M.; Mwachireya, S. Fish and sea urchin herbivory and competition in Kenyan coral reef lagoons: The role of reef management. *J. Exp. Mar. Biol. Ecol.* **1994**, *184*, 237–254. [CrossRef]
80. McClanahan, T.R. Recovery of a coral reef keystone predator, *Balistapus undulatus*, in East African marine parks. *Biol. Conserv.* **2000**, *94*, 191–198. [CrossRef]
81. McClanahan, T.R.; Obura, D. Status of Kenyan coral reefs. *Coast. Manag.* **1995**, *23*, 57–76. [CrossRef]
82. Watson, M.; Righton, D.; Austin, T.; Ormond, R. The effects of fishing on coral reef fish abundance and diversity. *J. Mar. Biol. Assoc. United Kingd.* **1996**, *76*, 229–233. [CrossRef]
83. McClanahan, T.R.; Mwaguni, S.; Muthiga, N.A. Management of the Kenyan coast. *Ocean. Coast. Manag.* **2005**, *48*, 901–931. [CrossRef]

*Article*

# Effect of Acute Seawater Temperature Increase on the Survival of a Fish Ectoparasite


**Mary O. Shodipo [1], Berilin Duong [2], Alexia Graba-Landry [3], Alexandra S. Grutter [2] and Paul C. Sikkel [4,5,***

[1]  Institute of Environmental and Marine Sciences, Silliman University, 6200 Dumaguete City, Negros Oriental, Philippines; mary.shodipo@gmail.com

[2]  School of Biological Sciences, The University of Queensland, St Lucia QLD 4072, Australia; berilin.duong@uq.net.au (B.D.); a.grutter@uq.edu.au (A.S.G.)

[3]  ARC Centre of Excellence for Coral Reef Studies, James Cook University, Townsville QLD 4811, Australia; alexia.grabalandry@my.jcu.edu.au

[4]  Department of Biological Sciences and Environmental Sciences Program, Arkansas State University, Jonesboro, AR 72467, USA

[5]  Water Research Group, Unit for Environmental Science and Management, North-West University, Potchefstroom 2520, South Africa

*  Correspondence: paul.sikkel@gmail.com


Received: 17 July 2020; Accepted: 27 September 2020; Published: 4 October 2020


**Abstract:** Extreme warming events that contribute to mass coral bleaching are occurring with increasing regularity, raising questions about their effect on coral reef ecological interactions. However, the effects of such events on parasite-host interactions are largely ignored. Gnathiid isopods are common, highly mobile, external parasites of coral reef fishes, that feed on blood during the juvenile stage. They have direct and indirect impacts on their fish hosts, and are the major food source for cleaner fishes. However, how these interactions might be impacted by increased temperatures is unknown. We examined the effects of acute temperature increases, similar to those observed during mass bleaching events, on survivorship of gnathiid isopod juveniles. Laboratory experiments were conducted using individuals from one species (*Gnathia aureamaculosa*) from the Great Barrier Reef (GBR), and multiple unknown species from the central Philippines. Fed and unfed GBR gnathiids were held in temperature treatments of 29 °C to 32 °C and fed Philippines gnathiids were held at 28 °C to 36 °C. Gnathiids from both locations showed rapid mortality when held in temperatures 2 °C to 3 °C above average seasonal sea surface temperature (32 °C). This suggests environmental changes in temperature can influence gnathiid survival, which could have significant ecological consequences for host-parasite-cleaner fish interactions during increased temperature events.



**Keywords:** Gnathiidae; Isopoda; coral reefs; climate change; ocean warming; coral bleaching; Great Barrier Reef; Coral Triangle


---

## 1. Introduction

Among the myriad anthropogenic impacts on the world's oceans, perhaps the most significant is the increase in temperature associated with production of greenhouse gases [1]. This warming is responsible for large-scale changes in circulation and storm activity through melting of glaciers, warming of air masses, and increased evaporation and salinity [1], and as such, warming may have an indirect effect on marine organisms. However, the majority of marine organisms are ectothermic, and are therefore dependent on environmental temperature to gain adequate energy for their own biological functions. The relationship between the performance of an ectotherm and temperature is non-linear, where performance gradually increases with temperature until it reaches a thermal optimum after which it rapidly declines (Thermal Performance Curve: [2]). Hence, the effect of increasing temperature on marine ectotherms may be more direct, affecting physiology and metabolism [3–7], which may have implications for growth, motor function, development, reproduction and behaviour [6–15], which in turn may impact species' abundance and distribution [16]. Therefore, increasing temperatures may be the most pervasive climate change factor influencing marine organisms [7,17,18]. Warming can subsequently impact entire ecological communities and ecosystems by differentially impacting individuals and functional traits [19–21].

Coral reefs are one of the most biodiverse ecosystems in the world [22,23]. Even though the rate of increase in sea surface temperature (SST) is 30% less in tropical oceans than the global average [24], coral reefs are also among the most sensitive ecosystems to changes in environmental conditions [22,25,26], and thus, are particularly at risk of thermal stress. Tropical ectotherms have a narrow thermal tolerance range, and their thermal optimum is close to their thermal maximum, as they have evolved under relatively stable thermal conditions [27,28]. As SSTs rise, corals and coral reef associated organisms are being subjected to higher temperatures (29–31 °C) for increasing periods of time [24]. As a consequence many tropical organisms are thought to be living at or close to their thermal limits [29]. SSTs are predicted to continue to rise over the coming years [1] and extreme warming events, resulting in global scale coral bleaching, are occurring with increasing regularity and severity [30–34], causing degradation of coral reef habitats [22,30,31,35,36].

In addition to the corals themselves, research on the effects of marine heatwaves has also focused heavily on fishes [37–48], which are also typically included in coral reef monitoring efforts. However, studies have almost completely ignored the myriad of small, cryptic, species, which make up a disproportionate amount of coral reef biodiversity [49–51]. One such group are parasites, which make up the largest consumer strategy globally [52] and comprise an estimated 40% of global biodiversity [53–56]. In addition to host behavior, physiology, and population dynamics, parasitic organisms have been shown to have impacts on interspecific interactions, energy flow, and the structure, ecology, and biodiversity of communities [55,57–60]. Parasites are particularly diverse on coral reefs [61] with an estimate of over 20,000 species on the Great Barrier Reef (GBR) alone [62]. However even with such a large presence in coral reef communities, they are significantly underrepresented in ecological studies [10,55,63]. Coral reef parasites are also ectothermic, and as such, may be affected by changes to their environmental temperature [10,11,64–67]. Some parasites are ectoparasitic and would be highly vulnerable to increased temperatures. Ectoparasites are also likely to be directly impacted by the temperature itself, in addition to being indirectly affected through changes in community structure due to temperature impacts on hosts [68] and other organisms. For example, the life cycle of a monogenean ectoparasite (*Neobenedinia*) was faster and the life span of their larvae (oncomiracidia) decreased as temperatures increased from 22 °C to 34 °C [69].

Gnathiid isopods are one of the most common ectoparasites in coral reef habitats [70–72]. They are small crustaceans, typically 1–3 mm long, that do not permanently live on their fish hosts [73,74]. In fact, with few exceptions, they associate only long enough to extract a blood meal and may therefore also be referred to as "micropredators" [75,76]. After feeding on tissue and blood from their fish host they return to the benthos to molt and progress to the next developmental stage [73,74]. They are only parasitic during their three juvenile stages, and no longer feed once they metamorphose into an adult. Gnathiids can have significant impacts on their hosts [66]. Direct effects include influencing behavior [77–83], physiology [84], and mortality [85]. Indeed, as few as one gnathiid can kill a young juvenile fish [83,86–90]. Indirect effects include transmission of blood-borne parasites [91] and wounds that can facilitate infection [92]. Gnathiids are also the most common items in the diet of many cleaner fishes, including *Labroides dimidiatus* [93,94], a species with far-reaching ecosystem effects [95–97]. Indeed, environmental perturbations, including a coral bleaching event with water temperatures reaching up to 30 °C, resulted in an 80% decline in *L. dimidiatus* at Lizard Island, GBR [48]. However, the processes leading to this decline remain unknown.

A long-term monitoring study of gnathiid isopods off Lizard Island, GBR, revealed a significant decrease in gnathiid abundance during extreme warm-water periods associated with bleaching events, compared with cooler periods in the same year or during non-bleaching years [65]. However, the cause of this decline was unclear. Sikkel et al. (2019) [65] hypothesized that the direct effects of temperature on gnathiid mortality may have partly contributed to the decline in gnathiid abundance. The aim of this study therefore, was to assess the direct effect of a rapid increase in seawater temperatures on mortality of shallow-reef gnathiid isopods. By conducting laboratory experiments on gnathiids in two coral reef regions subject to bleaching, GBR, Australia [31,32,98], and Philippines [99–101], we show that a rapid increase in temperature causes significant increases in mortality.

## 2. Materials and Methods

### 2.1. Study Sites

This study was conducted between January and February 2018 at the Lizard Island Research Station (LIRS), northern GBR and between July and October 2017 at the Silliman University–Institute of Environmental and Marine Sciences (SU-IEMS), Dumaguete City, Negros Oriental, Visayas, Philippines.

### 2.2. Gnathiid Collection

For the GBR study, gnathiids were obtained from a culture maintained at LIRS since 2001, which uses the continual availability of wrasse *Hemigymnus melapterus* (Labridae) as hosts [102]. The culture is outdoors and uses a flow-through seawater system that obtains water directly from the nearby reefs. The previous exposures of the experimental (and previous generations) of gnathiids would have, therefore, reflected similar temperatures to the ocean and land ones [102].

Gnathiids for the Philippines study were collected from the shallow fringing coral reefs (<10 m) of Cangmating reef (9°21′18.38″ N, 123°17′58.91″ E) and Agan-an reef (9°20′2.6″ N, 123°18′41.5″ E) in Sibulan and from Bantayan reef (9°19′49.22″ N, 123°18′43.43″ E) in Dumaguete City, all within Negros Oriental Province. The Bantayan reef has small patch reefs with inshore seagrass beds. Cangmating and Agan-an have larger patch reefs and inshore seagrass beds. Gnathiids are common at all three sites [103,104]. Gnathiids were collected using light traps, adapted from Artim et al. (2015) [89] and Artim and Sikkel (2016) [105]. The traps were set at dusk and retrieved the following morning and then transported by boat to the SU-IEMS laboratory where they were emptied into individual 10 L plastic buckets with aerators. The contents of each trap were filtered with a funnel and 55 μm plankton mesh. The gnathiids were then sorted using a stereoscope and placed in an aquarium (27 L) with fresh, filtered, aerated seawater. The species of gnathiids collected were unknown due to difficulty with species identification of the juvenile stages [73], and the fact that no species have yet been formally described from our Philippines study region.

### 2.3. Experimental Protocol

At both locations, gnathiid mortality was defined as the absence of detectable movement, even after disturbance (e.g., by moving the vial it was held in while viewing it under the microscope).

#### 2.3.1. Great Barrier Reef

Gnathiids, all belonging to the species *Gnathia aureamaculosa*, were collected from the culture in the morning and afternoon, and placed together into 75 mL holding containers filled with seawater.

They were collected by moving a black tray (35 × 25 × 5 cm) up the side of the gnathiid culture tank and were removed using a pipette. From the holding containers, gnathiids were individually transferred into 5 mL vials that were half-filled with seawater. These vials were then individually labelled. Collecting and processing took approximately 2 to 4 h, depending on the catch size of the day (ranging from 9 to 226 gnathiids). The daily number varied as a result of fluctuations in the number that were active, most likely due to normal high variation in their population dynamics [102].

A mixture of fed and unfed gnathiids was used, and it was not known how much time elapsed since the last feeding. Gnathiids were not fed for practical reasons. After processing the gnathiids, the vials were randomly allocated, in a balanced way (approximate equal number), to a temperature treatment and aquarium replicate combination; there were three aquarium replicates per temperature treatment. Vials were labelled with a unique number across all replicates. Only the lids of the vials were labelled, reducing any potential bias when viewing them under the microscope. It also made it easier to monitor and return them to their respective treatment and aquarium daily. Vials were held underwater in plastic baskets ($17 \times 17 \times 10$ cm), one for each treatment and replicate ($n = 9$ baskets). Baskets had four mesh ($1$ mm$^2$) windows ($12 \times 5$ cm) on the sides and one on the lid ($12 \times 12$ cm) to allow for flow of water. A dive weight was used to submerge the baskets. Aquaria were supplied with flow-through seawater, with seawater that was either chilled or heated in a sump under the aquarium benches and pumped up to the aquaria. Each bench had a different temperature treatment and held 10 aquaria (previously used for another experiment, see Graba-Landry et al. 2020 [106]). Three aquaria were randomly selected per bench and allocated to replicates.

We estimated the predicted ambient seawater temperature (29.25 °C $\pm$ 0.013 SE) based on the Australian Institute of Marine Science long-term average water temperature for February [107] (Figure S1). Actual average daily seawater temperature during the experiment was 29.0 °C $\pm$ 0.67 SE (February 1 to 23, 2018 available only). The temperature of the water that gnathiids had been maintained in throughout their lifetimes was not available. However, the temperature of incoming water from the station's holding tanks was on average 1.4 °C warmer than the ocean, when sampled at two sources at three times of the day (09:00, 15:00 and 21:00 h from 15 to 20 October, 2018) relative to the same period in the ocean [108].

Temperature was manipulated in an outdoor seawater flow-through system at LIRS using purpose built 1KW steel bar heaters and chillers (Teco®) in a header or sump tank. Each sump, one per temperature treatment, fed replicate 40 L tanks with the appropriate experimental flow-through seawater using 1000 L hr$^{-1}$ pumps (Eheim®) at a rate of approximately 1 L minute$^{-1}$. Tanks were wrapped in Insulbreak® insulation to stabilize water temperatures. Temperature ($\pm$0.1 °C) was also measured at 12:00 h daily from each of the nine tanks housing the gnathid cultures/vials using a portable temperature probe (Comark®) calibrated to 26 °C, 28 °C and 30 °C (National Association of Testing Authorities certified) to ensure temperature remained stable across treatments. Experimental temperatures at 12:00 h per treatment, averaged across the means of the three replicate aquaria, were: 29 °C: 29.5 °C $\pm$ 0.07 SE, 31 °C: 31.4 °C $\pm$ 0.02 SE, and 32 °C: 32.6 °C $\pm$ 0.02 SE. Temperatures were very similar among replicates within a treatment, (Figure S2). One calibrated temperature logger (HOBO Pendant temperature/light logger, UA-002-08) per treatment also recorded the temperature every 2 h throughout the course of the study to account for diurnal fluctuations in temperature [mean (SE) per treatment: 29 °C: 28.9(0.042); 31 °C: 31.2(0.035): 32 °C: 31.9(0.06); Figure S3].

Each day, one random basket from each treatment was removed (to reduce time exposed to air temperature). The vials from each basket were rinsed in freshwater and placed in a large tub (all treatments were examined together to avoid bias). Each vial was then examined under a dissection microscope to check for gnathiid mortality. Vials with alive gnathiids were sorted back into their respective treatments/replicates and placed back into the aquaria. This was repeated for each of the remaining baskets from each treatment-replicate combination. Gnathiids were monitored until all had died (except for four survivors, see Results for details). Vials with dead gnathiids were preserved for later to undertake headwidth measurements, by adding a few drops of formalin into the seawater.

### 2.3.2. Philippines

Unfed juvenile gnathiids were given one day to acclimatize in an aquarium after collection, before host fish, *Dascyllus trimaculatus* (Pomacentridae) and various species of Labridae, were placed in the aquarium overnight to allow them to feed. The gnathiids did not feed again for the duration of the experiment. The following day, fed, mobile and healthy-looking individuals were selected.

However, as the gnathiids' species and therefore the consequent size range for each stage was not known, they could not be separated by juvenile stage as in the GBR study. Instead they were sorted into two size classes (<2 mm and >2 mm, to account for any effect of size of the gnathiid on its molting rate and survival.

Gnathiids were then placed in 270 mL plastic containers, 5–10 gnathiids per container, with plankton mesh (55 µm) secured on the top, and the containers were submerged in one of five 27 L aerated aquaria, each with a different set temperature. Each container was labelled with the size class, treatment, trial and replicate. The first trial consisted of 5 temperature treatments, ambient (28 °C), 30 °C, 32 °C, 34 °C and 36 °C. A second trial was also conducted to obtain a finer resolution of the temperature effect, with treatments of 30 °C, 32 °C, 33 °C, 34 °C and 35 °C. Aquaria were individually heated gradually with 100W and 200W aquarium electric heaters (Venusaaqua®) over a 10 h period to their desired temperature. Temperature readings were taken daily with an aquarium-mounted digital thermometer (Doutop®) to ensure the desired water temperatures were maintained and to calculate the average temperature for each treatment per trial (Table S1a,b). Containers from each treatment were inspected daily for evidence of changes in gnathiid development and mortality. Dead gnathiids were removed from the containers and molted adult males were preserved in ethanol for future species identification. The experiment was concluded for each treatment when all gnathiids were dead or when 20 d had passed. One third of the water in each aquarium was removed daily and replaced with ambient temperature, filtered, fresh seawater. Over the 20 d duration of each replicate the aquaria used for each temperature treatment was alternated every week (once per replicate) to ensure there were no confounding effects associated with individual aquaria. Three replicates of each temperature treatment were run for each trial. Trials 1 ($n$ = 369 gnathiids) and 2 ($n$ = 318 gnathiids) had a range of 60–76 and 59–68 gnathiids per temperature treatment respectively (Table S2a,b).

The baseline ambient temperatures for Trials 1 (28 °C) and 2 (30 °C) were similar to the SST in the Bohol Sea, Philippines, which fluctuated by about 3 °C (about 27–30 °C) during 2017; the SST during the experiment (July–October 2017) averaged at 30 °C ± 0.04 SE. [109].

### 2.4. Statistical Analyses

#### 2.4.1. Great Barrier Reef

One live unfed gnathiid (stage three, 15 d), two adult males (alive, 15 d; dead, 29 d) and three adult females (alive, 15 and 24 d; dead, 16 d) were excluded from the data. These gnathiids were excluded because they were adults and thus their longevity would be different to that of the juveniles. The single juvenile that was still alive when we terminated the experiment was omitted for simplicity and consistency. We categorized the three juvenile stages based on their headwidth (stage one: 0.14–0.2, stage two: 0.21–0.24, stage three: 0.25–0.32 mm) [77]. Due to the large difference in sample size per unfed/fed status (based on the presence of an engorged gut), we conducted separate analyses for unfed ($n$ = 1133) and fed gnathiids ($n$ = 87).

To test whether survival of gnathiids differed among temperature treatments, we used a proportional hazards Cox mixed-effects model with temperature treatment and gnathiid juvenile stage as categorical fixed effects, aquarium as a random factor, and gnathiid headwidth as a covariate (Table 1 and Table S3). We used ambient temperature (29 °C) and juvenile stage one as the baselines for the analyses. We used the function "coxme" in the package "coxme" [110,111] and function "Anova" in the package "car" [112]. We tested the Cox model assumption of proportionality using the Global test statistic in the function "coxph" and "cox.zph" in the package "coxme" and graphically using a smoothed spline plot of the Shoenfeld residuals relative to time (see Tables S3 and S4 for results and Figures S4 and S5 for spline plots).

**Table 1. Great Barrier Reef;** Analysis of deviance table (Type II tests) for unfed gnathiid survival among temperature treatments and juvenile stages for Cox model. Bolded values are ones mentioned in main text. *** $p < 0.001$.

|  | Df | Chisq | Pr(>Chisq) |  |
|---|---|---|---|---|
| Temperature | 2 | 183.76 | **<0.0001** | *** |
| Headwidth | 1 | 1.87 | 0.1719 |  |
| Stage | 2 | 26.93 | <0.0001 | *** |
| Temperature × Headwidth | 2 | 0.07 | 0.9639 |  |
| Temperature × Stage | 4 | 3.06 | 0.5471 |  |
| Headwidth × Stage | 2 | 23.85 | **<0.0001** | *** |
| Temperature × Headwidth × Stage | 4 | 6.51 | 0.1642 |  |

### 2.4.2. Philippines

In the size class < 2 mm ($n = 168$ gnathiids) for Trial 1, five (3%) gnathiids molted into adult females and nine (5%) into males. In the size class < 2 mm ($n = 153$) for Trial 2, no gnathiids molted into adults. In contrast, for the larger size class > 2 mm for Trial 1 ($n = 201$), 20 (10%) gnathiids molted into females and 85 (42%) into males. In the size class > 2 mm ($n = 165$) for Trial 2, 37 (22%) gnathiids molted into females and 68 (41%) into males. In this study, newly metamorphosed males were first observed after day 1 in Trial 2 and day 2 in Trial 1, and no additional males appeared after day 5 in Trial 2 and day 7 in Trial 1. The mean number of days juvenile gnathiids molted into males for all treatments was 3.39 ± 0.75 and 3.23 ± 0.16 for Trials 1, and 2, respectively.

To test whether survival of gnathiids differed among temperature treatments, we used the same statistical methods as for the GBR data, with some modifications to the model. Temperature treatment was a fixed effect, and size class, life stage (male, female, or juvenile) and container (which the gnathiids were kept in) were treated as random effects. The ambient temperature of 28 °C was used as the baseline for analysis for Trial 1 and 30 °C for Trial 2. Assumptions of proportionality were met for both analyses (both Global tests: $p > 0.05$, see Tables S5 and S6 for results and Figures S6 and S7 for spline plots).

### 2.5. Ethics

All applicable international, national, and/or institutional guidelines for the care and use of animals were followed. All procedures performed in studies involving animals were in accordance with the ethical standards of Silliman University, Arkansas State University, The University of Queensland, and the Government of Australia and the Philippines.

## 3. Results

### 3.1. Great Barrier Reef

### 3.1.1. Unfed Gnathiids

From the 1220 gnathiids whose survival was followed over time, 93% were unfed individuals. For unfed individuals, the numbers of gnathiids per stage and per temperature was relatively even within each of the 29 °C, 31 °C, and 32 °C temperature treatments (stage one: 186, 198, 210; stage two: 73, 66, 71; stage three: 113, 113, 103, respectively). There was a significant effect of temperature on gnathiid survival ($p < 0.0001$, Table 1, Figure 1), due to a significantly lower survival at 32 °C compared with the 29 °C baseline temperature ($p = 0.016$, Table S7). The interaction between gnathiid headwidth and juvenile stage was significant ($p < 0.0001$, Table 1); when further explored separately by stage, the association was largely due to a weakly positive relationship between gnathiid survival and gnathiid headwidth in stage one ($p < 0.0001$, Table S9a).

### 3.1.2. Fed Gnathiids

For fed individuals, the numbers of gnathiids per juvenile stage and per temperature were also relatively even between the 29 °C, 31 °C, and 32 °C temperature treatments (stage one: 8, 6, 6; stage two: 17, 15, 13; stage three: 8, 7, 7, respectively). Three (3.4%) contained more blood than clear material (i.e., plasma) in their gut, the remainder had a clear gut. Of the 87 individuals followed, 60% had molted during the course of the study. Survival differed according to an interaction between temperature and juvenile stage ($p = 0.0085$, Table 2); when further explored separately by stage (Figure 2), the effect of temperature was significant for stage two ($p < 0.0001$), and three ($p = 0.0009$, Figure 2b,c, Table S10b,c), with the strongest effect of temperature being that between the baseline (29 °C) and the 32 °C treatments for stage three (Table S10c, Figure 2c). Survival differed according to an interaction between temperature and headwidth ($p = 0.0480$, Table 2); when further explored separately by temperature treatment, the effect of headwidth was largely due to non-significant weakly positive relationships between survival and headwidth at 29 °C ($p = 0.0758$) and 32 °C ($p = 0.0718$, Table S11a,c).

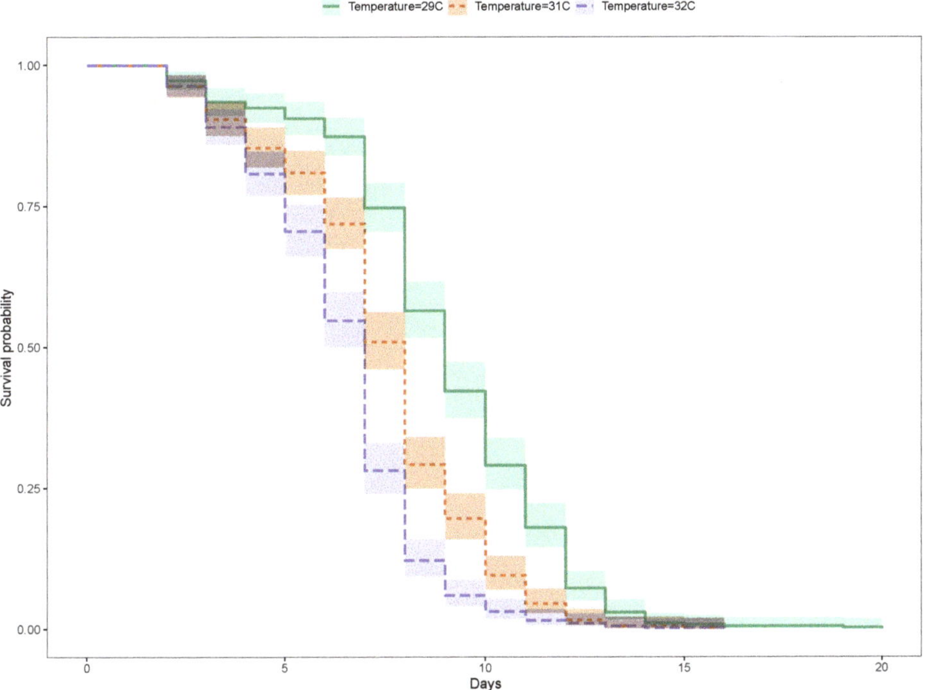


**Figure 1. Lizard Island Research Station**; Kaplan-Meier survival curves for unfed gnathiids per temperature treatment. Shaded areas are 95% confidence intervals.

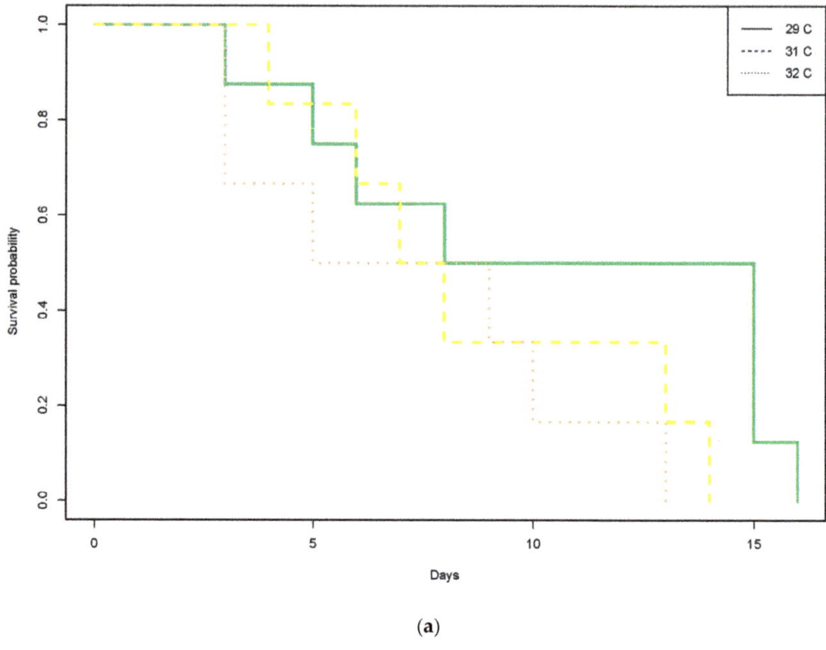

(**a**)

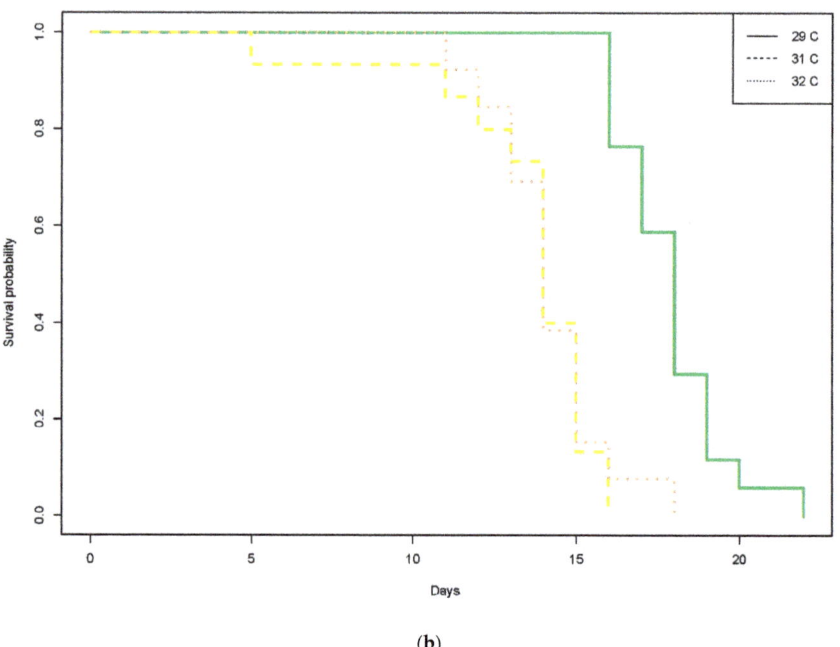

(**b**)

**Figure 2.** *Cont.*

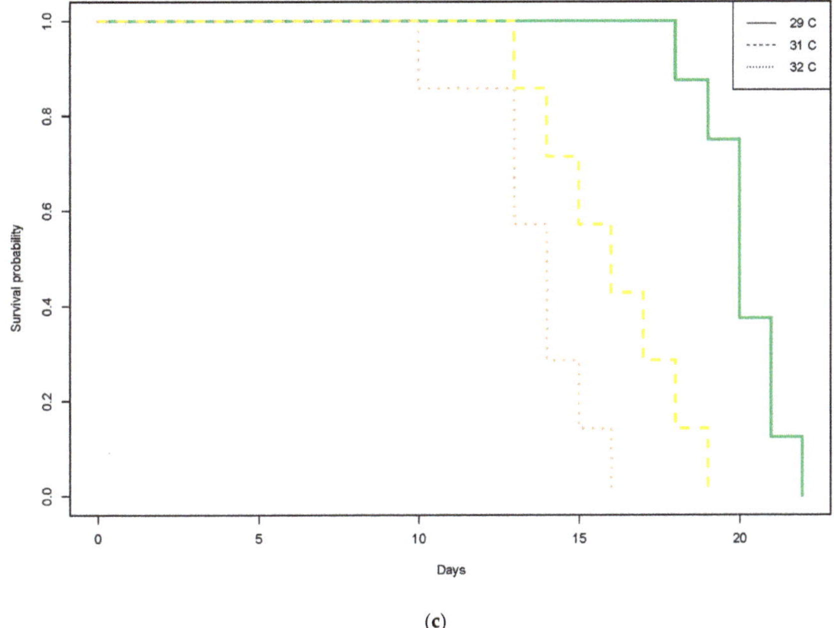


**Figure 2. Great Barrier Reef;** Kaplan-Meier survival curves for fed gnathiids per temperature treatment for (**a**) stage one, (**b**) stage two, and (**c**) stage three juveniles. For ease of interpretation, 95% confidence intervals are not included.

**Table 2. Great Barrier Reef;** Analysis of deviance table (Type II tests) for fed gnathiid survival among temperature treatments and juvenile stages for Cox model. Bolded values are ones mentioned in main text. ** $p < 0.01$, *** $p < 0.001$.

|  | Df | Chisq | Pr(>Chisq) |  |
|---|---|---|---|---|
| Temperature | 2 | 28.2063 | $7.50 \times 10^{-7}$ | *** |
| Headwidth | 1 | 1.9549 | 0.16206 |  |
| Stage | 2 | 21.3179 | $2.35 \times 10^{-5}$ | *** |
| Temperature × Headwidth | 2 | 6.0753 | **0.04795** | * |
| Temperature × Stage | 4 | 13.6392 | **0.00854** | ** |
| Headwidth × Stage | 2 | 3.0686 | 0.21561 |  |
| Temperature × Headwidth × Stage | 4 | 3.3548 | 0.5003 |  |

## 3.2. Philippines

### 3.2.1. Trial 1

There was a significant effect of temperature on gnathiid survival ($p < 0.0001$) in Trial 1, driven by lower survival curves for the 36 °C ($p < 0.0001$) and 32 °C ($p = 0.024$) treatments, compared with the 28 °C baseline temperature treatment (Table 3 and Table S12, Figure 3).

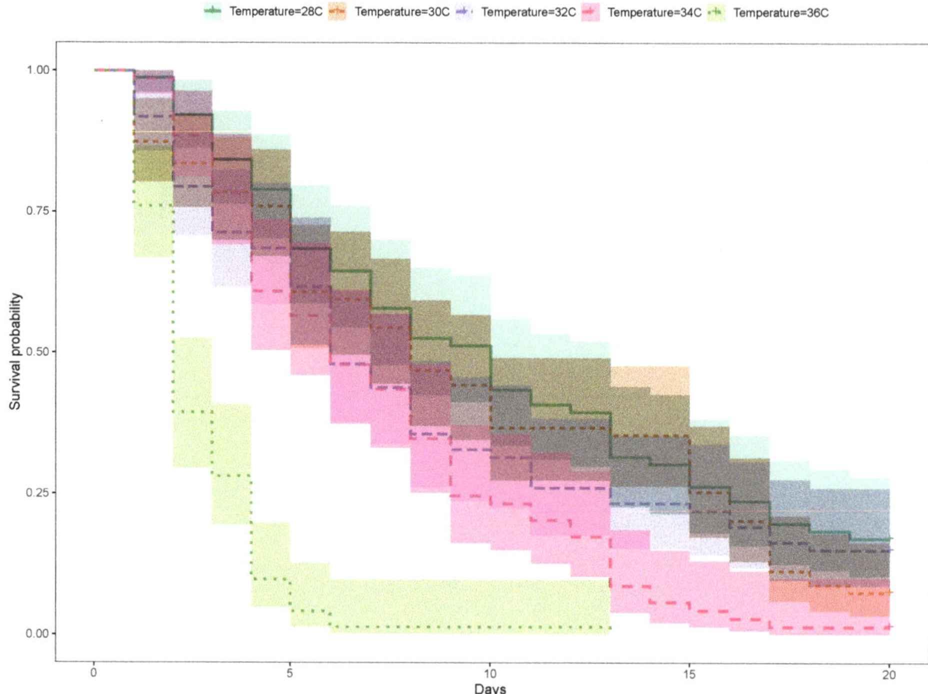


**Figure 3. Philippines;** Trial 1, Kaplan-Meier survival curves for gnathiids per temperature treatment. Shaded areas are 95% confidence intervals.

**Table 3. Philippines;** Analysis of deviance table (Type II tests) for Trial 1 and 2 gnathiid survival for Cox model. Bolded values are ones mentioned in main text.

| Trial | | Df | Chisq | Pr(>Chisq) |
|---|---|---|---|---|
| 1 | Temperature | 4 | 24.927 | **<0.0001** |
| 2 | Temperature | 4 | 8.4374 | **0.07681** |

### 3.2.2. Trial 2

Overall, the effect of temperature on gnathiid survival was not quite statistically significant ($p = 0.0768$, Table 3). Nevertheless, the 35 °C treatment had a much steeper survival curve than that for the 30 °C baseline (Figure 4) ($p = 0.0057$, Table S13).

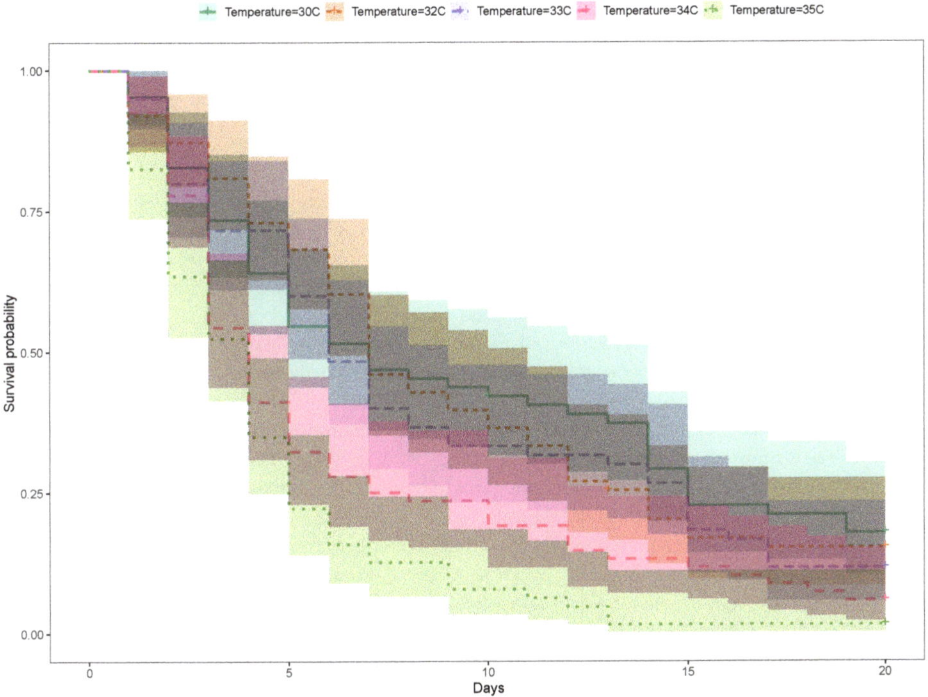



**Figure 4. Philippines;** Trial 2, Kaplan-Meier survival curves for gnathiids per temperature treatment. Shaded areas are 95% confidence intervals.

## 4. Discussion

With ocean temperatures predicted to rise 3 °C by the end of the century [113], the effects of ocean warming on coral reef organisms have received an increasing amount of attention. However, such studies largely ignore the cryptofauna that comprises most of coral reef's biodiversity and biomass, including parasites [67]. In the only long-term monitoring study of any marine parasitic crustacean, Sikkel et al. (2019) [65] reported that during extreme warm-water events in the GBR parasitic gnathiid isopod populations crashed. The findings reported here are consistent with their hypothesis that this may be attributable, in part, to a direct effect of temperature on gnathiid mortality. Such an effect of temperature on the larvae of a tropical ectoparasite has been shown for monogeneans on farmed tropical fish [69].

The present study is the first to examine effects of acute temperature increases on this common reef fish ectoparasite. In our study, gnathiids from both the GBR in Australia and Negros Oriental in the Philippines demonstrated rapid mortality in temperatures raised to above average SST, suggesting that environmental changes in temperature can influence gnathiid survival. In the Philippines, temperatures as little as 2 °C (i.e., 32 °C) above average seasonal SST (30 °C) caused significantly lower survival, with increasingly steep survival curves at 35 °C, with the steepest at 36 °C, where no gnathiids survived past five days. Unfed gnathiids on the GBR had lower survival at 32 °C compared with 29 °C, an effect which was consistent across all three juvenile stages. For fed gnathiids on the GBR, the effect of temperature was significant for juvenile stages two and three, with the strongest effect of temperature on stage three, also between the 32 °C and 29 °C treatments. It is, therefore, likely that gnathiids from both the Philippines and Australia may be living near their thermal limit, as small increases in temperature from the annual seasonal mean have resulted in increased mortality in organisms from both regions. These results indicate not only that an acute change of temperature to just 32 °C decreases the survival of gnathiids, but that the effect of increased temperature is greater on the larger juvenile stages.

In the GBR we found evidence that greater gnathiid headwidth, not just juvenile stage, increased gnathiid survival. For unfed gnathiids, there was a weakly positive relationship between gnathiid survival and gnathiid headwidth, but only in juvenile stage one. For fed gnathiids, there was also a weakly positive relationship between survival and headwidth, but only at 29 °C and 32 °C. Gnathiid length is correlated with headwidth [95,105] and thus likely with mass also. Both results suggest that even small increases in gnathiid size within a juvenile stage can increase gnathiid survival; these findings also supported our decision to include both headwidth and stage in the statistical model as being important factors to consider when modelling gnathiid juvenile survival. Such a difference in the thermal response related to size may be due to the increased metabolic demand caused by the increase in temperature, an effect which may lead to an energetic deficit for smaller individuals if enough food cannot be obtained, therefore, creating a metabolic mismatch between energy obtained versus energy required [114]. Alternatively, there may possibly be a higher baseline metabolism or higher growth rate at smaller sizes, which then slows down as they reach the maximum size for that stage, resulting in smaller sizes using up their reserves faster than larger sizes. Furthermore, the energetic demands of development may also differ among juvenile life-history stages. Thus increased metabolic demand for basal processes (such as cell maintenance) as a result of increases in temperature, coupled with differential energetic requirements for development may also explain the variation in the thermal response among juvenile stages in our study. Therefore, understanding the effect of increasing temperature on individual metabolism and survival also requires an understanding of food resources and availability [114].

Overall, these results suggest that even with a small increase of 2–3 °C above the normal ambient mean, raised temperature can ultimately lead to increased gnathiid mortality. However, there were some key differences in the experimental protocol between the GBR and Philippines studies that should be considered. First, only one species of gnathiid was used in the GBR experiment, compared to at least three (all undescribed, M.O.S. personal observation) species present in the pool of gnathiids used for the Philippines experiment. Second, because these species were unknown, we were unable to confidently separate juvenile gnathiids into their different stages, and so used size class as a proxy. Therefore, we cannot discount the possibility of some among-species and life-stage variation in thermal tolerance. Finally, in contrast to the Philippines, in the GBR study the time the gnathiids were last fed was unknown. This would account for much of the unexplained variation in survivorship in the analysis of the GBR data, as the variation in resources available to the gnathiid, in the form of a blood meal, would decrease over time since their last feeding event. In addition, it should be noted that, in both studies, the gnathiids were not fed for the duration of the experiment, and thus starvation may have been a contributing cause of mortality. While starvation may have influenced mortality of gnathiids among the treatments (as suggested by increased mortality over time in the ambient temperatures), the rates of mortality at higher temperatures were greater, with rapid mortality taking place very early on in the experiments (e.g., one to five days in the Phillipines). This supports the interpretation that increased temperatures influence gnathiid survival directly. It is of relevance that marine "heatwaves" (which are categorised as periods of abnormally high SST lasting for longer than five days [115]) have been predicted to become more frequent, longer and more severe [116,117]. Our observations of rapid gnathiid mortality even after just one day suggests there may be a decline in gnathiid survival from early on in a heatwave, so that gnathiid populations may be heavily impacted if there are more frequent and severe heatwaves in the future.

Our findings appear consistent with data for other tropical marine invertebrates, which have an upper thermal tolerance that is not far above normal sea temperature (reviewed in [4,118]). For example, in a meta-analysis on bivalves, and a study on porcelain crabs, tropical species were found to have upper thermal limits that were closer to the maximum temperature of their habitat than temperate species [118,119]. Tropical species of bivalves have also been shown to have a smaller thermal tolerance window than temperate species [118]. This is thought to be due to tropical marine organisms being more sensitive to changes in temperature as they have evolved under relatively invariable thermal conditions [27]. Other studies on marine invertebrates have also shown increased mortality with high SST [120,121], with hermatypic corals being particularly sensitive to increases, with SSTs needing to rise only a few degrees for bleaching to occur [25,26,31,122].

There are a number of studies investigating the potential impact of temperature increase associated with climate change on parasite communities and aquatic parasite-host interactions. The majority of these studies have been on endoparasitic trematodes from temperate regions [123–125]. Temperature was consistently observed to have a significant effect on the survival times of trematodes in their free-living juvenile stage, with survival rates decreasing as temperature increased (e.g., [126–133]). Similarly, temperature has been reported to have an effect on parasitic barnacles (rhizocephalans), with their prevalence decreasing at higher temperatures [68].

In one of the few other studies on ectoparasites, Conley and Curtis (1993) [134] found that, in temperatures of 8–20 °C, survival of copepodids was also inversely proportional to temperature. This same trend was observed in the survival rates of monogeneans, and isopods (Cymothoidae) in two studies in sub-tropical regions [135,136] and one study of monogeneans in a tropical region [69]. In all three studies, temperature treatments of 30 °C and above had the lowest survival rates [69,135,136]. Similar results were also observed with trematode cercariae from sub-tropical regions [137–139]. Summer temperatures for these lower latitudes parasites are in the range of 30–31 °C, which suggests that like gnathiids in the warmer months they are living close to their thermal limits.

Although, this study focused on effects of temperature on mortality, increased temperature can also have sub-lethal effects on marine organisms, impacting their ability to perform essential tasks [140–142]. Based on a review of the literature, Lough (2012) [24] suggested that temperatures between 30–32 °C may reflect a potential temperature threshold where a proportion of reef organisms' physiological processes are negatively impacted. Higher temperatures may also affect the ability of parasites with mobile life history stages (such as gnathiids) to successfully detect and associate with a host. To our knowledge there are no studies that specifically examine this. However, the ability to physically reach a host by swimming does appear to be influenced by temperature. For temperate parasitic copepods in their free-living stage, the duration of swimming activity was found to be inversely related to water temperature [134]. For newly emerged cercariae, swimming speed increased in higher temperatures (19–36 °C). However, the speed declined over time, with rate of decline increasing with temperature. This resulted in higher swimming speeds, but for shorter durations in water of 30 °C and above [143]. In a sub-tropical study, cercariae infectivity also increased with temperature with maximum infectivity occurring at 30 °C before declining at 36 °C and 40 °C [137]. This could be attributable to greater cercariae swimming activity [137]. Although, we did not quantify the effects of temperature on movement, in the Philippines study it was apparent that gnathiids moved more slowly and less frequently at temperatures of 32 °C and above, with movement decreasing further as temperature increased, and also with apparent effects greater for the larger size class (M.O.S. personal observation).

Elevated temperatures may also impact host physiology, behavior and survival in ways that impact the balance between parasite and host. For parasitic barnacles (rhizocephalans), the effects of temperature on infected host mortality (and consequent transmission) could threaten their survival, with models showing that just an increase of 2 °C in ambient temperature could cause local parasite eradication [68]. In contrast, reef fishes can live further away from their thermal limits than gnathiids were observed to do in this study and in some cases can tolerate temperatures of up to 34–40 °C, [3,15,37,47]. However, they can still experience sub-lethal effects with smaller increases in temperatures [3,6,12,14,15,142,144–146], which could also impact host-parasite interactions.

Large hosts, like many reef fishes, can also leave areas of warm water for cooler water, or leave habitat impacted by coral bleaching for other habitats [39,40,43–46,147], depriving gnathiids and other similar ectoparasites, like natatory-stage cymothoid isopods of hosts [148]. The potentially impaired physiological and swimming ability of the parasite, combined with direct effects on mortality and host availability, could result in a decline in parasite populations. However, the ability of some gnathiids to feed on invertebrate hosts [149,150], combined with weakened immune response for the smaller, less mobile, fish species could leave fish more susceptible to ectoparasites, and thus, compensate for the loss of larger hosts. Indeed, during the 2016–2017 mass bleaching event on the GBR, there was a significant decrease in the numbers of larger, more mobile host fishes in shallow areas, with only smaller, site-attached species remaining [48]. This could have also contributed significantly to the crash in gnathiid populations observed by Sikkel et al. (2019) [65]. However, it should also be noted that as gnathiids are mostly free-living and have a temporary association with their hosts, they too can potentially avoid higher water temperatures. This might happen passively by the gnathiid "hitching a ride" whilst feeding on their host, a process which can last from a few minutes up to a few hours [73]. The gnathiids may, thereby, be transferred to different locations [151,152]. However, as knowledge of the dispersal mechanisms of gnathiids, the infection rates of host fish, and fish movements after disturbances is limited, the proportion of the gnathiid populations that could transfer location with their hosts remains unknown.

Another indirect effect of increased SST may be effects of warming on predators of gnathiids and other ectoparasites' free-living stages. In particular, coral polyps are a major source of predation on juvenile gnathiids [153,154], and thus high coral mortality associated with warm-water events, combined with the loss of cleaner fish [48], which prey on ectoparasites [93], might increase living space and decrease predation on gnathiids. Indeed, once water cools following a bleaching event and most corals are dead, gnathiid populations appear to recover rapidly [65]. While, oceans are also experiencing increased acidification [1,155], Paula et al. (2020) [156] found no effect of acidification on the mortality of the same GBR gnathiid isopod as that studied here.

As parasites have a significant role in ecosystem function, changes in parasite abundance may pose consequences for ecological communities [157–159]. Therefore, while the diversity of coral reef parasites and their hosts makes it difficult to draw general conclusions on how warming events will impact parasite-host interactions, it remains important to further investigate parasite responces to both the direct and indirect effects of warming [67]. Future studies on gnathiids examining sublethal thermal effects on molting, physiology, locomotion, host-detecting mechanisms and reproductive performance will provide a more comprehensive understanding of effects of temperature on host-parasite interactions in coral reef systems.

**Supplementary Materials:** The following are available online at http://www.mdpi.com/2673-1924/1/4/16/s1. Table S1: Philippines; Average water temperature of aquaria for five treatments over three aquarium replicates, Table S2: Philippines; Sample size of larval gnathiid isopods < 2 mm and >2 mm in length in five different temperature treatments over three aquarium replicates, Table S3: Great Barrier Reef; Tests of proportionality, using function "cox.zph" in library "coxme" in R 3.2.5, for full model for unfed gnathiid survival among temperature treatments and juvenile stages, Table S4: Great Barrier Reef; Tests of proportionality, using function "cox.zph" in library "coxme", for full model of fed gnathiid survival among temperature treatments and juvenile stages, Table S5: Philippines; Tests of proportionality, using function "cox.zph" in library "coxme" in R 3.2.5, for full model for Trial 1 gnathiid survival among temperature treatments, Table S6: Philippines; Tests of proportionality, using function "cox.zph" in library "coxme" in R 3.2.5, for full model for Trial 2 gnathiid survival among temperature treatments, Table S7: Great Barrier Reef; Summary output for full model for unfed gnathiid survival among temperature treatments and juvenile stages for Cox model, Table S8: Great Barrier Reef; Summary output for full model for fed gnathiid survival among temperature treatments and juvenile stages for Cox model, Table S9: Great Barrier Reef; Analysis of deviance tables (Type II tests) and summary outputs for unfed gnathiid survival for separate Cox models for each juvenile stage, Table S10: Great Barrier Reef; Analysis of deviance tables (Type II tests) and summary outputs for fed gnathiid survival for separate Cox models for each juvenile stage, Table S11: Great Barrier Reef; Analysis of deviance tables (Type II tests) and summary outputs for fed gnathiid survival for separate Cox models for each temperature treatment, Table S12: Philippines; Summary output for full model for Trial 1gnathiid survival among temperature treatments for Cox model, Table S13: Philippines; Summary output for full model for Trial 2 gnathiid survival among temperature treatments for Cox model. Bolded values are ones mentioned in main text, Figure S1: Seawater temperature for Great Barrier Reef data, Figure S2: Great Barrier Reef; Temperatures, measured using a handheld device at 12:00 h, for three replicate aquaria per temperature treatment, Figure S3: Great Barrier Reef; Water temperatures in an aquarium over duration of study for each of the temperature treatments between February 1 and March 2 2018, Figure S4: Great Barrier Reef; Scaled Shoenfeld residual plot for full model testing unfed gnathiid survival relative to time (days), Figure S5: Great Barrier Reef;

Scaled Shoenfeld residual plot for full model testing fed gnathiid survival relative to time (days), Figure S6: Philippines; Scaled Shoenfeld residual plot for full model testing gnathiid survival relative to time (days) for Trial 1, Figure S7: Philippines; Scaled Shoenfeld residual plot for full model testing gnathiid survival relative to time (days), for Trial 2.

**Author Contributions:** Conceptualization, P.C.S. and A.S.G.; methodology, P.C.S., A.S.G., M.O.S. and B.D.; validation, A.S.G. and P.C.S.; formal analysis, A.S.G., M.O.S. and A.G.-L.; investigation, M.O.S. and B.D.; resources, P.C.S., A.S.G., M.O.S. and A.G.-L.; data curation, M.O.S. and B.D.; writing—original draft preparation, M.O.S., B.D., P.C.S. and A.S.G.; writing—review and editing, M.O.S., A.G.-L. and P.C.S.; visualization, A.S.G. and M.O.S.; supervision, P.C.S. and A.S.G.; project administration, A.S.G., P.C.S. and M.O.S.; funding acquisition, A.S.G., A.G.-L. and P.C.S. All authors have read and agreed to the published version of the manuscript.

**Funding:** This research was funded by the Australian Research Council (A00105175, A19937078, ARCFEL010G, DP0557058, DP120102415), Sea World Research and Rescue Foundation Australia (SWR/2/2012), and the US National Science Foundation (OCE-1536794, PC Sikkel, PI).

**Acknowledgments:** We thank the many volunteers and the Lizard Island Research Station (GBR) staff who helped maintain the gnathiid culture and provided equipment and facilities. We also thank Jessica Vorse, who conducted an earlier pilot study on the effects of temperature on gnathiid survival on the Great Barrier Reef; this was invaluable in the development of the final methodology implemented in the present study. We thank the municipality of Sibulan, and Dumaguete City, Negros Oriental, Philippines, for permission to conduct this study (0154-18 DA-BFAR). We also thank Hilconida P. Calumpong, Janet S. Estacion, Rene A. Abesamis, and the staff of the Silliman University Institute for Environmental and Marine Sciences for logistic support, equipment and use of facilities. We thank Jeremiah Gepaya and Lucille Jean Raterta for their field assistance and Dioscoro Inocencio for fish collections and field support. The Authors are also grateful to the three anonymous reviewers for their constructive comments.

**Conflicts of Interest:** The authors declare no conflict of interest.

## References

1.	Bindoff, N.L.; Cheung, W.W.L.; Kairo, J.G.; Arístegui, J.; Guinder, V.A.; Hilmi, N.; Jiao, N.; Karim, M.S.; Levin, L.; O'Donoghue, S.; et al. *Changing Ocean, Marine Ecosystems, and Dependent Communities*; ETH Zurich: Zurich, Switzerland, 2019; pp. 477–587.

2.	Huey, R.B.; Stevenson, R.D. Integrating Thermal Physiology and Ecology of Ectotherms: A Discussion of Approaches. *Am. Zool.* **1979**, *19*, 357–366. [CrossRef]

3.	Brett, J.R. 3. Temperature, 3.3. Animals, 3.32. Fishes-Functional Reponses. In *Marine Ecology: A Comprehensive, Integrated Treatise on Life in Oceans and Coastal Waters*; Kinne, O., Ed.; Wiley-Interscience: Hoboken, NJ, USA, 1984; Volume 1, pp. 515–560.

4.	Kinne, O. 3. Temperature, 3.3. Animals, 3.31. Invertebrates. In *Marine Ecology: A Comprehensive, Integrated Treatise on Life in Oceans and Coastal Waters*; Kinne, O., Ed.; Wiley-Interscience: Hoboken, NJ, USA, 1984; Volume 1, pp. 407–514.

5.	Newell, R.C.; Branch, G.M. The influence of temperature on the maintenance of metabolic energy balance in marine invertebrates. *Adv. Mar. Biol.* **1980**, *17*, 329–396. [CrossRef]

6.	Houde, E.D. Comparative growth, mortality, and energetics of marine fish larvae: Temperature and implied latitudinal effects. *Fish. Bull.* **1989**, *87*, 471–495.

7.	Schmidt-Nielsen, K. *Animal Physiology: Adaptation and Environment*; Cambridge University Press: Cambridge, UK, 1997.

8.	Marsden, I.D. Effect of temperature on the microdistribution of the isopod *Sphaeroma rugicauda* from a saltmarsh habitat. *Mar. Biol.* **1976**, *38*, 117–128. [CrossRef]

9.	McNamara, J.C.; Moreira, P.S.; Moreira, G.S. The effect of salinity on the upper thermal limits of survival and metamorphosis during larval development in *Macrobrachium amazonicum* (Heller) (Decapoda, Palaemonidae). *Crustaceana* **1986**, *50*, 231–238. [CrossRef]

10.	Marcogliese, D.J. The impact of climate change on the parasites and infectious diseases of aquatic animals. *Oie Rev. Sci. Tech.* **2008**, *27*, 467–484. [CrossRef]

11.	Marcogliese, D.J. The Distribution and Abundance of Parasites in Aquatic Ecosystems in a Changing Climate: More than Just Temperature. *Integr. Comp. Biol.* **2016**, *56*, 611–619. [CrossRef]

12.	Munday, P.L.; Jones, G.P.; Pratchett, M.S.; Williams, A.J. Climate change and the future for coral reef fishes. *Fish Fish.* **2008**, *9*, 261–285. [CrossRef]

13. Przeslawski, R.; Ahyong, S.; Byrne, M.; Wörheide, G.; Hutchings, P. Beyond corals and fish: The effects of climate change on noncoral benthic invertebrates of tropical reefs. *Glob. Chang. Biol.* **2008**, *14*, 2773–2795. [CrossRef]

14. Donelson, J.M.; Munday, P.L.; McCormick, M.I.; Pankhurst, N.W.; Pankhurst, P.M. Effects of elevated water temperature and food availability on the reproductive performance of a coral reef fish. *Mar. Ecol. Prog. Ser.* **2010**, *401*, 233–243. [CrossRef]

15. Zarco-Perello, S.; Pratchett, M.; Liao, V. Temperature-growth performance curves for a coral reef fish, *Acanthochromis polyacanthus*. *Galaxea J. Coral Reef Stud.* **2012**, *14*, 97–103. [CrossRef]

16. Pecl, G.T.; Araújo, M.B.; Bell, J.D.; Blanchard, J.; Bonebrake, T.C.; Chen, I.-C.; Clark, T.D.; Colwell, R.K.; Danielsen, F.; Evengård, B.; et al. Biodiversity redistribution under climate change: Impacts on ecosystems and human well-being. *Science* **2017**, *355*, eaai9214. [CrossRef] [PubMed]

17. *Animals and Temperature: Phenotypic and Evolutionary Adaptation*; Johnston, I.A.; Bennett, A.F. (Eds.) Cambridge University Press: Cambridge, UK, 2008.

18. Hoey, A.S.; Howells, E.; Johansen, J.L.; Hobbs, J.-P.A.; Messmer, V.; McCowan, D.M.; Wilson, S.K.; Pratchett, M.S. Recent Advances in Understanding the Effects of Climate Change on Coral Reefs. *Diversity* **2016**, *8*, 12. [CrossRef]

19. Fields, P.A.; Graham, J.B.; Rosenblatt, R.H.; Somero, G.N. Effects of expected global climate change on marine faunas. *Trends Ecol. Evol.* **1993**, *8*, 361–367. [CrossRef]

20. Jennings, S.; Brander, K. Predicting the effects of climate change on marine communities and the consequences for fisheries. *J. Mar. Syst.* **2010**, *79*, 418–426. [CrossRef]

21. Kordas, R.L.; Harley, C.D.G.; O'Connor, M.I. Community ecology in a warming world: The influence of temperature on interspecific interactions in marine systems. *J. Exp. Mar. Biol. Ecol.* **2011**, *400*, 218–226. [CrossRef]

22. Hoegh-Guldberg, O. Climate change, coral bleaching and the future of the world's coral reefs. *Mar. Freshw. Res.* **1999**. [CrossRef]

23. Bellwood, D.R.; Hughes, T.P. Regional-scale assembly rules and biodiversity of coral reefs. *Science* **2001**, *292*, 1532–1534. [CrossRef]

24. Lough, J.M. Small change, big difference: Sea surface temperature distributions for tropical coral reef ecosystems, 1950-2011. *J. Geophys. Res. Ocean.* **2012**, *117*. [CrossRef]

25. Hughes, T.P.; Kerry, J.T.; Baird, A.H.; Connolly, S.R.; Dietzel, A.; Eakin, C.M.; Heron, S.F.; Hoey, A.S.; Hoogenboom, M.O.; Liu, G.; et al. Global warming transforms coral reef assemblages. *Nature* **2018**, *556*, 492–496. [CrossRef]

26. Lough, J.M.; Anderson, K.D.; Hughes, T.P. Increasing thermal stress for tropical coral reefs: 1871-2017. *Sci. Rep.* **2018**, *8*, 1–8. [CrossRef] [PubMed]

27. Tewksbury, J.J.; Huey, R.B.; Deutsch, C.A. ECOLOGY: Putting the Heat on Tropical Animals. *Science* **2008**, *320*, 1296–1297. [CrossRef]

28. Sunday, J.M.; Bates, A.E.; Dulvy, N.K. Global analysis of thermal tolerance and latitude in ectotherms. *Proc. R. Soc. B: Biol. Sci.* **2011**, *278*, 1823–1830. [CrossRef] [PubMed]

29. Habary, A.; Johansen, J.L.; Nay, T.J.; Steffensen, J.F.; Rummer, J.L. Adapt, move or die - how will tropical coral reef fishes cope with ocean warming? *Glob. Chang. Biol.* **2017**, *23*, 566–577. [CrossRef]

30. Lough, J.M. 10th Anniversary Review: A changing climate for coral reefs. *J. Environ. Monit.* **2008**, *10*, 21–29. [CrossRef] [PubMed]

31. Hughes, T.P.; Kerry, J.T.; Álvarez-Noriega, M.; Álvarez-Romero, J.G.; Anderson, K.D.; Baird, A.H.; Babcock, R.C.; Beger, M.; Bellwood, D.R.; Berkelmans, R.; et al. Global warming and recurrent mass bleaching of corals. *Nature* **2017**, *543*, 373–377. [CrossRef]

32. Hughes, T.P.; Anderson, K.D.; Connolly, S.R.; Heron, S.F.; Kerry, J.T.; Lough, J.M.; Baird, A.H.; Baum, J.K.; Berumen, M.L.; Bridge, T.C.; et al. Spatial and temporal patterns of mass bleaching of corals in the Anthropocene. *Science* **2018**, *359*, 80–83. [CrossRef] [PubMed]

33. Oliver, J.K.; Berkelmans, R.; Eakin, C.M. Coral Bleaching in Space and Time. In *Coral Bleaching: Patterns, Processes, Causes and Consequences*; van Oppen, M.J.H., Lough, J.M., Eds.; Springer International Publishing: Cham, Switzerland, 2018; pp. 27–49. [CrossRef]

34. Skirving, W.J.; Heron, S.F.; Marsh, B.L.; Liu, G.; De La Cour, J.L.; Geiger, E.F.; Eakin, C.M. The relentless march of mass coral bleaching: A global perspective of changing heat stress. *Coral Reefs* **2019**, *38*, 547–557. [CrossRef]

35. Hughes, T.P.; Kerry, J.T.; Baird, A.H.; Connolly, S.R.; Chase, T.J.; Dietzel, A.; Hill, T.; Hoey, A.S.; Hoogenboom, M.O.; Jacobson, M.; et al. Global warming impairs stock–recruitment dynamics of corals. *Nature* **2019**, *568*, 387–390. [CrossRef]

36. Raymundo, L.J.; Burdick, D.; Hoot, W.C.; Miller, R.M.; Brown, V.; Reynolds, T.; Gault, J.; Idechong, J.; Fifer, J.; Williams, A. Successive bleaching events cause mass coral mortality in Guam, Micronesia. *Coral Reefs* **2019**, *38*, 677–700. [CrossRef]

37. Mora, C.; Ospína, A.F. Tolerance to high temperatures and potential impact of sea warming on reef fishes of Gorgona Island (tropical eastern Pacific). *Mar. Biol.* **2001**, *139*, 765–769. [CrossRef]

38. Booth, D.; Beretta, G. Changes in a fish assemblage after a coral bleaching event. *Mar. Ecol. Prog. Ser.* **2002**, *245*, 205–212. [CrossRef]

39. Spalding, M.D.; Jarvis, G.E. The impact of the 1998 coral mortality on reef fish communities in the Seychelles. *Mar. Pollut. Bull.* **2002**, *44*, 309–321. [CrossRef]

40. Jones, G.P.; McCormick, M.I.; Srinivasan, M.; Eagle, J.V. Coral decline threatens fish biodiversity in marine reserves. *Proc. Natl. Acad. Sci. USA* **2004**, *101*, 8251–8253. [CrossRef]

41. Ospina, A.F.; Mora, C. Effect of body size on reef fish tolerance to extreme low and high temperatures. *Environ. Biol. Fishes* **2004**, *70*, 339–343. [CrossRef]

42. Roessig, J.M.; Woodley, C.M.; Cech, J.J.; Hansen, L.J. Effects of global climate change on marine and estuarine fishes and fisheries. *Rev. Fish Biol. Fish.* **2004**, *14*, 251–275. [CrossRef]

43. Graham, N.A.J.; Wilson, S.K.; Jennings, S.; Polunin, N.V.C.; Bijoux, J.P.; Robinson, J. Dynamic fragility of oceanic coral reef ecosystems. *Proc. Natl. Acad. Sci. USA* **2006**, *103*, 8425–8429. [CrossRef]

44. Garpe, K.C.; Yahya, S.A.S.; Lindahl, U.; Öhman, M.C. Long-term effects of the 1998 coral bleaching event on reef fish assemblages. *Mar. Ecol. Prog. Ser.* **2006**, *315*, 237–247. [CrossRef]

45. Wilson, S.K.; Graham, N.A.J.; Pratchett, M.S.; Jones, G.P.; Polunin, N.V.C. Multiple disturbances and the global degradation of coral reefs: Are reef fishes at risk or resilient? *Glob. Chang. Biol.* **2006**, *12*, 2220–2234. [CrossRef]

46. Pratchett, M.S.; Munday, P.; Wilson, S.K.; Graham, N.A.J.; Cinner, J.; Bellwood, D.R.; Jones, G.P.; Polunin, N.V.C.; McClanahan, T.R. Effects Of Climate-Induced Coral Bleaching On Coral-Reef Fishes, Ecological And Economic Consequences. In *Oceanography and Marine Biology: An Annual Reviewiology*; Gibson, R.N., Atkinson, R.J.A., Gordon, J.D.M., Eds.; Taylor & Francis: Milton Park, UK, 2008; pp. 257–302.

47. Clark, T.D.; Roche, D.G.; Binning, S.A.; Speers-Roesch, B.; Sundin, J. Maximum thermal limits of coral reef damselfishes are size dependent and resilient to near-future ocean acidification. *J. Exp. Biol.* **2017**, *220*, 3519–3526. [CrossRef]

48. Triki, Z.; Wismer, S.; Levorato, E.; Bshary, R. A decrease in the abundance and strategic sophistication of cleaner fish after environmental perturbations. *Glob. Chang. Biol.* **2018**, *24*, 481–489. [CrossRef] [PubMed]

49. Plaisance, L.; Knowlton, N.; Paulay, G.; Meyer, C. Reef-associated crustacean fauna: Biodiversity estimates using semi-quantitative sampling and DNA barcoding. *Coral Reefs* **2009**, *28*, 977–986. [CrossRef]

50. Plaisance, L.; Caley, M.J.; Brainard, R.E.; Knowlton, N. The diversity of coral reefs: What are we missing? *PLoS ONE* **2011**, *6*, e25026. [CrossRef] [PubMed]

51. Brandl, S.J.; Tornabene, L.; Goatley, C.H.R.; Casey, J.M.; Morais, R.A.; Côté, I.M.; Baldwin, C.C.; Parravicini, V.; Schiettekatte, N.M.D.; Bellwood, D.R. Demographic dynamics of the smallest marine vertebrates fuel coral reef ecosystem functioning. *Science* **2019**, *364*, 1189–1192. [CrossRef] [PubMed]

52. De Meeûs, T.; Renaud, F. Parasites within the new phylogeny of eukaryotes. *Trends Parasitol.* **2002**, *18*, 247–250. [CrossRef]

53. Rohde, K. *Ecology of Marine Parasites*; University of Queensland Press: St Lucia, Australia, 1982; p. 245.

54. Dobson, A.; Lafferty, K.D.; Kuris, A.M.; Hechinger, R.F.; Jetz, W. Homage to linnaeus: How many parasites? How many hosts? *Light Evol.* **2009**, *2*, 63–82. [CrossRef]

55. Hatcher, M.J.; Dunn, A.M. *Parasites in Ecological Communities: From Interactions to Ecosystems*; Cambridge University Press: Cambridge, UK, 2011. [CrossRef]

56. Okamura, B.; Hartigan, A.; Naldoni, J. Extensive uncharted biodiversity: The parasite dimension. *Integr. Comp. Biol.* **2018**, *56*, 1132–1145. [CrossRef]

57. Hudson, P.J.; Dobson, A.P.; Lafferty, K.D. Is a healthy ecosystem one that is rich in parasites? *Trends Ecol. Evol.* **2006**, *21*, 381–385. [CrossRef]

58. Kuris, A.M.; Hechinger, R.F.; Shaw, J.C.; Whitney, K.L.; Aguirre-Macedo, L.; Boch, C.A.; Dobson, A.P.; Dunham, E.J.; Fredensborg, B.L.; Huspeni, T.C.; et al. Ecosystem energetic implications of parasite and free-living biomass in three estuaries. *Nature* **2008**, *454*, 515–518. [CrossRef]

59. Hatcher, M.J.; Dick, J.T.A.; Dunn, A.M. Diverse effects of parasites in ecosystems: Linking interdependent processes. *Front. Ecol. Environ.* **2012**, *10*, 186–194. [CrossRef]

60. Hatcher, M.J.; Dick, J.T.A.; Dunn, A.M. Parasites that change predator or prey behaviour can have keystone effects on community composition. *Biol. Lett.* **2014**, *10*. [CrossRef] [PubMed]

61. Rohde, K. Ecology and biogeography of marine parasites. *Adv. Mar. Biol.* **2002**, *43*, 1–86. [PubMed]

62. Rohde, K. Species diversity of parasites on the Great Barrier Reef. *Z. Für Parasitenkd.* **1976**, *50*, 93–94. [CrossRef]

63. Poulin, R.; Blasco-Costa, I.; Randhawa, H.S. Integrating parasitology and marine ecology: Seven challenges towards greater synergy. *J. Sea Res.* **2016**, *113*, 3–10. [CrossRef]

64. Marcogliese, D.J. Implications of climate change for parasitism of animals in the aquatic environment. *Can. J. Zool.* **2001**, *79*, 1331–1352. [CrossRef]

65. Sikkel, P.C.; Richardson, M.A.; Sun, D.; Narvaez, P.; Feeney, W.E.; Grutter, A.S. Changes in abundance of fish-parasitic gnathiid isopods associated with warm-water bleaching events on the northern Great Barrier Reef. *Coral Reefs* **2019**, *38*, 721–730. [CrossRef]

66. Sikkel, P.C.; Welicky, R.L. The Ecological Significance of Parasitic Crustaceans. In *Parasitic Crustacea*; Springer: London, UK, 2019; pp. 421–477.

67. Claar, D.C.; Wood, C.L. Pulse Heat Stress and Parasitism in a Warming World. *Trends Ecol. Evol.* **2020**, *35*, 704–715. [CrossRef]

68. Gehman, A.-L.M.; Hall, R.J.; Byers, J.E. Host and parasite thermal ecology jointly determine the effect of climate warming on epidemic dynamics. *Proc. Natl. Acad. Sci. USA* **2018**, *115*, 744–749. [CrossRef]

69. Brazenor, A.K.; Hutson, K.S. Effects of temperature and salinity on the life cycle of *Neobenedenia* sp. (Monogenea: Capsalidae) infecting farmed barramundi (*Lates calcarifer*). *Parasitol. Res.* **2015**, *114*, 1875–1886. [CrossRef]

70. Grutter, A.S. Spatial and temporal variations of the ectoparasites of seven reef fish species from Lizard Island and Heron Island, Australia. *Mar. Ecol. Prog. Ser.* **1994**, *115*, 21–30. [CrossRef]

71. Grutter, A.S. Parasite removal rates by the cleaner wrasse Labroides dimidiatus. *Mar. Ecol. Prog. Ser.* **1996**, *130*, 61–70. [CrossRef]

72. Grutter, A.S.; Poulin, R. Cleaning of Coral Reef Fishes by the Wrasse *Labroides dimidiatus*: Influence of Client Body Size and Phylogeny. *Copeia* **1998**, *1998*, 1447707. [CrossRef]

73. Smit, N.J.; Davies, A.J. The curious life-style of the parasitic stages of Gnathiid isopods. *Adv. Parasitol.* **2004**, *58*, 289–391. [CrossRef] [PubMed]

74. Tanaka, K. Life history of gnathiid isopods—Current knowledge and and future directions. *Plankton Benthos Res.* **2007**, *2*, 1–11. [CrossRef]

75. Lafferty, K.D.; Kuris, A.M. Trophic strategies, animal diversity and body size. *Trends Ecol. Evol.* **2002**, *17*, 507–513. [CrossRef]

76. Raffel, T.R.; Martin, L.B.; Rohr, J.R. Parasites as predators: Unifying natural enemy ecology. *Trends Ecol. Evol.* **2008**, *23*, 610–618. [CrossRef]

77. Grutter, A.S. Parasite infection rather than tactile stimulation is the proximate cause of cleaning behaviour in reef fish. *Proc. R. Soc. B Biol. Sci.* **2001**, *268*, 1361–1365. [CrossRef]

78. Grutter, A.S. Feeding ecology of the fish ectoparasite Gnathia sp. (Crustacea: Isopoda) from the Great Barrier Reef, and its implications for fish cleaning behaviour. *Mar. Ecol. Prog. Ser.* **2003**, *259*, 295–302. [CrossRef]

79. Sikkel, P.C.; Cheney, K.L.; Côté, I.M. In situ evidence for ectoparasites as a proximate cause of cleaning interactions in reef fish. *Anim. Behav.* **2004**, *68*, 241–247. [CrossRef]

80. Sikkel, P.C.; Herzlieb, S.E.; Kramer, D.L. Compensatory cleaner-seeking behavior following spawning in female yellowtail damselfish. *Mar. Ecol. Prog. Ser.* **2005**, *296*, 1–11. [CrossRef]

81. Cheney, K.L.; Côté, I.M. Mutualism or parasitism? The variable outcome of cleaning symbioses. *Biol. Lett.* **2005**, *1*, 162–165. [CrossRef] [PubMed]

82. Binning, S.A.; Roche, D.G.; Grutter, A.S.; Colosio, S.; Sun, D.; Miest, J.; Bshary, R. Cleaner wrasse indirectly affect the cognitive performance of a damselfish through ectoparasite removal. *Proc. R. Soc. B Biol. Sci.* **2018**, *285*. [CrossRef] [PubMed]

83. Sellers, J.C.; Holstein, D.M.; Botha, T.L.; Sikkel, P.C. Lethal and sublethal impacts of a micropredator on post-settlement Caribbean reef fishes. *Oecologia* **2019**, *189*, 293–305. [CrossRef] [PubMed]

84. Jones, C.M.; Grutter, A.S. Parasitic isopods (*Gnathia* sp.) reduce haematocrit in captive blackeye thicklip (Labridae) on the Great Barrier Reef. *J. Fish Biol.* **2005**, *66*, 860–864. [CrossRef]

85. Hayes, P.M.; Smit, N.J.; Grutter, A.S.; Davies, A.J. Unexpected response of a captive blackeye thicklip, *Hemigymnus melapterus* (Bloch), from Lizard Island, Australia, exposed to juvenile isopods *Gnathia aureamaculosa* Ferreira & Smit. *J. Fish Dis.* **2011**, *34*, 563–566. [CrossRef]

86. Grutter, A.S.; Pickering, J.L.; McCallum, H.; McCormick, M.I. Impact of micropredatory gnathiid isopods on young coral reef fishes. *Coral Reefs* **2008**, *27*, 655–661. [CrossRef]

87. Jones, C.M.; Grutter, A.S. Reef-based micropredators reduce the growth of post-settlement damselfish in captivity. *Coral Reefs* **2008**, *27*, 677–684. [CrossRef]

88. Penfold, R.; Grutter, A.S.; Kuris, A.M.; McCormick, M.I.; Jones, C.M. Interactions between juvenile marine fish and gnathiid isopods: Predation versus micropredation. *Mar. Ecol. Prog. Ser.* **2008**, *357*, 111–119. [CrossRef]

89. Artim, J.M.; Sellers, J.C.; Sikkel, P.C. Micropredation by gnathiid isopods on settlement stage reef fish in the eastern Caribbean Sea. *Bull. Mar. Sci.* **2015**, *91*, 479–487. [CrossRef]

90. Grutter, A.S.; Blomberg, S.P.; Fargher, B.; Kuris, A.M.; McCormick, M.I.; Warner, R.R. Size-related mortality due to gnathiid isopod micropredation correlates with settlement size in coral reef fishes. *Coral Reefs* **2017**, *36*, 549–559. [CrossRef]

91. Curtis, L.M.; Grutter, A.S.; Smit, N.J.; Davies, A.J. *Gnathia aureamaculosa*, a likely definitive host of *Haemogregarina balistapi* and potential vector for *Haemogregarina bigemina* between fishes of the Great Barrier Reef, Australia. *Int. J. Parasitol.* **2013**, *43*, 361–370. [CrossRef] [PubMed]

92. Honma, Y.; Chiba, A. Pathological changes in the branchial chamber wall of stingrays, *Dasyatis* spp., associated with the presence of juvenile gnathiids (Isopoda, Crustacea). *Fish Pathol.* **1991**, *26*, 9–16. [CrossRef]

93. Grutter, A.S. Spatiotemporal Variation and Feeding Selectivity in the Diet of the Cleaner Fish Labroides dimidiatus. *Copeia* **1997**, *1997*, 1447754. [CrossRef]

94. Grutter, A.S. Cleaning symbioses from the parasites' perspective. *Parasitology* **2002**, *124*, 65–81. [CrossRef]

95. Grutter, A.S.; Blomberg, S.P.; Box, S.; Bshary, R.; Ho, O.; Madin, E.M.P.; McClure, E.C.; Meekan, M.G.; Murphy, J.M.; Richardson, M.A.; et al. Changes in local free-living parasite populations in response to cleaner manipulation over 12 years. *Oecologia* **2019**, *190*, 783–797. [CrossRef]

96. Demairé, C.; Triki, Z.; Binning, S.A.; Glauser, G.; Roche, D.G.; Bshary, R. Reduced access to cleaner fish negatively impacts the physiological state of two resident reef fishes. *Mar. Biol.* **2020**, *167*, 48. [CrossRef]

97. Grutter, A.; Bejarano, S.; Cheney, K.; Goldizen, A.; Sinclair-Taylor, T.; Waldie, P. Effects of the cleaner fish Labroides dimidiatus on grazing fishes and coral reef benthos. *Mar. Ecol. Prog. Ser.* **2020**, *643*, 99–114. [CrossRef]

98. Hughes, T.P.; Kerry, J.T.; Simpson, T. Large-scale bleaching of corals on the Great Barrier Reef. *Ecology* **2018**, *99*, 501. [CrossRef]

99. Arceo, H.O.; Quibilan, M.C.; Aliño, P.M.; Lim, G.; Licuanan, W.Y. Coral bleaching in Philippine reefs: Coincident evidences with mesoscale thermal anomalies. *Bull. Mar. Sci.* **2001**, *69*, 579–593.

100. Raymundo, L.; Maypa, A. Recovery of the Apo Island Marine Reserve, Philippines, 2 years after the El Niño bleaching event. *Coral Reefs* **2002**, *21*, 260–261. [CrossRef]

101. Magdaong, E.T.; Fujii, M.; Yamano, H.; Licuanan, W.Y.; Maypa, A.; Campos, W.L.; Alcala, A.C.; White, A.T.; Apistar, D.; Martinez, R. Long-term change in coral cover and the effectiveness of marine protected areas in the Philippines: A meta-analysis. *Hydrobiologia* **2014**, *733*, 5–17. [CrossRef]

102. Grutter, A.S.; Feeney, W.E.; McClure, E.C.; Narvaez, P.; Smit, N.J.; Sun, D.; Sikkel, P.C. Practical methods for culturing parasitic gnathiid isopods. *Int. J. Parasitol.* **2020**, *50*, 825–837. [CrossRef] [PubMed]

103. Sikkel, P.C.; Tuttle, L.J.; Cure, K.; Coile, A.M.; Hixon, M.A. Low Susceptibility of Invasive Red Lionfish (*Pterois volitans*) to a Generalist Ectoparasite in Both Its Introduced and Native Ranges. *PLoS ONE* **2014**, *9*, e95854. [CrossRef] [PubMed]

104. Santos, T.R.N.; Sikkel, P.C. Habitat associations of fish-parasitic gnathiid isopods in a shallow reef system in the central Philippines. *Mar. Biodivers.* **2017**, *49*, 83–96. [CrossRef]

105. Artim, J.M.; Sikkel, P.C. Comparison of sampling methodologies and estimation of population parameters for a temporary fish ectoparasite. *Int. J. Parasitol. Parasites Wildl.* **2016**, *5*, 145–157. [CrossRef] [PubMed]

106. Graba-Landry, A.C.; Loffler, Z.; McClure, E.C.; Pratchett, M.S.; Hoey, A.S. Impaired growth and survival of tropical macroalgae (*Sargassum* spp.) at elevated temperatures. *Coral Reefs* **2020**, *39*, 475–486. [CrossRef]

107. Pacific Ocean: Great Barrier Reef: Sea Water Temperature Compared to Long Term Averages. Available online: http://data.aims.gov.au/aimsrtds/yearlytrends.xhtml (accessed on 13 November 2018).

108. Water Temperature at Lizard Island Graph. Australian Institute of Marine Science AIMS. Available online: http://data.aims.gov.au/aimsrtds/datatool.xhtml?from=2018-10-15&thru=2018-10-20&period=MONTH&aggregations=AVG&channels=3523 (accessed on 13 November 2018).

109. NOAA Coral Reef Watch Program. Available online: https://coralreefwatch.noaa.gov/satellite/vs/philippines.php (accessed on 5 January 2018).

110. Therneau, T.M.; Grambsch, P.M. *Modeling Survival Data: Extending the Cox Model*; Springer: New York, NY, USA, 2000.

111. Therneau, T. Coxme: Mixed Effects Cox Models. R Package Version 2.2-3. 2015. Available online: http://CRAN.R-project.org/package=coxme (accessed on 17 July 2020).

112. Fox, J.; Weisberg, S. Multivariate Linear Models in R. In *An R Companion to Applied Regression*, 2nd ed.; Thousand Oaks: Los Angeles, CA, USA, 2011; pp. 1–31.

113. Ganachaud, A.; Sen Gupta, A.S.; Orr, J.C.; Wijffels, S.E.; Ridgway, K.R.; Hemer, M.A.; Maes, C.; Steinberg, C.R.; Tribollet, A.D.; Qiu, B.; et al. Observed and expected changes to the tropical Pacific Ocean. In *Vulnerability of Tropical Pacific Fisheries and Aquaculture to Climate Change*; Bell, J.D., Johnson, J.E., Hobday, A.J., Eds.; Secretariat of the Pacific Community: Noumea, New Caledonia, 2011; pp. 101–187.

114. Huey, R.B.; Kingsolver, J.G. Climate Warming, Resource Availability, and the Metabolic Meltdown of Ectotherms. *Am. Nat.* **2019**, *194*, E140–E150. [CrossRef]

115. Smale, D.A.; Wernberg, T.; Oliver, E.C.J.; Thomsen, M.; Harvey, B.P.; Straub, S.C.; Burrows, M.T.; Alexander, L.V.; Benthuysen, J.A.; Donat, M.G.; et al. Marine heatwaves threaten global biodiversity and the provision of ecosystem services. *Nat. Clim. Chang.* **2019**, *9*, 306–312. [CrossRef]

116. Frölicher, T.L.; Fischer, E.M.; Gruber, N. Marine heatwaves under global warming. *Nature* **2018**, *560*, 360–364. [CrossRef]

117. Oliver, E.C.J.; Donat, M.G.; Burrows, M.T.; Moore, P.J.; Smale, D.A.; Alexander, L.V.; Benthuysen, J.A.; Feng, M.; Gupta, A.S.; Hobday, A.J.; et al. Longer and more frequent marine heatwaves over the past century. *Nat. Commun.* **2018**, *9*, 1324. [CrossRef]

118. Compton, T.J.; Rijkenberg, M.J.A.; Drent, J.; Piersma, T. Thermal tolerance ranges and climate variability: A comparison between bivalves from differing climates. *J. Exp. Mar. Biol. Ecol.* **2007**, *352*, 200–211. [CrossRef]

119. Stillman, J.H.; Somero, G.N. A comparative analysis of the upper thermal tolerance limits of eastern pacific porcelain crabs, genus *Petrolisthes*: Influences of latitude, vertical zonation, acclimation, and phylogeny. *Physiol. Biochem. Zool.* **2000**, *73*, 200–208. [CrossRef] [PubMed]

120. Attrill, M.J.; Kelmo, F.; Jones, M.B. Impact of the 1997-98 El Niño event on the coral reef-associated echinoderm assemblage from northern Bahia, northeastern Brazil. *Clim. Res.* **2004**, *26*, 151–158. [CrossRef]

121. Chan, B.K.K.; Morritt, D.; De Pirro, M.; Leung, K.M.Y.; Williams, G.A. Summer mortality: Effects on the distribution and abundance of the acorn barnacle *Tetraclita japonica* on tropical shores. *Mar. Ecol. Prog. Ser.* **2006**, *328*, 195–204. [CrossRef]

122. Jokiel, P.L.; Coles, S.L. Coral Reefs to elevated temperature. *Environ. Conserv.* **1990**, *8*, 155–162.

123. Mouritsen, K.N.; Poulin, R. Parasitism, climate oscillations and the structure of natural communities. *Oikos* **2002**, *97*, 462–468. [CrossRef]

124. Poulin, R. Global warming and temperature-mediated increases in cercarial emergence in trematode parasites. *Parasitology* **2006**, *132*, 143–151. [CrossRef]

125. Thieltges, D.W.; Fredensborg, B.L.; Studer, A.; Poulin, R. Large-scale patterns in trematode richness and infection levels in marine crustacean hosts. *Mar. Ecol. Prog. Ser.* **2009**, *389*, 139–147. [CrossRef]

126. Pechenik, J.A.; Fried, B. Effect of temperature on survival and infectivity of *Echinostoma trivolvis* cercariae: A test of the energy limitation hypothesis. *Parasitology* **1995**, *111*, 373–378. [CrossRef]

127. Lyholt, H.C.K.; Buchmann, K. Diplostomum spathaceum:effects of temperature and light on cercarial shedding and infection of rainbow trout. *Dis. Aquat. Org.* **1996**, *25*, 169–173. [CrossRef]

128. McCarthy, A.M. The influence of temperature on the survival and infectivity of the cercariae of *Echinoparyphium recurvatum* (Digenea: Echinostomatidae). *Parasitology* **1999**, *118*, 383–388. [CrossRef] [PubMed]

129. Toledo, R.; Muñoz-Antoli, C.; Pérez, M.; Esteban, J.G. Survival and infectivity of Hypoderaeum conoideum and Euparyphium albuferensis cercariae under laboratory conditions. *J. Helminthol.* **1999**, *73*, 177–182. [CrossRef] [PubMed]

130. Mouritsen, K.N. The *Hydrobia ulvae—Maritrema subdolum* association: Influence of temperature, salinity, light, water-pressure and secondary host exudates on cercarial emergence and longevity. *J. Helminthol.* **2002**, *76*, 341–347. [CrossRef] [PubMed]

131. Thieltges, D.W.; Rick, J. Effect of temperature on emergence, survival and infectivity of cercariae of the marine trematode *Renicola roscovita* (Digenea: Renicolidae). *Dis. Aquat. Org.* **2006**, *73*, 63–68. [CrossRef]

132. Leiva, N.V.; Manríquez, P.H.; Aguilera, V.M.; González, M.T. Temperature and pCO2 jointly affect the emergence and survival of cercariae from a snail host: Implications for future parasitic infections in the Humboldt Current system. *Int. J. Parasitol.* **2019**, *49*, 49–61. [CrossRef]

133. Franzova, V.A.; MacLeod, C.D.; Wang, T.; Harley, C.D.G. Complex and interactive effects of ocean acidification and warming on the life span of a marine trematode parasite. *Int. J. Parasitol.* **2019**, *49*, 1015–1021. [CrossRef]

134. Conley, D.C.; Curtis, M.A. Effects of temperature and photoperiod on the duration of hatching, swimming, and copepodid survival of the parasitic copepod *Salmincola edwardsii*. *Can. J. Zool.* **1993**, *71*, 972–976. [CrossRef]

135. Hirazawa, N.; Takano, R.; Hagiwara, H.; Noguchi, M.; Narita, M. The influence of different water temperatures on *Neobenedenia girellae* (Monogenea) infection, parasite growth, egg production and emerging second generation on amberjack *Seriola dumerili* (Carangidae) and the histopathological effect of this parasite on fish skin. *Aquaculture* **2010**, *299*, 2–7. [CrossRef]

136. Mahmoud, N.E.; Fahmy, M.M.; Abuowarda, M.M.; Zaki, M.; Ismail, E.M.; Ismael, E.S. Mediterranean sea fry; a source of isopod infestation problem in Egypt with reference to the effect of salinity and temperature on the survival of *Livoneca redmanii* (Isopoda: Cymothoidae) juvenile stages. *J. Egypt. Soc. Parasitol.* **2019**, *49*, 235–242.

137. Evans, N.A. The influence of environmental temperature upon transmission of the cercariae of *Echinostoma liei* (Digenea: Echinostomatidae). *Parasitology* **1985**, *90*, 269–275. [CrossRef]

138. Lo, C.T.; Lee, K.M. Pattern of Emergence and the Effects of Temperature and Light on the Emergence and Survival of Heterophyid Cercariae (*Centrocestus formosanus* and *Haplorchis pumilio*). *J. Parasitol.* **1996**, *82*, 347. [CrossRef] [PubMed]

139. Fried, B.; Ponder, E.L. Effects of temperature on survival, infectivity and in vitro encystment of the cercariae of *Echinostoma caproni*. *J. Helminthol.* **2003**, *77*, 235–238. [CrossRef] [PubMed]

140. Tucker, C.S.; Sommerville, C.; Wootten, R. The effect of temperature and salinity on the settlement and survival of copepodids of *Lepeophtheirus salmonis* (Krøyer, 1837) on Atlantic salmon, *Salmo salar* L. *J. Fish Dis.* **2000**, *23*, 309–320. [CrossRef]

141. Peck, L.S.; Webb, K.E.; Bailey, D.M. Extreme sensitivity of biological function to temperature in Antarctic marine species. *Funct. Ecol.* **2004**, *18*, 625–630. [CrossRef]

142. Pörtner, H.O.; Knust, R. Climate change affects marine fishes through the oxygen limitation of thermal tolerance. *Science* **2007**, *315*, 95–97. [CrossRef] [PubMed]

143. Meyrowitsch, D.; Christensen, N.; Hindsbo, O. Effects of temperature and host density on the snail-finding capacity of cercariae of *Echinostoma caproni* (Digenea: Echinostomatidae). *Parasitology* **1991**, *102*, 391–395. [CrossRef]

144. Cech, J.J.; Moyle, P.B. *Fishes: An Introduction to Ichthyology*, 5th ed.; Prentice Hall: Upper Saddle River, NJ, USA, 2004.

145. Gagliano, M.; McCormick, M.I.; Meekan, M.G. Temperature-induced shifts in selective pressure at a critical developmental transition. *Oecologia* **2007**, *152*, 219–225. [CrossRef]

146. Nilsson, G.E.; Crawley, N.; Lunde, I.G.; Munday, P.L. Elevated temperature reduces the respiratory scope of coral reef fishes. *Glob. Chang. Biol.* **2009**, *15*, 1405–1412. [CrossRef]

147. Nay, T.J.; Johansen, J.L.; Habary, A.; Steffensen, J.F.; Rummer, J.L. Behavioural thermoregulation in a temperature-sensitive coral reef fish, the five-lined cardinalfish (*Cheilodipterus quinquelineatus*). *Coral Reefs* **2015**, *34*, 1261–1265. [CrossRef]

148. Jones, C.M.; Miller, T.L.; Grutter, A.S.; Cribb, T.H. Natatory-stage cymothoid isopods: Description, molecular identification and evolution of attachment. *Int. J. Parasitol.* **2008**, *38*, 477–491. [CrossRef]

149. Shodipo, M.O.; Gomez, R.D.C.; Welicky, R.L.; Sikkel, P.C. Apparent kleptoparasitism in fish—parasitic gnathiid isopods. *Parasitol. Res.* **2018**, *118*, 653–655. [CrossRef] [PubMed]

150. Nicholson, M.D.; Artim, J.D.; Hendrick, G.C.; Packard, A.J.; Sikkel, P.C. Fish-Parasitic Gnathiid Isopods Metamorphose Following Invertebrate-Derived Meal. *J. Parasitol.* **2019**, *105*, 793. [CrossRef] [PubMed]

151. Ota, Y.; Hoshino, O.; Hirose, M.; Tanaka, K.; Hirose, E. Third-stage larva shifts host fish from teleost to elasmobranch in the temporary parasitic isopod, *Gnathia trimaculata* (Crustacea; Gnathiidae). *Mar. Biol.* **2012**, *159*, 2333–2347. [CrossRef]

152. Sikkel, P.C.; Welicky, R.L.; Artim, J.M.; McCammon, A.M.; Sellers, J.C.; Coile, A.M.; Jenkins, W.G. Nocturnal migration reduces exposure to micropredation in a coral reef fish. *Bull. Mar. Sci.* **2017**, *93*, 475–489. [CrossRef]

153. Artim, J.M.; Sikkel, P.C. Live coral repels a common reef fish ectoparasite. *Coral Reefs* **2013**, *32*, 487–494. [CrossRef]

154. Artim, J.M.; Nicholson, M.D.; Hendrick, G.C.; Brandt, M.; Smith, T.; Sikkel, P.C. Abundance of a Cryptic Generalist Parasite Reflects Degradation of an Ecosystem. *Ecosphere* **2020**, in press.

155. Dupont, S.; Pörtner, H. Get ready for ocean acidification. *Nature* **2013**, *498*, 429. [CrossRef]

156. Paula, J.R.; Otjacques, E.; Hildebrandt, C.; Grutter, A.S.; Rosa, R. Ocean Acidification Does Not Affect Fish Ectoparasite Survival. *Oceans* **2020**, *1*, 3. [CrossRef]

157. Carlson, C.J.; Burgio, K.R.; Dougherty, E.R.; Phillips, A.J.; Bueno, V.M.; Clements, C.F.; Castaldo, G.; Dallas, T.A.; Cizauskas, C.A.; Cumming, G.S.; et al. Parasite biodiversity faces extinction and redistribution in a changing climate. *Sci. Adv.* **2017**, *3*. [CrossRef]

158. Carlson, C.J.; Cizauskas, C.A.; Burgio, K.R.; Clements, C.F.; Harris, N.C. The More Parasites, the Better? *Science* **2013**, *342*, 1041. [CrossRef]

159. Cizauskas, C.A.; Carlson, C.J.; Burgio, K.R.; Clements, C.F.; Dougherty, E.R.; Harris, N.C.; Phillips, A.J. Parasite vulnerability to climate change: An evidence-based functional trait approach. *R. Soc. Open Sci.* **2017**, *4*. [CrossRef] [PubMed]

*Article*

# Severe Heat Stress Resulted in High Coral Mortality on Maldivian Reefs following the 2015–2016 El Niño Event

Pia Bessell-Browne [1], Hannah E. Epstein [2], Nora Hall [3], Patrick Buerger [4,5] and Kathryn Berry [6,*]

[1]   CSIRO Oceans and Atmosphere, Castray Esplanade, Hobart, TAS 7000, Australia; pia.bessell-browne@csiro.au

[2]   Department of Microbiology, Oregon State University, Corvallis, OR 97331, USA; epsteinh@oregonstate.edu

[3]   College of Science and Engineering, James Cook University, Townsville, QLD 4811, Australia; Noramhall@gmail.com

[4]   CSIRO Synthetic Biology Future Science Platform, Land & Water, Black Mountain, ACT 2601, Australia; patrick.buerger@unimelb.edu.au

[5]   School of BioSciences, The University of Melbourne, Parkville, VIC 3010, Australia

[6]   Department of Fisheries and Oceans, Institute of Ocean Sciences, Sidney, BC V8L 4B2, Canada

*   Correspondence: kathrynbberry@gmail.com

**Abstract:** Coral cover worldwide has been declining due to heat stress caused by climate change. Here we report the impacts of the 2015–2016 El Niño mass coral bleaching event on the coral cover of reefs located on central and northern atolls of the Maldives. We surveyed six reef sites in the Alifu Alifu (Ari) and Baa (South Maalhosmadulu) Atolls using replicate 20 m benthic photo transects at two depths per reef site. Live and recently dead coral cover identified from images differed between reef sites and depth. Recently dead corals on average made up 33% of the coral assemblage at shallow sites and 24% at deep sites. This mortality was significantly lower in massive corals than in branching corals, reaching an average of only 6% compared to 41%, respectively. The best predictors of live coral cover were depth and morphology, with a greater percentage of live coral at deep sites and in massive corals. The same predictors best described the prevalence of recently dead coral, but showed inverse trends to live coral. However, there was high variability among reef sites, which could be attributed to additional local stressors. Coral bleaching and resulting coral mortalities, such as the ones reported here, are of particular concern for small island nations like the Maldives, which are reliant on coral reefs.

**Keywords:** Maldives; Indian Ocean; El Niño; mass coral bleaching; coral mortality; benthic cover

**Citation:** Bessell-Browne, P.; Epstein, H.E.; Hall, N.; Buerger, P.; Berry, K. Severe Heat Stress Resulted in High Coral Mortality on Maldivian Reefs following the 2015–2016 El Niño Event. *Oceans* **2021**, *2*, 233–245. https://doi.org/10.3390/oceans2010014

Academic Editor: Rupert Ormond

Received: 28 October 2020
Accepted: 22 February 2021
Published: 3 March 2021

**Publisher's Note:** MDPI stays neutral with regard to jurisdictional claims in published maps and institutional affiliations.



## 1. Introduction

Elevated seawater temperatures resulting from climate change are causing widespread coral bleaching events across tropical regions of the world [1–3]. Over the past 30 years the frequency and severity of these bleaching events have increased and they are now occurring on an unprecedented scale [1,4]. In 2015–2016, a severe El Niño event led to widespread coral bleaching across the Pacific and Indian Oceans [1,2,4–8], with some regions experiencing repeated heat stress events throughout this period (e.g., Chagos Archipelago [9], Great Barrier Reef (GBR) [2,10]). During the El Niño, some areas of the Pacific experienced extreme heat stress, reaching over 20 degree heating weeks (DHWs) [5], a metric representing accumulated heat stress over a 90-day window in comparison to historical maximum monthly means (MMMs) [10]. Although not quite as extreme, DHWs from the Western and Eastern Indian Ocean during the same period still greatly exceeded the bleaching threshold, with records ranging from four to 11 [11–13].

Coral bleaching occurs when corals expel their photosynthetic endosymbiotic algae (Symbiodiniaceae), often as a result of temperature stress, leaving a white skeleton visible through translucent tissue [14]. Corals may recover over time or coral mortality may occur

165

if the algae cells are not regained and temperature stress conditions persist. Susceptibility to bleaching and mortality can vary among coral species due to a variety of factors, which include coral morphology [15,16], history of stress or temperature exposure [17,18], associated algal symbiont type [19–21], and the coral microbiome [22]. This variability in susceptibility can cause not only a loss in overall coral cover, but also a reduction in both the biodiversity and functional diversity of coral assemblages [23]. In severe cases bleaching and subsequent mass mortality events result in decreased larval supply and connectivity, which can result in phase shifts that are often from coral to macroalgal dominated reefs [24]. Other alternate states include sponge- or urchin-dominated reefs [25,26]. It becomes difficult for corals to re-populate reefs and reverse such phase shifts (reviewed in [27]), resulting in obvious changes to benthic and reef-associated pelagic assemblages across the reef ecosystem.

Mass coral bleaching events impact not only the health of the coral reef ecosystem, but also the human populations that may rely on artisanal fisheries, coastal protection, and ecotourism. Coral mortality can be particularly concerning for small island nations because reefs represent a lifeline that provides food and livelihood. Loss of coral in the ecosystem can decrease catches of artisanal reef fisheries and discourage tourism, ultimately impacting the food security and economy of local communities [28,29].

One such small island nation is the Republic of Maldives, located in the central equatorial Indian Ocean. The Maldives is made up of over 1000 islets situated within 26 atolls. It is the largest group of coral reefs in the Indian Ocean, hosting approximately 1100 reef-associated fish species and 250 species of coral [30]. The economy of the Maldives benefits from marine resources through tourism and ecosystem services such as coastal protection, food, and construction materials [31]. Thus, changes to the state of Maldivian reef health is of concern to the nation. During the 2015–2016 El Niño event, reefs in the Maldives experienced between four and 11 DHWs, suggesting considerable heat stress [32]. This resulted in widespread bleaching across the region [7,33–35], with bleaching observed in 73% of corals from 0–13 m depth [12]. Here we document the effects of the 2015–2016 global bleaching event on the survival and mortality of coral communities in the northern atolls of the Maldives.

## 2. Results

Six coral reef sites were monitored in the Alifu Alifu and Baa Atolls in the Republic of Maldives (Figure 1) to assess post-bleaching coral mortality. The percentage of live coral cover at each coral reef site ($n = 6$) and depth was variable and ranged between 3% and 26% of substrate cover in December 2016, following the peak of the heating event that occurred in May of the same year (Figure 2). Of all the sites, live coral cover was lowest at Hurasdhoo, with an average of $3 \pm 4\%$ (mean $\pm$ standard error, $n = 51$) at deep sites (Figure 2). No records were available for this site at shallow depths. All other sites had similar live coral cover at shallow depths, ranging from $13 \pm 8\%$ ($n = 34$) at Rasdhoo to $26 \pm 13\%$ ($n = 51$) at Dhonfanu (Figure 2). Live coral cover at depth had a similar range, from $13 \pm 14\%$ ($n = 51$) at Nika Point to $27 \pm 12\%$ ($n = 51$) at Dhonfanu (Figures 2 and 3).

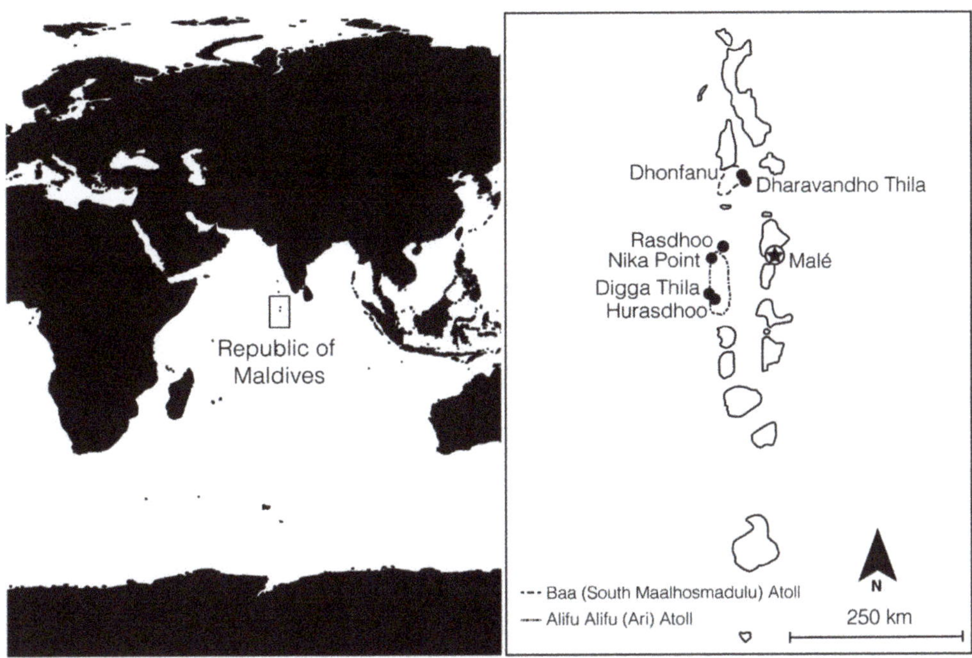



**Figure 1.** Reef site locations within the northern atolls (Baa and Alifu Alifu Atolls) of the Republic of Maldives. Sampling locations are marked with a black dot on the map.

Recently dead coral, classified as coral skeletons covered with turf algae but with the skeleton still intact, was also abundant at each of the sites surveyed (Figure 2). At shallow depths, this was as high as $54 \pm 19\%$ ($n = 52$) of the substrate cover at Digga Thila (Figure 2). The percentage of recently dead coral at shallow transects at all other sites ranged from $12 \pm 8\%$ ($n = 50$) at Nika Point to $40 \pm 14\%$ ($n = 51$) at Dhonfanu (Figure 2). Of the deep transects, the percentage of recently dead coral was highest at Hurasdhoo, averaging $41 \pm 25\%$ ($n = 51$), although this was highly variable between transects and replicates (Figure 2).

Very few coral colonies were still bleached, with translucent live tissue covering the skeleton, at the time of data collection seven months after peak heat stress. The total substrate covered by bleached corals ranged from $0.1 \pm 0.7\%$ ($n = 54$) to $2.9 \pm 7.6\%$ ($n = 54$) across all reef sites and depths.

Modelling of variables influencing the percentage cover of live coral suggested that the best model fit was obtained when including additive effects of reef site, depth, and morphology (Table 1). When accounting for these variables, live coral cover was estimated to be considerably lower at Hurasdhoo (3% live cover, with confidence intervals (CI) of 2–5%), compared with all other reef sites, over which coral cover ranged from 18% (15–22% CI) to 29% (25–34% CI) (Figure 4a). Mean live coral cover was found to be higher at deep rather than shallow sites, estimated to be 23% (19–27% CI) and 16% (13–19% CI), respectively (Figure 4b). In addition, mean live coral cover was found to be greater in massive corals than branching corals, estimated to be 33% (27–39% CI) in massive corals and 6% (5–7% CI) in branching corals (Figure 4c).

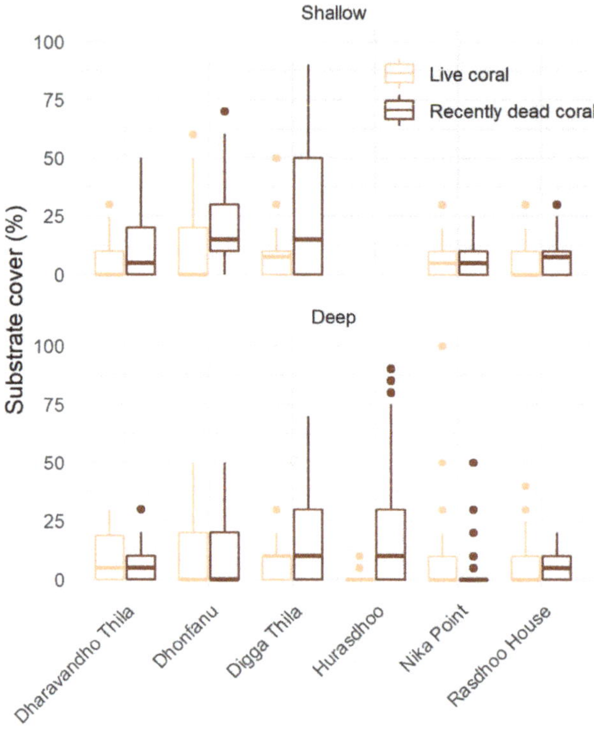

Figure 2. Box and whisper plot showing the percentage of substrate cover of live corals and recently dead corals at both shallow and deep transects at each of the six reef sites surveyed.

**Figure 3.** Extensive stands of dead corals (*Acropora* spp.) in the Maldives, where reefs exhibited high levels of coral mortality in December 2016 following the 2015–2016 El Niño mass coral bleaching event ((**a**) and (**b**)).

**Table 1.** Generalised linear mixed effect model (GLMM) fit statistics investigating the most important variables influencing both (**a**) live coral and (**b**) recently dead coral. Shown are the number of parameters ($n$), Akaike information criteria (AICc), δ AICc, and AICc weight ($\omega_i$). The best fitting model is highlighted in bold.

| Coral Cover | Model | $n$ | AICc | δ AICc | $\omega_i$ |
|---|---|---|---|---|---|
| (a) Live coral | **Reef site + depth + morphology** | **11** | **1013.3** | **0** | **1** |
| | Reef site + morphology | 10 | 1036.7 | 23.31 | 0 |
| | Depth + morphology | 6 | 1153 | 139.64 | 0 |
| | Morphology | 5 | 1156.4 | 143.05 | 0 |
| | Reef site + depth | 10 | 1523.5 | 510.12 | 0 |
| | Reef site | 9 | 1537.7 | 524.31 | 0 |
| | Depth | 5 | 1635.1 | 621.79 | 0 |
| | Null | 4 | 1637 | 623.64 | 0 |
| (b) Recently dead coral | **Reef site + depth + morphology** | **11** | **1034.7** | **0** | **1** |
| | Reef site + morphology | 10 | 1056.2 | 21.53 | 0 |
| | Depth + morphology | 6 | 1093 | 58.25 | 0 |
| | Morphology | 5 | 1096.5 | 61.81 | 0 |
| | Reef site + depth | 10 | 1177.8 | 143.12 | 0 |
| | Reef site | 9 | 1193.5 | 158.75 | 0 |
| | Depth | 5 | 1236.9 | 202.18 | 0 |
| | Null | 4 | 1238.1 | 203.41 | 0 |

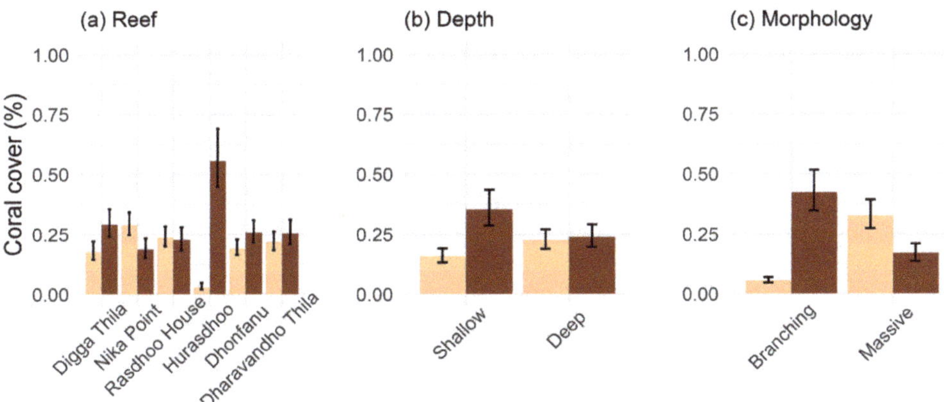


**Figure 4.** Predicted percentage of substrate cover of both live coral and recently dead coral by (**a**) reef, (**b**) depth, and (c) morphology, with 95% confidence intervals.

Modelling the cover of recently dead coral also suggested the best fitting model included additive effects of reef site, depth, and morphology (Table 1b). The percentage of dead coral cover at each reef site varied. The highest percentage was found at Hurasdhoo, with 56% (±45–69%) recently dead coral cover, whereas the remainder of the reef sites ranged from 19% (±15–23%) to 29% (±24–36%) (Figure 4a). Overall, there were more recently dead corals at shallow depths, with estimates of 35% (±28–43%) compared to 24% (±20–29%) at deep sites (Figure 4b). Estimates of recently dead massive corals were also lower than those for branching corals, with massive estimates of 17% (±14–21%) and branching estimates of 42% (±34–52%) (Figure 4c). The percentage of cover of recently dead corals was inverse to the trends observed for live coral cover (Figure 4).

## 3. Discussion

### 3.1. Bleaching and Mortality

During the 2016 El Niño the Maldives experienced a severe bleaching event with the overall mean percentage of bleached corals (across 71 sites and depths ranging from 0–13 m) reaching 73%, with some sites exhibiting close to 100% bleaching [12]. Although the peak of the bleaching event occurred between mid-April and mid-June 2016 and some mortality was seen at this time, many colonies remained bleached for several months. Live mean coral cover was reduced to approximately 6–20% across the central and southern Maldives [33–35], with both lagoonal and oceanic reefs impacted [35]. Higher mortality was observed at reef sites in proximity to higher local population sizes, suggesting that these sites were more strongly impacted by human pressures (e.g., North and South Malé Atolls [35]). Coral reef sites investigated in the present study were exposed to comparatively lower human pressures (Baa and Alifu Alifu Atolls), but still exhibited extensive mortality (Figures 3 and 4), with mean live coral cover reduced to 3–27%. The observed bleaching-induced mortality followed similar patterns as previously recorded bleaching severity in the area (see [7,12,34]); mortality was highest in shallow depths (~30% mortality) and among branching corals (~40% mortality), especially *Acropora* spp.

Bleaching severity is considered to vary spatially (e.g., [36,37]) and to be influenced by microhabitat features [38]. Additionally, coral morphology is a factor and branching corals are more susceptible to bleaching and mortality during heat stress in comparison to massive corals [15,39,40]. Branching structures can help optimize light dispersion and reflection that may be important for maintaining colony health at depth or in normal light conditions. However, for colonies in shallow habitats where irradiance is naturally higher, this structure may cause supra-optimal irradiance conditions during heatwaves, accelerating or intensifying bleaching [15,39,40]. In contrast, massive corals have less structural complexity and have been found to have a higher resistance to bleaching, taking longer to bleach and experiencing less mortality [41]. Although branching corals did experience higher bleaching-induced mortality than massive corals in the present study, these mortality levels (approximately 40%) were comparably lower than branching coral mortality in other areas of the Maldives in the aftermath of the 2016 El Niño event. North Malé (Kaafu) Atoll experienced 80% mortality of *Acropora* spp., the dominant branching coral genus in the Maldives [12], and southern sheltered reefs exhibited 75% mortality in branching corals [33]. In a similar pattern, the Chagos Archipelago, situated south of the Maldives, lost 86% of *Acropora* spp. cover (from 2–14%) during the 2015–2016 thermal anomalies, shifting to a community dominated by *Porites* spp. post bleaching [9].

Interestingly, although branching corals are generally considered more susceptible than massive corals on mildly bleaching reefs, this pattern breaks down when reefs are exposed to severe bleaching: Some corals that are normally classified as thermotolerant (e.g., many massive species) can nevertheless become bleached under severe heat stress (e.g., [1]). This may suggest that the Maldives was not hit as hard by this El Niño event as some reefs in other areas around the globe. In support of this, although DHWs in the Maldives reached 11, which exceeds the proposed eight DHW mass bleaching threshold, other reefs around the globe reached much higher DHWs: up to 16 on the GBR and over 20 in the central Pacific [1,5]. Mortality in some of these areas was nearly 100% [42]. The mortality recorded here on Maldivian reef sites still exceeded 50%, and although this is less than what some of the hardest hit reefs globally experienced during the 2015–2016 El Niño, this remains a mass mortality event that can have major consequences for reef recovery.

### 3.2. Potential for Recovery of Maldivian Reefs

Maldivian coral reefs experienced severe bleaching in 1998 [43,44] and again in 2016 [12], and between these dates also endured a series of mild bleaching events in 2003, 2005, 2007, and 2010 [45–47]. During this recovery period, Maldivian reefs faced other disturbances, such as a tsunami in 2004 [48] and outbreaks of crown-of-thorns starfish (*Acanthaster planci*), both of which contributed to further coral mortality and influenced recovery capacity in the impacted localized areas [49]. Following the 1998 bleaching event, which resulted in mass mortality (90%) of hard corals [50] and a near total disappearance of tabular coral [51], Maldivian reefs underwent a recovery trajectory that spanned the next 18 years. Initial mortality of acroporids (up to 90% [51]) led to a shift in assemblage composition of coral cover, from *Acropora*-dominated to *Porites*-dominated, for the first eight-plus years of recovery [43,49,52,53]. By 2014, 15 years post bleaching, reefs in the region resembled pre-1998 bleaching assemblages, however, with an absence of species that had yet to recover [46,49]. After the 1998 bleaching event the constructional capacity (i.e., bioconstruction potential, carbonate deposition, and reef accretion) of Maldivian reefs was severely affected, with two to three years required for renewed carbonate deposition and 14–16 years required before accretion rates were high enough to ensure constratal (low-relief) growth [54].

Recovery of Maldivian reefs post 1998 bleaching event was slower than rates reported in the neighbouring Chagos Archipelago [49,55,56], which lacks local human threats and recovered in 10 years [9]. Slower recovery was likely related to the increased level of human pressures that Maldivian reefs experience (e.g., overexploitation of resources, pollution, and coastal development) that influence the success of coral reef recovery, overall coral reef health, and resilience (e.g., [57,58]). Environmental degradation associated with human-induced pressures have increased in the Maldives, particularly related to local population growth, coastal development, land reclamation works, and tourism [59–62]. Whereas local populations reached 344,023 in 2014, increases in the numbers of tourists grew by 400% between 1997 and 2019, reaching 1.5 million visitors in 2019 [35].

Following the 2016 bleaching event, coral mortality was greater at coral reefs exposed to increased human pressures (e.g., North and South Malé [35]). For example, reefs in proximity to land reclamation projects experienced four-fold increases in sedimentation loads and exhibited significant reductions in live coral cover post bleaching event in comparison to reefs not impacted by both land reclamation activities and bleaching [62]. Land reclamation and associated dredging (Figure 5a) and dumping activities contribute to increased suspended sediments and deposited sediment loads (i.e., sedimentation). Increased suspended sediments reduce light levels required by corals to photosynthesize [63–65], but even low levels of sediment or other particulate deposition reduces suitable reef substrate for coral recruitment [66,67]. Bleached corals are less capable of removing sediments from their tissue surface [68], resulting in prolonged smothering and increased energy expenditure to clear sediments [69,70]. Recent investigations into cumulative impacts of increased suspended sediments and bleaching events indicate that these impacts can be synergistic when suspended sediments are at high concentrations, resulting in greater mortality than would be observed with either individual event [71].

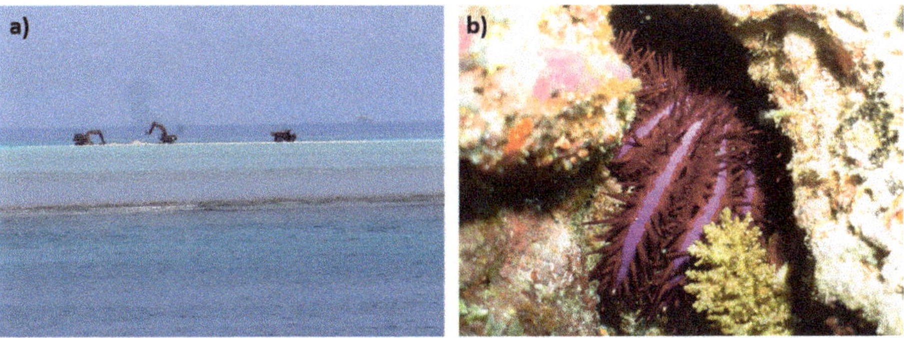


**Figure 5.** Maldivian coral reefs are exposed to local disturbances such as coastal development and land reclamation projects (**a**) and crown-of-thorns starfish that prey on corals (**b**). These additional pressures and thermal stress events may contribute to cumulative pressures on coral reef health and influence reef resilience and recovery.

Although land reclamation activities were not taking place near the coral reef sites surveyed in the present study, two crown-of-thorns starfish (Figure 5b) were observed at Hurasdhoo, which exhibited the highest coral mortality. Coral mortality caused by bleaching can be compounded and/or exacerbated by crown-of-thorns starfish, which favor branching corals as prey [34], and likely contributed to the higher reported mortality at this reef site.

The 2016 bleaching event resulted in lower coral mortality than the 1998 event and initial recovery has been faster [54]. By May 2017 hard coral cover had increased to a mean of approximately 30% (at eight reef sites surveyed across the atolls of Ari Felidhoo, Gaafu Alifu (Suvadiva), North Malé, South Malé, Rasdhoo, and Thoddoo), however, *Acropora* spp. had yet to show evidence of recovery at that time [54]. Current research shows that recovery trajectories are following different trends post 1998 and post 2016 bleaching events [54]. These differences are not surprising given that local human pressures have increased while pre-bleaching coral assemblage and reef health differed. Though faster recovery rates are positive for these reefs, the projected increases in both the frequency and intensity of bleaching events, lengthy recovery timeframes, and local disturbances suggest that Maldivian reefs may not have time to recover before the next severe bleaching event [35,54]. It is therefore critical that policies and management actions are created and implemented that focus on reducing local and regional stressors. Effective management initiatives will be crucial in facilitating the future resilience of both Maldivian reefs and the livelihoods of those that depend on them.

## 4. Materials and Methods

### 4.1. Field Surveys

Coral mortality surveys were conducted from 2–8 December, 2016, approximately 7 months after the peak thermal stress occurred [7]. Six coral reef sites were monitored in the Alifu Alifu and Baa Atolls (Figure 1) in the Republic of Maldives using modified Reef Check International protocols [72]. In brief, within each reef, coral cover was surveyed on both the reef crest and reef slope of lagoon reefs. Three 20 m transect lines were surveyed at each depth, with 5 m between each transect (for a total of $n = 6$ transects per reef site, 60 m length per depth). Shallow transects were placed along a 4–5.5 m contour (average depth = 4.9 m $\pm$ 0.2), and deep transects were placed at a 9–11 m contour (average depth = 10.7 m $\pm$ 0.3). Photos were taken at 1 m interval points along each transect from approximately 1 m above the substrate and 0.5 m in each direction from the interval point. The shallow transect of one reef, Hurasdhoo House Reef, was omitted from analysis due to poor image quality.

### 4.2. Image Analyses

Photographs taken along each transect were assessed to determine the percentage of cover of benthic organisms and substrate. Each photograph contained approximately one square meter of the substrate. A grid was placed over each image to aid in the calculation of substrate proportions. In every photograph each component of the benthic substrate was recorded as a percentage of the total area and rounded to the nearest 5%. The categories that were used to determine benthic substrate included the 10 Reef Check International substrate categories, including hard coral, soft coral, recently dead coral, nutrient indicator algae, sponge, rock, rubble, sand, silt, and other [72].

Only data in the hard coral category were analysed for the purpose of this study. Insufficient numbers of some growth forms in the photographs limited the morphological categories to two: branching and massive, within the first of which corymbrose corals were grouped with branching and within the second of which sub-massive corals were grouped with massive (see Table 2).

**Table 2.** Categories used to characterize benthic substrate during image analysis.

| Category | Description |
| --- | --- |
| Live hard coral | Live coral that is pigmented. Split into three morphology categories. |
| Bleached hard coral | Live coral that appears white or fluorescent in color and has lost normal pigment. Tissue has not been colonised by any algae and is therefore considered to still be alive. |
| Recently dead hard coral | Coral colonies that have recently died. Skeletons had a light covering of turf algae that distinguish them from translucent tissue in bleached coral and pigmented tissue in live coral. Algae cover was not well established and there was no breakdown of skeletal structure, suggesting coral mortality was recent. |

Each of the hard coral categories were grouped by growth form and health status, including live branching hard coral (BC), massive hard coral (MC), and hard coral of other growth forms (HC). Those corals that were visibly bleached with translucent tissue covering their skeleton were classified as bleached branching corals (BBC), bleached massive corals (BBM), and bleached corals of other growth forms (BHC). Those corals that were recently dead were also classified into their respective growth forms, including recently dead branching corals (RDBC), recently dead massive corals (RDMC), and other recently dead corals (RDHC).

Each of the hard coral categories used in the analysis are illustrated in Figure 6. These photographs illustrate the specification of recently dead coral in both massive and branching colonies as opposed to coral that has been dead for an extended period that has since turned to rock or rubble (Figure 6). Examples are included from both shallow and deep transects across various reef sites (Figure 6).

*4.3. Statistical Analyses*

To assess the impact of reef site, depth, and morphology on the percentage of cover of live and recently dead coral, data were analysed using generalised linear mixed models (GLMM, R version 3.6.3 [73]). The influence of reef, morphology, and depth was investigated using a full subsets model selection approach [74]. Following convention, the simplest model with the lowest Akaike information criterion was considered optimal [74]. A null model with only the intercept and random effect was included to determine whether any of the included variables were useful predictors of coral cover. The percentage of live or recently dead hard coral was modelled in relation to reef site, coral morphology, and transect depth, with transect included as a random effect. The model used a Tweedie distribution with a log link and was modelled using the glmmTMB function within the glmmTMB package [75].

**Figure 6.** Examples of different coral health categories. The legend on each image indicates the site name (see Figure 1), transect depth (either deep or shallow), and number (T1, T2, or T3), along with the recorded coral health category. Coral health abbreviations include live branching coral (BC), bleached branching coral (BBC), recently dead branching coral (RDBC), live massive coral (MC), and recently dead massive coral (RDMC).

**Author Contributions:** K.B. and N.H. designed the study. P.B.-B., K.B. and P.B. processed the data. P.B.-B. and H.E.E. analysed the data. All authors contributed to the writing and review of the manuscript. There are no competing or conflicting interests. All authors have read and agreed to the published version of the manuscript.

**Funding:** This research received no external funding.

**Data Availability Statement:** The data presented in this study are available on request from the corresponding author.

**Acknowledgments:** This research was conducted with the support of The Hydrous NGO. We thank Hassan Hameez, Abdul Sattar Hassan, Kate Malmgren, Rick Miskiv, and Erika Woolsey for their help with field surveys and trip logistics.

**Conflicts of Interest:** The authors declare no conflict of interest.

## References

1. Hughes, T.P.; Kerry, J.T.; Álvarez-Noriega, M.; Álvarez-Romero, J.G.; Anderson, K.D.; Baird, A.H.; Babcock, R.C.; Beger, M.; Bellwood, D.R.; Berkelmans, R.; et al. Global Warming and Recurrent Mass Bleaching of Corals. *Nature* **2017**, *543*, 373–377. [CrossRef] [PubMed]
2. Hughes, T.P.; Anderson, K.D.; Connolly, S.R.; Heron, S.F.; Kerry, J.T.; Lough, J.M.; Baird, A.H.; Baum, J.K.; Berumen, M.L.; Bridge, T.C.; et al. Spatial and Temporal Patterns of Mass Bleaching of Corals in the Anthropocene. *Science* **2018**, *359*, 80–83. [CrossRef]
3. Heron, S.F.; Maynard, J.A.; Van Hooidonk, R.; Eakin, C.M. Warming Trends and Bleaching Stress of the World's Coral Reefs 1985–2012. *Sci. Rep. Nat. Publ. Group Lond.* **2016**, *6*, 38402. [CrossRef] [PubMed]
4. Eakin, C.M.; Sweatman, H.P.A.; Brainard, R.E. The 2014–2017 Global-Scale Coral Bleaching Event: Insights and Impacts. *Coral Reefs* **2019**, *38*, 539–545. [CrossRef]

5.    Barkley, H.C.; Cohen, A.L.; Mollica, N.R.; Brainard, R.E.; Rivera, H.E.; DeCarlo, T.M.; Lohmann, G.P.; Drenkard, E.J.; Alpert, A.E.; Young, C.W.; et al. Repeat Bleaching of a Central Pacific Coral Reef over the Past Six Decades (1960–2016). *Commun. Biol.* **2018**, *1*, 177. [CrossRef]

6.    Monroe, A.A.; Ziegler, M.; Roik, A.; Röthig, T.; Hardenstine, R.S.; Emms, M.A.; Jensen, T.; Voolstra, C.R.; Berumen, M.L. In Situ Observations of Coral Bleaching in the Central Saudi Arabian Red Sea during the 2015/2016 Global Coral Bleaching Event. *PLoS ONE* **2018**, *13*, e0195814. [CrossRef]

7.    Cowburn, B.; Moritz, C.; Grimsditch, G.; Solandt, J. Evidence of Coral Bleaching Avoidance, Resistance and Recovery in the Maldives during the 2016 Mass-Bleaching Event. *Mar. Ecol. Prog. Ser.* **2019**, *626*, 53–67. [CrossRef]

8.    McClanahan, T.R.; Darling, E.S.; Maina, J.M.; Muthiga, N.A.; 'agata, S.D.; Jupiter, S.D.; Arthur, R.; Wilson, S.K.; Mangubhai, S.; Nand, Y.; et al. Temperature Patterns and Mechanisms Influencing Coral Bleaching during the 2016 El Niño. *Nat. Clim. Change* **2019**, *9*, 845–851. [CrossRef]

9.    Head, C.E.I.; Bayley, D.T.I.; Rowlands, G.; Roche, R.C.; Tickler, D.M.; Rogers, A.D.; Koldewey, H.; Turner, J.R.; Andradi-Brown, D.A. Coral Bleaching Impacts from Back-to-Back 2015–2016 Thermal Anomalies in the Remote Central Indian Ocean. *Coral Reefs* **2019**, *38*, 605–618. [CrossRef]

10.   Raymundo, L.J.; Burdick, D.; Hoot, W.C.; Miller, R.M.; Brown, V.; Reynolds, T.; Gault, J.; Idechong, J.; Fifer, J.; Williams, A. Successive Bleaching Events Cause Mass Coral Mortality in Guam, Micronesia. *Coral Reefs* **2019**, *38*, 677–700. [CrossRef]

11.   Le Nohaïc, M.; Ross, C.L.; Cornwall, C.E.; Comeau, S.; Lowe, R.; McCulloch, M.T.; Schoepf, V. Marine Heatwave Causes Unprecedented Regional Mass Bleaching of Thermally Resistant Corals in Northwestern Australia. *Sci. Rep.* **2017**, *7*, 14999. [CrossRef]

12.   Ibrahim, N.; Mohamed, M.; Basheer, A.; Haleem, I.; Nistharan, F.; Schmidt, A.; Naeem, R.; Abdulla, A.; Grimsditch, G. *Status of Coral Bleaching in the Maldives 2016*; Maldives Marine Research Centre: Malé, Maldives, 2017; pp. 1–47.

13.   Cerutti, J.M.B.; Burt, A.J.; Haupt, P.; Bunbury, N.; Mumby, P.J.; Schaepman-Strub, G. Impacts of the 2014–2017 Global Bleaching Event on a Protected Remote Atoll in the Western Indian Ocean. *Coral Reefs* **2020**, *39*, 15–26. [CrossRef]

14.   Glynn, P.W. Widespread Coral Mortality and the 1982–83 El Niño Warming Event. *Environ. Conserv.* **1984**, *11*, 133–146. [CrossRef]

15.   Marcelino, L.A.; Westneat, M.W.; Stoyneva, V.; Henss, J.; Rogers, J.D.; Radosevich, A.; Turzhitsky, V.; Siple, M.; Fang, A.; Swain, T.D.; et al. Modulation of Light-Enhancement to Symbiotic Algae by Light-Scattering in Corals and Evolutionary Trends in Bleaching. *PLoS ONE* **2013**, *8*, e61492. [CrossRef] [PubMed]

16.   Smith, H.; Epstein, H.; Torda, G. The Molecular Basis of Differential Morphology and Bleaching Thresholds in Two Morphs of the Coral Pocillopora Acuta. *Sci. Rep.* **2017**, *7*, 10066. [CrossRef]

17.   Grottoli, A.G.; Warner, M.E.; Levas, S.J.; Aschaffenburg, M.D.; Schoepf, V.; McGinley, M.; Baumann, J.; Matsui, Y. The Cumulative Impact of Annual Coral Bleaching Can Turn Some Coral Species Winners into Losers. *Glob. Chang. Biol.* **2014**, *20*, 3823–3833. [CrossRef] [PubMed]

18.   Safaie, A.; Silbiger, N.J.; McClanahan, T.R.; Pawlak, G.; Barshis, D.J.; Hench, J.L.; Rogers, J.S.; Williams, G.J.; Davis, K.A. High Frequency Temperature Variability Reduces the Risk of Coral Bleaching. *Nat. Commun.* **2018**, *9*, 1–12. [CrossRef]

19.   Berkelmans, R.; van Oppen, M.J.H. The Role of Zooxanthellae in the Thermal Tolerance of Corals: A 'Nugget of Hope' for Coral Reefs in an Era of Climate Change. *Proc. R. Soc. B Biol. Sci.* **2006**, *273*, 2305–2312. [CrossRef] [PubMed]

20.   Boulotte, N.M.; Dalton, S.J.; Carroll, A.G.; Harrison, P.L.; Putnam, H.M.; Peplow, L.M.; van Oppen, M.J. Exploring the Symbiodinium Rare Biosphere Provides Evidence for Symbiont Switching in Reef-Building Corals. *ISME J.* **2016**, *10*, 2693–2701. [CrossRef]

21.   Carballo-Bolaños, R.; Denis, V.; Huang, Y.-Y.; Keshavmurthy, S.; Chen, C.A. Temporal Variation and Photochemical Efficiency of Species in Symbiodinaceae Associated with Coral Leptoria Phrygia (Scleractinia; Merulinidae) Exposed to Contrasting Temperature Regimes. *PLoS ONE* **2019**, *14*, e0218801. [CrossRef] [PubMed]

22.   Ziegler, M.; Grupstra, C.G.B.; Barreto, M.M.; Eaton, M.; BaOmar, J.; Zubier, K.; Al-Sofyani, A.; Turki, A.J.; Ormond, R.; Voolstra, C.R. Coral Bacterial Community Structure Responds to Environmental Change in a Host-Specific Manner. *Nat. Commun. Lond.* **2019**, *10*, 1–11. [CrossRef]

23.   Cadotte, M.W. The New Diversity: Management Gains through Insights into the Functional Diversity of Communities. *J. Appl. Ecol.* **2011**, *48*, 1067–1069. [CrossRef]

24.   Hughes, T.P.; Baird, A.H.; Bellwood, D.R.; Card, M.; Connolly, S.R.; Folke, C.; Grosberg, R.; Hoegh-Guldberg, O.; Jackson, J.B.C.; Kleypas, J.; et al. Climate Change, Human Impacts, and the Resilience of Coral Reefs. *Science* **2003**, *301*, 929–933. [CrossRef]

25.   Norström, A.; Nyström, M.; Lokrantz, J.; Folke, C. Alternative States on Coral Reefs: Beyond Coral–Macroalgal Phase Shifts. *Mar. Ecol. Prog. Ser.* **2009**, *376*, 295–306. [CrossRef]

26.   Chaves-Fonnegra, A.; Riegl, B.; Zea, S.; Lopez, J.V.; Smith, T.; Brandt, M.; Gilliam, D.S. Bleaching Events Regulate Shifts from Corals to Excavating Sponges in Algae-Dominated Reefs. *Glob. Chang. Biol.* **2018**, *24*, 773–785. [CrossRef] [PubMed]

27.   Graham, N.A.; Bellwood, D.R.; Cinner, J.E.; Hughes, T.P.; Norström, A.V.; Nyström, M. Managing Resilience to Reverse Phase Shifts in Coral Reefs. *Front. Ecol. Environ.* **2013**, *11*, 541–548. [CrossRef]

28.   Pratchett, M.S.; Hoey, A.S.; Wilson, S.K. Reef Degradation and the Loss of Critical Ecosystem Goods and Services Provided by Coral Reef Fishes. *Curr. Opin. Environ. Sustain.* **2014**, *7*, 37–43. [CrossRef]

29.   Rogers, A.; Blanchard, J.L.; Mumby, P.J. Fisheries Productivity under Progressive Coral Reef Degradation. *J. Appl. Ecol.* **2018**, *55*, 1041–1049. [CrossRef]

30. Naseer, A.; Hatcher, B.G. Inventory of the Maldives? Coral Reefs Using Morphometrics Generated from Landsat ETM+ Imagery. *Coral Reefs* **2004**, *23*, 161–168. [CrossRef]
31. Agardy, T.; Hicks, F.; Nistharan, F.; Fisam, A.; Schmidt, A.; Grimsditch, G. *Ecosystem Services Assessment of North Ari Atoll Maldives*; IUCN: Gland, Switzerland, 2017.
32. Liu, G.; Rauenzahn, J.; Heron, S.; Eakin, C.M.; Skirving, W.; Christensen, T.; Strong, A.; Li, J. NOAA Coral Reef Watch 50 Km Satellite Sea Surface Temperature-Based Decision Support System for Coral Bleaching Management. *NOAA Tech. Rep. NESDIS* **2013**, *143*, 41.
33. Perry, C.T.; Morgan, K.M. Post-Bleaching Coral Community Change on Southern Maldivian Reefs: Is There Potential for Rapid Recovery? *Coral Reefs* **2017**, *36*, 1189–1194. [CrossRef]
34. Pisapia, C.; Burn, D.; Pratchett, M.S. Changes in the Population and Community Structure of Corals during Recent Disturbances (February 2016-October 2017) on Maldivian Coral Reefs. *Sci. Rep.* **2019**, *9*, 1–12. [CrossRef]
35. Montefalcone, M.; Morri, C.; Bianchi, C.N. Influence of Local Pressures on Maldivian Coral Reef Resilience Following Repeated Bleaching Events, and Recovery Perspectives. *Front. Mar. Sci.* **2020**, *7*, 587. [CrossRef]
36. Penin, L.; Adjeroud, M.; Schrimm, M.; Lenihan, H.S. High Spatial Variability in Coral Bleaching around Moorea (French Polynesia): Patterns across Locations and Water Depths. *C. R. Biol.* **2007**, *330*, 171–181. [CrossRef]
37. Done, T.; Wooldridge, S. Learning to Predict Large-Scale Coral Bleaching from Past Events: A Bayesian Approach Using Remotely Sensed Data, in-Situ Data, and Environmental Proxies. *Coral Reefs* **2004**, *23*, 96–108. [CrossRef]
38. Hoogenboom, M.O.; Frank, G.E.; Chase, T.J.; Jurriaans, S.; Álvarez-Noriega, M.; Peterson, K.; Critchell, K.; Berry, K.L.E.; Nicolet, K.J.; Ramsby, B.; et al. Environmental Drivers of Variation in Bleaching Severity of Acropora Species during an Extreme Thermal Anomaly. *Front. Mar. Sci.* **2017**, *4*, 376. [CrossRef]
39. Kaniewska, P.; Anthony, K.R.N.; Hoegh-Guldberg, O. Variation in Colony Geometry Modulates Internal Light Levels in Branching Corals, Acropora Humilis and Stylophora Pistillata. *Mar. Biol.* **2008**, *155*, 649–660. [CrossRef]
40. Swain, T.D.; DuBois, E.; Gomes, A.; Stoyneva, V.P.; Radosevich, A.J.; Henss, J.; Wagner, M.E.; Derbas, J.; Grooms, H.W.; Velazquez, E.M.; et al. Skeletal Light-Scattering Accelerates Bleaching Response in Reef-Building Corals. *BMC Ecol.* **2016**, *16*, 10. [CrossRef]
41. Baird, A.H.; Marshall, P. Mortality, Growth and Reproduction in Scleractinian Corals Following Bleaching on the Great Barrier Reef. *Mar. Ecol. Prog. Ser.* **2002**, *237*, 133–141. [CrossRef]
42. Vargas-Ángel, B.; Huntington, B.; Brainard, R.E.; Venegas, R.; Oliver, T.; Barkley, H.; Cohen, A. El Niño-Associated Catastrophic Coral Mortality at Jarvis Island, Central Equatorial Pacific. *Coral Reefs* **2019**, *38*, 731–741. [CrossRef]
43. McClanahan, T.R. Bleaching Damage and Recovery Potential of Maldivian Coral Reefs. *Mar. Pollut. Bull.* **2000**, *40*, 587–597. [CrossRef]
44. Edwards, A.J.; Clark, S.; Zahir, H.; Rajasuriya, A.; Naseer, A.; Rubens, J. Coral Bleaching and Mortality on Artificial and Natural Reefs in Maldives in 1998, Sea Surface Temperature Anomalies and Initial Recovery. *Mar. Pollut. Bull.* **2001**, *42*, 7–15. [CrossRef]
45. Tkachenko, K.S. The Northernmost Coral Frontier of the Maldives: The Coral Reefs of Ihavandippolu Atoll under Long-Term Environmental Change. *Mar. Environ. Res.* **2012**, *82*, 40–48. [CrossRef]
46. Morri, C.; Montefalcone, M.; Lasagna, R.; Gatti, G.; Rovere, A.; Parravicini, V.; Baldelli, G.; Colantoni, P.; Bianchi, C.N. Through Bleaching and Tsunami: Coral Reef Recovery in the Maldives. *Mar. Pollut. Bull.* **2015**, *98*, 188–200. [CrossRef]
47. McClanahan, T.R.; Muthiga, N.A. Community Change and Evidence for Variable Warm-Water Temperature Adaptation of Corals in Northern Male Atoll, Maldives. *Mar. Pollut. Bull.* **2014**, *80*, 107–113. [CrossRef] [PubMed]
48. Goffredo, S.; Piccinetti, C.; Zaccanti, F. Tsunami Survey Expedition: Preliminary Investigation of Maldivian Coral Reefs Two Weeks after the Event. *Environ. Monit. Assess.* **2007**, *131*, 95–105. [CrossRef] [PubMed]
49. Pisapia, C.; Burn, D.; Yoosuf, R.; Najeeb, A.; Anderson, K.D.; Pratchett, M.S. Coral Recovery in the Central Maldives Archipelago since the Last Major Mass-Bleaching, in 1998. *Sci. Rep.* **2016**, *6*, 34720. [CrossRef]
50. Bianchi, C.N.; Morri, C.; Pichon, M.; Benzoni, F.; Colantoni, P.; Baldelli, G.; Sandrini, M. Dynamics and Pattern of Coral Recolonization Following the 1998 Bleaching Event in the Reefs of the Maldives. In Proceedings of the 10th International Coral Reef Symposium, Okinawa, Japan, 28 June–2 July 2004; Japanese Coral Reef Society: Tokyo, Japan, 2004; pp. 30–37.
51. Schuhmacher, H.; Loch, K.; Loch, W.; See, W.R. The Aftermath of Coral Bleaching on a Maldivian Reef—a Quantitative Study. *Facies* **2005**, *51*, 80–92. [CrossRef]
52. Goreau, T.; McClanahan, T.; Hayes, R.; Strong, A. Conservation of Coral Reefs after the 1998 Global Bleaching Event. *Conserv. Biol.* **2000**, *14*, 5–15. [CrossRef]
53. Lasagna, R.; Albertelli, G.; Giovannetti, E.; Grondona, M.; Milani, A.; Morri, C.; Bianchi, C.N. Status of Maldivian Reefs Eight Years after the 1998 Coral Mass Mortality. *Chem. Ecol.* **2008**, *24*, 67–72. [CrossRef]
54. Montefalcone, M.; Morri, C.; Bianchi, C.N. Long-Term Change in Bioconstruction Potential of Maldivian Coral Reefs Following Extreme Climate Anomalies. *Glob. Chang. Biol.* **2018**, *24*, 5629–5641. [CrossRef]
55. Sheppard, C.R.C.; Harris, A.; Sheppard, A.L.S. Archipelago-Wide Coral Recovery Patterns since 1998 in the Chagos Archipelago, Central Indian Ocean. *Mar. Ecol. Prog. Ser.* **2008**, *362*, 109–117. [CrossRef]
56. Graham, N.A.J.; McClanahan, T.R.; MacNeil, M.A.; Wilson, S.K.; Polunin, N.V.C.; Jennings, S.; Chabanet, P.; Clark, S.; Spalding, M.D.; Letourneur, Y.; et al. Climate Warming, Marine Protected Areas and the Ocean-Scale Integrity of Coral Reef Ecosystems. *PLoS ONE* **2008**, *3*, e3039. [CrossRef] [PubMed]

57. Carpenter, K.E.; Abrar, M.; Aeby, G.; Aronson, R.B.; Banks, S.; Bruckner, A.; Chiriboga, A.; Cortés, J.; Delbeek, J.C.; DeVantier, L.; et al. One-Third of Reef-Building Corals Face Elevated Extinction Risk from Climate Change and Local Impacts. *Science* **2008**, *321*, 560–563. [CrossRef]

58. Cowburn, B.; Samoilys, M.A.; Obura, D. The Current Status of Coral Reefs and Their Vulnerability to Climate Change and Multiple Human Stresses in the Comoros Archipelago, Western Indian Ocean. *Mar. Pollut. Bull.* **2018**, *133*, 956–969. [CrossRef]

59. Jaleel, A. The Status of the Coral Reefs and the Management Approaches: The Case of the Maldives. *Ocean Coast. Manag.* **2013**, *82*, 104–118. [CrossRef]

60. Nepote, E.; Bianchi, C.N.; Chiantore, M.; Morri, C.; Montefalcone, M. Pattern and Intensity of Human Impact on Coral Reefs Depend on Depth along the Reef Profile and on the Descriptor Adopted. *Estuar. Coast. Shelf Sci.* **2016**, *178*, 86–91. [CrossRef]

61. Cowburn, B.; Moritz, C.; Birrell, C.; Grimsditch, G.; Abdulla, A. Can Luxury and Environmental Sustainability Co-Exist? Assessing the Environmental Impact of Resort Tourism on Coral Reefs in the Maldives. *Ocean Coast. Manag.* **2018**, *158*, 120–127. [CrossRef]

62. Pancrazi, I.; Ahmed, H.; Cerrano, C.; Montefalcone, M. Synergic Effect of Global Thermal Anomalies and Local Dredging Activities on Coral Reefs of the Maldives. *Mar. Pollut. Bull.* **2020**, *160*, 111585. [CrossRef]

63. Bessell-Browne, P.; Negri, A.P.; Fisher, R.; Clode, P.L.; Duckworth, A.; Jones, R. Impacts of Turbidity on Corals: The Relative Importance of Light Limitation and Suspended Sediments. *Mar. Pollut. Bull.* **2017**, *117*, 161–170. [CrossRef]

64. Bessell-Browne, P.; Fisher, R.; Duckworth, A.; Jones, R. Mucous Sheet Production in Porites: An Effective Bioindicator of Sediment Related Pressures. *Ecol. Indic.* **2017**, *77*, 276–285. [CrossRef]

65. Jones, R.; Bessell-Browne, P.; Fisher, R.; Klonowski, W.; Slivkoff, M. Assessing the Impacts of Sediments from Dredging on Corals. *Mar. Pollut. Bull.* **2016**, *102*, 9–29. [CrossRef]

66. Ricardo, G.F.; Jones, R.J.; Nordborg, M.; Negri, A.P. Settlement Patterns of the Coral Acropora Millepora on Sediment-Laden Surfaces. *Sci. Total Environ.* **2017**, *609*, 277–288. [CrossRef] [PubMed]

67. Berry, K.L.E.; Hoogenboom, M.O.; Brinkman, D.L.; Burns, K.A.; Negri, A.P. Effects of Coal Contamination on Early Life History Processes of a Reef-Building Coral, Acropora Tenuis. *Mar. Pollut. Bull.* **2017**, *114*, 505–514. [CrossRef] [PubMed]

68. Bessell-Browne, P.; Negri, A.P.; Fisher, R.; Clode, P.L.; Jones, R. Cumulative Impacts: Thermally Bleached Corals Have Reduced Capacity to Clear Deposited Sediment. *Sci. Rep.* **2017**, *7*, 2716. [CrossRef]

69. Fitt, W.K.; McFarland, F.K.; Warner, M.E.; Chilcoat, G.C. Seasonal Patterns of Tissue Biomass and Densities of Symbiotic Dinoflagellates in Reef Corals and Relation to Coral Bleaching. *Limnol. Oceanogr.* **2000**, *45*, 677–685. [CrossRef]

70. Porter, J.W.; Fitt, W.K.; Spero, H.J.; Rogers, C.S.; White, M.W. Bleaching in Reef Corals: Physiological and Stable Isotopic Responses. *Proc. Natl. Acad. Sci. USA* **1989**, *86*, 9342–9346. [CrossRef] [PubMed]

71. Fisher, R.; Bessell-Browne, P.; Jones, R. Synergistic and Antagonistic Impacts of Suspended Sediments and Thermal Stress on Corals. *Nat. Commun.* **2019**, *10*, 2346. [CrossRef] [PubMed]

72. Hill, J.; Loder, J. *Reef Check Australia Survey Methods*; Reef Check Foundation Ltd.: Queensland, Australia, 2013; p. 18.

73. R Core Team. *R: A Language and Environment for Statistical Computing*; R Foundation for Statistical Computing: Vienna, Austria, 2020.

74. Burnham, K.P.; Anderson, D.R. *Model Selection and Multimodel Inference: A Practical Information-Theoretic Approach*, 2nd ed.; Springer: New York, NY, USA, 2002. [CrossRef]

75. Brooks, M.E.; Kristensen, K.; van Benthem, K.J.; Magnusson, A.; Berg, C.W.; Nielsen, A.; Skaug, H.J.; Mächler, M.; Bolker, B.M. GlmmTMB Balances Speed and Flexibility Among Packages for Zero-Inflated Generalized Linear Mixed Modeling. *R J.* **2017**, *9*, 378–400. [CrossRef]

*Article*

# An Experimental Approach to Assessing the Roles of Magnesium, Calcium, and Carbonate Ratios in Marine Carbonates

**Claire E. Reymond * and Sönke Hohn**

Leibniz Center for Tropical Marine Research (ZMT), Fahrenheitstr. 6, 28359 Bremen, Germany;
soenke.hohn@googlemail.com
* Correspondence: claire.reymond@gmail.com

**Abstract:** Marine biomineralization is a globally important biological and geochemical process. Understanding the mechanisms controlling the precipitation of calcium carbonate [$CaCO_3$] within the calcifying fluid of marine organisms, such as corals, crustose coralline algae, and foraminifera, presents one of the most elusive, yet relevant areas of biomineralization research, due to the often-impenetrable ability to measure the process in situ. The precipitation of $CaCO_3$ is assumed to be largely controlled by the saturation state [$\Omega$] of the extracellular calcifying fluid. In this study, we mimicked the typical pH and $\Omega$ known for the calcifying fluid in corals, while varying the magnesium, calcium, and carbonate concentrations in six chemo-static growth experiments, thereby mimicking various dissolved inorganic carbon concentration mechanisms and ionic movement into the extracellular calcifying fluid. Reduced mineralization and varied $CaCO_3$ morphologies highlight the inhibiting effect of magnesium regardless of pH and $\Omega$ and suggests the importance of strong magnesium removal or calcium concentration mechanisms. In respect to ocean acidification studies, this could allow an explanation for why specific marine calcifiers respond differently to lower saturation states.

**Keywords:** marine biomineralization; inorganic mineralization; coral reefs; ocean acidification (OA); omega; dissolved inorganic carbon (DIC); extracellular calcifying fluid (ECF)

**Citation:** Reymond, C.E.; Hohn, S. An Experimental Approach to Assessing the Roles of Magnesium, Calcium, and Carbonate Ratios in Marine Carbonates. *Oceans* **2021**, *2*, 193–214. https://doi.org/10.3390/oceans2010012

Academic Editor: Peter Schupp

Received: 30 August 2020
Accepted: 22 February 2021
Published: 3 March 2021

**Publisher's Note:** MDPI stays neutral with regard to jurisdictional claims in published maps and institutional affiliations.

## 1. Introduction

Calcium carbonate ($CaCO_3$) is the most important biogenic mineral, in terms of quantity, global distribution, and diversity [1]. The production of $CaCO_3$ provides a number of ecological goods and services, such as shoreline protection and habitat structures. For example, coral reefs are one of the most important living bioconstructions of $CaCO_3$ [2] harboring one-quarter to one-third of all marine species [3], and thus serving to be socially and economically important [4]. Unfortunately, future projections show marine biomineralization will become severely impacted by ocean acidification (OA) due to the reduction of carbonate ion concentrations in the oceans [5,6].

Corals calcify extracellularly in a fluid that is separated from the seawater by at least two cell layers [7,8] and rely on a number of active and passive ionic exchanges. For example, calcium ions are actively transported into the extracellular calcifying fluid (ECF) by the epithelium cells of the coral polyp [9,10] while protons are removed [11], establishing favorable conditions for the precipitation of $CaCO_3$ [12]. Similarly, carbon either diffuses into the ECF as carbon dioxide [$CO_2$] or is actively transported into the ECF in the form of bicarbonate [13,14]. Some coral species can calcify in ocean water that is undersaturated with respect to aragonite [15], whereas other species cease to grow and vanish [16,17], which demonstrates a range of biological controls governing the mineralization process. Therefore, to understand which marine calcifiers will be affected by future reduction in ocean saturation states and to estimate its implications for the global carbon cycle, we need to explore a range of possible ECF scenarios.

The significance of biologically-induced and biologically-influenced mineralization is irrefutable. For example, the skeletal organic matrix [SOM] within corals is considered a major factor controlling the precipitation of $CaCO_3$. A number of studies have reported that the SOM contains not only acid-rich proteins (e.g., sulphated proteoglycans), but also assemblages of adhesion and structural proteins, which together are thought to provide a template for aragonite precipitation [10,18–20]. Additionally, the dissolution and precipitation of $CaCO_3$ in aqueous solution is largely dependent on abiotic factors relating to the saturation state ($\Omega$) of the ECF [21,22], which is defined by the product of the dissolved ions forming the mineral divided by the stoichiometric solubility product, $K_{sp}^{*}$ (Equation (1)).

$$\Omega = [Ca^{2+}]^{*}[CO_3^{2-}]/K_{sp}^{*}, \tag{1}$$

As expressed above, $\Omega$ is an extremely useful indicator of the equilibrium or disequilibrium of a solution with a mineral surface. When the ion product, $[Ca^{2+}]^{*}[CO_3^{2-}]$, equals the solubility product, $K_{sp}^{*}$, the saturation state equals one and the system is in equilibrium. If the saturation state is below one because the ion product is lower than the solubility product, the solution is undersaturated and the mineral dissolves. A saturation state higher than one indicates supersaturation where the product of the ion concentrations is greater than the solubility product. In this case, it is thermodynamically viable that dissolved ions precipitate into a crystal structure [22]. The observation that the precipitation rate of $CaCO_3$ increases with an increasing saturation state [21,23–27] has led to the development of empirical relationships (Equation (2)) that describe the calcification rate, G, as a function of the saturation state, where *k* is the reaction rate constant and n is the empirical reaction order.

$$G = k^{*}[\Omega - 1]^{n}, \tag{2}$$

This equation has been applied to predict calcification rates in corals [28] and has successfully been used to simulate the dynamics of ion concentrations in the calcifying fluid of corals and coccolithophores [14,28–30]. However, Equation (2) appears to ignore the theoretical basis shown in Figure 1, which emphasizes how the product of varying calcium and carbonate ion concentrations can obtain the same $\Omega$ value. This is also supported by [31], which demonstrated how the rate of calcite precipitation differed due to the ratio of calcium to carbonate despite having the same oversaturated $\Omega$. Although $\Omega$ is a good predictor of dissolution and precipitation of $CaCO_3$, it does exclude the possibility that ion concentrations differ while obtaining the same $\Omega$ and could therefore account for observational variations among marine calcifiers.

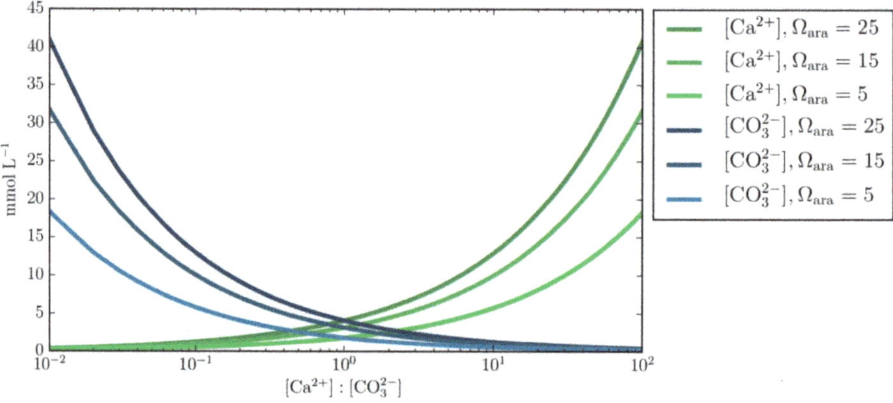


**Figure 1.** Varying concentrations of calcium and carbonate ions $[Ca^{2+}:CO_3^{2-}]$ at fixed $\Omega_{ara}$ of 25, 15, and 5. This demonstrates the underlining principle that calcium and carbonate ion concentrations can obtain the same $\Omega_{ara}$ value at different Ca:CO$_3$ ratios, therefore questioning the empirical equation that prescribes the calcification rate as a function of $\Omega_{ara}$ alone.

Based on this rational, we decided to incubate coral skeleton fragments under six controlled abiotic chemo-static scenarios. By emulating previously measured ECF conditions, we kept all the solutions oversaturated in respect to $\Omega_{ara}$ = 10, pH 8.7, and maintained a typical tropical temperature of 25 °C. Experiment 1a recreated a magnesium [Mg] free condition (strong Mg removal activity) with a high Ca:CO$_3$ ratio (e.g., no dissolved inorganic carbon (DIC) concentrating mechanism and a weak proton removal from the ECF), while experiment 1b recreated a Mg-free solution with a low Ca:CO$_3$ ratio (e.g., mimicking a DIC concentrating mechanism resulting in DIC concentrations three times greater than ambient seawater and a strong proton removal from the ECF resulting in elevated total alkalinity ($_T$A) four times greater than ambient seawater). Experiment 2a recreated a medium Mg scenario (representing concentrations half that of the modern seawater) with a high Ca:CO$_3$ ratio, while experiment 2b recreated a medium Mg scenario with a low Ca:CO$_3$ ratio. Experiment 3a recreated a high Mg scenario (equal to that of modern seawater, i.e., no active removal of ions from the ECF) with a high Ca:CO$_3$ ratio, while experiment 3b recreated a high Mg scenario with a low Ca:CO$_3$ ratio.

## 2. Materials and Methods

### 2.1. Preparation of the Seed Material

There are a range of methodological approaches used to study CaCO$_3$ precipitation, previous studies have used powdered calcite 3–7 μm diameter as the seeding material [32], or Iceland spar [31], living specimens, e.g., [33,34], or synthetic crystals [35]. We used bioclastic fragments of *Stylophora pistillata* to add a potentially realistic coral aragonite crystal structure and investigate if active ion transport, as mediated by the coral calcifying tissue, suffices to drive coral calcification. The ion transporters of the tissue are simulated via the pumped fluids. Our experiment, therefore, aimed to mimic natural processes. However, the only biological component that was not included in the experiments were the organic molecules. This approach is also comparable to a recently published study that did include organic molecules in the incubations [36]. The seeding material for all experiments was obtained from aquarium grown *Stylophora pistillata* (Leibniz Center for Tropical Marine Research [ZMT], Bremen) and followed the methods of [36]. The coral skeleton fragments were cleaned for 48 h with hydrogen peroxide (H$_2$O$_2$ 30%) to remove any soluble components and organic tissue. The coral skeleton fragments were then rinsed in Millipore® water, dried at 40 °C for 24 h, afterwards ground in a planetary ball mill (PM100, Retsch®) for 1 min, and dry sieved (1–200 μm). Individual bioclastic fragments were then hand-picked under a light microscope and selected based on uniform size and shape. These bioclasts are considered rough and represent a typical biogenic skeleton structure. The heterogenetic nature of coral skeletal structure adds a potentially realistic portrayal of the crystal surface adjustment to the ECF but also adds natural variability that occurs in all treatments. Afterwards, each bioclastic fragment was placed in an individual Eppendorf® Safe-Lock 0.5-mL microcentrifuge tube filled with ethanol and placed in an ultrasonic bath for 5 min to remove residual powder and again dried at 40 °C for 24 h. A by-product of this cleaning procedure could result in an increase of the micro-porosity of the bioclastic fragments, by the removal of organic material or breakage. Each fragment was weighed before and after the incubations on a Mettler Toledo® scale with a 1-μg precision (room humidity 30% and temperature 22 °C). As the size and weight of each bioclastic fragment was not perfectly uniform (0.364–1.449 g; Table A1), all bioclastic fragments were evenly distributed among treatments. The initial and end weights, and standardized daily weight increases can be found in the Appendix A (Table A1). To understand the difference in precipitation rate, a two-way factorial analysis of variance (ANOVA), least square (LS), and Tukey–Kramer honest significance difference (HSD) test of the standardized mean weight change between the six experimental scenarios were performed using the software JMP version 9.0. Microstructure formed during each experimental scenario was identified using a scanning electron microscope (Tescan Vega 3 XMU SEM, ZMT) back-scatter electron (BSE) images. Crystal structures of individual CaCO$_3$ polymorphs (vaterite, calcite and aragonite) were analyzed under the Raman microscope at the Alfred Wegener Institute for Polar and Marine Research (AWI) in Bremerhaven, Germany, with the help of Dr. Gernot Nehrke. Due to the uneven surface of the incubated crystals, we did not perform a mapping of the whole crystal but focused on individual crystal structures to qualitatively identify the polymorphs with the Raman spectrum (Figure A1).

### 2.2. Experimental Setup

Ten custom-built incubation chambers made of Teflon [31,36] were used to conduct three cross-factor experiments with two Ca:CO$_3$ scenarios (high and low) in parallel with three Mg concentrations (0 mM, 26.5 mM, and 53 mM) in series (Figure 2; Table 1). Each chamber was attached with Tygon and Marprene® tubing to two 1-L Tedlar® gas sampling bags filled with either a calcium chloride [CaCl$_2$] or sodium bicarbonate [NaHCO$_3$], the preparation of the stock solution is detailed below. Five replicates were run for each treatment. The volume of each incubation chambers was 0.25 mL. The initial flow rate of the solutions into the chambers was accelerated to quickly fill the incubation chambers and then reduced to a constant flow (10 µL min$^{-1}$) via a 24-channel peristaltic pump (Ismatec®). The seed material was placed in each of the incubation chambers and the experiments were run between 32–70 days in a temperature-controlled Rumed® climate cabinet maintained at a constant 25 °C ($\pm$0.5 °C). The variation in experimental duration was due to unexpected health and safety issues of the authors not being allowed into the laboratory.

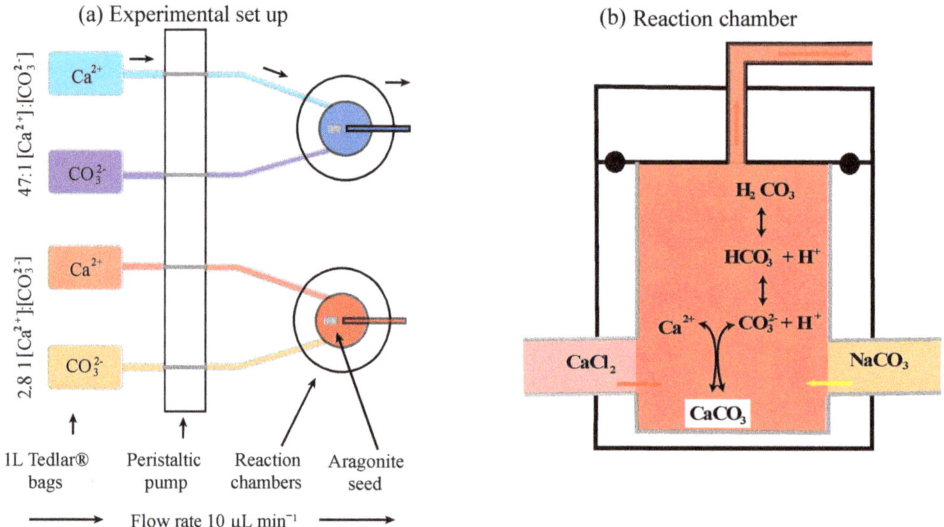


**Figure 2.** Schematic design of the experimental conditions showing (**a**) the separate stock solutions of Ca and CO$_3$ passing through the peristaltic pump and mixing in the reaction chambers, and (**b**) the physicochemical conversions of CaCO$_3$, bicarbonate [HCO$_3^-$] and CO$_3$ occurring during mineral precipitation and dissolution phases. Each of the six experimental scenarios is outlined in Table 1.

### 2.3. Preparation of Stock Solutions

All experiment stock solutions were prepared with Millipore® water, which was initially boiled to drive out dissolved CO$_2$ and then kept in a constant N$_2$ atmosphere to prevent CO$_2$ in-gassing. For all stock solutions pH was measured with a WTW-Multi 3430 Set K pH senor and calibrated with the pH 4 and 10 buffers at 25 °C. Among all the experiments temperature, salinity, and pH remained constant at 25 °C, 36, and 8.7, respectively. The aragonite saturation state, $\Omega_{ara}$ in all incubations was 10, with a saturation index, $SI_{ara} = \log(\Omega_{ara})$, of 2.8, which should induce aragonite precipitation. These parameters represent conservative estimates of realistic scenarios for the coral ECF [11,37]. The aquatic properties chosen for the stock solutions in these experiments reflect the ECF parameters known for *Galaxea fascicularis*, but may not be representative for other coral species, e.g., [38,39]. The full details of the quantity of chemical compounds used for each experiment can be found in Table 2. Each chemical compound was weighed on a Mettler Toledo® scale with a 1 µg precision (room humidity 30% and temperature 22 °C). Concentrations of $CaCl_2$ and magnesium chloride [$MgCl_2$] in the calcium stock solution and $NaHCO_3$ in the carbonate stock solution were double the target concentrations for calcium and carbonate ions because the fluids enter the incubation chambers at a 1:1 ratio and dilute each other's concentration by half (quantiles are given in Table 2A). After adding all the necessary chemical compounds to the stock solutions, they were transferred into 1-L Tedlar® gas sampling bags [36] and put into the climate cabinet at constant 25 °C ($\pm$0.5 °C). This study would improve greatly if microsensors were installed in the incubation chambers to monitor the real time chemistry. Unfortunately, our approach relies on the calculated parameters inside the chambers similar to the work of [31,36].

The calcium stock solutions were prepared by dissolving $CaCl_2$ in 5-L of carbon-free Millipore® water. For the experiments containing magnesium, $MgCl_2$ was added to the stock solution of $CaCl_2$. The Mg concentrations were chosen to represent a strong ion removal mechanism (0 mM, control Mg treatment), a medium ion removal mechanism [26.5 mM, equivalent to half the concentration in present day seawater], and a weak ion removal mechanism (53 mM, equivalent to the concentration in present day seawater). The amount of sodium chloride [NaCl] was then adjusted to maintain a final salinity of 36. Neither carbon nor alkalinity was present in the $CaCl_2$ stock solution (pH = 7), therefore maintaining a zero DIC and $_TA$ concentration.

The carbonate stock solutions were prepared by dissolving $NaHCO_3$ in 5-L of carbon-free Millipore® water. Sodium hydroxide [NaOH] was added via titration to adjust $_TA$ and to reach a pH of 8.716. The pH of the carbonate stock solution was 8.716 because when it mixes with the $CaCl_2$ solution (pH = 7) in the incubation chamber the pH will adjust to 8.700 because DIC and $_TA$ are known to mix conservatively [40]. DIC and $_TA$ were calculated for equilibrium carbonate chemistry in NaCl using the dissociation constants of [41].

**Table 1.** Solutions setup for incubation experiments. Calcium, magnesium, and DIC concentrations are arranged by the amounts of $CaCl_2$, $MgCl_2$, and $NaHCO_3$ added to the solutions. Temperature and salinity remained constant at 25 °C and 36 g kg$^{-1}$, respectively. Carbonate concentrations and $_TA$ are calculated for equilibrium carbonate chemistry in NaCl using the dissociation constants of [41].

| Exp. | $Ca^{2+}:CO_3^{2-}$ mol:mol | $Mg^{2+}:Ca^{2+}$ mol:mol | $Ca^{2+}$ mM | $CO_3^{2-}$ µM | $Mg^{2+}$ mM | $\Omega_{ara}$ | pH | DIC µM | $_TA$ µM |
|------|------|------|------|------|------|------|------|------|------|
| 1a | 47 | 0 | 10.6 | 226 | 0 | 10 | 8.7 | 1777 | 2440 |
| 1b | 2.8 | 0 | 2.6 | 926 | 0 | 10 | 8.7 | 7270 | 9885 |
| 2a | 47 | 2.5 | 10.6 | 226 | 26.5 | 10 | 8.7 | 1777 | 2440 |
| 2b | 2.8 | 10.2 | 2.6 | 926 | 26.5 | 10 | 8.7 | 7270 | 9885 |
| 3a | 47 | 5 | 10.6 | 226 | 53 | 10 | 8.7 | 1777 | 2440 |
| 3b | 2.8 | 20.4 | 2.6 | 926 | 53 | 10 | 8.7 | 7270 | 9885 |

Although the concentration of calcium in the ECF is known to vary, previous studies have recorded values between 9–15 mM from cold-water corals 9–12.3 mM with a mean of 9.9 mM; [42] and the tropical corals *Pocillopora damicornis* and *Acropora youngei* range between ca. 9–15 mM; [43]. The target value of 10.6 mM calcium, was chosen for the incubations with a high Ca:CO$_3$ (47:1) stoichiometry because it represents conditions measured with microelectrodes in the ECF of *Galaxea fascicularis* 9–11 mM; [11]. Although the target value of 2.6 mM calcium is perhaps unrealistically low, it was chosen for the incubations with a low Ca:CO$_3$ (2.8:1) stoichiometry to emulate a strong proton removal from the ECF (resulting in elevated $_TA$ four times greater than ambient seawater), as well as a DIC concentrating mechanism (three times greater DIC than ambient seawater) as proposed by a number of authors [28,44–46]. These Ca:CO$_3$ stoichiometries were also chosen to maintain constant pH and $\Omega_{ara}$ between the treatments.

## 3. Results

### 3.1. Precipitation Rates

A highly significant interactive effect of Mg ion concentration and Ca:CO$_3$ concentration was observed (Figure 3; Table 2; F$_{[2,24]}$ = 150.924, $p$ < 0.001). When Mg was included into the aquatic solution, neither a significant weight change of the CaCO$_3$ seed nor a difference between the two Ca:CO$_3$ scenarios were observed (Table 2; Table A1). Conversely, both Mg-free Ca:CO$_3$ scenarios had significant weight increases. The Mg-free 47:1 Ca:CO$_3$ scenario had a calcification rate four times that of the Mg-free 2.8:1 Ca:CO$_3$ scenario (Figure 3, points labelled A and B). The average weight increase ($\pm$ SD) in the Mg$^-$free 47:1 Ca:CO$_3$ treatment was 1.017 ($\pm$ 0.130] mg d$^{-1}$ and 0.229 ($\pm$ 0.061) mg d$^{-1}$ in the Mg-free 2.8:1 Ca:CO$_3$ treatment. The high amount of newly formed CaCO$_3$ measured in the Mg-free 47:1 Ca:CO$_3{}^{2-}$ is partly explained by spontaneous nucleation, which was only observed in this scenario. Slight dissolution was observed under the intermediate (26.5 mM] and high (53 mM) Mg scenario with a 47:1 Ca:CO$_3$ concentration ($-$0.001 mg d$^{-1}$ $\pm$ 0.001 and $-$0.002 mg d$^{-1}$ $\pm$ 0.003, respectively). The intermediate and high Mg scenario with a 2.8:1 Ca:CO$_3$ concentration had no significant weight changes (0.000 mg d$^{-1}$ $\pm$ 0.001 and 0.002 mg d$^{-1}$ $\pm$ 0.007, respectively), the Ca:CO$_3$ concentration had no effect on the growth rate when Mg ion concentration was equal to or half that found in the ambient ocean, indicating that in this situation Mg has a stronger inhibiting effect towards calcification than Ca:CO$_3$ concentrations.

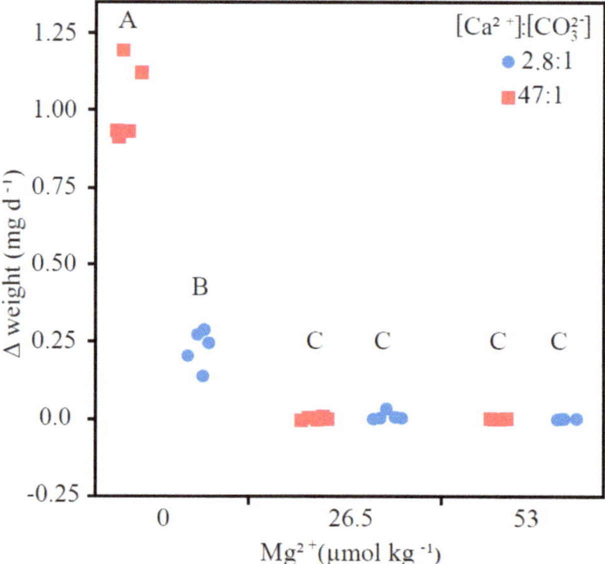


**Figure 3.** Scatter plot showing the standardized total weight change per day [mg d$^{-1}$] of each experimental scenario. The data shows a clear inhibiting effect of Mg and 2.8:1 Ca:CO$_3$ stoichiometry towards inorganic mineralization. Treatments not connected by the same alphabetical symbol (A, B, or C) are significantly different as shown in Table 2; F$_{[2,24]}$ = 150.924, $p$ < 0.001.

**Table 2.** Two-way factorial ANOVA and least square (LS) means difference. The effect test showed significant interaction between Ca:CO$_3$ and Mg concentration. Tukey–Kramer HSD comparisons indicates a significant difference between Mg-free and high Ca:CO$_3$, and all other treatments, as well as Mg-free and low Ca:CO$_3$ and all other treatments.

| Source | DF | SS | MS | F-Ratio | $p > F$ |
|---|---|---|---|---|---|
| Model | 5 | 4.142 | 0.828 | 240.390 | <0.001 |
| Mg$^{2+}$ | 2 | 2.592 | | 375.980 | <0.001 |
| Ca$^{2+}$:CO$_3$$^{2-}$ | 1 | 0.510 | | 148.143 | <0.001 |
| Mg$^{2+}$ * Ca$^{2+}$:CO$_3$$^{2-}$ | 2 | 1.040 | | 150.924 | <0.001 |
| Error | 24 | 0.083 | 0.003 | | |
| Total Error | 29 | 4.224 | | | |

### 3.2. Mineralogy and Crystal Morphology

Back-scatter electron (BSE) images of the CaCO$_3$ surfaces show distinct morphological differences between the six treatments. The Mg-free (control Mg treatment) 47:1 Ca:CO$_3$ incubations formed homogeneous and heterogeneous nucleation in the form of crosshatched vaterite pre-spherical and laminated cubed calcite (Figure 4a). In the Mg-free 2.8:1 scenario, laminated cube calcite precipitated, along with amorphic calcium carbonate (ACC) or a stable prenucleation calcium carbonate cluster c.f. [47], and presumably unfinished calcite transforming from proto-vaterite precursors (Figure 4b). Despite the absence of a measurable weight increase from both the Mg addition scenarios with the 2.8:1 Ca:CO$_3$, newly formed aragonite needles were visibly precipitated on top of the seeding material (Figure 4d,f). Well-defined and abundant acicular crystals were precipitated in random directions and from multiple centers of nucleation as well as from cemented CaCO$_3$ (Figure 4f). In the scenario with high Mg concentration (equal to present day seawater, 53 mM) and at a 2.8:1 Ca:CO$_3$, CaCO$_3$, cements were also observed along with dissolution pits in a needle form (Figure 4f). In the scenario with lower Mg concentration (equal to half the present-day seawater, 26.5 mM) with a 47:1 Ca:CO$_3$, both ACC and dissolution pits were observed, in addition to Mg-calcite (Figure 4c). Conversely, in the scenario with the high Mg concentration and 47:1 Ca:CO$_3$, cements primarily formed along with dissolution pits, and low-relief aragonite needles within the seed material crevices (Figure 4e).

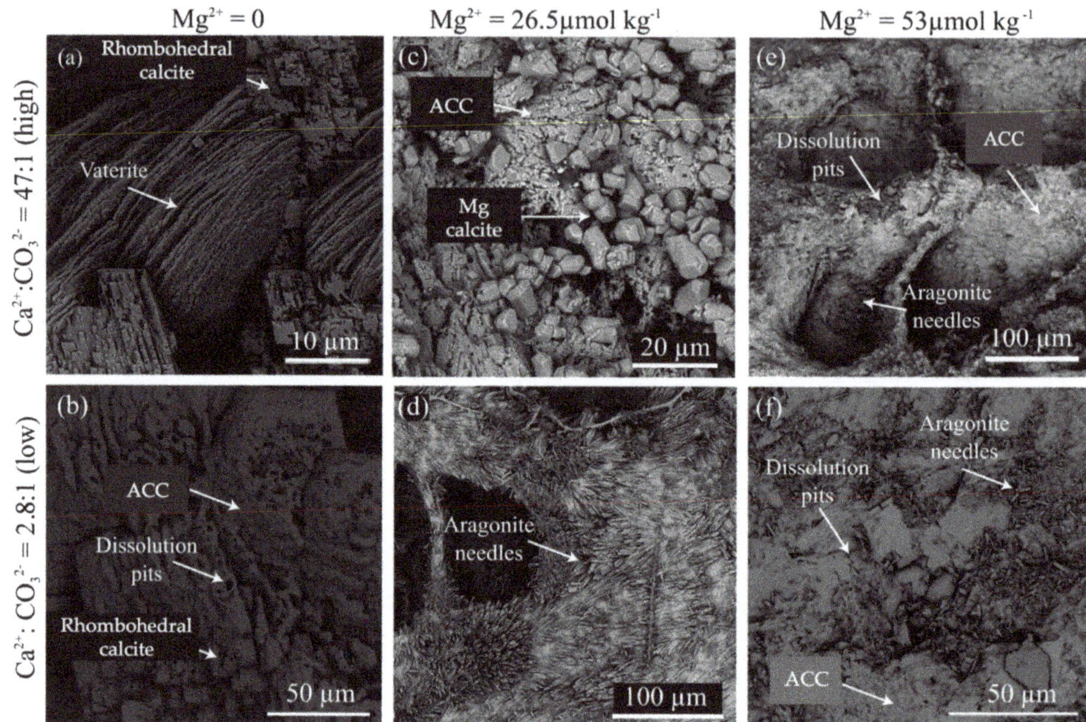



**Figure 4.** Representative SEM images of the precipitated CaCO₃ from each experimental scenario. This shows the influence of Ca:CO₃ stoichiometry, as well as the Mg concentration on the CaCO₃ morphology. The Mg-free incubations with a 47:1 Ca:CO₃ (**a**) resulted in crosshatched vaterite and layered rhombohedral calcite. Mg-free incubations (control Mg treamtent) with a 2.8:1 Ca:CO₃ stoichiometry (**b**) resulted in an intermediate form of ACC together with rhombohedral calcite and some signs of dissolution. Where the Mg concentration was half that of the modern seawater values (26.5 mM) and with a 47:1 Ca:CO₃ stoichiometry (**c**) resulted in a Mg-calcite structure c.f. [48]. Mg concentration of 26.5 mM with a 2.8:1 Ca:CO₃ stoichiometry (**d**) resulted in the formation of aragonite needles. In the scenario where the Mg concentration was 53 mM (equivalent to modern seawater values) no matter if the Ca:CO₃ stoichiometry was 2.8:1 or 47:1 (**e**,**f**) very little new CaCO₃ material precipitated, instead we can see ACC with low relief aragonite needles or unconnected aragonite needles and dissolution pits.

## 4. Discussion

This study compartmentalizes hypothetical abiotic conditions of the ECF, with the aim to gain a broader understanding of the chemical mechanisms relating to biomineralization among tropical marine calcifiers. We show that despite the same ion product of Ca and CO₃, calcification rates vary with different Ca:CO₃ ratios and Mg concentrations. In agreement with [31,36], this study emphasizes the importance of considering the ratio of Ca:CO₃ when estimating the $\Omega_{ara}$ within the ECF of marine calcifiers as exemplified in Figure 1. It is worth noting that in these experiments calcite and vaterite were also precipitated. The calculated $\Omega_{ara}$ in all experiments was 10, $\Omega_{calcite}$ was 15.15. Even though $\Omega_{vaterite}$ has been reported to be lower than $\Omega_{ara}$ and $\Omega_{calcite}$ it is not possible to calculate $\Omega_{vaterite}$ because this requires knowledge of the solubility product [49,50]. Our results show that high calcification rates, are not possible when the Mg concentration is equal to or half that of present-day oceanic concentrations, unless it is counterbalanced by a number of additional factors such as Ca:CO₃ stoichiometry, temperature, $\Omega_{ara}$, proton pumping, or organic molecules. This study infers that the specific conditions required for CaCO₃ precipitation among marine calcifiers is positively amplified by the organism. It is difficult to fully understand the process that controls biomineralization without further in situ ionic measurements from the ECF or more in vitro experiments.

### 4.1. Comparing Low and High Ca:CO$_3$ Scenarios

Generalized CaCO$_3$ precipitation models, as described by [23,28], present disagreement regarding the values for the coefficients n and k [24,25,51], which overestimate the calcification flux at low carbonate concentrations, e.g., when the ECF becomes DIC limited. It is well documented that biomineralization requires elevated $\Omega$ and in doing so implies DIC concentrating mechanisms [28,44–46]. However, direct DIC measurements from the ECF within tropical corals indicate concentrations similar to that of ambient seawater [37], which may result from high DIC consumption during calcification. As seen in previous studies, calcification among tropical corals can be maintained by elevating the Ca ions in the ECF to compensate decreasing seawater pH [43], while decreasing the strontium:calcium [Sr:Ca] and borate [B(OH)$_4$:CO$_3{}^{2-}$] ratios [52]. When comparing the low Ca:CO$_3$ (mimicking strong proton removal from the ECF resulting in elevated $_T$A four times greater than ambient seawater, and a DIC concentrating mechanism resulting in DIC three times greater than ambient seawater) with the high Ca:CO$_3$ scenario [ambient seawater DIC and $_T$A] at elevated pH = 8.7 and $\Omega_{ara}$ = 10 and 0 Mg [control Mg treatment], calcite precipitation rates were three times greater in the ambient seawater treatment than in the DIC concentrating mechanism treatment (Figure 3). This implies that ambient DIC is sufficient to induce calcification provided that homeostasis is maintained in the ECF and the DIC withdrawal from calcification is balanced by ionic flow rates in and out of the ECF [44–46].

### 4.2. The Connection between Mg and Calcification

Marine carbonate-producing organisms exert strong biogenic control to promote calcification within their ECF. This biogenic control is evident by the various mineralogy types and microstructures found among marine carbonate-producing organisms [53]. Calcite is preferentially precipitated as a function of lower temperatures and/or Mg:Ca ratios [54], in addition to the preferential substitution of Ca for Mg, e.g., high-Mg calcite [25]. For example, previous studies have shown coralline algae [33], scleractinian corals [34], and juvenile scleractinian coral [55] can produce calcite when the Mg:Ca ratio of seawater is <2 (e.g., Cretaceous calcitic seas) but at a slower rate. A recent study also found the presence of Mg ions to inhibit not only calcite nucleation during crystal formation but also aragonite [56]. Similarly, it has been shown that strontium also inhibits precipitation rates as a direct correlation with the aqueous calcium activity, thus preventing the attachment of calcium ions to the reactive sites [57,58]. Aragonite microstructure has also been shown to vary as a function of calcification rate, from rapidly formed granular centers of calcification to slower formed fibrous needles [59,60] as determined by the fractionation of $\delta^{18}$O and $\delta^{13}$C isotopes [61,62] and Mg:Ca ratios [59,60,63].

### 4.3. Polynucleation and Spontaneous Nucleation

The ratio of calcium to carbonate clearly matters within the ECF, as it has been shown to describe the rate and morphology of CaCO$_3$ (Figures 3 and 4). Precipitation pathways can be either direct or sequential depending on the free energy available on the surface as determined by pre-nucleation clusters (PNC), growth, and transformation [64–67]. Polynucleation occurred in all scenarios in this study, which led to a complex situation increasing the number of active sites on the surface layer, and therefore a stronger dependence on supersaturation than solely the layer-by-layer mineralization process [31]. Further complications arise because the rate-determining step may change with time as the number of defects and the relative dimensions of the crystal faces become modified during precipitation. Therefore, there are often deviations from the idealized kinetic models, as there may be a number of mechanisms operating in concert [66–68]. It is interesting to note that spontaneous nucleation appeared in the experiment with high excess of calcium ions relative to carbonate ions (i.e., the ambient DIC and TA seawater treatment) but not in the low $Ca:CO_3$ treatment (i.e., the DIC concentrating mechanism scenario), however previous studies have shown that PNC usually form in a low $Ca:CO_3$ solution, equivalent to the binding of ions during crystal formation [47]. This may be due to metastable conditions under which precipitation of the mineral is delayed despite the solution being oversaturated in respect to $\Omega_{ara}$ [67].

The higher calcification rates in the Mg-free and 47:1 $Ca:CO_3$ scenario were obtained primarily due to spontaneous nucleation within the incubation chamber leading to much higher precipitation and thus greater reactive surface area. Previous calcite precipitation experiments in supersaturated ($\Omega_{calcite}$ 5, 16; pH = 10; T = 20 °C) conditions did not produce spontaneous nucleation and showed an optimum precipitation rate when $Ca:CO_3$ = 1:1 [31]. However, there are differences between this study and [31], one of which is the use of $NaHCO_3$ to prepare the carbonate solution in this study instead of $K_2CO_3$. This together with a temperature difference of 5 °C can potentially explain the variation between our observations and [31].

### 4.4. CaCO_3 Polymorphs

In the SEM images (Figure 4), the Mg-free high $Ca:CO_3$ scenario we see ACC, metastable inter-crosshatched vaterite pre-spheres, and rhombohedral calcite blocks. The sequential dissolution and re-precipitation mechanism can be explained via the kinetic rate, which is primarily controlled by the surface area of the crystal [69]. The mixture of vaterite and calcite suggests that calcite mineralization is the rate-determining step. The substitution of Mg into the ACC will however precipitate directly into calcite without the intermediate vaterite phase [70,71] as seen in the 2.5 Mg high $Ca:CO_3$ scenario (Figure 4c). Under the present-day Mg:Ca ratio, aragonite dominates the kinetics of nucleation due to the calcite nucleation barrier being greater than metastable aragonite [72], which explains the lack of calcite in the high-Mg scenario. However, nucleation and precipitation in both the high-Mg scenarios where close to zero, implying that the aragonite seeding material was in equilibrium with the solution as shown by the dominance of ACC (Figure 4d–f) and dissolution pits in the shape of aragonite needles (Figure 4f).

The $Ca:CO_3$ ratio as well as the Mg concentration affected the $CaCO_3$ polymorph precipitated from the oversaturated solutions (Figure 4). In the Mg-free incubations, we obtained crosshatched vaterite and layered rhombohedral calcite in the high $Ca:CO_3$ scenario (Figure 4a] and an intermediate form of ACC together with rhombohedral calcite in the low $Ca:CO_3$ scenario (Figure 4b). In the incubations with 26.5 mM Mg, we obtained an unconnected Mg-calcite in the high $Ca:CO_3$ scenario (Figure 4c) and aragonite needles in the low $Ca:CO_3$ scenario (Figure 4d). With a 53 mM Mg concentration, representing normal seawater conditions, very little new material precipitated, most of which were ACC with sparse low relief aragonite needles (Figure 4e) or unconnected aragonite needles with dissolution pits in the form of needles (Figure 4f). Varying the $Ca:CO_3$, while keeping the Mg concentration fixed, changes the Mg:Ca ratio, which may have driven the differences in polymorphs shown in Figure 4c,d. Overall, the variety of polymorphs precipitated at a pH of 8.7 and an $\Omega_{ara}$ of 10 demonstrates that $\Omega$ alone does not control the precipitation process, as also suggested by [73]. Therefore, caution should be applied when inferring saturation state from the crystal morphology [60], particularly if other factors, e.g., Mg concentrations, temperature, or DIC, are not known.

Even though this study removes the organic aspect of biomineralization, organic molecules have been shown to act as a template to facilitate or induce crystallization [20,74–77] due to their strong binding potential with calcium ions [78,79]. The source of the organics is likely a combination of polyp-derived SOM and seawater-derived SOM as demonstrated from a comparison of coral skeletons and abiotic aragonite [80]. However, the presence of SOM or coral mucus in oversaturated solutions has also been shown to inhibit the nucleation of $CaCO_3$ [81] or pose no effect towards the rate of calcification [31]. Rather, organic molecules appear to influence the $CaCO_3$ polymorph that precipitates from an oversaturated solution [31,77,82]. This suggests that organic molecules have a greater influence on the processes at the crystal surface that leads to the formation of a crystal structure, but not the kinetic processes, which transports the ions to the crystal surface.

An interesting observation from this study is that the aragonite needles precipitated in synthetic seawater (observed in the 26.5 mM Mg with a 2.8:1 $Ca:CO_3$ treatment] with no added biomolecules have a similar morphological appearance to synthetic aragonite experiments made from natural seawater, presumably with some residual organic carbon [60]. This could suggest that coral aragonite crystals may precipitate abiogenically after being initially nucleated, since abiotic systems that lack biomolecule templates altogether show similar morphologies.

*4.5. Implications for Coral Reef Calcifiers*

The concentrations of Mg or $PO_4$, which actively influence crystallization [21,23,24,67,83], are not well known for the ECF among marine organisms. Several studies assume the Mg concentrations in the ECF to be the same as in seawater and thus imply very high $\Omega_{ara}$ (>20) in order to explain the high precipitation rates as observed in corals [11,23,24,70–79,81,84]. Pioneer studies, which utilized various techniques, are largely in agreement with the range of $\Omega_{ara}$ in the ECF. For example, based on microsensor measurements, $\Omega_{ara}$ ranges from 11–25.5 [11,12,38,46], with the exception of 3.2 in the dark [11], 11–12.3 from Raman spectra [43,85], 11–25 inferred from $\delta^{11}B$ isotopes [44,86], 11.1–17.3 predicted by X-ray diffraction-based crystallographic estimates [87], and previously reviewed by [38] to range between 10.16–38.31. While, previous studies have reported the pH in the ECF to be 0.5–0.2 units higher than ambient seawater [46] and that homoeostasis can be maintained within the ECF regardless of varying external seawater pH [88]. Thus, implying a wide range of plausible $Ca:CO_3$ and Ca:Mg ratios within the ECF which enable $CaCO_3$ precipitation.

Measured Mg:Ca ratios from coral skeletons are between 1.5–5.5 mmol/mol [89–92] and from inorganic aragonite has between ~8.5–10 mmol/mol [59], while inorganic calcite has between 30–140 mmol/mol [93], demonstrating the importance of a Mg removal mechanism to facilitate the rate and morphology of calcification. Our results show that high calcification rates observed in corals are not possible when the Mg ion concentration in the ECF is equal to or half that of present day oceanic concentrations (Mg:Ca > 2.5). The high calcification rates observed in this study suggest a mechanism for active removal of inhibiting ions such as Mg from the coral ECF or a Ca concentration mechanism as suggested previously [43]. These points stress the well held belief that biomineralization is a highly complex and biologically mediated process, orchestrated by the secretion of organic molecules [94] and active ion transport [29].

Heterogeneous nucleation is largely inferred by the presence of biomolecules such as acid-rich proteins (e.g., sulphated proteoglycans) and various adhesion and structural proteins [10,18,19] are considered vital for the promotion and functioning of $CaCO_3$ structures. Additionally, the presence of SOM is known to influence the $CaCO_3$ crystal polymorph precipitated from over saturated solutions [82]. Recent experiments [36] confirmed an inhibiting role of coral organic molecules towards rate but not form of $CaCO_3$ [81]. For instance, the role of an ACC precursor phase is likely initiated by a series of controlled biomineralization mechanisms [95,96], particularly for polymorphic calcifying marine organisms. Heterogeneous nucleation has been observed in a range of marine calcifiers such as barnacles [97], echinoderms [95,98], coralline algae [99], foraminifera [100], and corals [43,96].

## 5. Conclusions

To understand the nuances of how coral reef calcifiers can adapt to global change, such as ocean acidification, we need to better understand the ionic composition at the site of calcification. Unfortunately, in the short-term tropical calcifying organisms show little acclimatization potential to ocean acidification [101] particularly when coupled with thermal stress [102], but there are few examples of resistance by altering the ionic concentrations in the ECF, for example Ca [43]. Additionally, this study considered the influence of various Ca:CO$_3$ stoichiometry and Mg concentrations on the precipitation rates and morphology of CaCO$_3$ in a homeostatic experiment. Although there is still a need to conduct more experiments covering a range of other possible scenarios, we believe that our findings are highly relevant within the field of coral reef research for the following reasons:

1. Varying concentrations of calcium and carbonate ions at fixed $\Omega_{ara}$ demonstrates the underlining principal that calcium and carbonate ion concentrations can obtain the same $\Omega_{ara}$ value at different Ca:CO$_3$ stoichiometry and questions the generalized applicability of the empirical equation that prescribes the calcification rate as a function of $\Omega_{ara}$ alone.

2. As shown, calcifying fluid stoichiometry alters the precipitation rate and morphology of CaCO$_3$ at a constant $\Omega$ and pH. Therefore, our findings suggest caution should be applied when inferring saturation state from the crystal morphology, particularly if other factors e.g., Mg, temperature, or DIC are not known.

3. When comparing a strong proton removal scenario and a DIC concentrating mechanism to a scenario with ambient seawater pH and DIC conditions, calcite precipitation rates were three times greater in the ambient seawater conditions. Implying ambient seawater pH and DIC within the calcifying fluid is sufficient to induce calcification provided homeostasis is maintained.

4. Mg exerts a stronger effect on the instability of CaCO$_3$ than Ca:CO$_3$ stoichiometry, in which Mg incorporation locally disturbs the coordination environment in the aragonite structure [87,103]. These differences emphasize the importance of Mg removal from the calcifying fluid. Future studies are recommended to additionally monitor the Mg concentration in the calcifying fluid along with the carbon chemistry.

**Author Contributions:** C.E.R. and S.H., contributed equally to all aspects of this study from the conceptualization, methodology, validation, analysis, drafting, writing, and editing, visualization, project administration, and funding acquisition of this study. All authors have read and agreed to the published version of the manuscript.

**Funding:** This project was funded by the Leibniz Center for Tropical Marine Research [ZMT, Bremen] who is supported by the Ministerium für Kultur und Wissenschaft des Landes Nordrhein-Westfalen, the Regierende Bürgermeister von Berlin-inkl. Wissenschaft und Forschung, and the Bundesministerium für Bildung und Forschung.

**Acknowledgments:** Special thanks to Matthias Birkicht for analytical support and Gernot Nehrke for use of the CRM to identify the crystal polymorphs. We greatly appreciate the constructive comments from the three anonymous reviewers.

**Conflicts of Interest:** The authors declare no conflict of interest. The funders had no role in the design of the study; in the collection, analyses, or interpretation of data; in the writing of the manuscript, or in the decision to publish the results.

## Appendix A

**Table A1.** Initial and final seed weights, including the duration of the experiment and the standardized weight change. Experimental identification corresponds to the aqueous conditions detailed in Table 1. The seed weight values are listed in grams, the experimental identification code is listed as Exp., the individual $CaCO_3$ bioclast identification is listed under Seed, the initial weight and end weight of the seed material correspond to column $t_0$ and $t_e$, the number of days in the experimental conditions is in column D (variation in experimental duration was due to unexpected health and safety issues), and the standardized weight change per day (g d$^{-1}$) is in column D$\Delta$.

| Exp. | Seed | $t_0$ | $t_e$ | D | D$\Delta$ |
|---|---|---|---|---|---|
| 1a | 1Mg-Free | 0.480 | 30.290 | 32 | 0.932 |
| 1a | 2Mg-Free | 0.657 | 36.500 | 32 | 1.120 |
| 1a | 3Mg-Free | 0.883 | 30.620 | 32 | 0.929 |
| 1a | 4Mg-Free | 1.057 | 30.220 | 32 | 0.911 |
| 1a | 5Mg-Free | 1.202 | 39.370 | 32 | 1.193 |
| 1b | 6Mg-Free | 0.591 | 9.320 | 32 | 0.273 |
| 1b | 7Mg-Free | 0.862 | 5.260 | 32 | 0.137 |
| 1b | 8Mg-Free | 0.983 | 7.480 | 32 | 0.203 |
| 1b | 9Mg-Free | 1.064 | 10.300 | 32 | 0.289 |
| 1b | 10Mg-Free | 1.218 | 9.070 | 32 | 0.245 |
| 2a | 1Mg | 0.212 | 0.310 | 70 | 0.001 |
| 2a | 2Mg | 0.715 | 0.570 | 70 | −0.002 |
| 2a | 3Mg | 0.777 | 0.649 | 70 | −0.002 |
| 2a | 4Mg | 0.924 | 0.718 | 70 | −0.003 |
| 2a | 5Mg | 1.013 | 1.006 | 70 | 0.000 |
| 2b | 6Mg | 0.423 | 0.322 | 70 | −0.001 |
| 2b | 7Mg | 0.766 | 13.003 | 70 | 0.175 |
| 2b | 8Mg | 0.818 | 0.793 | 70 | 0.000 |
| 2b | 9Mg | 0.972 | 0.944 | 70 | 0.000 |
| 2b | 10Mg | 1.336 | 1.381 | 70 | 0.001 |
| 3a | 1Mg + | 0.364 | 0.360 | 38 | 0.000 |
| 3a | 2Mg + | 0.816 | 0.643 | 38 | −0.005 |
| 3a | 3Mg + | 0.947 | 0.902 | 38 | −0.001 |
| 3a | 4Mg + | 1.150 | 1.146 | 38 | 0.000 |
| 3a | 5Mg + | 1.274 | 1.212 | 38 | −0.002 |
| 3b | 6Mg + | 0.668 | 0.534 | 38 | −0.004 |
| 3b | 7Mg + | 0.918 | 0.922 | 38 | 0.000 |
| 3b | 8Mg + | 1.069 | 1.064 | 38 | 0.000 |
| 3b | 9Mg + | 1.220 | 1.237 | 38 | 0.000 |
| 3b | 10Mg + | 1.449 | 1.485 | 38 | 0.001 |

**Table A2.** The quantity of compounds (mg/5 L) added to the stock solution to obtain the experimental parameters outlined in Table 1.

| Exp. | Calcium Stock (mg/5 L) | | | Carbonate Stock (mg/5 L) | |
|---|---|---|---|---|---|
| | $CaCl_2$ | $MgCl_2$ | NaCl | $NaHCO_3$ | NaCl |
| 1a | 15.583 | 0.000 | 164.416 | 1.493 | 178.507 |
| 1b | 3.809 | 0.000 | 176.191 | 6.107 | 173.893 |
| 2a | 15.583 | 53.874 | 110.542 | 1.493 | 178.507 |
| 2b | 3.809 | 53.874 | 122.316 | 6.107 | 173.893 |
| 3a | 15.583 | 107.749 | 56.667 | 1.493 | 178.507 |
| 3b | 3.809 | 107.749 | 68.442 | 6.107 | 173.893 |

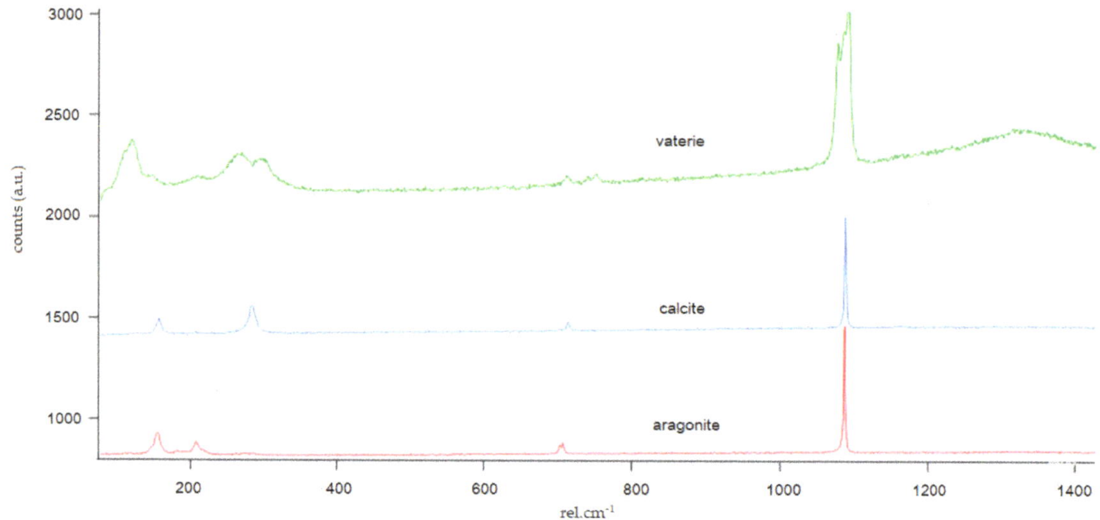

**Figure A1.** CaCO$_3$ polymorphs (aragonite, calcite, and vaterite) were identified via Raman spectroscopy, which were first morphologically identified by SEM on the incubated crystals. These Raman spectra were used to qualitatively identify the crystal structures found on the seeding crystals after incubation.

**Script A1. Below are the calculations used to modify the artificial seawater for the six experiments.**

```
# script for calculating salts and acids to set up solutions for experiment
import numpy as np
import scipy.optimize as opt
#*****************************************************************************************************
# some functions
def KstarW(tempK,salt): #after Millero_1995 p.670 Eq.63 and OA best practices guide
a0 = -1.384726e4
a1 = 1.489652e2
a2 = -2.36521e1
b0 = 1.1867e2
b1 = -5.977
b2 = 1.0495
g = -1.615e-2
lnKWT = a0/tempK + a1 + a2*np.log(tempK)
fT = b0/tempK + b1 + b2*np.log(tempK)
kstarw = np.exp(lnKWT + fT*(salt**0.5) + g*salt)
return kstarw
def Kstar1(tempK,salt): #after OA best practices guide
kstar1 = 10**(-3633.86/tempK + 61.2172-9.67770*np.log(tempK) + 0.011555*salt-0.0001152*salt**2.)
return kstar1
def Kstar2(tempK,salt): #after OA best practices guide
kstar2 = 10**(-471.78/tempK-25.9290 + 3.16967*np.log(tempK) + 0.01781*salt-0.0001122*salt**2.)
return kstar2
def Ksp_ara(tempK,salt):
ksp_ara = -171.945-0.077993*tempK + 2903.293/tempK + 71.595*np.log10(tempK) + (-0.068393 +0.0017276*tempK
+ 88.135/tempK)*salt**0.5-0.10018*salt + 0.0059415*salt**1.5 #mol2 kg-2
return 10.0**ksp_ara
def Kstar0(tempK,salt): # after Weiss, R. F., Marine Chemistry 2:203-215, 1974. (taken from CO2sys)
TempK100 = tempK/100.0;
```

```python
    lnK0 = −60.2409 + 93.4517/TempK100 + 23.3585 * np.log(TempK100) + salt *(0.023517 - 0.023656 * TempK100 +
0.0047036 * TempK100 **2.)
    K0 = np.exp(lnK0) # this is in mol/kg-SW/atm
    return K0
def Ksp_cal(tempK,salt):
    ksp_cal = −171.9065−0.077993*tempK + 2839.319/tempK + 71.595*np.log10(tempK) + (−0.77712 +0.0028426*tempK
+ 178.34/tempK)*salt**0.5−0.07711*salt + 0.0041249*salt**1.5 #mol2 kg−2
    return 10.0**ksp_cal
#*********************************************************************************************************
# molar masses
m_Na = 22.98977 # [g/mol]
m_Ca = 40.078 # [g/mol]
m_Mg = 24.3050 # [g/mol]
m_Cl = 35.4527 # [g/mol]
m_C = 12.0107 # [g/mol]
m_O = 15.9994 # [g/mol]
m_H = 1.00794
m_H2O= 2*m_H + m_O
m_CaCl2 = m_Ca + 2.0*m_Cl + 2.0*m_H2O # [g/mol]
m_MgCl2 = 203.30 # [g/mol]
m_NaCl = m_Na + m_Cl # [g/mol]
m_Na2CO3 = 2.0*m_Na + m_C + 3.0*m_O # [g/mol]
m_NaHCO3 = m_Na + m_H + m_C + 3*m_O
#*********************************************************************************************************
# constant forcing and salt matrix
Temperature = 25.0 # Celsius
Salinity = 36.0 # we use g per kg
TK = 273.15 + Temperature
Sal = Salinity
K0F = Kstar0(TK,Sal)
K1F = Kstar1(TK,Sal)
K2F = Kstar2(TK,Sal)
KWF = KstarW(TK,Sal)
Ksp = Ksp_ara(TK,Sal)
Kspcal = Ksp_cal(TK,Sal)
#*********************************************************************************************************
#!!!!!!!!!!!!!!!!!!!!!!!!!!!!!!!!!!!!!!!!!!!!!!!!!!!!!!!!!!!!!!!!!!!!!!!!!!!!!!!!!!!!!!!!!!!!!!!!!!!!!!!!
# input variables:
# target values (in the experiment):
pH_chamber = 8.7
Omega_chamber = 10.0
Mg_chamber = 53.0e−3/1.0 # mol kg−1
#Ca_chamber = 10.60e−3 # mol kg−1
# or
Stoichiometry = 1.0/1.0 # mol Ca : mol CO3
Ca_chamber = np.sqrt(Omega_chamber*Ksp*Stoichiometry)
H_chamber = 10.0**(-pH_chamber)
#!!!!!!!!!!!!!!!!!!!!!!!!!!!!!!!!!!!!!!!!!!!!!!!!!!!!!!!!!!!!!!!!!!!!!!!!!!!!!!!!!!!!!!!!!!!!!!!!!!!!!!!!
#———————————————————————————————————————————————
# calculated values
# for the calculation of calcium and carbonate ions, I use two equations:
# Stoichiometry = Calcium/CO3
# Omega = Calcium*CO3/Ksp
## now I solve for CO3
```

```python
# CO3 = Calcium/Stoichiometry
# CO3 = Omega*Ksp/Calcium
# Calcium/Stoichiometry = Omega*Ksp/Calcium # *Calcium *Stoichiometry
# Calcium**2 = Omega*Ksp*Stoichiometry
#------------------------------------------------------------------
# once I know Calcium the rest is as follows
# (Chamber values)
CO3_chamber = Omega_chamber*Ksp/Ca_chamber # mol kg-1
DIC_chamber = CO3_chamber*(K1F*H_chamber + H_chamber*H_chamber + K1F*K2F)/(K1F*K2F)
CAlk_chamber = DIC_chamber*K1F*(H_chamber + 2.0*K2F)/(H_chamber*H_chamber + K1F*H_chamber + K1F*K2F)
OH_chamber = KWF/H_chamber
TA_chamber = CAlk_chamber + OH_chamber - H_chamber
Omegacal = Ca_chamber*CO3_chamber/Kspcal
print 'Omega calcite = ', Omegacal
#*****************************************************************************************************
# output
# (Chamber values)
print ' '
print 'expected values:'
print 'Ca =', Ca_chamber*1e3, 'e-3 mol kg-1'
print 'Mg =', Mg_chamber*1e3, 'e-3 mol kg-1'
print 'CO3 =', CO3_chamber*1e6, 'e-6 mol kg-1'
print 'pH =', pH_chamber
print 'DIC =', DIC_chamber*1e6, 'e-6 mol kg-1'
print 'TA =', TA_chamber*1e6, 'e-6 mol kg-1'
print 'Omega =', Omega_chamber
print 'stoichiometry =', Ca_chamber/CO3_chamber,': 1', '(mol Ca : mol CO3)'
print 'Cai paper: Omega=', 10.6e-3*600.0e-6/Ksp
#*****************************************************************************************************
# (Bag values)
print ' '
print 'the amounts of salts needed are:'
g_CaCl2 = Ca_chamber*m_CaCl2*2.0 # times two because the concentrations will be diluted in the chamber
g_MgCl2 = Mg_chamber*m_MgCl2*2.0 # times two because the concentrations will be diluted in the chamber
print g_CaCl2*5.0, 'g CaCl2 per 5 Liters'
print g_MgCl2*5.0, 'g MgCl2 per 5 Liters'
# (Bag values)
g_NaHCO3 = DIC_chamber*m_NaHCO3*2.0# times two because the concentrations will be diluted in the chamber
print g_NaHCO3*5.0, 'g NaHCO3 per 5 Liters'
print 'and to adjust salinity in the solutions we need:'
g_NaCl_1 = Salinity-g_CaCl2-g_MgCl2 # since Salinity is defined as g/kg
print g_NaCl_1*5.0, 'g NaCl per 5 Liters in the CaCl2 bag:'
g_NaCl_2 = Salinity-g_NaHCO3 # since Salinity is defined as g/kg
print g_NaCl_2*5.0, 'g NaCl per 5 Liters in the NaHCO3 bag:'
print 'remark:'
print '(Although I believe that the amount of carbonate might be overestimated and that measured Salinity might
actually be lower. However, we can test this with a calibrated salinity electrode.)'
#*****************************************************************************************************
print ' '
print ' '
print 'and the pH of the solutions will be:'
C0 = DIC_chamber*2.0
print 'DIC = ', C0
pKW = -np.log10(KWF)
```

```python
    pKS1 = -np.log10(K1F)
    pKS2 = -np.log10(K2F)
    pKB1 = pKW-pKS1
    pKB2 = pKW-pKS2
    print 'pKS1=', pKS1, 'pKS2=', pKS2
    #pH_NaHCO3 = 0.5*(pKS1-np.log10(C0))
    pH_NaHCO3 = pKW-0.5*(pKB1-np.log10(C0)) # this does not work because the approximation that the acid is only
a 1 proton acid is too crude.
    #pH_NaHCO3 = 0.5*(pKS1 + pKS2) # for an amphoter
    print 'approximated pH for NaHCO3=', pH_NaHCO3
    #**********************************************************************************************
    # now calculate the real pH
    # molarity of base or acid to add
    mol_base = 0.04 # mol L-1
    mol_acid = 10.172/1000.0 # mol L-1
    K0 = K0F
    K1 = K1F
    K2 = K2F
    KW = KWF
    # the function of the H+ concentration is an equation of fourth order and has to be solved numerically
    # I use fmin to solve it:
    # (Bag values)
    DIC_CO3 = DIC_chamber*2.0
    # the equation differs if you use NaHCO3 because the charge balance is slightly different:
    def H_func_NaHCO3(H):
    val = (KW/H + H*DIC_CO3/(H*H/(K2*K1) + H/K2 + 1.0)/K2 + 2.0*DIC_CO3/(H*H/(K2*K1) + H/K2 + 1.0)-H-
DIC_CO3)**2.0
    #print val
    return val
    # take a good guess from the approximation
    H_init = 10**(-pH_NaHCO3)
    H_opt = opt.fmin(H_func_NaHCO3, H_init, xtol = 0.1, ftol = 0.000001, maxiter = None, maxfun = None, full_output
= 0, disp = 1, retall = 0, callback = None)
    print 'print H for NaHCO3:'
    print 'H for NaHCO3 = ', H_opt
    pH_NaHCO3_opt = -np.log10(H_opt)
    print 'print pH for NaHCO3:'
    print 'the resulting pH of the NaHCO3 solution is', pH_NaHCO3_opt
    #**********************************************************************************************
    # we will not adjust the pH of the CaCl2 bag !!!
    # the milliQ is cooked and has a pH of 7
    # adding CaCl2 does not add alkalinity, it might have a small effect on the pH due to CaOH and CaOH2
    # we assume TA_Ca = 0.0; DIC_Ca = 0.0; and pH_Ca = 7.0
    # (Bag values) Calcium Bag:
    DIC_Ca = 0.0
    TA_Ca =0.0
    #**********************************************************************************************
    # now I have to calculate the required TA of the CO3 bag, which is double the TA in the chamber
    TA_CO3 = TA_chamber*2.0
    DIC_CO3 = DIC_chamber*2.0
    print 'TA CO3=', TA_CO3*1e6
    print 'DIC CO3=', DIC_CO3*1e6
    # from this I have to derive the required pH of the solution
    diff_TA = TA_CO3-DIC_CO3 # this is the amount of alkalinity that has to be added via NaOH
```

```
# so, basically, I have the amount of NaOH that has to be added (at least in theory)
print 'diff TA =', diff_TA*1e6
# (Bag values) Carbonate Bag:
#H_CO3 = 10.0**(-pH_target)
#OH_CO3 = KWF/H_CO3
#CAlk_CO3 = DIC_CO3*K1F*(H_CO3 + 2.0*K2F)/(H_CO3*H_CO3 + K1F*H_CO3 + K1F*K2F)
#TA_CO3 = CAlk_CO3 + OH_CO3 - H_CO3
#0 = DIC_CO3*K1F*(H_CO3 + 2.0*K2F)/(H_CO3*H_CO3 + K1F*H_CO3 + K1F*K2F) + KWF/H_CO3 - H_CO3 - TA_CO3
# the equation differs if you use NaHCO3 because the charge balance is slightly different:
def H_func_CO3bag(H):
val = (TA_CO3 - (DIC_CO3*K1F*(H + 2.0*K2F)/(H*H + K1F*H + K1F*K2F) + KWF/H - H))**2.0
#print val
return val
# take a good guess from the approximation
H_init = 10**(-pH_chamber)
H_opt = opt.fmin(H_func_CO3bag, H_init, xtol = 1e-12, ftol = 1e-12, maxiter=None, maxfun=None, full_output=0, disp=1, retall=0, callback=None)
print 'print target H for the NaHCO3 bag:'
print 'H for NaHCO3=', H_opt
pH_CO3_opt=-np.log10(H_opt)
print 'print target pH for the NaHCO3 bag:'
print 'the resulting pH of the NaHCO3 solution is', pH_CO3_opt
# test if this pH results the correct TA
CAlk_opt = DIC_CO3*K1F*(H_opt + 2.0*K2F)/(H_opt*H_opt + K1F*H_opt + K1F*K2F)
OH_opt = KWF/H_opt
TA_opt = CAlk_opt + OH_opt - H_opt
print 'TA CO3', TA_CO3*1e6, 'minus TA opt', TA_opt*1e6, '=', (TA_CO3-TA_opt)*1e6
#*****************************************************************************************************
# now titrate the NaHCO3 bag to the target pH
# and the same for the other solution
# (Bag values)
pH_start = 7.968 #pH_NaHCO3_opt
pH_target= pH_CO3_opt
#amount of acid needed (first calculated without buffering capacity)
mol_NaOH=10.0**(-pH_start)−10.0**(-pH_target)
# I have to calculate the amount of protons that are consumed also by the buffering system of the carbonate chemistry, and this on top of the pH change without the buffering capacity.
H_start=10.0**(-pH_start)
H_end=10.0**(-pH_target)
OH_start=KWF/H_start
OH_end=KWF/H_end
CO3_start=DIC_CO3*K1F*K2F/(K1F*H_start + H_start*H_start + K1F*K2F)
HCO3_start=DIC_CO3*K1F*H_start/(K1F*H_start + H_start*H_start + K1F*K2F)
H2CO3_start=DIC_CO3-HCO3_start-CO3_start
CO3_end=DIC_CO3*K1F*K2F/(K1F*H_end + H_end*H_end + K1F*K2F)
HCO3_end=DIC_CO3*K1F*H_end/(K1F*H_end + H_end*H_end + K1F*K2F)
H2CO3_end=DIC_CO3-HCO3_end-CO3_end
print 'CO3 from:', CO3_start*1e6, 'to', CO3_end*1e6, '\mu mol'
print 'HCO3 from:', HCO3_start*1e6, 'to', HCO3_end*1e6, '\mu mol'
print 'H2CO3 from:', H2CO3_start*1e6, 'to', H2CO3_end*1e6, '\mu mol'
H_diff=(2*H2CO3_start+HCO3_start+H_start-OH_start)-(2*H2CO3_end+HCO3_end+H_end-OH_end)
print 'mol NaOH needed to adjust pH_NaHCO3 from', pH_start, 'to', pH_target, 'is:', H_diff, 'compared to:', mol_NaOH, 'without considering the buffer capacity'
```

```
print 'at a molarity of ', mol_base, 'this requires ', H_diff/mol_base*1e3*5.0, 'ml of base for 5 L solution'
#**********************************************************************************************
# added for review:
# calculate Omega calcite for comparison
Omegacal=Ca_chamber*CO3_chamber/Kspcal
print 'Omega calcite=', Omegacal
Omegaara=Ca_chamber*CO3_chamber/Ksp
print 'Omega aragonite=', Omegaara
```

## References


1. Lowenstam, H.A.; Weiner, S. *On Biomineralization*; Oxford University Press: New York, NY, USA, 1989; p. 336.
2. Spalding, M.D.; Ravilious, C.; Green, E.P. *World Atlas of Coral Reefs*; University of California Press: Berkeley, CA, USA, 2001; p. 424.
3. Knowlton, N.; Brainard, R.E.; Fisher, R.; Moews, M.; Plaisance, L.; Caley, M.J. Coral Reef Biodiversity. In *Life in the World's Oceans: Diversity, Distribution, and Abundance*; Wiley online Books; McIntyre, A.D., Ed.; Blackwell Publishing Ltd.: Hoboken, NJ, USA, 2010; pp. 65–78.
4. Nystrom, M.; Folke, C.; Moberg, F. Coral reef disturbance and resilience in a human-dominated environment. *Trends Ecol. Evol.* **2000**, *15*, 413–417. [CrossRef]
5. Teng, H.H.; Dove, P.M.; Orme, C.A.; de Yoreo, J.J. Thermodynamics of calcite growth: Baseline for understanding biomineral formation. *Science* **1998**, *282*, 724–727. [CrossRef] [PubMed]
6. Hoegh-Guldberg, O.; Mumby, P.J.; Hooten, A.J.; Steneck, R.S.; Greenfield, P.; Gomez, E.; Harvell, C.D.; Sale, P.F.; Edwards, A.J.; Caldeira, K.; et al. Coral Reefs Under Rapid Climate Change and Ocean Acidification. *Science* **2007**, *318*, 1737–1742. [CrossRef] [PubMed]
7. Allemand, D.; Ferrierpages, C.; Furla, P.; Houlbreque, F.; Puverel, S.; Reynaud, S.; Tambutté, E.; Tambutté, S.; Zoccola, D. Biomineralisation in reef-building corals: From molecular mechanisms to environmental control. *Comptes Rendus Palevol.* **2004**, *3*, 453–467. [CrossRef]
8. Tambutté, S.; Holcomb, M.; Ferrier-Pagés, C.; Reynaud, S.; Tambutté, É.; Zoccola, D.; Allemand, D. Coral biomineralization: From the gene to the environment. *J. Exp. Mar. Bio Ecol.* **2011**, *408*, 58–78. [CrossRef]
9. Al-Horani, F.A.; Ferdelman, T.; Al-Moghrabi, S.M.; De Beer, D. Spatial distribution of calcification and photosynthesis in the scleractinian coral Galaxea fascicularis. *Coral. Reefs.* **2005**, *24*, 173–180. [CrossRef]
10. Tambutté, E.; Allemand, D.; Zoccola, D.; Meibom, A.; Lotto, S.; Caminiti, N.; Segonds, N. Observations of the tissue-skeleton interface in the scleractinian coral Stylophora pistillata. *Coral. Reefs.* **2007**, *26*, 517–529.
11. Al-Horani, F.A.; Al-Moghrabi, S.M.; De Beer, D. The mechanism of calcification and its relation to photosynthesis in the scleractinian coral *Galaxea fascicularis*. *Mar. Biol.* **2003**, *142*, 419–426. [CrossRef]
12. Ries, J.B. A physicochemical framework for interpreting the biological calcification response to $CO_2$ induced ocean acidification. *Geochim. Cosmochim. Acta.* **2011**, *75*, 4053–4064. [CrossRef]
13. Furla, P.; Allemand, D.; Orsenigo, M.N. Involvement of $H^+$-ATPase and carbonic anhydrase in inorganic carbon uptake for endosymbiont photosynthesis. *Am. J. Physiol. Regul. Integr. Comp. Physiol.* **2000**, *278*, R870–R881. [CrossRef] [PubMed]
14. Hohn, S.; Merico, A. Modelling coral polyp calcification in relation to ocean acidification. *Biogeosciences* **2012**, *9*, 4441–4454. [CrossRef]
15. Venn, A.A.; Tambutté, E.; Holcomb, M.; Laurent, J.; Allemand, D.; Tambutté, S. Impact of seawater acidification on pH at the tissue–skeleton interface and calcification in reef corals. *Proc. Natl. Acad. Sci. USA* **2013**, *110*, 1634–1639. [CrossRef]
16. Fabricius, K.E.; De'ath, G.; Noonan, S.; Uthicke, S. Ecological effects of ocean acidification and habitat complexity on reef-associated macroinvertebrate communities. *Proc. R. Soc. B Biol. Sci.* **2014**, *81*, 2013479. [CrossRef]
17. Fabricius, K.E.; Langdon, C.; Uthicke, S.; Humphrey, C.; Noonan, S.; De 'ath, G.; Okazaki, R.; Muehllehner, N.; Glas, M.S.; Lough, J.M. Losers and winners in coral reefs acclimatized to elevated carbon dioxide concentrations. *Nat. Clim. Chang.* **2011**, *1*, 165–169. [CrossRef]
18. Muscatine, L.; Tambutte, E.; Allemand, D. Morphology of coral desmocytes, cells that anchor the calicoblastic epithelium to the skeleton. *Coral Reefs.* **1997**, *16*, 205–213. [CrossRef]
19. Mass, T.; Drake, J.L.; Peters, E.C.; Jiang, W.; Falkowski, P.G. Immunolocalization of skeletal matrix proteins in tissue and mineral of the coral *Stylophora pistillata*. *Proc. Natl. Acad. Sci. USA* **2014**, *111*, 12728–12733. [CrossRef]
20. Cuif, J.-P.; Dauphin, Y. The Environment Recording Unit in coral skeletons—a synthesis of structural and chemical evidences for a biochemically driven, stepping-growth process in fibres. *Biogeosciences* **2005**, *2*, 61–73. [CrossRef]
21. Lippmann, F. *Sedimentary Carbonate Minerals*; Springer: Berlin/Heidelberg, Germany, 1973; p. 228.
22. Lasaga, A.C. *Kinetic Theory in the Earth Sciences*; Princeton University Press: Princeton, NJ, USA, 2014; p. 822.
23. Burton, E.A.; Walter, L.M. Relative precipitation rates of aragonite and Mg calcite from seawater: Temperature or carbonate ion control? *Geology* **1987**, *15*, 111–114. [CrossRef]
24. Burton, E.A.; Walter, L.M. The role of pH in phosphate inhibition of calcite and aragonite precipitation rates in seawater. *Geochim. Cosmochim. Acta* **1990**, *54*, 797–808. [CrossRef]

25. Zuddas, P.; Mucci, A. Kinetics of calcite precipitation from seawater: I. A classical chemical kinetics description for strong electrolyte solutions. *Geochim. Cosmochim. Acta* **1994**, *58*, 4353–4362. [CrossRef]

26. Reddy, M.M.; Nancollas, G.H. The crystallization of calcium carbonate. *IV. The effect of magnesium, strontium and sulfate ions. J. Cryst. Growth* **1976**, *35*, 33–38. [CrossRef]

27. Gattuso, J.P.; Frankignoulle, M.; Bourge, I.; Romaine, S.; Buddemeier, R.W. Effect of calcium carbonate saturation of seawater on coral calcification. *Glob. Planet Chang.* **1998**, *18*, 37–46. [CrossRef]

28. McCulloch, M.; Falter, J.; Trotter, J.; Montagna, P. Coral resilience to ocean acidification and global warming through pH up-regulation. *Nat. Clim. Chang.* **2012**, *2*, 623–627. [CrossRef]

29. Hohn, S.; Merico, A. Quantifying the relative importance of transcellular and paracellular ion transports to coral polyp calcification. *Front Earth Sci.* **2015**, *2*, 1–11. [CrossRef]

30. Holtz, L.M.; Wolf-Gladrow, D.; Thoms, S. Simulating the effects of light intensity and carbonate system composition on particulate organic and inorganic carbon production in Emiliania huxleyi. *J. Theor. Biol.* **2015**, *372*, 192–204. [CrossRef]

31. Nehrke, G.; Reichart, G.J.; Van Cappellen, P.; Meile, C.; Bijma, J. Dependence of calcite growth rate and Sr partitioning on solution stoichiometry: Non-Kossel crystal growth. *Geochim. Cosmochim. Acta* **2007**, *71*, 2240–2249. [CrossRef]

32. Hartley, G.; Mucci, A. The influence of $P_{CO2}$ on the partitioning of magnesium in calcite overgrowth precipitated from artificial seawater at 25° and 1 atm total pressure. *Geochim. Cosmochim. Acta* **1996**, *60*, 315–324. [CrossRef]

33. Ries, J.B. Aragonitic Algae in Calcite Seas: Effect of Seawater Mg/Ca Ratio on Algal Sediment Production. *J. Sediment Res.* **2006**, *76*, 515–523. [CrossRef]

34. Ries, J.B.; Stanley, S.M.; Hardie, L.A. Scleractinian corals produce calcite, and grow more slowly, in artificial Cretaceous seawater. *Geology* **2006**, *34*, 525–528. [CrossRef]

35. Kim, Y.-Y.; Ganesan, K.; Yang, P.; Kulak, A.N.; Borukhin, S.; Pechook, S.; Ribeiro, L.; Kröger, R.; Eichhorn, S.J.; Armes, S.P.; et al. An artificial biomineral formed by incorporation of copolymer micelles in calcite crystals. *Nat. Mater.* **2011**, *10*, 890–896. [CrossRef] [PubMed]

36. Hohn, S.; Reymond, C.E. Coral calcification, mucus, and the origin of skeletal organic molecules. *Coral Reefs.* **2019**, *38*, 973–984. [CrossRef]

37. Cai, W.-J.; Ma, Y.; Hopkinson, B.M.; Grottoli, A.G.; Warner, M.E.; Ding, Q.; Hu, X.; Yuan, X.; Schoepf, V.; Xu, H.; et al. Microelectrode characterization of coral daytime interior pH and carbonate chemistry. *Nat. Commun.* **2016**, *4*, 11144. [CrossRef]

38. Raybaud, V.; Tambutté, S.; Ferrier-Pagès, C.; Reynaud, S.; Venn, A.A.; Tambutté, É.; Nival, P.; Allemand, D. Computing the carbonate chemistry of the coral calcifying medium and its response to ocean acidification. *J. Theor. Biol.* **2017**, *424*, 26–36. [CrossRef] [PubMed]

39. Sevilgen, D.S.; Venn, A.A.; Hu, M.Y.; Tambutté, E.; de Beer, D.; Planas-Bielsa, V.; Tambutté, S. Full in vivo characterization of carbonate chemistry at the site of calcification in corals. *Sci. Adv.* **2019**, *5*, 7447. [CrossRef] [PubMed]

40. Wolf-Gladrow, D.A.; Zeebe, R.E.; Klaas, C.; Körtzinger, A.; Dickson, A.G. Total alkalinity: The explicit conservative expression and its application to biogeochemical processes. *Mar. Chem.* **2007**, *106*, 287–300. [CrossRef]

41. Millero, F.; Huang, F.; Graham, T.; Pierrot, D. The dissociation of carbonic acid in NaCl solutions as a function of concentration and temperature. *Geochim. Cosmochim. Acta* **2007**, *71*, 46–55. [CrossRef]

42. Schleinkofer, N.; Raddatz, J.; Freiwald, A.; David Evans, D.; Beuck, L.; Rüggeberg, A.; Liebetrau, V. Environmental and biological controls on Na/Ca ratios in scleractinian cold-water corals. *Biogeosciences* **2019**, *16*, 3565–3582. [CrossRef]

43. DeCarlo, T.M.; Comeau, S.; Cornwall, C.E.; McCulloch, M.T. Coral resistance to ocean acidification linked to increased calcium at the site of calcification. *Proc. R. Soc. B Biol. Sci.* **2018**, *285*, 1878. [CrossRef] [PubMed]

44. Allison, N.; Cohen, I.; Finch, A.A.; Erez, J.; Tudhope, A.W. Corals concentrate dissolved inorganic carbon to facilitate calcification. *Nat. Commun.* **2014**, *5*, 5741. [CrossRef] [PubMed]

45. de Nooijer, L.J.; Spero, H.J.; Erez, J.; Bijma, J.; Reichart, G.J. Biomineralization in perforate Foraminifera. *Earth Sci. Rev.* **2014**, *135*, 48–58. [CrossRef]

46. Venn, A.A.; Tambutté, E.; Holcomb, M.; Allemand, D.; Tambutté, S. Live tissue imaging shows reef corals elevate pH under their calcifying tissue relative to seawater. *PLoS ONE* **2011**, *6*, e20013.65. [CrossRef]

47. Gebauer, D.; Völkel, A.; Cölfen, H. Stable prenucleation calcium carbonate clusters. *Science* **2009**, *322*, 1819–1822. [CrossRef] [PubMed]

48. Ihli, J.; Clark, J.N.; Kanwal, N.; Kim, Y.-Y.; Holden, M.A.; Harder, R.J.; Tang, C.C.; Ashbrook, S.E.; Robinson, I.K.; Meldrum, F.C. Visualization of the effect of additives on the nanostructures of individual bio-inspired calcite crystals. *Chem. Sci.* **2019**, *10*, 1176–1185. [CrossRef]

49. Plummer, L.N.; Busenberg, E. The solubilities of calcite, aragonite and vaterite in CO2-H2O solutions between 0 and 90 °C, and an evaluation of the aqueous model for the system CaCO3-CO2-H2O. *Geochim. Cosmochim. Acta* **1982**, *46*, 1011–1040. [CrossRef]

50. De Visscher, A.; Vanderdeelen, J. Estimation of the Solubility Constant of Calcite, Aragonite, and Vaterite at 25°C Based on Primary Data Using the Pitzer Ion Interaction Approach. *Monatshefte fur Chemie* **2003**, *134*, 769–775. [CrossRef]

51. Zhong, S.; Mucci, A. Calcite precipitation in seawater using a constant addition technique: A new overall reaction kinetic expression. *Geochim. Cosmochim. Acta* **1993**, *57*, 1409–1417.

52. Allison, N.; Finch, A.A. δ11B, Sr, Mg and B in a modern Porites coral: The relationship between calcification site pH and skeletal chemistry. *Geochim. Cosmochim. Acta* **2010**, *74*, 1790–1800. [CrossRef]

53. Opdyke, B.N.; Wilkinson, B.H. Paleolatitude distribution of Phanerozoic marine ooids and cements. *Palaeogeogr Palaeoclim Palaeoecol* **1990**, *78*, 135–148. [CrossRef]

54. Morse, J.W.; Wang, Q.; Tsio, M.Y. Influences of temperature and Mg: Ca ratio on $CaCO_3$ precipitates from seawater. *Geology* **1997**, *25*, 85–87. [CrossRef]

55. Higuchi, T.; Fujimura, H.; Yuyama, I.; Harii, S.; Agostini, S.; Oomori, T. Biotic control of skeletal growth by scleractinian corals in aragonite–calcite seas. *PLoS ONE* **2014**, *9*, e91021. [CrossRef]

56. Boon, M.; Rickard, W.D.; Rohl, A.L.; Jones, F. Stabilization of Aragonite: Role of $Mg^{2+}$ and Other Impurity Ions. *Cryst. Growth Des.* **2020**, *20*, 5006–5017. [CrossRef]

57. Bracco, J.N.; Grantham, M.C.; Stack, A.G. Calcite Growth Rates as a Function of Aqueous Calcium-to-Carbonate Ratio, Saturation Index, and Inhibitor Concentration: Insight into the Mechanism of Reaction and Poisoning by Strontium. *Cryst. Growth Des.* **2012**, *12*, 3540–3548. [CrossRef]

58. Gebrehiwet, T.A.; Redden, G.D.; Fujita, Y.; Beig, M.S.; Smith, R.W. The Effect of the $CO_3{}^{2-}$ to $Ca^{2+}$ Ion activity ratio on calcite precipitation kinetics and $Sr^{2+}$ partitioning. *Geochem. Trans.* **2012**, *13*, 1–11. [CrossRef]

59. Gaetani, G.A.; Cohen, A.L. Element partitioning during precipitation of aragonite from seawater: A framework for understanding paleoproxies. *Geochim. Cosmochim. Acta* **2006**, *70*, 4617–4634. [CrossRef]

60. Holcomb, M.; Cohen, A.L.; Gabitov, R.; Hutter, J. Compositional and morphological features of aragonite precipitated experimentally from seawater and biogenically by corals. *Geochim. Cosmochim. Acta* **2009**, *73*, 166–179. [CrossRef]

61. Adkins, J.F.; Boyle, E.A.; Curry, W.B.; Lutringer, A. Stable isotopes in deep-sea corals and a new mechanism for "vital effects". *Geochim. Cosmochim. Acta* **2003**, *67*, 1129–1143. [CrossRef]

62. Rollion-Bard, C.; Chaussidon, M.; France-Lanord, C. pH control on oxygen isotopic composition of symbiotic corals. *Earth Planet Sci. Lett.* **2003**, *215*, 275–288. [CrossRef]

63. Watson, E.B. A conceptual model for near-surface kinetic controls on the trace- element and stable isotope composition of abiogenic calcite crystals. *Geochim. Cosmochim. Acta* **2004**, *68*, 1473–1488. [CrossRef]

64. Boistelle, R.; Astier, J.P. Crystallization mechanisms in solution. *J. Cryst. Growth.* **1988**, *90*, 14–30. [CrossRef]

65. Davey, R.J.; Schroeder, S.L.M.; Ter Horst, J.H. Nucleation of organic crystals—A molecular perspective. *Angew Chemie Int. Ed.* **2013**, *52*, 2167–2179. [CrossRef]

66. Mann, S. *Biomineralization: Principles and Concepts in Bioinorganic Materials Chemistry*; Oxford University Press: Oxford, UK, 2001; p. 210.

67. Gebauer, D.; Kellermeier, M.; Gale, J.D.; Bergström, L.; Cölfen, H. Pre-nucleation clusters as solute precursors in crystallisation. *Chem. Soc. Rev.* **2014**, *43*, 2348–2371. [CrossRef]

68. Bonucci, E. Role of collagen fibrils in calcification. In *Calcification in Biological Systems*; Bonucci, E., Ed.; CRC Press: Boca Raton, FL, USA, 1992; pp. 19–39.

69. Rodriguez-Blanco, J.D.; Shaw, S.; Benning, L.G. The kinetics and mechanisms of amorphous calcium carbonate [ACC] crystallization to calcite, viavaterite. *Nanoscale* **2011**, *3*, 265–271. [CrossRef] [PubMed]

70. Wray, J.; Daniels, F. Precipitation of calcite and aragonite. *J. Am. Chem. Soc.* **1957**, *79*, 2031–2034. [CrossRef]

71. Ogino, T.; Suzuki, T.; Sawada, K. The formation and transformation mechanism of calcium carbonate in water. *Geochemica Cosmochem Acta.* **1987**, *51*, 2757–2767. [CrossRef]

72. Sun, W.; Jayaraman, S.; Chen, W.; Persson, K.A.; Ceder, G. Nucleation of metastable aragonite CaCO3 in seawater. *Proc Natl. Acad. Sci. USA* **2015**, *112*, 3199–3204. [CrossRef]

73. Cyronak, T.; Schulz, K.G.; Jokiel, P.L. The Omega myth: What really drives lower calcification rates in an acidifying ocean. *ICES J. Mar. Sci.* **2016**, *73*, 558–562. [CrossRef]

74. Allemand, D.; Tambutte, E.; Girard, J.; Jaubert, J.; Tambutté, E.; Girard, J.; Jaubert, J. Organic matrix synthesis in the scleractinian coral Stylophora pistillata: Role in biomineralization and poten- tial target of the organotin tributyltin. *J. Exp. Biol.* **1998**, *201*, 2001–2009. [PubMed]

75. Watanabe, T.; Fukuda, I.; China, K.; Isa, Y. Molecular analyses of protein components of the organic matrix in the exoskeleton of two scleractinian coral species. *Comp. Biochem. Physiol. B Biochem. Mol. Biol.* **2003**, *136*, 767–774. [CrossRef]

76. Helman, Y.; Natale, F.; Sherrell, R.M.; LaVigne, M.; Starovoytov, V.; Gorbunov, M.Y.; Falkowski, P.G. Extracellular matrix production and calcium carbonate precipitation by coral cells in vitro. *Proc. Natl. Acad. Sci. USA* **2008**, *105*, 54–58. [CrossRef] [PubMed]

77. Mass, T.; Drake, J.L.; Haramaty, L.; Kim, J.D.; Zelzion, E.; Bhattacharya, D.; Falkowski, P.G. Cloning and characterization of four novel coral acid-rich proteins that precipitate carbonates *in vitro*. *Curr. Biol.* **2003**, *23*, 1126–1131. [CrossRef]

78. Kretsinger, R.H. Calcium-Binding Proteins. *Ann. Rev. Biochem.* **1976**, *45*, 239–266. [CrossRef] [PubMed]

79. Carafoli, E. Intracellular Calcium Homeostasis. *Ann. Rev. Biochem.* **1987**, *56*, 395–433. [CrossRef]

80. DeCarlo, T.M.; Ren, H.; Farfan, G.A. The Origin and Role of Organic Matrix in Coral Calcification: Insights from Comparing Coral Skeleton and Abiogenic Aragonite. *Front Mar. Sci.* **2018**, *5*, 170. [CrossRef]

81. Marin, F.; Smith, M.; Isa, Y.; Muyzer, G.; Westbroek, P. Skeletal matrices, muci, and the origin of invertebrate calcification. *Proc. Natl. Acad. Sci. USA* **1996**, *93*, 1554–1559. [CrossRef]

82. Westbroek, P.; Marin, F. A marriage of bone and nacre. *Nature* **1998**, *392*, 861–862. [CrossRef] [PubMed]

83. Kawano, J.; Sakuma, H.; Nagai, T. Incorporation of $Mg^{2+}$ in surface $Ca^{2+}$ sites of aragonite: An ab initio study. *Prog. Earth Planet Sci.* **2015**, *2*, 7. [CrossRef]

84. Cohen, A.L.; McConnaughey, T.A. Geochemical Perspectives on Coral Mineralization. *Rev. Mineral. Geochem.* **1990**, *4*, 151–187.

85. DeCarlo, T.M.; D'Olivo, J.P.; Foster, T.; Holcomb, M.; Becker, T.; McCulloch, M.T. Coral calcifying fluid aragonite saturation states derived from Raman spectroscopy. *Biogeosciences* **2017**, *14*, 5253–5269. [CrossRef]

86. McCulloch, M.; Trotter, J.; Montagna, P.; Falter, J.; Dunbar, R.; Freiwald, A.; Försterra, G.; López Correa, M.; Maier, C.; Rüggeberg, A.; et al. Resilience of cold-water scleractinian corals to ocean acidification: Boron isotopic systematics of pH and saturation state up-regulation. *Geochim. Cosmochim. Acta.* **2012**, *87*, 21–34. [CrossRef]

87. Farfan, G.A.; Cordes, E.E.; Waller, R.G.; DeCarlo, T.M.; Hansel, C.M. Mineralogy of Deep-Sea Coral Aragonites as a Function of Aragonite Saturation State. *Front. Mar. Sci.* **2018**, *5*, 473. [CrossRef]

88. Georgiou, L.; Falter, J.; Trotter, J.; Kline, D.I.; Holcomb, M.; Dove, S.G.; Hoegh-Guldberg, O.; Mcculloch, M. pH homeostasis during coral calcification in a free ocean $CO_2$ enrichment [FOCE] experiment, Heron Island reef flat, Great Barrier Reef. *Proc. Natl. Acad. Sci. USA* **2015**, *112*, 13219–13224. [CrossRef]

89. Mitsuguchi, T.; Matsumoto, E.; Abe, O.; Uchida, T.; Isdale, P.J. Mg/Ca Thermometry in Coral Skeletons. *Science.* **1996**, *274*, 961–963. [CrossRef] [PubMed]

90. Gaetani, G.A.; Cohen, A.L.; Wang, Z.; Crusius, J. A biomineralization approach to developing climate proxies. *Geochim. Cosmochim. Acta* **2011**, *75*, 1920–1932. [CrossRef]

91. Wei, G.; Sun, M.; Li, X.; Nie, B. Mg/Ca, Sr/Ca and U/Ca ratios of a porites coral from Sanya Bay, Hainan Island, South China Sea and their relationships to sea surface temperature. *Palaeogeogr Palaeoclimatol Palaeoecol.* **2000**, *162*, 59–74. [CrossRef]

92. Marchitto, T.M.; Bryan, S.P.; Doss, W.; McCulloch, M.T.; Montagna, P. A simple biomineralization model to explain Li, Mg, and Sr incorporation into aragonitic foraminifera and corals. *Earth Planet Sci. Lett.* **2018**, *481*, 20–29. [CrossRef]

93. De Choudens-Sánchez, V.; González, L.A. Calcite and aragonite precipitation under controlled instantaneous supersaturation: Elucidating the role of CaCO3 saturation state and Mg/Ca ratio on calcium carbonate polymorphism. *J. Sediment. Res.* **2009**, *79*, 363–376. [CrossRef]

94. Falini, G.; Fermani, S.; Goffredo, S. Seminars in Cell & Developmental Biology Coral biomineralization: A focus on intra-skeletal organic matrix and calcification. *Semin. Cell Dev. Biol.* **2015**, *46*, 17–26. [PubMed]

95. Politi, Y.; Arad, T.; Klein, E.; Weiner, S.; Addadi, L. Sea urchin spine calcite forms via a transient amorphous calcium carbonate phase. *Science* **2004**, *306*, 1161–1164. [CrossRef]

96. Mass, T.; Giuffre, A.J.; Sun, C.-Y.; Stifler, C.A.; Frazier, M.J.; Neder, M.; Tamura, N.; Stan, C.V.; Marcus, M.A.; Gilbert, P.U.P.A. Amorphous calcium carbonate particles form coral skeletons. *Proc. Natl. Acad. Sci. USA* **2017**, *114*, E7670–E7678. [CrossRef]

97. Checa, A.G.; Macías-Sánchez, E.; Rodríguez-Navarro, A.B.; Sánchez-Navas, A.; Lagos, N.A. Origin of the biphase nature and surface roughness of biogenic calcite secreted by the giant barnacle *Austromegabalanus psittacus*. *Sci. Rep.* **2020**, *10*, 16784. [CrossRef]

98. Politi, Y.; Metzler, R.A.; Abrecht, M.; Gilbert, B.; Wilt, F.H.; Sagi, I.; Addadi, L.; Weiner, S.; Gilbert, P.U.P.A. Transformation mechanism of amorphous calcium carbonate into calcite in sea urchin larval spicule. *Proc. Natl. Acad. Sci. USA* **2008**, *105*, 17362–17366. [CrossRef] [PubMed]

99. Nash, M.C.; Diaz-Pulido, G.; Harvey, A.S.; Adey, W. Coralline algal calcification: A morphological and process-based understanding. *PLoS ONE* **2019**, *14*, e0221396. [CrossRef]

100. Jacob, D.E.; Wirth, R.; Agbaje, O.B.A.; Branson, O.; Eggins, S.M. Planktic foraminifera form their shells via metastable carbonate phases. *Nat. Commun.* **2017**, *8*, 1265. [CrossRef] [PubMed]

101. Comeau, S.; Cornwall, C.E.; DeCarlo, T.M.; Doo, S.S.; Carpenter, R.C.; McCulloch, M.T. Resistance to ocean acidification in coral reef taxa is not gained by acclimatization. *Nat. Clim. Chang.* **2019**, *9*, 477–483. [CrossRef]

102. Guillermic, M.; Cameron, L.P.; De Corte, I.; Misra, S.; Bijma, J.; De Beer, D.; Reymond, C.E.; Westphal, H.; Ries, J.B.; Robert, A.; et al. Thermal stress reduces pocilloporid coral resilience to ocean acidification by impairing control over calcifying fluid chemistry. *Sci. Adv.* **2021**, *7*, eaba9958. [CrossRef]

103. Soldati, A.L.; Jacob, D.; Glatzel, P.; Swarbrick, J.C.; Geck, J. Element substitution by living organisms: The case of manganese in mollusc shell aragonite. *Sci. Rep.* **2016**, *6*, 22514. [CrossRef] [PubMed]

Brief Report

# The Effects of the UV-Blocker Oxybenzone (Benzophenone-3) on Planulae Swimming and Metamorphosis of the Scyphozoans *Cassiopea xamachana* and *Cassiopea frondosa*

**William K. Fitt [1],* and Dietrich K. Hofmann [2]**

1    Odum School of Ecology, University of Georgia, Athens, GA 30602, USA
2    Department of Zoology and Neurobiology, Ruhr-Universität, 44781 Bochum, Germany;
     Dietrich.Hofmann@ruhr-uni-bochum.de
*    Correspondence: fitt@uga.edu

Received: 7 July 2020; Accepted: 21 September 2020; Published: 24 September 2020

**Abstract:** Benzophenones are UV-blockers found in most common sunscreens. The ability of Scyphozoan planula larvae of *Cassiopea xamachana* and *C. frondosa* to swim and complete metamorphosis in concentrations 0–228 µg/L benzophenone-3 (oxybenzone) was tested. Planulae of both species swam in erratic patterns, 25–30% slower, and experienced significant death ($p < 0.05$) in the highest concentrations of oxybenzone tested, whereas the larvae exhibited normal swimming patterns and no death in ≤2.28 µg/L oxybenzone. In addition, metamorphosis decreased 10–30% over 3 days for both species maintained in 228 µg/L oxybenzone. These effects do not involve symbiotic dinoflagellates, as planulae larvae of *Cassiopea* sp. are aposymbiotic. It is concluded that oxybenzone can have a detrimental impact on these jellyfish.

**Keywords:** *Cassiopea xamachana*; *C. frondosa*; Scyphozoan; planulae; settlement; metamorphosis; oxybenzone

---

## 1. Introduction

Benzophenones are organic compounds used in a variety of personal-care products for their abilities to absorb UV-B wavelengths coming from the sun. Oxybenzone avobenzone, octisalate, octocrylene, homosalate, and octinoxate are all used in conventional sunscreens and other topical, personal-care products to prevent damage from the sun's UV rays. Between 6000 and 14,000 tons of sunscreen lotion, much of which contains between 1 and 10% benzophenone-3 (oxybenzone), are estimated to be released onto coral reef areas each year [1]. Concern over oxybenzone arose when a study reported that 1–2% was absorbed by the skin immediately after application [2]. Application to the epidermis is not the only mode for absorption. A German study reported traces of oxybenzone in breast milk of mammals e.g., [3]. Since oxybenzone is a photo-toxicant, its negative effects are activated and exacerbated by light [4]. Soon after reacting with light oxybenzone goes through rapid oxidation causing inactivation of antioxidants and a negative impact on the skin's overall homeostasis [5]. Oxybenzone can induce photoactivated and non-photo-activated contact dermatitis, contact cheilitis, uticaria, and anaphylactoid allergic reactions e.g., [6]. In addition, oxybenzone, and other sunscreen chemicals, act to suppress the immune system [7].

Benzophenones put approximately 40% of coral reefs located along coastal areas, with at least 10% of reefs overall, at risk of exposure [1]. Oxybenzone has been found to cause environmental concerns such as bleaching of symbiotic dinoflagellates from corals [8,9], failure of larvae to settle [10], and increased mortality of corals [9–11] and fish [12,13]. Research has shown correlations

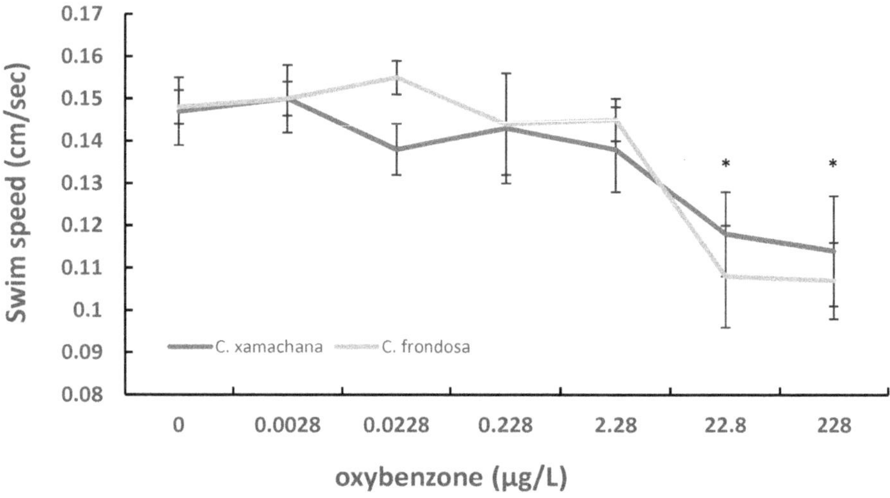

**Figure 2.** The relationship between average (± s.d.) swim speed (cm/sec) of larvae of *Cassiopea xamachana* and *C. frondosa* and the concentration of oxybenzone. * = significantly different (ANOVA, $p \leq 0.05$) from 0, 0.00228, 0.0228, 0.228 µg/L oxybenzone.

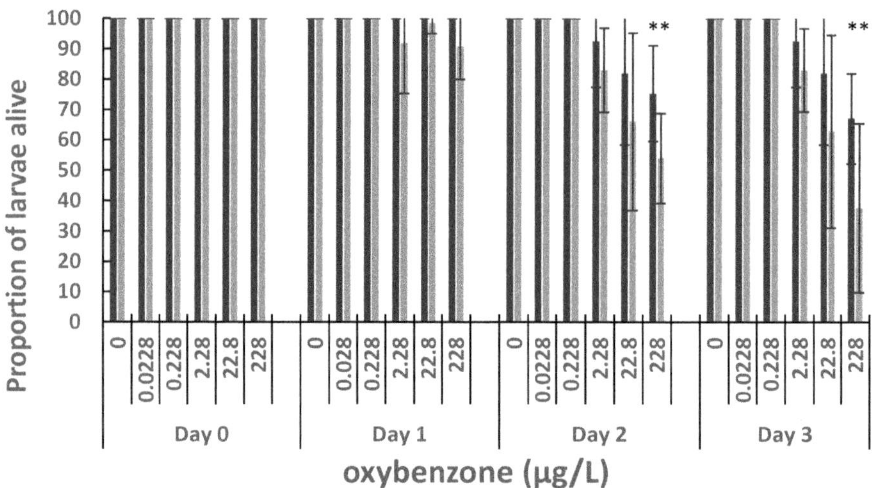

**Figure 3.** The average number of larvae remaining in the wells after each time period per concentration of oxybenzone. Dark bars: *Cassiopea xamachana*, Light bars: *Cassiopea frondosa*. * = significantly different (ANOVA, $p \leq 0.05$) from controls.

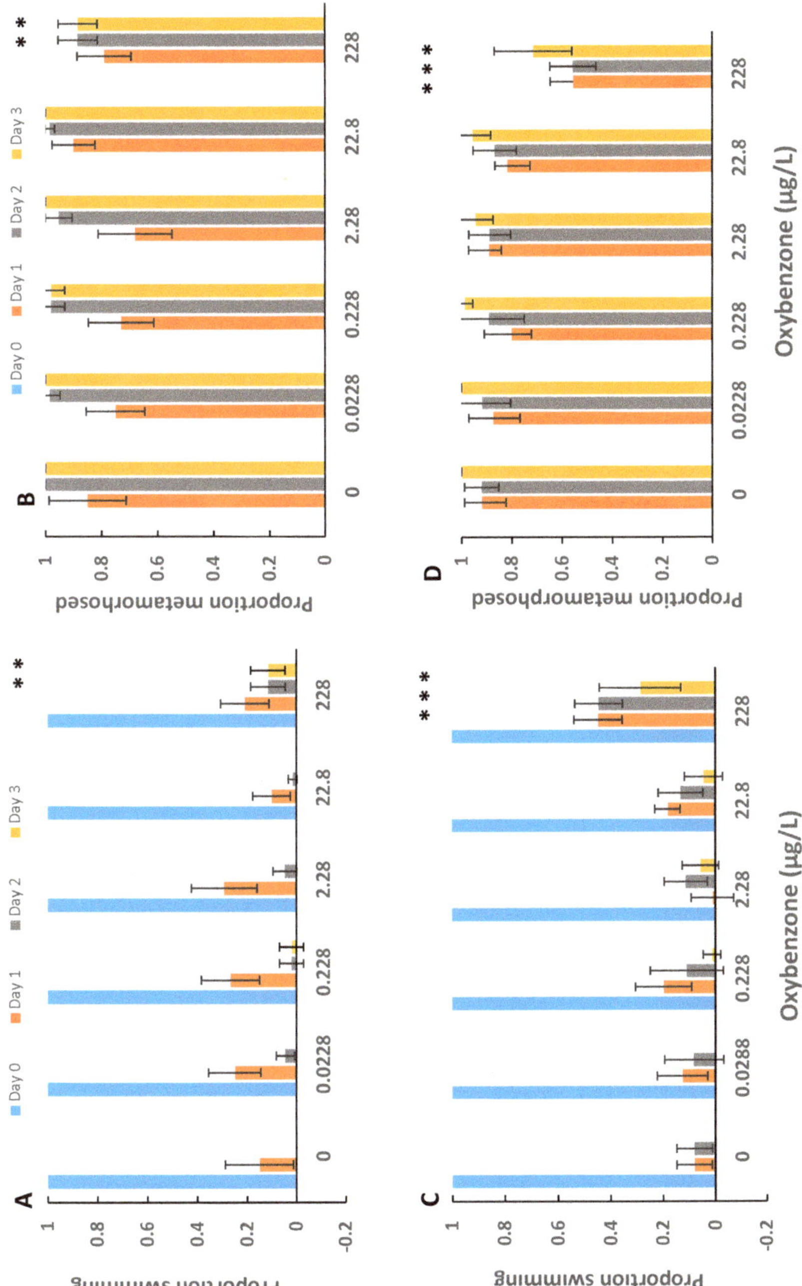

**Figure 4.** Proportion of swimming (**A,C**) or metamorphosed (**B,D**) larvae *Cassiopea* sp. larvae in varying concentrations of oxybenzone over the course of 3 days. **A**: swimming *C. xamachana*, **B**: metamorphosed *C. xamachana*, **C**: swimming *C. frondosa*, **D**. metamorphosed *C. frondosa*. * = animals maintained in 228 μg/L oxybenzone significantly different ($p \leq 0.05$, ANOVA) from the controls.

## 4. Discussion

Oxybenzone has been found to damage and deform planulae larvae, which could explain why the motile skills of planulae larvae of *C. xamachana* and *C. frondosa* were partially inhibited (Figure 2) [4,14]. Larvae in higher concentrations of oxybenzone (228, 22.8 µg/L) were swimming in irregular patterns, some spun in circles in the same spot, and many of them died (Figure 3). The disorientation of the surviving larvae would most likely inhibit their ability to settle and metamorphose (Figure 4). Although the response of the two species is very similar, planulae of *C. frondosa* appear to be more sensitive to high concentrations of oxybenzone than *C. xamachana* (Figures 3 and 4). Since there was no death in the control group of larvae, nor those maintained in 0.0228–2.28 µg/L oxybenzone, it appears that the low amount of DMSO used in the experiments did not detrimentally affect the planulae.

Planulae and newly metamorphosed scyphistomae (= polyps) are aposymbiotic in *Cassiopea* sp., unlike the normally symbiotic planulae of the coral *Stylophora pistillata* [4,14]. Many of the larval responses seen by Downs [4,14] were thought to be partially due to the effects on the symbiotic algae. *Cassiopea* sp. can acquire Symbiodiniaceae, soon after the new polyps develop a mouth. Therefore, Symbiodiniaceae were not involved in the responses of planulae larvae of *Cassiopea* sp. to oxybenzone. The symbionts enable the polyps of *Cassiopea* sp. to strobilate, turning into a medusa (Figure 1), although the mechanism of the symbiotic interaction is not known [17].

Sunscreens washed or flushed into the ocean during tourist season are probably having a negative effect on corals and jellyfish [18]. Downs et al. [14] found that the upper concentration varied in Hawaii (0.8–19.2 µg/L) and the US Virgin Islands (75–1400 µg/L), up to 6 times higher than the top concentration of oxybenzone used in the current experiments. The lethal concentration of oxybenzone that kills half of planulae of *S. pistillata* (LC50) in the light for an 8- and 24-h exposure was 3100 µg/L and 139 µg/L, respectively [14].

Planulae normally use their swimming ability to investigate substrates to settle on. *Cassiopea xamachana* and *C. frondosa* normally settle on specific substrates in their environment [19]. Oxybenzone can be attributed to the decreased motility and settlement/metamorphosis of *C. xamachana* and *C. frondosa* larvae, and even death at the higher concentrations used, posing a threat to the survival of these species.

**Author Contributions:** For Methodology, experimental set-up, original draft preparation W.K.F., methodology, experimental set-up, reading and editing D.K.H. All authors have read and agreed to the published version of the manuscript.

**Funding:** This research received no external funding.

**Acknowledgments:** The authors acknowledge the student research of Alex Amalfitano, Anna Schramski, Sarah Gardner, Ethan Turner, and Rebecca Farley. The Key Largo Marine Research Laboratory (contribution #167) provided housing and access to the jellyfish *Cassiopea* sp.

**Conflicts of Interest:** The authors declare no conflict of interest.

## References

1. Shaath, N.A.; Shaath, M. Recent sunscreen market trends. In *Sunscreens, Regulations and Commercial Development*; Shaath, N.A., Ed.; Taylor & Francis: Boca Raton, FL, USA, 2005; Volume 3, pp. 929–940.
2. French, J.E. NTP technical report on the toxicity studies of 2-hydroxy-4-methoxybenzophenone (CAS No. 131-57-7) administered topically and in dosed feed to F344/N Rats and B6C3F1 mice. *Toxic. Rep. Ser.* **1992**, *21*, 1–14. [PubMed]
3. Hany, J.; Nagel, R. Detection of sunscreen agents in human breast milk. *Dtsch. Lebensm. Rundsch.* **1995**, *91*, 341–345.
4. Downs, C.A.; Kramarsky-Winter, E.; Segal, R.; Fauth, J.; Knutson, S.; Bronstein, O.; Ciner, F.R.; Jeger, R.; Lichtenfeld, Y.; Woodley, C.M.; et al. Toxic pathological effects of the sunscreen UV filter, oxybenzone (benzophenone-3), on coral planulae and cultured primary cells and its environmental contamination in Hawaii and the U.S. Virgin Islands. *Arch. Environ. Contam. Toxicol.* **2016**, *70*, 265–288.

5. Schallreuter, K.U.; Wood, J.M.; Farwell, D.W.; Moore, J.; Edwards, H.G.M. Oxybenzone oxidation following solar irradiation of skin: Photoprotection versus antioxidant inactivation. *J. Investig. Derm.* **1996**, *106*, 583–586. [CrossRef] [PubMed]

6. Huang, Y.; Wnag, P.; Law, J.C.; Zhao, Y.; Wei, Y.; Zhou, Y.; Zang, Y.; Shi, H.; Leung, K.S. Organic UV filter exposure and pubertal development: A prospective follow-up study of urban Chinese adolescents. *Environ. Int.* **2020**, *143*. [CrossRef] [PubMed]

7. Frikeche, J. Research on the immunosuppressive activity of ingredients contained in sunscreens. *Arch. Dermatol. Res.* **2015**, *307*, 211–218. [CrossRef] [PubMed]

8. Danovaro, R.; Bongiorni, L.; Corinaldesi, C.; Giovannelli, D.; Damiani, E.; Astolfi, P.; Greci, L.; Pusceddu, A. Sunscreens cause coral bleaching by promoting viral infections. *Environ. Health Perspect.* **2008**, *116*, 337–340. [CrossRef] [PubMed]

9. He, T.; Tsui, M.M.P.; Tan, C.J.; Ma, C.Y.; Yui, S.K.F.; Wang, L.H.; Chen, T.H.; Fan, T.Y.; Lam, P.K.S.; Murphy, M.B. Toxicological effects of two organic ultraviolet filters and a related commercial sunscreen product in adult corals. *Environ. Pollut.* **2019**, *245*, 462–471. [CrossRef] [PubMed]

10. He, T.; Tsui, M.M.P.; Tan, C.J.; Ng, K.Y.; Guo, F.W.; Wang, L.H.; Fan, T.Y.; Lam, P.K.S.; Murphy, M.B. Comparative toxicities of four benzophenone ultraviolet filters to two life stages of two coral species. *Sci. Total Environ.* **2019**, *651*, 2391–2399. [CrossRef]

11. Wijgerde, T.; van Ballegooijen, M.; Nijland, R.; van der Loos, L.; Kwadijk, C.; Osinga, R.; Murk, A.; Slijkerman, D. Adding insult to injury: Effects of chronic oxybenzone exposure and elevated temperature on two reef-building corals. *Sci. Total Environ.* **2020**, *733*, 139130. [CrossRef]

12. DiNardo, J.C.; Downs, C.A. Dermatological and environmental toxicological impact of the sunscreen ingredient oxybenzone/benzophenone-3. *J. Cosmet. Derm.* 2017. [CrossRef] [PubMed]

13. Blüthgen, N.; Zucchi, S.; Fent, K. Effects of the UV filter benzophenone-3 (oxybenzone) at low concentrations in zebrafish (Danio rerio). *Toxicol. Appl. Pharm.* **2012**, *263*, 184–194. [CrossRef] [PubMed]

14. Downs, C.A.; Kramarsky-Winter, E.; Fauth, J.E.; Segal, R.; Bronstein, O.; Jeger, R.; Lichtenfeld, Y.; Woodley, C.M.; Pennington, P.; Kushmaro, A.; et al. Toxicological Effects of the Sunscreen UV Filter, Benzophenone-2, on Planulae and In Vitro Cells of the Coral. *Stylophora Pist. Ecotoxicol.* **2014**, *23*, 175. [CrossRef] [PubMed]

15. Raffa, R.B.; Pergolizzi, J.V.; Taylor, R.; Kitzen, J.M. Sunscreen bans: Coral reefs and skin cancer. *J. Clin. Pharm. Ther.* **2019**, *44*, 134–139. [CrossRef] [PubMed]

16. Banning Personal Care Products. In Proceedings of the Hawaii State Legislature, SB 260, Honolulu, HI, USA, January 2017.

17. Hofmann, D.K.; Fitt, W.K.; Fleck, J. Checkpoints in the life-cycle of Cassiopea spp.: Control of metagenesis and metamorphosis in a tropical jellyfish. *Int. J. Dev. Biol.* **1996**, *40*, 331–338. [PubMed]

18. Tsui, M.M.P.; Lam, J.C.; Ng, T.Y.; Murphy, M.B.; Lam, P.K.S. Occurrence, distribution and fate of organic UV filters in coral communities. *Environ. Sci. Technol.* **2017**, *51*, 4182–4190. [CrossRef] [PubMed]

19. Fleck, J.; Fitt, W.K. Degrading Mangrove Leaves of Rhizophora mangle Linne Provide a Natural Cue for Settlement and Metamorphosis of the Upside Down Jellyfish Cassiopea xamachana Bigelow. *J. Exp. Mar. Biol. Ecol.* **1999**, *234*, 83–94. [CrossRef]

*Article*

# The Condition of Four Coral Reefs in Timor-Leste before and after the 2016–2017 Marine Heatwave

**Catherine J. S. Kim** [1,*], **Chris Roelfsema** [2], **Sophie Dove** [1] **and Ove Hoegh-Guldberg** [1]

[1] ARC Centre of Excellence for Coral Reef Studies, School of Biological Sciences, The University of Queensland, Brisbane, QLD 4072, Australia; sophie@uq.edu.au (S.D.); oveh@uq.edu.au (O.H.-G.)

[2] Remote Sensing Research Center, The University of Queensland, Brisbane, QLD 4072, Australia; c.roelfsema@uq.edu.au

* Correspondence: catherine.kim37@gmail.com; Tel.: +61-7-3365-2118

**Abstract:** El Niño Southern Oscillation global coral bleaching events are increasing in frequency, yet the severity of mass coral bleaching is not geographically uniform. Based in Timor-Leste, the present project had two major objectives: (1) assess the baseline of reefs and coral health at four sites and (2) explore water quality and climate-related changes in ocean temperatures on these understudied reef systems. The impacts of climate change were surveyed on coral reefs before and after the 2016–2017 global underwater heatwave, (principally by following coral mortality). Temperature loggers were also deployed between surveys, which were compared to Coral Reef Watch (CRW) experimental virtual station sea surface temperature (SST). CRW is an important and widely used tool; however, we found that the remotely sensed SST was significantly warmer (>1 °C) than in situ temperature during the austral summer accruing 5.79-degree heating weeks. In situ temperature showed no accumulation. There were significant differences in coral cover, coral diversity, and nutrient concentrations between sites and depths, as well as a low prevalence of disease recorded in both years. Change in coral cover between surveys was attributed to reef heterogeneity from natural sources and localized anthropogenic impacts. Timor-Leste has both pristine and impacted reefs where coral cover and community composition varied significantly by site. Degradation was indicative of impacts from fishing and gleaning. The comparison of in situ temperature and remotely sensed SST indicated that bleaching stress in Timor-Leste is potentially mitigated by seasonal coastal upwelling during the Northwest monsoon season. As a climate refugium, the immediate conservation priority lies in the mitigation of localized anthropogenic impacts on coral reefs through increasing the management of expanding human-related sedimentation and fishing.

**Keywords:** coral reefs; Coral Triangle; ENSO; coral bleaching; temperature; stable isotope; coral disease; coral health; nutrients; Indonesian ThroughFlow

**Citation:** Kim, C.J.S.; Roelfsema, C.; Dove, S.; Hoegh-Guldberg, O. The Condition of Four Coral Reefs in Timor-Leste before and after the 2016–2017 Marine Heatwave. *Oceans* **2021**, *3*, 147–171. https://doi.org/10.3390/oceans3020012

Academic Editor: Rupert Ormond

Received: 30 October 2020
Accepted: 7 March 2022
Published: 8 April 2022

**Publisher's Note:** MDPI stays neutral with regard to jurisdictional claims in published maps and institutional affiliations.



## 1. Introduction

Timor-Leste is a developing country with limited infrastructure following decades of war. It is one of six member states of the Coral Triangle (CT), the global center of marine biodiversity (numbers of species), housing 29% of the world's coral reefs [1,2]. Much of this diversity, however, is under threat due to a range of growing local and global stresses [1,3–5]. Globally, climate change-induced coral bleaching via ocean warming and coral disease are among the main threats facing coral reefs [3,6]. Mass coral bleaching events driven by anomalous increases in sea surface temperature (SST) maintained over time, or marine heatwaves, have been occurring with increasing frequency [7]. The El Niño Southern Oscillation (ENSO) associated marine heatwave in 2016–2017 was the longest and most intense in history with global, but patchy, impacts on coral reefs [8,9]. Comparatively few reports of the bleaching event exist in the CT, with one report from Sulawesi, Indonesia attributing coral mortality in shallow reef flat zones to ENSO-related sea level fall [10]. The

CT arguably has the most to lose in terms of loss of biodiversity and resources associated with reefs that support the 360 million people who live in the region [11,12].

Coral reefs within the CT are highly threatened by local impacts. In Timor-Leste, 92% of reefs are at high or very high risk due to fishing pressure, watershed-based pollution, coastal development, and pollution from marine activities (shipping, oil, and gas extraction) [11]. While dynamite fishing has decreased since the Indonesian occupation [13], subsistence fishing is important to livelihoods and food security. There are an estimated 5000 fishers on the narrow, productive shelf that supports coral reefs [14,15]. Additionally, gleaning, or harvesting invertebrates from intertidal flats for consumption, known locally as *meti*, is commonly practiced by women and children and has its own significant impact [16–18]. Agricultural practices consist of small-scale, subsistence farming without the use of non-organic fertilizers and pesticides [19]; nevertheless, watershed-based pollution is widespread due to an estimated 24% decrease in forest cover from 1972 to 1999 caused by slash and burn agriculture, logging, and consumption for fuel [20–22]. With planned development in these sectors [11,20–22], further understanding of natural and anthropogenic pressures affecting reefs is crucial in order to properly manage coral reefs for nature and people alike.

### 1.1. Disease in the Context of Coral Reef Health

Coral disease has been a major contributor to the decline in corals in the Caribbean [23], as well as reefs in the Indo-Pacific [3,24–26]. By contrast, there have been few studies of coral disease in the CT (Table S1). In this study, diseases were defined as syndromes caused by pathogens. It is important to document lesions, or morphologic abnormalities, predation, physical breakages (storms, anchors), and aggressive interactions that may result in tears or breaks in the tissue, partial mortality, stress to the coral host, and abiotic diseases which we collectively refer to as compromised health. Abiotic diseases, such as coral bleaching, are not caused by microbial agents [27]. These are some of the many indicators of coral reef condition (loosely defined as coral health). Disease can be endemic and highly visible [23], or present in low frequency in any given population [26]. Tracking disease and other signs of compromised health through time can be paired with other datasets (such as herbivore biomass, hurricane incidence, environmental parameters, etc.) and key physiological parameters such as growth rates, fecundity, and community composition of reefs [28]. Many coral diseases have been linked to increasing ocean temperatures, nutrient pollution, sedimentation, and fishing [29–32]. At most sites in Timor-Leste, these types of measurements are absent, highlighting the importance of the present study as a crucial baseline on the conditions of important marine resources.

### 1.2. Water Quality and Coral Reefs along the North Coast of Timor-Leste

Pollution arising from disturbed coastal regions and watersheds poses a serious threat to coral reefs globally. This type of pollution includes a wide range of compounds such as agrochemicals (pesticides), inorganic nutrients (nitrate, ammonia, and phosphate), soils and sediments, and fossil fuel residues that flow from disturbed landscapes. Many of these compounds negatively affect coral physiology by reducing calcification rates, fecundity, fertilization success, and larval development [33]. This can degrade reef communities reducing coral cover, community composition, diversity, and structural complexity [34,35]. High levels of marine pollution can increase the prevalence and severity of disease and susceptibility to bleaching [36–40]. Dissolved inorganic nitrogen (DIN = ammonium + nitrate + nitrite) measurements on reefs are generally <1.5 µM (individual species ammonium, nitrate, nitrite < 1 µM) with even lower phosphate concentrations (<0.3 µM; Table S2) [37,41–47]. A greater prevalence of disease has been associated with elevated concentrations of DIN from anthropogenic sources (fertilizer, sewage pollution, etc.) and phosphate ranging from 3.6 µM to 25.6 µM and 0.3 µM to 0.4 µM, respectively [41,42,45–47].

The isotopic signature of nutrients such as nitrogen can often serve as a tracer for different sources of coastal pollution, with different forms of input having different impacts (sewage can increase pathogen concentrations) and solutions [32,48–57]. Stable isotope analyses of nitrogen stored in macroalgae can provide a nutrient signal integrated over time versus water sampling, which is highly variable over space and time [58]. Generally, $\delta^{15}N$ signatures in algae associated with urban wastewater are >10‰ [59–62]. Natural and synthetic fertilizers display a large range from −4‰ to +4‰ of $\delta^{15}N$ values while nitrogen fixation typically has a negative $\delta^{15}N$ signature between −2 and 0‰ [63]. Upwelling can have variable $\delta^{15}N$ values ranging from 5 to12‰ [50,64–70]. Both fertilizer use and waste infrastructure are expected to be developed as described in the national strategic development plan [19].

### 1.3. Global Impacts—Ocean Warming, Mass Coral Bleaching, and Mortality

The US National Oceanic and Atmospheric Administration (NOAA) Coral Reef Watch virtual station in Timor-Leste (CRWTL) reported anomalous warming between the two survey periods of November 2015 and July 2017. Between January and May 2016, and again in January and February 2017, the water temperature of the region attained degree heating weeks (DHWs) between 4 and 8 °C-weeks [71]. This DHW range has been associated with 30–80% bleaching [72,73], suggesting that corals may have bleached twice within the 20 months between sampling intervals. Surviving corals, however, would have had four to five months to recover before resurveying in July 2017. Typically, mortality is not expected below DHW of 8 °C-weeks [74], although this is variable between species [75,76]. Corals that have experienced a recent thermal event that is sufficiently warm to cause temporary bleaching in some corals, may nonetheless be vulnerable to disease or other signs of compromised health [3,77–79]. Additionally, corals may endure sublethal effects for months after the event as they attempt to rebuild energy reserves [4,80].

### 1.4. Aims

The aims of the present study were two-fold. The first aim was to investigate the state and health of coral reefs along the north coast of Timor-Leste, as measured by the presence of coral disease and other signs of compromised health. The second aim was to explore the global and local impacts on Timorese reefs through water quality measurements, temperature data, and surveys before and after the 2016–2017 global bleaching event. This was achieved through repeated coral health surveys, seawater nutrient and nitrogen stable isotope analyses of macroalgae to assess nutrients, and in situ and remotely sensed temperature data.

## 2. Materials and Methods

### 2.1. Study Site

Timor-Leste is a small island country located inside the southern edge of the CT and between Australia and Indonesia (Figure 1). The country gained its independence in 2002 following nearly 25 years of Indonesian occupation. It lies within the Indonesian ThroughFlow (ITF), a major oceanographic feature connecting the Pacific and Indian Oceans. The ITF transports an estimated 4.9 Sverdrups ($10^6$ m$^3$/s) of water through the Ombai Strait along the north coast (Figure 1) [81–85].

This study was undertaken along the coast of Dili, Timor-Leste to complement a growing body of coral reef science undertaken in the area. Previous indications of reef health in this area have typically been anecdotes from surveys with other objectives. Dili, the capital (8°33′ S and 125°34′ E), houses a quarter of the country's population, with 252,884 people recorded in the 2015 Census [86]. The Northwest monsoon season extends from December to May, when there is more rain and greater runoff to the coast [87]. The seasonal Comoro River runs through Dili, with flows ranging from less than 0.5 m$^3$/s from July to November, to 12.3 m$^3$/s in March [20]. Timorese waters are affected by semi-diurnal tides with a range of 1–2 m along the north coast [88]. The present study was conducted in two, three-week field trips that occurred in the dry season in 15–27 November 2015 and

15–29 June 2017. The dry, Northwest monsoon season offers safer surveying conditions but would also limit terrestrial run-off inputs such as nutrients.

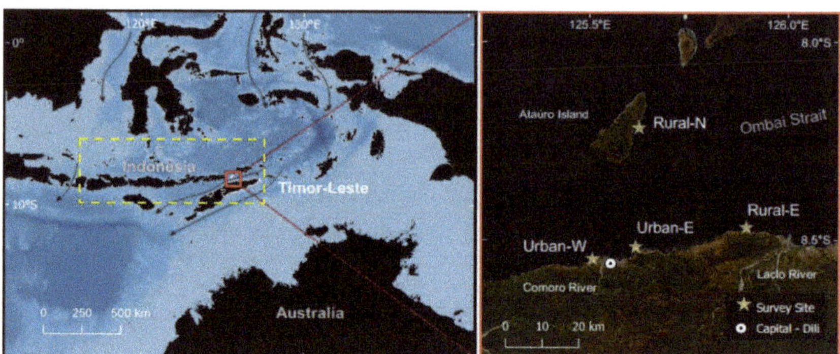


**Figure 1.** Timor-Leste lies between Indonesia and Australia within the Indonesian ThroughFlow (ITF). Arrows indicate inflows and outflows of the ITF adapted from [81,82]. The yellow dashed box indicates the Lesser Sunda Islands (adapted from [85]). Enlarged map on the right shows the location of the survey sites in Timor-Leste near the capital of Dili. Survey sites are Rural-N on Ataúro Island in the Ombai Strait, Rural-E 40 km east of Dili, and Urban-W and Urban-E flanking Dili. The highly seasonal Comoro River can be seen just east of Urban-W (in Dili) and the Laclo River east of Rural-E. The four sites were sampled on two occasions in November 2015 and June 2017. Jaco Island lies at the easternmost point of the country.

Surveys were conducted at four sites; two sites flanking Dili were representative of reefs under urban influences ("Urban-W" with 4993.0 people/km$^2$; "Urban-E" with 449.8 people/km$^2$) and two sites were representative of reefs under rural influences ("Rural-N" 30.7 people/km$^2$, and "Rural-E" with 34.9 people/km$^2$ [86]; Figure 1). Sites were chosen for logistics and to complement NOAA data collected between 2012 and 2016 [89].

## 2.2. Coral Community Composition and Coral Health Surveys

To assess benthic cover and coral health, we deployed 15 m line intercept transects (LIT) [90] and 15 m × 2 m belt transects, respectively [28]. At each of the four sites, three transects were laid at both 5 m (reef flat) and 10 m (reef slope) depths, for a total of 24 transects across all sites. The first of three transects at each site and depth was chosen haphazardly, with the subsequent transects at least 5 m away from the start/end of other transects on the appropriate depth contour. For the LIT surveys, all benthos under the 15 m tape was categorized into a major benthic category (hard coral, soft coral, substrate/sand, macroalgae, turf algae, cyanobacteria, and crustose coralline algae [CCA]). LIT transects were conducted by two volunteers per survey trip who were at least ReefCheck certified. On the coral health belt transects performed by the first author, every coral colony within the 15 × 2 m transect area was identified to genus and assessed visually for coral disease and signs of potentially compromised coral health, such as overgrowth by macroalgae, turf and cyanobacteria overgrowth, encrusting invertebrates (sponges, tunicates, flatworm infestation), burrowing invertebrates (gastropods, bivalves, crustaceans, etc.), signs of predation (fish and *Drupella* spp. snails), signs of bleaching (partial or total loss of algal symbionts appearing white), signs of coral response (pigmentation, mucus), and physical damage (sedimentation, breakage) as per protocols developed by the Global Environment Facility and World Bank Coral Disease Working Group (Figure 2 and Figure S1; Table S3) [28]. High prevalence or increasing numbers of burrowing invertebrates or overgrowth of corals by turf algae and invertebrates could be indicative of reef degradation due to anthropogenic disturbances [91]. No *Acanthaster* spp. sea stars or feeding scars were observed during surveys. Any uncertain diagnoses were photographed for later consultation. The prevalence of disease and compromised health was calculated as the number of corals affected by the disease/compromised health category divided by the total number of coral colonies in the transect [28]. The GPS coordinates at the start of each transect were recorded and used for the second survey in July 2017 (Table S4).

**Figure 2.** Examples of disease and compromised health recorded on surveys undertaken in Timor-Leste between 15–27 November 2015. Photos outlined in red represent close-ups of adjacent photos (**b1,c1**). (**a**) WS–White Syndrome band of distinct tissue loss on tabulate acroporids with white skeleton abutting live tissue with exposed skeleton gradually colonized by turf algae, (**a1**) exposed coral skeleton caused by coral tissue loss from WS, (**b**) bleached tissue displaying white, living tissue lacking symbionts, (**c**) flatworm infestation on a *Fungia* coral, (**d**) turf algae overgrowth on a massive *Porites* coral, and (**e**) cyanobacterial (purple) overgrowth on a reef affecting more than one genus (i.e., *Fungia* and branching montiporids). See Figure S1 for other compromised states and Table S3 for more information.

### 2.3. Measurement of Nutrient Concentrations and Stable Isotope Ratios

Seawater samples were collected for measuring the concentration of inorganic nutrients as an indicator of nutrient pollution. Three replicate 100 mL seawater samples were collected on each transect at a depth of 0.5 m above the benthos, after benthic surveys were completed (Table S4), kept on ice until filtered through a 0.22 µm pore membrane filter, and stored frozen. Seawater samples were analyzed within four months for ammonium, nitrite, nitrate, and phosphate using flow injection analysis at the Advanced Water Management Center (The University of Queensland). Nitrite had mostly zero values and was combined with nitrate for analyses.

Macroalgal samples were collected for stable isotope analysis to explore the origin of inorganic nitrogen. Three replicates of *Halimeda* spp. and of *Chlorodesmis* spp. macroalgae (approximately 5 g dry weight) were collected when available on each transect, rinsed, and air-dried for transport. In the laboratory, the samples were re-dried at 60 °C for a minimum of 24 h before homogenization using a mortar and pestle and subsequently analyzed at the Cornell University Stable Isotope Laboratory (Finnigan MAT Delta Plus isotope ratio mass spectrometer) for $\delta^{15}$N.

### 2.4. In Situ and Satellite Temperature Data

Calibrated HOBO pendant temperature loggers (Onset Computer Corporation, Bourne, MA, USA) were deployed at each site and depth in November 2015. Temperature was recorded every 30 min. All were collected in June 2017 except those from Rural-E which could not be retrieved. Remotely sensed satellite SST data from the NOAA's CRWTL were downloaded from August 2015 through August 2017. This product uses 5 km$^2$ resolution to predict bleaching stress across an entire jurisdiction such as Timor-Leste instead of producing values at every pixel [71].

### 2.5. Statistical Analyses

All analyses were conducted in R version 4.0.4 [92] and PRIMER7 [93,94]. Three-way repeated measures permutational multivariate analysis of variance (PERMANOVA) with 9999 permutations were conducted to test for significant effects between sites (Rural-N, Rural-E, Urban-W, Urban-E), depths (5 m, 10 m), and years (2015, 2017) on a Bray–Curtis similarity matrix of benthic cover categories, prevalence of disease and compromised health, and a Bray–Curtis similarity matrix of the number of colonies per coral genera (the count of coral genera per belt transect), all square root transformed [93,94]. All PERMANOVAs were also tested for homogeneity of dispersion akin to the homogeneity of variance in univariate tests [93]. Principal Coordinates Analysis (PCoA) was run on the same similarity matrix of coral genera to visualize coral community structure. Repeated measures analysis of variance (ANOVA; Anova in the car, emmeans, and nlme R packages) [95–97] was used to test the transformed hard coral abundance, the categories of disease and compromised health (only the bleached category was transformed), and the transformed Shannon diversity index of coral genera for significant effects between sites, depths, and years. All transformations were square root. A repeated measures ANOVA was also conducted on the log-transformed number of acroporids per transect between site, depth, and year. Normality was visually inspected (hist, qqplot, qqnorm, and leveneTest in the car package).

Nutrient data were only collected in 2015, with which a two-way ANOVA with factors site, and depth was performed on the seawater nutrient data, including DIN (transformations: log—$NH_4^+$ and DIN; square root—$NO_3^- + NO_2^-$). A two-way ANOVA was used to test for significant differences in $\delta^{15}$N for each of the two genera of algae, *Halimeda* spp. and *Chlorodesmis* spp., with the factors site and depth. Only three samples of *Chlorodesmis* spp. were collected on a single transect at Rural-E and these were removed from the analysis. Variables were visually inspected for normality and tested for homogeneity of variance using Levene's test (leveneTest). Percent nitrogen was log-transformed for *Halimeda* spp. Post hoc tests were conducted (multcomp and emmeans R packages) for *Halimeda* spp. and *Chlorodesmis* spp.

The monthly mean temperature was calculated from the 24 h daily maximum temperature obtained from both datasets (in situ temperature logger data and remotely sensed CRWTL data). A one-way repeated measures ANOVA was used to test temperature logger data (pooled by site) to test for differences in the monthly means between sites. A two-way repeated ANOVA assessed seasonal (summer January–March, fall April–June, winter July–September, spring October–December) and methodological (in situ loggers, remotely sensed SST from CRWTL) differences between the monthly temperature means. Both analyses employed random intercept models with residual autocorrelation structures to account for temporal autocorrelation of individual temperature loggers and CRWTL measurements (nlme R package). Post hoc tests were conducted using the emmeans R package. To assess levels of thermal stress, remotely sensed DHWs were retrieved from CRWTL online [71].

## 3. Results

### 3.1. Coral Cover and Community Composition at Four Sites

A total of 9521 corals from 51 genera were counted within the 1440 m$^2$ of the 15 m × 2 m belt transects in 2015 and 2017. LIT surveys indicated that benthic composition was significantly different between a three-way interaction, where variables affect benthic composition in conjunction with each other, between year, site, and depth [three-way repeated measures PERMANOVA, pseudo-$F$(3,47) = 2.6888, $p$(perm) = 0.0109]. All sites, except Urban-W, were significantly different between survey years at 10 m while only Urban-W and Rural-N changed at 5 m [$p$(perm) < 0.05]. Dispersion, or variability, was also significant for year [$F$(1,46) = 5.9165, $p$(perm) = 0.0205], and site [$F$(1,44) = 6.7038, $p$(perm) = 0.0012]. Urban-W was significantly more variable in benthic composition than all sites [$p$(perm) < 0.05]. Live hard coral cover (henceforth coral cover) was significantly different with a three-way interaction between year, site, and depth [three-way repeated measures ANOVA $\chi^2$(3) = 18.6751, $p$ = 0.0003]. In both years, Urban-W at 5 m had the lowest coral cover (mean ± SE; 4.8 ± 1.8% in 2015 and 4.5 ± 1.5% in 2017) and Rural-N 5 m (58.2 ± 1.7%) and Rural-N 10 m (56.9 ± 3.3%) had the highest coral cover, respectively, in 2015 and 2017 (Figure 3). Overall, coral cover was ~25% higher at the rural sites (37.3 ± 5.3%) than at the urban sites (12.9 ± 3.8%).

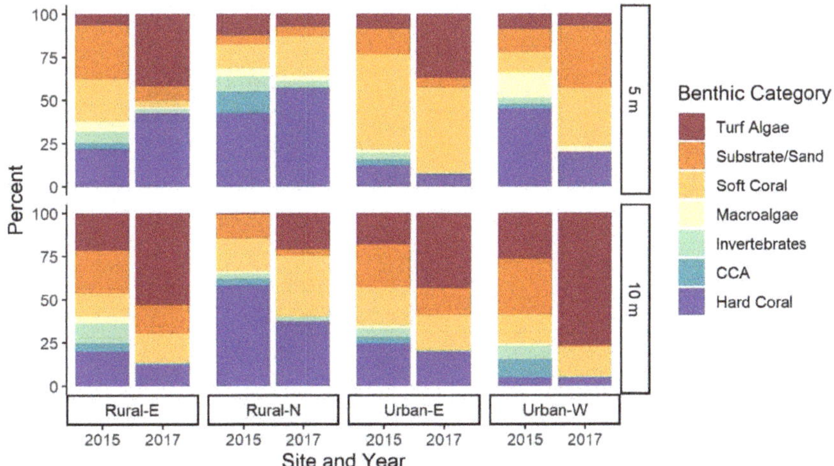


**Figure 3.** Benthic composition cover from 15 m line intercept transects by site (Rural-N, Rural-E, Urban-W, Urban-E) and depth (5 and 10 m) for November 2015 and June 2017 surveys in Timor-Leste. Major categories include: Hard Coral; CCA—crustose coralline algae; Invertebrates—mobile invertebrates; Macroalgae; Soft Coral; Substrate/Sand; and Turf algae.

Coral community composition, as measured by the abundance of individual coral genera from belt transects, also differed significantly by a site and depth interaction [three-way repeated measures PERMANOVA pseudo-$F(3,47)$ = 3.2546, $p$(perm) = 0.0001]. Hard coral diversity, calculated using the Shannon diversity index of hard coral genera counted on belt transects, showed significant site and depth differences [three-way repeated-measures ANOVA $\chi^2(3)$ = 16.1668, $p$ = 0.0010]. Urban-W had the lowest coral diversity of all sites at 10 m but had comparable diversity at 5 m depth (Shannon index 1.7 $\pm$ 0.2; 18 $\pm$ 2 genera) (Figure S2). The maximum genus richness of 33 $\pm$ 2 was present at Rural-N with consistently high (>40%) coral cover in both survey years. This site also had significantly more tabulate [three-way repeated measures ANOVA $\chi^2(3)$ = 31.5895, $p$ < 0.0001] at both depths than all other sites with 21.1 $\pm$ 0.7 colonies per transect ($p$ < 0.05). Remaining sites averaged less than five acroporid colonies per transect. There was a significant three-way interaction between site, depth, and year for branching acroporids [three-way repeated measures ANOVA $\chi^2(1)$ = 7.8254, $p$ = 0.0498]. Branching Acroporids decreased at all sites and depths between survey years except for Rural-N and Urban-W at 10 m (Figure S3). Although 51 genera were found across the belt transects over all four sites, only a few genera dominated the reef, namely *Porites* (2015 = 17.4%, 2017 = 13.0%), *Fungia* (2015 = 13.7%, 2017 = 19.0%), and *Montipora* (2015 = 12.9%, 2017 = 13.4%).

Community composition measured from belt transects was also significantly different with a year and site interaction [pseudo-$F(3,47)$ = 2.1713, $p$(perm) = 0.0002]. Urban-E was the only site with comparable coral community between years, and all sites were significantly different within years [$p$(perm) < 0.05]. In Figure 4, sites were generally distributed along axis two of the PCoA with Rural-N most positively associated with tabulate acroporids, *Galaxea*, *Goniopora*, *Montipora*, and *Stylophora*, while differences between depth were aligned along axis one with more *Pocillopora*, *Platygyra*, and massive *Porites* corals on the 5 m transects. Dispersion, or variability in community composition was also significantly different by site [$F(3,44)$ = 15.7770, $p$(perm) = 0.0001] and depth [$F(1,46)$ = 5.6388, $p$(perm) = 0.0433]. Specifically, the dispersion (spread of points) was significantly greater at 10 m and highest at Urban-W and lowest at Rural-N (Figure 4).

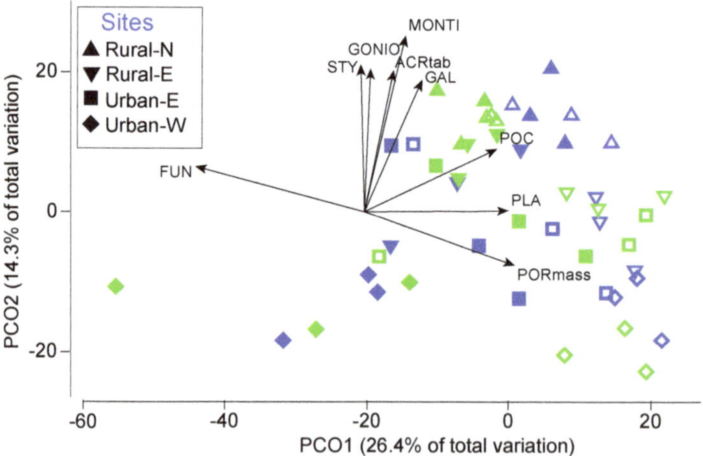


**Figure 4.** Principal Coordinates Analysis biplot of square root transformed Bray–Curtis similarity matrix of the hard coral genera counted on belt transects. Empty and solid markers indicating 5 and 10 m depths, respectively. Color indicates survey year: blue-2015 and green-2017. Vectors represent top nine coral genera/morphology correlated to axes as calculated by Person's correlation. Abbreviations are coral genera/morphology as follows: ACRtab—*Acropora* tabulate; FUN—Fungiids; GAL—*Galaxea*; MONTI—*Montipora*; GONIO—*Goniopora*; PLA—*Platygyra*; POC—*Pocillopora*; PORmass—*Porites* massive; and STY—*Stylophora*.

*3.2. Prevalence of Coral Disease and Indicators of Compromised Health*

Overall, most hard corals at the sites surveyed appeared healthy as recorded on the belt transects. Those categorized as "healthy" made up 65.7 ± 2.9% of corals surveyed averaged over both years, with a low (<1%) prevalence of diseases and a 33.7 ± 1.7% prevalence of other compromised states. In 2015, there was 0.9 ± 0.2% prevalence of WS at Rural-N on *Acropora* spp., which made up 44.9% of all diseases recorded (Figure 5). In the same year, Rural-N also had the highest prevalence of Growth Anomalies (GAs; 0.6 ± 0.2%). There was one case of unconfirmed Trematodiasis, which requires microscopic confirmation of the larval trematode. In 2017, disease prevalence was lower with the highest prevalence of WS (0.5 ± 0.1%) at Rural-N again but Urban-W recording the most GAs (0.6 ± 0.3%). All cases of WS were documented on acroporids in 2017, while GAs were less host-specific and found on nine genera across the two years. In 2015, the number of branching acroporids was comparable between depths at each site while in 2017 the overall number of this coral morphology decreased, more branching acroporids were found at 10 m (Figure S3).

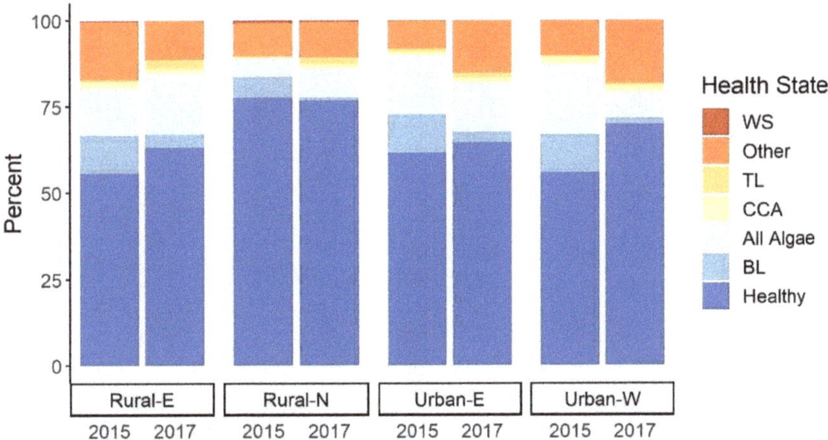


**Figure 5.** Prevalence of disease and states of compromised health on corals recorded in 15 × 2 m belt transect surveys at four sites (Rural-N, Rural-E, Urban-W, Urban-E), pooled by depth in Timor-Leste from November 2015 and June 2017. Abbreviations as follows: All Algae—combined macroalgae, turf, and cyanobacteria overgrowth; BL—Bleaching; CCA—Crustose coralline algae overgrowth; Other: combined pigmentation, predation, invertebrate infestation/overgrowth, burrowing invertebrates; TL—Unexplained tissue loss; WS—White Syndrome.

Prevalence of disease and compromised health categories from belt transects varied significantly by two interactions: year and site [three-way repeated measures PERMANOVA, pseudo-$F(1,47)$ = 3.8234; $p$ = 0.0005] and site and depth [pseudo-$F(3,47)$ = 2.3313; $p$ = 0.0150]. Rural-N had the lowest prevalence of disease and compromised health among all sites in both 2015 (22.4 ± 0.8%) and 2017 (23.2 ± 2.1%). However, Rural-N was the only site where the prevalence of compromised health and disease increased between survey years. Despite this, Rural-N was also characterized by the highest percentage of healthy corals (77.2 ± 1.1%) [three-way ANOVA $\chi^2(3)$ = 30.7576, $p$ < 0.0001}, significantly higher than all other sites ($p$ < 0.05). This site also had the lowest prevalence of algal overgrowth on corals in 2015 (5.3 ± 1.2%) and the lowest amount of coral bleaching in both years (6.0 ± 0.9% in 2015, 0.8 ± 0.2% in 2017; Figure 5).

### 3.3. Water Quality

Nutrients and Stable Isotopes

Seawater nutrient levels and N stable isotopes of macroalgae were assessed simultaneously to obtain an indication of land-based pollution. Nutrients were not elevated (>10 μM DIN) at the urban sites compared to the rural sites, though there were significant site and depth interactions [two-way MANOVA $F(3,63)$ = 3.208, Pillai = 0.398, $p$ = 0.0012]. Combined nitrate and nitrite, and phosphate were responsible for these interactions [two-way ANOVA $NO_3^- + NO_2^-$: $F(3,63)$ = 10.8991, $p$ < 0.0001; $PO_4^{3-}$: $F(3,63)$ = 4.5597, $p$ = 0.0059]. Rural-N 10 m had significantly higher combined nitrate and nitrite ($NO_3^- + NO_2^-$: 1.05 ± 0.07 μM) and phosphate ($PO_4^{3-}$ 0.15 ± 0.01 μM; Table S6) than all other sites at 10 m, but similar levels of both nutrients at 5 m compared to remaining sites (Figure 6). Variation in DIN was marginally significant with a site and depth interaction [two-way ANOVA $F(3,63)$ = 2.7769, $p$ = 0.0484], but there were no significant differences in the pairwise test ($p$ < 0.05; Figure 6).

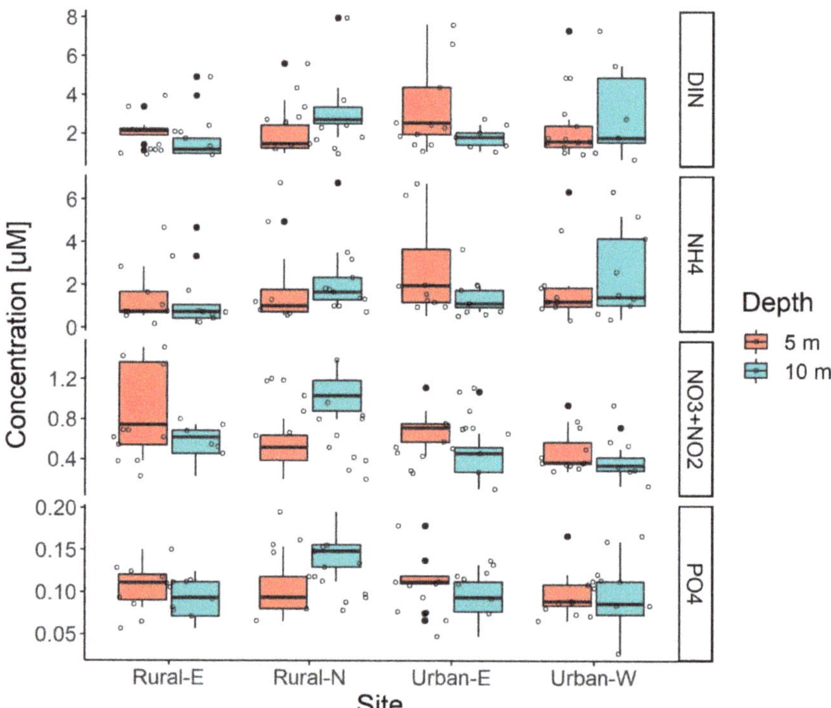


**Figure 6.** Seawater nutrient concentrations (top to bottom: dissolved inorganic nitrogen [DIN], $NH_4^+$, $NO_3^- + NO_2^-$, $PO_4^{3-}$) sampled in triplicate on each transect at the four sites (Urban-W, Urban-E, Rural-N, Rural-E), two depths (5 m and 10 m), and three transects per depth in Timor-Leste in November 2015. Bold line is the median, box ends are the first and third quartile, lines are 95% confidence interval of the median, and points are Tukey's outliers. Gray points indicate samples. Note differences in scales.

Stable isotope values were consistent across sites, with no elevated values at the urban sites compared to the rural sites. Delta $^{15}N$ stable isotopes had a significant site difference for both algae species. Urban-E had significantly lower $\delta^{15}N$ in both algae species (Table 1).

**Table 1.** Delta $^{15}$N stable isotope ANOVA results of algae sampled at the four sites (Urban-W, Urban-E, Rural-N, Rural-E), two depths (5 m and 10 m), and three transects per depth in Timor-Leste in 2015. Starred values are significant results with mean, standard error, and post hoc groupings (a, b, or ab) presented per site with "-" indicating no samples collected.

| Algae | Effect | df | F-Value | *p*-Value | Rural-E | Rural-N | Urban-E | Urban-W |
|---|---|---|---|---|---|---|---|---|
| *Halimeda* spp. | Site | 3 | 3.8199 | 0.0121 * | 4.26‰ | 4.31‰ | 4.03‰ | 4.26‰ |
| | Depth | 1 | 0.5442 | 0.4624 | ±0.06 | ±0.07 | ±0.06 | ±0.07 |
| | Site x Depth | 3 | 1.3801 | 0.2531 | ab | b | a | b |
| *Chlorodesmis* spp. | Site | 1 | 10.0028 | 0.0064 * | 4.57‰ | - | 4.11‰ | 4.47‰ |
| | Depth | 1 | 0.1747 | 0.6819 | ±0.15 | | ±0.08 | ±0.10 |
| | Site x Depth | 1 | 2.4127 | 0.1412 | ab | | a | b |

### 3.4. Temperature and the Prevalence of Bleaching

The average temperature difference between the 5 and 10 m temperature loggers across the three sites was 0.3 °C (±0.1 °C) and thus the loggers were pooled by site for further testing of site differences. The monthly means of the temperature recorded by the in situ loggers were not significantly different by site [one-way ANOVA: $\chi^2(2) = 0.1277$, $p = 0.9382$] (Figure 7).

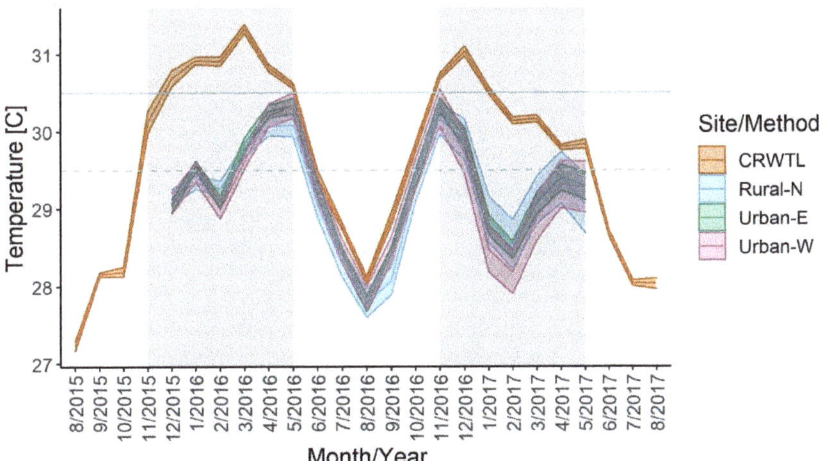


**Figure 7.** Mean temperature by month ± 2 standard error of the daily maximum temperature for remotely sensed sea surface temperature (5 km$^2$ pixels) from Coral Reef Watch Timor-Leste (CRWTL) and in situ temperature loggers pooled by depth (*n* = 6). Loggers sampled every 30 min between November 2015 and July 2017, the period between benthic surveys. The gray shading indicates the annual monsoon season from November to May. The dashed blue line is the maximum monthly mean (MMM, 29.5 °C) from the CRWTL data and the solid blue line is MMM + 1 °C (30.5 °C), the bleaching threshold for accumulation of degree heating weeks.

Comparison of the monthly means of all in situ temperature loggers with the monthly means of the CRWTL SSTs showed they were significantly different by a season and method interaction [two-way ANOVA: $\chi^2(3) = 8.2054$, $p = 0.0420$]. Pairwise tests for this interaction revealed that the CRWTL satellite-derived SST values were not significantly higher than the in situ logger temperatures within the same season (Summer CRWTL:Summer in situ, $p = 0.4991$; Fall CRWTL:Fall in situ, $p = 0.5562$; Spring CRWTL:Spring in situ, $p = 0.2291$; Winter CRWTL:Winter in situ, $p = 0.0712$). However, the elevated CRTWL temperatures during the austral summer (Jan–Mar, CRWTL = 30.7 ± 0.2 °C and in situ = 29.1 ± 0.1 °C) and austral spring (Oct–Dec, CRWTL = 30.5 °C and in situ = 29.7 °C) were ecologically significant, 1.6 °C and 0.8 °C warmer, respectively, than in situ temperatures for the same seasons. During the summer, the 1.6 °C difference between methods meant that CRWTL SSTs were above the MMM + 1 °C bleaching threshold while in situ temperatures were below the MMM. Thus, according the CRWTL, DHWs were accumulated over this period while in situ temperature remained below the bleaching threshold with no DHW accumulation (Figure 8).

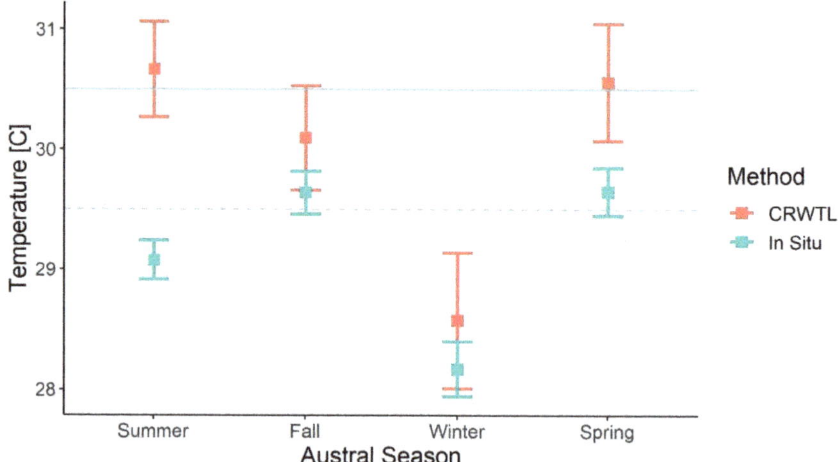


**Figure 8.** Plot of the significant season × method interaction between the Coral Reef Watch Timor-Leste virtual station (CRWTL) remotely sensed sea surface temperature and in situ temperature logger data collected between November 2015 and July 2017. Divergence in temperature data between the two methods was greatest during the austral summer. The dashed blue line indicates the maximum monthly mean (MMM, 29.5 °C) and the solid blue line is MMM + 1 °C (30.5 °C), the bleaching threshold for the accumulation of degree heating weeks. CRWTL was above the MMM + 1 °C in both the austral summer and spring seasons while the in situ temperature remained below this threshold. Error bars represent 95% confidence intervals.

During the 639 days between surveys, major heat stress events occurred. CRWTL indicated there were 190 days (30.2%) of bleaching warning (0 < DHW < 4) and 161 days (25.2%) of bleaching alert 1 (4 <= DHW < 8). The accumulation of DHWs was limited to November 2015 through July 2016 (224 days) and November 2016 through March 2017 (119 days). This corresponded to the months where the CRWTL monthly averaged temperatures were greater than the maximum monthly mean (MMM; Figure 7), the warmest month of the twelve-monthly mean SST climatology values from the CRWTL data. The accumulation of DHWs during 2015–2016 was almost 8 months, nearly twice as long as the DHW accumulation during 2016–2017. The in situ temperature data, however, never reached the MMM + 1 °C threshold for bleaching, and based on these data there would be no accumulation of DHWs.

There was a three-way interaction between site, depth, and year on bleaching prevalence [three-way repeated-measures ANOVA, $\chi^2(3) = 19.662$, $p = 0.0002$] (Figure 5). All sites at each depth showed a decrease in the prevalence of coral bleaching from one survey to the next ($p < 0.05$; Figure 5), which is expected as the second survey was conducted at the onset of austral winter. However, only Rural-E at 10 m (13.4 ± 0.7% and 2.7 ± 1.2% in 2015 and 2017, respectively) and Urban-W at 5 m (17.4 ± 1.6% and 1.8 ± 1.3%) had significant decreases in the prevalence of bleaching.

## 4. Discussion

This study provides a baseline for the condition of four outer reef slope communities that are located within ~40 km of the capital Dili, in Timor-Leste. Insights are provided for two key questions posed at the outset of this study. Firstly, coral cover and health was variable and we infer that small-scale fishing and gleaning are influencing reef health at one of the urban sites, and, second, local impacts are a greater immediate threat to Timorese reefs compared to ocean warming for the time being.

### 4.1. Coral Community Composition and Human Impacts

The underlying coral community composition was variable across the four surveyed sites. There is some evidence that increased human population density correlates with damaged reefs [98–101] and the rural sites did have greater percent coral cover compared to the urban sites, a difference which could be an indication of better reef condition. Site-level differences, however, seemed to play a greater role than rural versus urban classifications. Rural-N was distinct in having the highest coral cover and a diversity dominated by acroporids, a finding comparable to the biodiversity assessment of the same site in the 2012 Rapid Marine Assessment [102]. This site was also the only barrier reef surveyed, barrier reef formations being uncommon along the steep bathymetry of the north coast and harder to access. Anecdotally, it was observed that site-specific factors, such as ease of access to the reef, appear to be associated with indications of reduced reef health, such as reduced coral cover and coral diversity, and with an increase in compromised health states.

In addition to accessibility, geography, seasonal changes in precipitation, land-use, accumulated wave exposure, and storm exposure are likely to affect the reef community. Although these factors are important in shaping coral reefs, they were outside the scope of the study. Rural-N may be less subject to sedimentation than the other three sites as Ataúro Island does not have any significant rivers on it. Although Rural-N was the only barrier reef, all sites had steep reef slopes which is characteristic of Timorese reefs [103]. Despite the distance from large rivers for the remaining sites, coastal construction and the addition of culverts under roads could provide inputs of sediments especially at Rural-E. Large storms and waves that would have destructive effects on reefs are uncommon along the north coast [87]. The regions comprising Rural-N and Urban-E have similar wave exposure regimes of 0.5 m maximum wave height which is likely not responsible for the low coral cover compared to Rural-N (Figure 3) [104]. Temperature differences likely have a negligible influence on community composition as the temperature logger data were consistent between the three sites (Figure 7). This is with the caveat only three sites were measured; however, two years of NOAA temperature logger data measured from October 2012 to October 2014 at sites corresponding to Rural-N, Rural-E, and Urban-E were also very consistent [104]. More extreme differences in temperature are likely to shape coral reefs. This leads to the conclusion that localized human impacts play an important role in the site differences observed on these coral reefs.

Fishing is playing an increasingly significant role in Timor-Leste. Observations of extensive rubble slopes at Urban-W suggest this damage may be due to blast fishing, although the damage did not appear to be recent [102]. Gleaning is largely overlooked and an important means of food security in Timor-Leste with most (>80%) of Timor-Leste households in coastal communities gleaning [17,105,106]. Increased gleaning could also be a sign of diminishing fishing returns [107] or economic crises [108] and could result in degraded coral reef flats [109–111].

Urban-W site had the most fishing activity and showed the greatest signs of blast fishing impacts observed during the fieldwork. The subdistrict of Dom Alexio encompassing this site has the highest human population density adjacent to any of the four sites, with 4993.0 people/km$^2$ compared to 449.8 people/km$^2$ at Urban-E and 79.3 people/km$^2$ nationally. Both rural sites had less than half (<35 people/km$^2$) of the national population density. At Urban-W, the low coral cover at 5 m and the low diversity at 10 m could be attributed to the high subsistence and recreational (swimming) usage at the site. During surveys, women were observed gleaning for invertebrates on the low tide, small children were playing in the surf and on the reef flat, and men were net-fishing from small boats (Figure S4). While distance to the nearest river may be a sensible explanation for community-level differences between Rural-N and Rural-E, relative ease of access in a densely populated area differentiated Urban-W from Urban-E.

### 4.2. The Health of Coral Reefs along the North Coast of Timor-Leste

The prevalence of disease and compromised coral health was expected to be greater at urban sites with larger nutrient input and greater $\delta^{15}$N values at the shallow 5 m transects. Contrary to expectations, disease was highest at Rural-N at 5 m, with levels of WS at ~1% in both survey years. The low levels of disease detected in the current study agree with previous surveys [102,112], although no previous studies were specifically quantifying disease and compromised health. WS was the main pathology consistently observed during surveys. The WS documented at Rural-N was likely an infectious disease [26] and in the Indo-Pacific, WS is known to target acroporids [26,29,113]. Signs of WS spreading between acroporid corals were observed in the field. The pathogen causing WS at Rural-N is unknown but was likely *Vibrio* spp. a genus of bacteria that have been associated with diseases of multiple organisms including corals and humans [114–120]. Additionally, there was a positive association between host abundance and disease prevalence. This follows the classic density-dependent host–pathogen relationship [90–92]. In this study, all but one case of WS were found on acroporids. Rural-N had the highest density of Acroporids and 13 of the 17 recorded WS cases in 2015 and all 10 cases in 2017 were documented at this site [24,29,121,122]. The few cases of coral tissue mortality at other sites could have been from other causes such as unidentified predation; however, the pattern of distribution of corals affected by WS at Rural-N indicated that it was caused by an infectious pathogen.

There was likely coral mortality caused by the WS, inferred from the proportion of dead coral on some colonies (Figure 2a) [123,124], but this was likely not responsible for the decrease in coral cover in Rural-N. WS has the capacity to significantly decrease coral cover through mortality, which is associated with a much higher disease prevalence (>30%) [79]. Additionally, prevalence of WS did not differ between depths (Figure 5) and there was an increase in coral cover at Rural-N 10 m; thus, the changes in coral cover could be attributed to spatial heterogeneity recorded on non-permanent transects. WS recorded here is likely typical background levels of disease comparable to other CT locations, not an outbreak. The prevalence of WS, however, should continue to be monitored [122,125–128]. The low prevalence of coral disease in the CT supports the disease-diversity hypothesis which predicts that higher host species diversity will decrease the severity of outbreaks of a specialist pathogen [129–131]. The majority (>50%) of cases of WS were on tabulate acroporids, which are known to be the most susceptible to this syndrome. Four different acroporid morphologies were documented with WS during surveys likely encompassing different acroporid species with variable resistance to WS.

WS is a dynamic disease and can occur in outbreaks devastating acroporid populations [26,132] and thus altering overall coral community structure [132]. WS outbreaks have been linked to sediment plumes from dredging, terrestrial runoff, and elevated ocean temperature [29,30,133,134]. This is especially relevant given the recent global bleaching event and expected increase in the prevalence and severity of marine diseases given continued ocean warming [135]. A significant relationship between WS and coral bleaching co-infection was found on the GBR during the 2016–2017 global bleaching event. *Acropora* colonies that exhibited both WS and bleaching had seven times more tissue loss than solely bleached colonies [79]. Cooler temperatures could have been a protective factor against outbreaks of WS in Timor-Leste given the cooler subsurface temperatures on reefs compared to SST during the wet season. This phenomenon also coincides with the yearly ocean temperature maximum which is when corals would be most prone to bleaching. Increased sedimentation from catchments, however, is a continued threat as watersheds in Timor-Leste are degraded [20,136]. Future work assessing the downstream impacts of sedimentation on reefs and coral health is warranted.

The number of indicators of compromised health exceeded the prevalence of disease at surveyed sites. Rural-N at 5 m had the highest prevalence of non-coral invertebrate overgrowth (Figure 5); this could be explained by greater coral cover eliciting more coral-invertebrate interactions, as the cover of invertebrates was comparable between all sites. The infestation of flatworms was found at all sites, except Urban-W 10 m, with a prevalence similar to that reported from Indonesia [122] including some severe cases (Figure 2c). Although their role in coral reef environments is not well understood, flatworms consume coral mucus, reduce heterotrophic feeding, and at high densities inhibit photosynthesis [137–139]. There was also a notable absence of turf overgrowth at Rural-N, while the remaining sites had high levels which could be indicative of depauperate herbivore communities or elevated nutrients at these locations [140,141]. Competitive interactions between corals and other organisms such as boring barnacles, CCA overgrowth, and turf overgrowth were more commonly found on genera with massive morphologies such as *Platygyra*, *Montastrea*, and massive *Porites*.

*4.3. Water Quality and Sources of Nutrients in Timor-Leste*

The nutrient concentrations plus stable isotope ratios were more indicative of oceanic processes (upwelling, internal waves, etc.) than of terrestrially derived nutrient pollution. The levels of inorganic nutrients found in this study were not indicative of nutrient pollution. These values were comparable to nutrients measured in the Laclo river (~12 km east of Rural-E) in 2006 (See Supplementary Materials) [22]. Overall, nutrients were not found to be significantly elevated at surveyed sites. The sampling was also undertaken during the dry season (Mar to Nov) which would limit nutrient inputs from land and capturing elevated signals from point source pollution can be difficult.

Contrary to what was expected, combined nitrate and nitrite and phosphate averages were highest at Rural-N at 10 m; this could be a sign of upwelling nutrient-rich water [66,68,102,103,142]. Another source of nutrients at this depth could derive from submarine groundwater discharge [143]. An ephemeral bloom of cyanobacteria overgrowth was found at Urban-W 10 m in 2015 (6.1% prevalence), which can be a sign of elevated nutrients or other disturbances (ship strikes, etc.) [144–146]; however, seawater nutrients and stable isotope values were not elevated at this site suggesting the cause was not nutrient related. Although $NH_4^+$ was not significantly different between survey sites with the exception Rural-E 5 m, the range of 1.32 to 2.69 µM was greater than values between 0.3 µM and 2.2 µM [43,44,53] previously recorded for reefs in the Indo-Pacific.

The stable isotope data were consistent across sites and depths (range 2.5–5.5‰ excluding outliers) falling within the range of pristine oceanic (2–3‰) [48,147] and upwelling areas (5–6‰) [50,64–70]. The data were not indicative of $\delta^{15}N$ sewage enrichment typically ranging from 8 to 22‰ [49,52,61,148,149]. The mean $\delta^{15}N$ was significantly higher for the *Chlorodesmis* spp. at Urban-W as compared to Urban-E (Table 1). However, no *Chlorodesmis* spp. was found at Rural-N. *Halimeda* spp. were more abundant at sites and calcareous algae are good integrators of nitrogen over weeks to months versus days with fleshy macroalgae [60]. Similar values were recorded for both algae collected across sites and depths which indicates that the influx of nitrogen had been stable across several months. This is likely due to sampling being undertaken at the end of the dry season when there is little terrestrial runoff. There were a few outlier data points with much higher (12.17‰, 15.12‰) and lower (−6.79‰) $\delta^{15}N$ values recorded in *Halimeda* samples, and these could be indicative of localized inputs on a scale of tens of meters of nutrients such as fish waste or groundwater discharge. Previous studies have demonstrated that macroalgal $\delta^{15}N$ signatures decrease with depth where there is land-based pollution [50,52,57,150]. The influence of upwelling is less clear as both $\delta^{15}N$ depletion and enrichment have been reported with upwelling [50,64,68,151].

In summary, assigning direct links between the condition of coral reefs and the source of nutrients is difficult. The mean $\delta^{15}N$ values of algae sampled were higher than those reported from the open ocean. However, given the timing of collection, it is unlikely our sampling captured the effects of terrestrial run-off or of potential sewage pollution. Additionally, nutrients can be absorbed rapidly by biota and thus differences between pristine and polluted sites may not be readily apparent. Significant seasonal differences have been demonstrated for stable isotope values in macroalgae [51] so that further seasonal investigations are needed to elucidate the source of nutrients in nearshore waters.

*4.4. Elevated Temperature and the Prevalence of Bleaching from Thermal Stress*

The surveys in the present study were conducted immediately before the austral summer during the 2015 ENSO event which triggered mass bleaching globally [9]. The CRWTL virtual monitoring station indicated that the temperature began rising above the maximum monthly mean (MMM) in November 2015; however, care must be taken in interpreting such data given that the satellite only measures the temperature of the first 10–20 μm of the ocean [152] compared to loggers placed at 5 and 10 m. Satellite temperature products can be misleading in nearshore waters where pixels encompassing mostly land would be omitted. Timorese reefs are very steep and close to the coast and likely not to be included in satellite temperature products [103].

Timor-Leste appears to have experienced lower levels of bleaching compared to some other reef regions such as the Northern Great Barrier Reef (NGBR), one of the most severely affected by bleaching in 2016. The CRWTL accumulated DHWs on 55% of the days between survey periods compared 49% of days during the same time in the NGBR according to Coral Reef Watch data. However, the magnitude of DHWs in the NGBR reached 13.59 °C-weeks, more than double the 5.79 °C-weeks maximum in Timor-Leste. Comparison of in situ bleaching surveys and DHWs on the GBR indicated that 2–3 °C-weeks are associated with low levels of bleaching, >4 °C-weeks with 30–40% corals bleached, and >8 °C-weeks with a mean of 70–90% of corals bleached [72,73]. The bleaching severity of the NGBR was greater than 60% for all surveyed reefs in 2016. Although there are no data on the extent or severity of bleaching on reefs in Timor-Leste, DHW data would project mass coral bleaching in Timor-Leste of around 30–40%.

Local dive operators in Timor-Leste reported mass coral bleaching at Jaco Island, the easternmost point of the country, at the end of March. By the end of May, bleaching was affecting the majority (estimated 90%) of *Goniopora* spp. on Ataúro Island (Figure S5a), massive *Porites* spp. from 5–18 m at Jaco Island (Figure S5b), and staghorn acroporids at shallower depths in the same area. Bleaching reportedly began at shallow depths and progressively affected corals at greater depths (T. Crean, personal communication, 31 May 2016). The timing of the observed bleaching matched the in situ temperature logger timeline in which the mean monthly temperatures exceeded the MMM in March 2016. The in situ temperatures never exceeded the MMM + 1 °C threshold for DHW accumulation and mass bleaching (Figure 7). The range of temperature recorded by the loggers during December 2015 was from 27 °C to almost 31 °C, indicating reefs did experience elevated temperatures, but not for prolonged periods. The in situ mean temperature began to creep over the MMM and close the gap with the CRWTL data in March and April 2016. The in situ temperature approached MMM + 1 °C in May of 2016 five months after the CRWTL temperatures had been above the bleaching threshold (Figure 7). The in situ data are limited to the Dili and Ataúro Island areas and may not be representative of temperature regimes in the Jaco Island region. Even so, anecdotal reports that most bleaching occurred in May 2016 on both Ataúro and Jaco Islands matches the temperature timeline of the in situ temperature.

Based on the comparison of in situ temperature logger data in Timor-Leste and the satellite-derived SST, CRWTL overestimates the bleaching stress in-country. This is likely due to upwelling at the study sites during the Northwest monsoon (NWM) [85]. The north coast of the Lesser Sunda Islands is a NWM upwelling zone with intensity increasing eastward. Timor-Leste is one of the easternmost islands in the Lesser Sundas (Figure 1) and strong westerly winds (>4 m/s) promote offshore Ekman Mass Transport (3–5 m$^2$/s) on the north coast facilitating coastal upwelling. There were inconsistencies identified between SST and other oceanographic metrics analyzed in the region which was attributed to local oceanographic context and the influence of the Indonesian ThroughFlow (ITF) [85]. The ITF is strongest in the eastern Lesser Sundas [83–85,153] with the strongest currents during the austral summer/NWM which may promote mixing of the water column [154]. Vertical profiles also show upwelled less dense water masses approaching the surface around February for a short time period (~1 month) [85] which could be too short to influence remotely sensed SST measurements such as CRWTL. Additionally, ENSO was found to play a dominant role in interannual variability of NWM upwelling with decreased wind speeds and upwelling during El Niño [85,155,156].

The confluence of coastal upwelling during the season of the annual ocean temperature maximum indicates Timor-Leste could serve as a climate refugium for coral reefs against climate change-induced ocean warming as identified in other reef regions [157,158]. However, as discussed, coral bleaching did occur in Timor-Leste during the 2016–2017 marine heatwave, only not to the same extent as other reef regions. As such, cooling from upwelling would provide temporary respite as predicted warming of 2 °C from climate change would push Timorese reefs over the bleaching threshold [157]. There are negative impacts associated with upwelling such as hypercapnic ($CO_2$-rich) upwelled waters impeding the calcification and growth of corals [89,157,159]. This lower calcification rate could affect the ability of Timorese reefs to cope with sea-level rise and recovery from disturbances. There is, however, evidence that calcifying organisms can withstand seasonal increase in acidity through increased heterotrophic feeding [160,161]. Additionally, a study in the Eastern Tropical Pacific found that increased resilience to coral bleaching will potentially outweigh negative impacts to coral physiological from upwelling [157] especially as oceans warm and ENSO events become more frequent and extreme [162,163]. Further research on the complex oceanography of the region, variability of ENSO in a changing climate, and interactions between environmental (light, temperature, $CO_2$, salinity, etc.) and biological (disease, heterotrophy, calcification, etc.) parameters are required to understand and manage the country's marine resources.

## 5. Conclusions

The present study set out to understand the nature of both local and global threats to the relatively understudied coral reefs of Timor-Leste. Baseline information on these systems is limited despite the current and future importance of these marine resources to Timor-Leste. Coral reefs on the north coast of Timor-Leste are characterized by high coral cover, as much as 58.2 ± 6.4%. There is concern, however, that sites close to the urban areas of the capital city, Dili, are showing signs of degradation, since there is <5% hard coral cover at 5 m depth at one of the two urban sites. Coral disease and excess nutrients were not identified as significant causes of reef degradation at the sites surveyed in this work, although these aspects of reef health should continue to be monitored. Sites were affected by coral bleaching during a marine heatwave between the two surveys. However, in situ water temperatures were significantly lower than the CRWTL measurements, in line with previous oceanographic work identifying seasonal coastal upwelling in the region [155]. If this is so, healthy shallow reefs in such locations may serve as a climate refugia against ocean warming [157,158] which is corroborated by a global analysis including Timor-Leste as one of the 50 reef regions that are less vulnerable to climate change relative to other reefs [164]. As such, both community and national level coral reef management such as customary law (*tara bandu*) and climate mitigation policies, respectively, are necessary to ensure biodiversity is maintained while supporting coastal communities in a changing climate [165]. Although tackling climate change at an international level is still important for Timorese coral reefs, coral reef conservation efforts in-country should focus on mitigation of localized anthropogenic impacts such as sedimentation and fishing.

**Supplementary Materials:** The following supporting information can be downloaded at: https://www.mdpi.com/article/10.3390/oceans2010012/s1, Tables S1–S8; Figures S1–S5. Supplemental materials includes supporting tables, figures, discussion on relevant nutrient data, and images. References [166–200] are cited in the Supplementary Materials.

**Author Contributions:** Conceptualization, C.J.S.K. and S.D.; methodology, C.J.S.K. and S.D.; formal analysis, C.J.S.K. and S.D.; investigation, C.J.S.K.; resources, C.J.S.K.; data curation, C.J.S.K.; writing—original draft preparation, C.J.S.K.; writing—review and editing, C.J.S.K., S.D. and O.H.-G.; visualization, C.J.S.K.; supervision, S.D., C.R. and O.H.-G.; project administration, C.J.S.K.; funding acquisition, C.J.S.K. and O.H.-G. All authors have read and agreed to the published version of the manuscript.

**Funding:** This research was funded by the ARC Laureate FL 120100066 to O.H.-G.; the Society of Conservation Biology Small Grant Award to C.J.S.K.; and the Winifred Violet Scott Trust to C.J.S.K.

**Institutional Review Board Statement:** Not applicable.

**Informed Consent Statement:** Not applicable.

**Data Availability Statement:** The data presented in this study are openly available in The University of Queensland eSpace data repository at doi: 10.48610/7278446, record number UQ:7278446. The R code and associated data used for analysis can also be found in a publicly accessible GitHub repository here: https://github.com/seaCatKim/Timor_surveys20152017.

**Acknowledgments:** We gratefully acknowledge the following sources of funding in support of this research: the Australian Research Council, the Society of Conservation Biology, and the Winifred Violet Scott Trust. We thank the Ministry of Agriculture and Fisheries in Timor-Leste; Conservation International Timor-Leste; volunteers during fieldwork; and the XL CSS project logistical team. We also thank the Compass Boating & Diving and Aquatica crews. Samples were exported from Timor-Leste under export permit No. 455 and imported to Australia under AQIS import permit IP15000663. Finally, we thank those who contributed to the discussions and feedback on drafts.

**Conflicts of Interest:** The authors declare no conflict of interest. The funders had no role in the design of the study; in the collection, analyses, or interpretation of data; in the writing of the manuscript, or in the decision to publish the results.

## References

1. Burke, L.; Reytar, K.; Spalding, M.; Perry, A. *Reefs at Risk Revisited*; World Resources Institute: Washington, DC, USA, 2011; p. e0116200.
2. Veron, J.E.N.; DeVantier, L.M.; Turak, E.; Green, A.L.; Kininmonth, S.; Stafford-Smith, M.; Peterson, N. Delineating the Coral Triangle. *Galaxea J. Coral Reef Stud.* **2009**, *11*, 91–100. [CrossRef]
3. Harvell, D.; Jordán-Dahlgren, E.; Merkel, S.; Rosenberg, E.; Raymundo, L.; Smith, G.; Weil, E.; Willis, B. Coral Disease, Environmental Drivers, and the Balance Between Coral and Microbial Associates. *Oceanography* **2007**, *20*, 172–195. [CrossRef]
4. Hoegh-Guldberg, O.; Mumby, P.J.; Hooten, A.J.; Steneck, R.S.; Greenfield, P.; Gomez, E.; Harvell, C.D.; Sale, P.F.; Edwards, A.J.; Caldeira, K.; et al. Coral Reefs Under Rapid Climate Change and Ocean Acidification. *Science* **2007**, *318*, 1737–1742. [CrossRef] [PubMed]
5. Jackson, J.B.C.; Kirby, M.X.; Berger, W.H.; Bjorndal, K.A.; Botsford, L.W.; Bourque, B.J.; Bradbury, R.H.; Cooke, R.; Erlandson, J.; Estes, J.A.; et al. Historical Overfishing and the Recent Collapse of Coastal Ecosystems. *Science* **2001**, *293*, 629–637. [CrossRef]
6. Hoegh-Guldberg, O. Climate change, coral bleaching and the future of the world's coral reefs. *Mar. Freshw. Res.* **1999**, *50*, 839–866. [CrossRef]
7. Heron, S.F.; Maynard, J.A.; Van Hooidonk, R.; Eakin, C.M. Warming Trends and Bleaching Stress of the World's Coral Reefs 1985–2012. *Sci. Rep.* **2016**, *6*, 38402. [CrossRef]
8. Eakin, C.M.; Liu, G.; Gomez, A.M.; De La Cour, J.L.; Heron, S.F.; Skirving, W.J.; Geiger, E.F.; Marsh, B.L.; Tirak, K.V.; Strong, A.E.; et al. The Witch Is Dead (?)—Three Years of Global Coral Bleaching 2014–2017. *Reef Encount.* **2017**, *32*, 33–38.
9. Hughes, T.P.; Anderson, K.D.; Connolly, S.R.; Heron, S.F.; Kerry, J.T.; Lough, J.M.; Baird, A.H.; Baum, J.K.; Berumen, M.L.; Bridge, T.C.; et al. Spatial and temporal patterns of mass bleaching of corals in the Anthropocene. *Science* **2018**, *359*, 80–83. [CrossRef]
10. Ampou, E.E.; Johan, O.; Menkes, C.E.; Niño, F.; Birol, F.; Ouillon, S.; Andréfouët, S. Coral mortality induced by the 2015–2016 El-Niño in Indonesia: The effect of rapid sea level fall. *Biogeosciences* **2017**, *14*, 817–826. [CrossRef]
11. Burke, L.; Reytar, K.; Spalding, M.; Perry, A. *Reefs at Risk Revisited in the Coral Triangle*; World Resources Institute: Washington, DC, USA, 2012; pp. 1–72.
12. ADB. *State of the Coral Triangle: Timor-Leste*; Asian Development Bank: Mandaluyong City, Philippines, 2014; p. 57.
13. Macaulay, J. Timor Leste: Newest and Poorest of Asian Nations. *Geography* **2003**, *88*, 40–46.

14. Barbosa, M.; Booth, S. Timor-Leste's Fisheries Catches (1950–2009): Fisheries under Different Regimes. In *Fisheries Catch Reconstructions: Islands, Part I*; Zeller, D., Harper, S., Eds.; Fisheries Centre, University of British Columbia: Vancouver, BC, Canada, 2009; Volume 17, pp. 39–52.

15. Kingsbury, D.; Soares, D.B.; Harris, V.; Fox, J.J.; Bateman, S.; Bergin, A. A Reliable Partner: Strengthening Australia—Timor-Leste Relations. *Aust. Strateg. Policy Inst. Ltd.* **2011**, *39*, 68.

16. McWilliam, A. Perspectives on Customary Marine Tenures in East Timor. *Asia Pac. J. Anthropol.* **2002**, *3*, 6–32. [CrossRef]

17. Tilley, A.; Burgos, A.; Duarte, A.; Lopes, J.D.R.; Eriksson, H.; Mills, D. Contribution of women's fisheries substantial, but overlooked, in Timor-Leste. *AMBIO* **2021**, *50*, 113–124. [CrossRef] [PubMed]

18. Grantham, R.; Álvarez-Romero, J.G.; Mills, D.J.; Rojas, C.; Cumming, G.S. Spatiotemporal determinants of seasonal gleaning. *People Nat.* **2021**, *3*, 376–390. [CrossRef]

19. RDTL. *Timor-Leste Strategic Development Plan 2011–2030*; Republica Democratica de Timor-Leste: Dili, Timor-Leste, 2011; pp. 1–215.

20. JICA. *The Study on Community-Based Integrated Watershed Management in Laclo and Comoro River Basins in the Democratic Republic of Timor-Leste*; Japan International Cooperation Agency: Tokyo, Japan, 2010; pp. 1–43.

21. Sandlund, O.T.; Bryceson, I.; de Carvalho, D.; Rio, N.; da Silva, J.; Silva, M.I. *Assessing Environmental Needs and Priorities in Timor-Leste*; United Nations Development Programme (UNDP) Commissioned Paper; UNDP: New York, NY, USA, 2001.

22. Alongi, D.M.; Amaral, A.; de Carvalho, N.; McWilliam, A.; Rouwenhorst, J.; Tirendi, F.; Trott, L.; Wasson, R.J. *River Catchments and Marine Productivity in Timor Leste: Caraulun and Laclo Catchments; South and North Coasts—Final Report*; Ministry of Agriculture & Fisheries, Government of Timor Leste: Dili, Timor-Leste, 2012.

23. Aronson, R.; Precht, W.F. White-band disease and the changing face of Caribbean coral reefs. *Hydrobiologia* **2001**, *460*, 25–38. [CrossRef]

24. Myers, R.; Raymundo, L. Coral disease in Micronesian reefs: A link between disease prevalence and host abundance. *Dis. Aquat. Org.* **2009**, *87*, 97–104. [CrossRef]

25. Weil, E.; Irikawa, A.; Casareto, B.; Suzuki, Y. Extended geographic distribution of several Indo-Pacific coral reef diseases. *Dis. Aquat. Org.* **2012**, *98*, 163–170. [CrossRef]

26. Willis, B.L.; Page, C.A.; Dinsdale, E.A. Coral Disease on the Great Barrier Reef. In *Coral Health and Disease*; Springer: Berlin/Heidelberg, Germnay, 2004; pp. 69–104. [CrossRef]

27. Beeden, R.; Willis, B.L.; Page, C.A.; Weil, E. *Underwater Cards for Assessing Coral Health on Indo-Pacific Reefs*; Coral Reef Targeted Research and Capacity Building for Management Program, Currie Communications: Melbourne, Australia, 2008.

28. Raymundo, L.J.; Couch, C.S.; Bruckner, A.W.; Harvell, C.D. *Coral Disease Handbook Guidelines for Assessment: Guidelines for Assessment Monitoring and Management*; Currie Communications: Melbourne, Australia, 2008; ISBN 978-1-921317-01-9.

29. Bruno, J.F.; Selig, E.R.; Casey, K.; Page, C.A.; Willis, B.L.; Harvell, C.D.; Sweatman, H.; Melendy, A.M. Thermal Stress and Coral Cover as Drivers of Coral Disease Outbreaks. *PLoS Biol.* **2007**, *5*, e124. [CrossRef]

30. Pollock, F.J.; Lamb, J.; Field, S.N.; Heron, S.; Schaffelke, B.; Shedrawi, G.; Bourne, D.G.; Willis, B.L. Sediment and Turbidity Associated with Offshore Dredging Increase Coral Disease Prevalence on Nearby Reefs. *PLoS ONE* **2014**, *9*, e102498. [CrossRef]

31. Raymundo, L.J.; Halford, A.R.; Maypa, A.P.; Kerr, A.M. Functionally diverse reef-fish communities ameliorate coral disease. *Proc. Natl. Acad. Sci. USA* **2009**, *106*, 17067–17070. [CrossRef]

32. Yoshioka, R.M.; Kim, C.J.; Tracy, A.M.; Most, R.; Harvell, C.D. Linking sewage pollution and water quality to spatial patterns of Porites lobata growth anomalies in Puako, Hawaii. *Mar. Pollut. Bull.* **2016**, *104*, 313–321. [CrossRef] [PubMed]

33. Fabricius, K.E. Effects of terrestrial runoff on the ecology of corals and coral reefs: Review and synthesis. *Mar. Pollut. Bull.* **2005**, *50*, 125–146. [CrossRef] [PubMed]

34. Aronson, R.B.; Macintyre, I.G.; Wapnick, C.M.; O'Neill, M.W. Phase Shifts, Alternative States, and the Unprecedented Convergence of Two Reef Systems. *Ecology* **2004**, *85*, 1876–1891. [CrossRef]

35. Cleary, D.F.R.; Suharsono; Hoeksema, B.W. Coral diversity across a disturbance gradient in the Pulau Seribu reef complex off Jakarta, Indonesia. In *Marine, Freshwater, and Wetlands Biodiversity Conservation*; Hawksworth, D.L., Bull, A.T., Eds.; Springer: Dordrecht, The Netherlands, 2007; pp. 285–306.

36. Baker, D.; MacAvoy, S.; Kim, K. Relationship between water quality, Δ15N, and aspergillosis of Caribbean sea fan corals. *Mar. Ecol. Prog. Ser.* **2007**, *343*, 123–130. [CrossRef]

37. Thurber, R.L.V.; Burkepile, D.E.; Fuchs, C.; Shantz, A.; McMinds, R.; Zaneveld, J.R. Chronic nutrient enrichment increases prevalence and severity of coral disease and bleaching. *Glob. Change Biol.* **2013**, *20*, 544–554. [CrossRef] [PubMed]

38. Voss, J.D.; Richardson, L.L. Nutrient enrichment enhances black band disease progression in corals. *Coral Reefs* **2006**, *25*, 569–576. [CrossRef]

39. Wagner, D.; Kramer, P.; Van Woesik, R. Species composition, habitat, and water quality influence coral bleaching in southern Florida. *Mar. Ecol. Prog. Ser.* **2010**, *408*, 65–78. [CrossRef]

40. Wooldridge, S.A.; Done, T.J. Improved water quality can ameliorate effects of climate change on corals. *Ecol. Appl.* **2009**, *19*, 1492–1499. [CrossRef]

41. Amato, D.W.; Bishop, J.M.; Glenn, C.R.; Dulai, H.; Smith, C.M. Impact of Submarine Groundwater Discharge on Marine Water Quality and Reef Biota of Maui. *PLoS ONE* **2016**, *11*, e0165825. [CrossRef]

42. Dinsdale, E.A.; Pantos, O.; Smriga, S.; Edwards, R.; Angly, F.; Wegley, L.; Hatay, M.; Hall, D.; Brown, E.; Haynes, M.; et al. Microbial Ecology of Four Coral Atolls in the Northern Line Islands. *PLoS ONE* **2008**, *3*, e1584. [CrossRef]

43. Osawa, Y.; Fujita, K.; Umezawa, Y.; Kayanne, H.; Ide, Y.; Nagaoka, T.; Miyajima, T.; Yamano, H. Human impacts on large benthic foraminifers near a densely populated area of Majuro Atoll, Marshall Islands. *Mar. Pollut. Bull.* **2010**, *60*, 1279–1287. [CrossRef]

44. Smith, J.; Smith, C.; Hunter, C. An experimental analysis of the effects of herbivory and nutrient enrichment on benthic community dynamics on a Hawaiian reef. *Coral Reefs* **2001**, *19*, 332–342. [CrossRef]

45. Aeby, G.S.; Williams, G.J.; Franklin, E.C.; Kenyon, J.; Cox, E.F.; Coles, S.; Work, T.M. Patterns of Coral Disease across the Hawaiian Archipelago: Relating Disease to Environment. *PLoS ONE* **2011**, *6*, e20370. [CrossRef] [PubMed]

46. Bruno, J.F.; Petes, L.E.; Harvell, C.D.; Hettinger, A. Nutrient enrichment can increase the severity of coral diseases. *Ecol. Lett.* **2003**, *6*, 1056–1061. [CrossRef]

47. Kaczmarsky, L.; Richardson, L.L. Do elevated nutrients and organic carbon on Philippine reefs increase the prevalence of coral disease? *Coral Reefs* **2011**, *30*, 253–257. [CrossRef]

48. Costanzo, S.; O'Donohue, M.; Dennison, W.; Loneragan, N.; Thomas, M. A New Approach for Detecting and Mapping Sewage Impacts. *Mar. Pollut. Bull.* **2001**, *42*, 149–156. [CrossRef]

49. Dailer, M.L.; Ramey, H.L.; Saephan, S.; Smith, C.M. Algal $\Delta15N$ values detect a wastewater effluent plume in nearshore and offshore surface waters and three-dimensionally model the plume across a coral reef on Maui, Hawai'i, USA. *Mar. Pollut. Bull.* **2012**, *64*, 207–213. [CrossRef]

50. Lapointe, B.E.; Barile, P.J.; Littler, M.M.; Littler, D.S. Macroalgal blooms on southeast Florida coral reefs: II. Cross-shelf discrimination of nitrogen sources indicates widespread assimilation of sewage nitrogen. *Harmful Algae* **2005**, *4*, 1106–1122. [CrossRef]

51. Lapointe, B.E.; Barile, P.J.; Matzie, W.R. Anthropogenic Nutrient Enrichment of Seagrass and Coral Reef Communities in the Lower Florida Keys: Discrimination of Local versus Regional Nitrogen Sources. *J. Exp. Mar. Biol. Ecol.* **2004**, *308*, 23–58. [CrossRef]

52. Lin, H.-J.; Wu, C.; Kao, S.; Kao, W.; Meng, P. Mapping anthropogenic nitrogen through point sources in coral reefs using $\Delta15N$ in macroalgae. *Mar. Ecol. Prog. Ser.* **2007**, *335*, 95–109. [CrossRef]

53. Moynihan, M.; Baker, D.M.; Mmochi, A.J. Isotopic and microbial indicators of sewage pollution from Stone Town, Zanzibar, Tanzania. *Mar. Pollut. Bull.* **2012**, *64*, 1348–1355. [CrossRef] [PubMed]

54. Redding, J.E.; Myers-Miller, R.L.; Baker, D.M.; Fogel, M.; Raymundo, L.J.; Kim, K. Link between sewage-derived nitrogen pollution and coral disease severity in Guam. *Mar. Pollut. Bull.* **2013**, *73*, 57–63. [CrossRef] [PubMed]

55. Savage, C.; Elmgren, R. Macroalgal (fucus vesiculosus) $\Delta15n$ values trace decrease in sewage influence. *Ecol. Appl.* **2004**, *14*, 517–526. [CrossRef]

56. Sutherland, K.P.; Porter, J.W.; Turner, J.W.; Thomas, B.J.; Looney, E.E.; Luna, T.P.; Meyers, M.K.; Futch, J.C.; Lipp, E.K. Human sewage identified as likely source of white pox disease of the threatened Caribbean elkhorn coral, Acropora palmata. *Environ. Microbiol.* **2010**, *12*, 1122–1131. [CrossRef]

57. Umezawa, Y.; Miyajima, T.; Kayanne, H.; Koike, I. Significance of groundwater nitrogen discharge into coral reefs at Ishigaki Island, southwest of Japan. *Coral Reefs* **2002**, *21*, 346–356. [CrossRef]

58. Fry, B.; Baltz, D.M.; Benfield, M.C.; Fleeger, J.W.; Gace, A.; Haas, H.L.; Quiñones-Rivera, Z.J. Stable isotope indicators of movement and residency for brown shrimp (Farfantepenaeus aztecus) in coastal Louisiana marshscapes. *Estuaries* **2003**, *26*, 82–97. [CrossRef]

59. Dailer, M.L.; Knox, R.S.; Smith, J.E.; Napier, M.; Smith, C.M. Using $\Delta15N$ values in algal tissue to map locations and potential sources of anthropogenic nutrient inputs on the island of Maui, Hawai'i, USA. *Mar. Pollut. Bull.* **2010**, *60*, 655–671. [CrossRef] [PubMed]

60. Gartner, A.; Lavery, P.; Smit, A. Use of d15N signatures of different functional forms of macroalgae and filter-feeders to reveal temporal and spatial patterns in sewage dispersal. *Mar. Ecol. Prog. Ser.* **2002**, *235*, 63–73. [CrossRef]

61. Heaton, T. Isotopic studies of nitrogen pollution in the hydrosphere and atmosphere: A review. *Chem. Geol. Isot. Geosci. Sect.* **1986**, *59*, 87–102. [CrossRef]

62. Tucker, J.; Sheats, N.; Giblin, A.; Hopkinson, C.; Montoya, J. Using stable isotopes to trace sewage-derived material through Boston Harbor and Massachusetts Bay. *Mar. Environ. Res.* **1999**, *48*, 353–375. [CrossRef]

63. Montoya, J.P.; Carpenter, E.J.; Capone, D.G. Nitrogen fixation and nitrogen isotope abundances in zooplankton of the oligotrophic North Atlantic. *Limnol. Oceanogr.* **2002**, *47*, 1617–1628. [CrossRef]

64. Huang, H.; Li, X.B.; Titlyanov, E.A.; Ye, C.; Titlyanova, T.V.; Guo, Y.P.; Zhang, J. Linking macroalgal $\Delta15N$-values to nitrogen sources and effects of nutrient stress on coral condition in an upwelling region. *Bot. Mar.* **2013**, *56*, 471–480. [CrossRef]

65. Lamb, K.; Swart, P.; Altabet, M. Nitrogen and Carbon Isotopic Systematics of the Florida Reef Tract. *Bull. Mar. Sci.* **2012**, *88*, 119–146. [CrossRef]

66. Leichter, J.J.; Wankel, S.; Paytan, A.; Hanson, K.; Miller, S.; Altabet, M.A. Nitrogen and oxygen isotopic signatures of subsurface nitrate seaward of the Florida Keys reef tract. *Limnol. Oceanogr.* **2007**, *52*, 1258–1267. [CrossRef]

67. Sigman, D.M.; Altabet, M.A.; McCorkle, D.C.; Francois, R.; Fischer, G.E. The $\Delta15N$ of nitrate in the Southern Ocean: Nitrogen cycling and circulation in the ocean interior. *J. Geophys. Res. Earth Surf.* **2000**, *105*, 19599–19614. [CrossRef]

68. Firstater, F.N.; Hidalgo, F.J.; Lomovasky, B.; Tarazona, J.; Flores, G.; Iribarne, O.O. Coastal upwelling may overwhelm the effect of sewage discharges in rocky intertidal communities of the Peruvian coast. *Mar. Freshw. Res.* **2010**, *61*, 309–319. [CrossRef]

69. Radice, V.Z.; Hoegh-Guldberg, O.; Fry, B.; Fox, M.D.; Dove, S.G. Upwelling as the major source of nitrogen for shallow and deep reef-building corals across an oceanic atoll system. *Funct. Ecol.* **2019**, *33*, 1120–1134. [CrossRef]

70. Radice, V.Z.; Fry, B.; Dove, S.G.; Hoegh-Guldberg, O. Biogeochemical variability and trophic status of reef water column following a coral bleaching event. *Coral Reefs* **2021**, *40*, 1–7. [CrossRef]

71. NOAA Coral Reef Watch NOAA Coral Reef Watch Version 3.0 Daily Gobal 50km Satellite Virtual Station Time Series Data for Timor-Leste, 1 August 2015–31 August 2017. Available online: https://coralreefwatch.noaa.gov/product/vs/timeseries/coral_triangle.php#timor_leste (accessed on 15 September 2017).

72. Hughes, T.P.; Kerry, J.T.; Álvarez-Noriega, M.; Álvarez-Romero, J.G.; Anderson, K.D.; Baird, A.H.; Babcock, R.C.; Beger, M.; Bellwood, D.R.; Berkelmans, R.; et al. Global warming and recurrent mass bleaching of corals. *Nature* **2017**, *543*, 373–377. [CrossRef]

73. Strong, A.E.; Liu, G.; Skirving, W.; Eakin, C.M. NOAA's Coral Reef Watch program from satellite observations. *Ann. GIS* **2011**, *17*, 83–92. [CrossRef]

74. Liu, G.; Strong, A.E.; Skirving, W.; Arzayus, L.F. Overview of NOAA Coral Reef Watch Program's Near-Real-Time Satellite Global Coral Bleaching Monitoring Activities. In Proceedings of the 10th International Coral Reef Symposium, Okinawa, Japan; 2006; pp. 1783–1793.

75. Baird, A.H.; Marshall, P.A. Mortality, growth and reproduction in scleractinian corals following bleaching on the Great Barrier Reef. *Mar. Ecol. Prog. Ser.* **2002**, *237*, 133–141. [CrossRef]

76. Marshall, P.A.; Baird, A.H. Bleaching of corals on the Great Barrier Reef: Differential susceptibilities among taxa. *Coral Reefs* **2000**, *19*, 155–163. [CrossRef]

77. Harvell, C.D.; Kim, K.; Burkholder, J.M.; Colwell, R.R.; Epstein, P.R.; Grimes, D.J.; Hofmann, E.E.; Lipp, E.K.; Osterhaus, A.D.M.E.; Overstreet, R.M.; et al. Emerging Marine Diseases–Climate Links and Anthropogenic Factors. *Science* **1999**, *285*, 1505–1510. [CrossRef] [PubMed]

78. Harvell, C.D.; Mitchell, C.E.; Ward, J.R.; Altizer, S.; Dobson, A.P.; Ostfeld, R.S.; Samuel, M.D. Climate Warming and Disease Risks for Terrestrial and Marine Biota. *Science* **2002**, *296*, 2158–2162. [CrossRef] [PubMed]

79. Brodnicke, O.B.; Bourne, D.G.; Heron, S.F.; Pears, R.J.; Stella, J.S.; Smith, H.A.; Willis, B.L. Unravelling the links between heat stress, bleaching and disease: Fate of tabular corals following a combined disease and bleaching event. *Coral Reefs* **2019**, *38*, 591–603. [CrossRef]

80. van der Zande, R.M.; Achlatis, M.; Bender-Champ, D.; Kubicek, A.; Dove, S.; Hoegh-Guldberg, O. Paradise lost: End-of-century warming and acidification under business-as-usual emissions have severe consequences for symbiotic corals. *Glob. Change Biol.* **2020**, *26*, 2203–2219. [CrossRef]

81. Gordon, A.; Sprintall, J.; Van Aken, H.; Susanto, R.D.; Wijffels, S.; Molcard, R.; Ffield, A.; Pranowo, W.; Wirasantosa, S. The Indonesian throughflow during 2004–2006 as observed by the INSTANT program. *Dyn. Atmos. Oceans* **2010**, *50*, 115–128. [CrossRef]

82. Taufiqurrahman, E.; Wahyudi, A.J.; Masumoto, Y. The Indonesian Throughflow and its Impact on Biogeochemistry in the Indonesian Seas. *ASEAN J. Sci. Technol. Dev.* **2020**, *37*, 29–35. [CrossRef]

83. Sprintall, J.; Wijffels, S.; Molcard, R.; Jaya, I. Direct estimates of the Indonesian Throughflow entering the Indian Ocean: 2004–2006. *J. Geophys. Res. Earth Surf.* **2009**, *114*, 114. [CrossRef]

84. Susanto, D.; Wei, Z.; Adi, R.; Zhang, Q.; Fang, G.; Fan, B.; Supangat, A.; Agustiadi, T.; Li, S.; Trenggono, M.; et al. Oceanography Surrounding Krakatau Volcano in the Sunda Strait, Indonesia. *Oceanography* **2016**, *29*, 264–272. [CrossRef]

85. Wirasatriya, A.; Susanto, R.D.; Kunarso, K.; Jalil, A.R.; Ramdani, F.; Puryajati, A.D. Northwest monsoon upwelling within the Indonesian seas. *Int. J. Remote Sens.* **2021**, *42*, 5433–5454. [CrossRef]

86. RDTL. *Population and Housing Census 2015 Preliminary Results*; Democratic Republic of Timor-Leste (RDTL): Dili, Timor-Leste, 2015; p. 39.

87. DNMG; BOM; CSIRO. *Current and Future Climate of Timor-Leste*; Timor-Leste National Directorate of Meteorology and Geophysics: Dili, Timor-Leste, 2015; pp. 1–8.

88. United Nations Development Program; Democratic Republic of Timor-Leste. *National Coastal Vulnerability Assessment and Designing of Integrated Coastal Management and Adaptation Strategic Plan for Timor-Leste*; United Nations Development Programme: Dili, Timor-Leste, 2018; p. 73.

89. PIFSC. *Interdisciplinary Baseline Ecosystem Assessment Surveys to Inform Ecosystem-Based Management Planning in Timor-Leste: Final Report*; NOAA Pacific Islands Fisheries Science Center: Honolulu, HI, USA, 2017; p. 234.

90. English, S.; Wilkinson, C.; Baker, V. *Survey Manual for Tropical Marine Resources*; Australian Institute of Marine Science: Townsville, Australia, 1997.

91. Zvuloni, A.; Armoza-Zvuloni, R.; Loya, Y. Structural deformation of branching corals associated with the vermetid gastropod Dendropoma maxima. *Mar. Ecol. Prog. Ser.* **2008**, *363*, 103–108. [CrossRef]

92. The R Core Team. *R: A Language and Environment for Statistical Computing*; R Foundation for Statistical Computing: Vienna, Austria, 2020.

93. Anderson, M.; Gorley, R.N.; Clarke, K. *PERMANOVA+ for PRIMER: Guide to Software and Statistical Methods*; PRIMER-E Ltd.: Plymouth, UK, 2008.

94. Clarke, K.R.; Gorley, R.N. *Getting Started with PRIMER V7*; PRIMER-E: Plymouth, UK, 2015.

95. Fox, J.; Weisberg, S. *An {R} Companion to Applied Regression*; Sage: Thousand Oaks, CA, USA, 2019.

96. Lenth, R. Emmeans: Estimated Marginal Means, Aka Least-Squares Means. 2020. Available online: https://cran.r-project.org/web/packages/emmeans/index.html (accessed on 6 March 2022).

97. Pinheiro, J.; Bates, D.; Debroy, S.; Sarkar, D.; R Core Team. *Nlme: Linear and Nonlinear Mixed Effects Models*; R Core Team: Vienna, Austria, 2020.

98. Brown, K.; Bender-Champ, D.; Bryant, D.E.; Dove, S.; Hoegh-Guldberg, O. Human activities influence benthic community structure and the composition of the coral-algal interactions in the central Maldives. *J. Exp. Mar. Biol. Ecol.* **2017**, *497*, 33–40. [CrossRef]

99. Bruno, J.F.; Valdivia, A. Coral reef degradation is not correlated with local human population density. *Sci. Rep.* **2016**, *6*, 29778. [CrossRef]

100. Smith, J.E.; Brainard, R.; Carter, A.; Grillo, S.; Edwards, C.; Harris, J.; Lewis, L.; Obura, D.; Rohwer, F.; Sala, E.; et al. Re-evaluating the health of coral reef communities: Baselines and evidence for human impacts across the central Pacific. *Proc. R. Soc. B Boil. Sci.* **2016**, *283*, 20151985. [CrossRef]

101. Wedding, L.M.; Lecky, J.; Gove, J.M.; Walecka, H.R.; Donovan, M.K.; Williams, G.J.; Jouffray, J.-B.; Crowder, L.B.; Erickson, A.; Falinski, K.; et al. Advancing the integration of spatial data to map human and natural drivers on coral reefs. *PLoS ONE* **2018**, *13*, e0189792. [CrossRef]

102. Erdmann, M.V.; Mohan, C. (Eds.) *A Rapid Marine Biological Assessment of Timor-Leste, RAP Bulletin of Biological Assessment 66*; Coral Triangle Support Partnership, Conservation International: Dili, Timor-Leste, 2013; ISBN 978-1-934151-56-3.

103. Boggs, G.; Edyvane, K.; de Carvalho, N.; Penny, S.; Rouwenhorst, J.; Brocklehurst, P.; Cowie, I.; Barreto, C.; Amaral, A.; Monteiro, J.; et al. *Marine and Coastal Habitat Mapping in Timor Leste (North Coast)—Final Report*; Ministry of Agriculture & Fisheries, Government of Timor Leste: Dili, Timor-Leste, 2012.

104. Kim, C.J.S. Drivers of Coral Reef Composition, Cryptic Marine Biodiversity, and Coral Health along the North Coast of Timor-Leste. Ph.D. Thesis, The University of Queensland, St. Lucia, Australia, 2021.

105. Teh, L.S.L.; Teh, L.C.L.; Sumaila, U.R. A Global Estimate of the Number of Coral Reef Fishers. *PLoS ONE* **2013**, *8*, e65397. [CrossRef]

106. Da Costa, M.D.; Lopes, M.; Ximenes, A.; Ferreira, A.D.R.; Spyckerelle, L.; Williams, R.; Nesbitt, H.; Erskine, W. Household food insecurity in Timor-Leste. *Food Secur.* **2013**, *5*, 83–94. [CrossRef]

107. Cesar, H.; Burke, L.; Pet-Soede, L. *The Economics of Worldwide Coral Reef Degradation*; Cesar Environmental Economics Consulting, Arnhem, and WWF-Netherlands: Arnhem, The Netherlands, 2003; p. 23.

108. Gillett, R. *Fisheries in the Economies of the Pacific Island Countries and Territories*; Asia Development Bank: Mandaluyong, Philippines, 2009.

109. Andréfouët, S.; Guillaume, M.M.M.; Delval, A.; Rasoamanendrika, F.M.A.; Blanchot, J.; Bruggemann, J.H. Fifty years of changes in reef flat habitats of the Grand Récif of Toliara (SW Madagascar) and the impact of gleaning. *Coral Reefs* **2013**, *32*, 757–768. [CrossRef]

110. Ashworth, J.S.; Ormond, R.F.; Sturrock, H.T. Effects of reef-top gathering and fishing on invertebrate abundance across take and no-take zones. *J. Exp. Mar. Biol. Ecol.* **2004**, *303*, 221–242. [CrossRef]

111. Woodland, D.; Hooper, J. The effect of human trampling on coral reefs. *Biol. Conserv.* **1977**, *11*, 1–4. [CrossRef]

112. Ayling, A.M.; Ayling, A.L.; Edyvane, K.S.; Penny, S.; de Carvalho, N.; Fernandes, A.; Amaral, A.L. Preliminary Biological Resource Survey of Fringing Reefs in the Proposed Nino Konis Santana Marine Park, Timor-Leste. *Rep. North. Territ. Dep. Nat. Resour. Environ. Arts. Palmerst. North. Territ.* **2009**, *830*.

113. Maynard, J.A.; Van Hooidonk, R.; Eakin, C.M.; Puotinen, M.; Garren, M.; Williams, G.J.; Heron, S.; Lamb, J.; Weil, E.; Willis, B.L.; et al. Projections of climate conditions that increase coral disease susceptibility and pathogen abundance and virulence. *Nat. Clim. Change* **2015**, *5*, 688–694. [CrossRef]

114. Amaro, C.; Biosca, E.G. Vibrio vulnificus biotype 2, pathogenic for eels, is also an opportunistic pathogen for humans. *Appl. Environ. Microbiol.* **1996**, *62*, 1454–1457. [CrossRef] [PubMed]

115. Cervino, J.M.; Hayes, R.L.; Polson, S.W.; Polson, S.C.; Goreau, T.J.; Martinez, R.J.; Smith, G.W. Relationship of Vibrio Species Infection and Elevated Temperatures to Yellow Blotch/Band Disease in Caribbean Corals. *Appl. Environ. Microbiol.* **2004**, *70*, 6855–6864. [CrossRef]

116. Linkous, D.A.; Oliver, J.D. Pathogenesis of Vibrio vulnificus. *FEMS Microbiol. Lett.* **1999**, *174*, 207–214. [CrossRef]

117. Milton, D.L.; Norqvist, A.; Wolf-Watz, H. Cloning of a metalloprotease gene involved in the virulence mechanism of Vibrio anguillarum. *J. Bacteriol.* **1992**, *174*, 7235–7244. [CrossRef]

118. Sussman, M.; Willis, B.L.; Victor, S.; Bourne, D.G. Coral Pathogens Identified for White Syndrome (WS) Epizootics in the Indo-Pacific. *PLoS ONE* **2008**, *3*, e2393. [CrossRef]

119. Ushijima, B.; Videau, P.; Burger, A.H.; Shore, A.; Runyon, C.M.; Sudek, M.; Aeby, G.S.; Callahan, S.M. Vibrio coralliilyticus Strain OCN008 Is an Etiological Agent of Acute Montipora White Syndrome. *Appl. Environ. Microbiol.* **2014**, *80*, 2102–2109. [CrossRef]

120. Ushijima, B.; Smith, A.; Aeby, G.S.; Callahan, S.M. Vibrio owensii Induces the Tissue Loss Disease Montipora White Syndrome in the Hawaiian Reef Coral Montipora capitata. *PLoS ONE* **2012**, *7*, e46717. [CrossRef]

121. Aeby, G.S.; Ross, M.; Williams, G.J.; Lewis, T.D.; Work, T.M. Disease dynamics of Montipora white syndrome within Kaneohe Bay, Oahu, Hawaii: Distribution, seasonality, virulence, and transmissibility. *Dis. Aquat. Org.* **2010**, *91*, 1–8. [CrossRef] [PubMed]

122. Haapkylä, J.; Unsworth, R.; Seymour, A.; Melbourne-Thomas, J.; Flavell, M.; Willis, B.; Smith, D. Spatio-temporal coral disease dynamics in the Wakatobi Marine National Park, South-East Sulawesi, Indonesia. *Dis. Aquat. Org.* **2009**, *87*, 105–115. [CrossRef] [PubMed]

123. Aeby, G.S. Outbreak of coral disease in the Northwestern Hawaiian Islands. *Coral Reefs* **2005**, *24*, 481. [CrossRef]

124. Roff, G.; Hoegh-Guldberg, O.; Fine, M. Intra-colonial response to Acroporid "white syndrome" lesions in tabular *Acropora* spp. (Scleractinia). *Coral Reefs* **2006**, *25*, 255–264. [CrossRef]

125. Haapkylä, J.; Seymour, A.; Trebilco, J.; Smith, D. Coral disease prevalence and coral health in the Wakatobi Marine Park, south-east Sulawesi, Indonesia. *J. Mar. Biol. Assoc. UK* **2007**, *87*, 403–414. [CrossRef]

126. Johan, O.; Bengen, D.G.; Zamani, N.P.; Suharsono; Sweet, M. The Distribution and Abundance of Black Band Disease and White Syndrome in Kepulauan Seribu, Indonesia. *HAYATI J. Biosci.* **2015**, *22*, 105–112. [CrossRef]

127. Muller, E.M.; Raymundo, L.J.; Willis, B.L.; Haapkylä, J.; Yusuf, S.; Wilson, J.R.; Harvell, D.C. Coral Health and Disease in the Spermonde Archipelago and Wakatobi, Sulawesi. *J. Indones. Coral Reefs* **2012**, *1*, 147–159.

128. Muller, E.M.; Van Woesik, R. Caribbean coral diseases: Primary transmission or secondary infection? *Glob. Chang. Biol.* **2012**, *18*, 3529–3535. [CrossRef]

129. Aeby, G.S.; Bourne, D.G.; Wilson, B.; Work, T. Coral Diversity and the Severity of Disease Outbreaks: A Cross-Regional Comparison of Acropora White Syndrome in a Species-Rich Region (American Samoa) with a Species-Poor Region (Northwestern Hawaiian Islands). *J. Mar. Biol.* **2011**, *2011*, 490198. [CrossRef]

130. Elton, C.S. *The Biology of Invasions by Animals and Plants*; John Wiley and Sons: New York, NY, USA, 1958.

131. Plank, J.E. *Plant Diseases-Epidemics and Control*; Academic Press: New York, NY, USA, 1963.

132. Hobbs, J.-P.A.; Frisch, A.J.; Newman, S.; Wakefield, C.B. Selective Impact of Disease on Coral Communities: Outbreak of White Syndrome Causes Significant Total Mortality of Acropora Plate Corals. *PLoS ONE* **2015**, *10*, e0132528. [CrossRef]

133. Sheridan, C.; Grosjean, P.; Leblud, J.; Palmer, C.V.; Kushmaro, A.; Eeckhaut, I. Sedimentation rapidly induces an immune response and depletes energy stores in a hard coral. *Coral Reefs* **2014**, *33*, 1067–1076. [CrossRef]

134. Heron, S.F.; Willis, B.L.; Skirving, W.J.; Eakin, C.M.; Page, C.A.; Miller, I.R. Summer Hot Snaps and Winter Conditions: Modelling White Syndrome Outbreaks on Great Barrier Reef Corals. *PLoS ONE* **2010**, *5*, e12210. [CrossRef]

135. Altizer, S.; Ostfeld, R.S.; Johnson, P.T.J.; Kutz, S.; Harvell, C.D. Climate Change and Infectious Diseases: From Evidence to a Predictive Framework. *Science* **2013**, *341*, 514–519. [CrossRef] [PubMed]

136. Alongi, D.M.; da Silva, M.; Wasson, R.J.; Wirasantosa, S. Sediment discharge and export of fluvial carbon and nutrients into the Arafura and Timor Seas: A regional synthesis. *Mar. Geol.* **2013**, *343*, 146–158. [CrossRef]

137. Barneah, O.; Brickner, I.; Hooge, M.; Weis, V.; LaJeunesse, T.; Benayahu, Y. Three party symbiosis: Acoelomorph worms, corals and unicellular algal symbionts in Eilat (Red Sea). *Mar. Biol.* **2007**, *151*, 1215–1223. [CrossRef]

138. Naumann, M.S.; Mayr, C.; Struck, U.; Wild, C. Coral mucus stable isotope composition and labeling: Experimental evidence for mucus uptake by epizoic acoelomorph worms. *Mar. Biol.* **2010**, *157*, 2521–2531. [CrossRef]

139. Wijgerde, T.; Schots, P.; Van Onselen, E.; Janse, M.; Karruppannan, E.; Verreth, J.; Osinga, R. Epizoic acoelomorph flatworms impair zooplankton feeding by the scleractinian coral Galaxea fascicularis. *Biol. Open* **2013**, *2*, 10–17. [CrossRef]

140. McClanahan, T. Primary succession of coral-reef algae: Differing patterns on fished versus unfished reefs. *J. Exp. Mar. Biol. Ecol.* **1997**, *218*, 77–102. [CrossRef]

141. Vermeij, M.J.A.; Van Moorselaar, I.; Engelhard, S.; Hörnlein, C.; Vonk, S.M.; Visser, P.M. The Effects of Nutrient Enrichment and Herbivore Abundance on the Ability of Turf Algae to Overgrow Coral in the Caribbean. *PLoS ONE* **2010**, *5*, e14312. [CrossRef]

142. Leichter, J.J.; Stewart, H.L.; Miller, S.L. Episodic nutrient transport to Florida coral reefs. *Limnol. Oceanogr.* **2003**, *48*, 1394–1407. [CrossRef]

143. Risk, M.J.; Lapointe, B.; Sherwood, O.A.; Bedford, B.J. The use of $\Delta15N$ in assessing sewage stress on coral reefs. *Mar. Pollut. Bull.* **2009**, *58*, 793–802. [CrossRef] [PubMed]

144. Burgett, J. *Summary of Algal Community Changes Observed on the Southwest Arm of Rose Atoll from 1995–2002*; USFWS: Honolulu, HI, USA, 2012.

145. Thacker, R.; Ginsburg, D.; Paul, V. Effects of herbivore exclusion and nutrient enrichment on coral reef macroalgae and cyanobacteria. *Coral Reefs* **2001**, *19*, 318–329. [CrossRef]

146. Littler, M.M.; Littler, D.S.; Brooks, B. Harmful algae on tropical coral reefs: Bottom-up eutrophication and top-down herbivory. *Harmful Algae* **2006**, *5*, 565–585. [CrossRef]

147. Titlyanov, E.A.; Kiyashko, S.I.; Titlyanova, T.V.; Van Huyen, P.; Yakovleva, I.M. Identifying nitrogen sources for macroalgal growth in variously polluted coastal areas of southern Vietnam. *Bot. Mar.* **2011**, *54*, 367–376. [CrossRef]

148. Costanzo, S.D.; Udy, J.; Longstaff, B.; Jones, A. Using nitrogen stable isotope ratios ($\Delta15N$) of macroalgae to determine the effectiveness of sewage upgrades: Changes in the extent of sewage plumes over four years in Moreton Bay, Australia. *Mar. Pollut. Bull.* **2005**, *51*, 212–217. [CrossRef]

149. Thornber, C.S.; Kinlan, B.; Graham, M.H.; Stachowicz, J.J. Population ecology of the invasive kelp Undaria pinnatifida in California: Environmental and biological controls on demography. *Mar. Ecol. Prog. Ser.* **2004**, *268*, 69–80. [CrossRef]

150. Smith, J.; Runcie, J.; Smith, C. Characterization of a large-scale ephemeral bloom of the green alga Cladophora sericea on the coral reefs of West Maui, Hawai'i. *Mar. Ecol. Prog. Ser.* **2005**, *302*, 77–91. [CrossRef]

151. Fourqurean, J.W.; Moore, T.O.; Fry, B.; Hollibaugh, J.T. Spatial and Temporal Variation in C:N:P Ratios, Δ15N, and Δ13C of Eelgrass Zostera Marina as Indicators of Ecosystem Processes, Tomales Bay, California, USA. *Mar. Ecol. Prog. Ser.* **1997**, *157*, 147–157. [CrossRef]
152. Liu, G.; Rauenzahn, J.L.; Heron, S.; Eakin, C.M.; Skirving, W.J.; Christensen, T.R.L.; Strong, A.E.; Li, J. *NOAA Coral Reef Watch 50 Km Satellite Sea Surface Temperature-Based Decision Support System for Coral Bleaching Management*; NOAA/NERDIS: College Park, MD, USA, 2013.
153. Sprintall, J.; Gordon, A.L.; Wijffels, S.; Feng, M.; Hu, S.; Koch-Larrouy, A.; Phillips, H.; Nugroho, D.; Napitu, A.; Pujiana, K.; et al. Detecting Change in the Indonesian Seas. *Front. Mar. Sci.* **2019**, *6*, 257. [CrossRef]
154. Shinoda, T.; Han, W.; Metzger, E.J.; Hurlburt, H.E. Seasonal Variation of the Indonesian Throughflow in Makassar Strait. *J. Phys. Oceanogr.* **2012**, *42*, 1099–1123. [CrossRef]
155. Wirasatriya, A.; Prasetyawan, I.B.; Triyono, C.D.; Muslim; Maslukah, L. Effect of ENSO on the variability of SST and Chlorophyll-a in Java Sea. *IOP Conf. Ser. Earth Environ. Sci.* **2018**, *116*, 012063. [CrossRef]
156. Dewi, Y.W.; Wirasatriya, A.; Sugianto, D.N.; Helmi, M.; Marwoto, J.; Maslukah, L. Effect of ENSO and IOD on the Variability of Sea Surface Temperature (SST) in Java Sea. In *IOP Conference Series: Earth and Environmental Science*; IOP Publishing: Bristol, UK, 2020; Volume 530, p. 012007.
157. Randall, C.J.; Toth, L.T.; Leichter, J.J.; Maté, J.L.; Aronson, R.B. Upwelling buffers climate change impacts on coral reefs of the eastern tropical Pacific. *Ecology* **2020**, *101*, e02918. [CrossRef] [PubMed]
158. Glynn, P.W.; Maté, J.L.; Baker, A.C.; Calderón, M.O. Coral Bleaching and Mortality in Panama and Ecuador during the 1997–1998 El Niño–Southern Oscillation Event: Spatial/Temporal Patterns and Comparisons with the 1982–1983 Event. *Bull. Mar. Sci.* **2001**, *69*, 79–109.
159. Feely, R.A.; Sabine, C.L.; Hernandez-Ayon, J.M.; Ianson, D.; Hales, B. Evidence for Upwelling of Corrosive "Acidified" Water onto the Continental Shelf. *Science* **2008**, *320*, 1490–1492. [CrossRef] [PubMed]
160. Leichter, J.; Genovese, S. Intermittent upwelling and subsidized growth of the scleractinian coral Madracis mirabilis on the deep fore-reef slope of Discovery Bay, Jamaica. *Mar. Ecol. Prog. Ser.* **2006**, *316*, 95–103. [CrossRef]
161. Rixen, T.; Jiménez, C.; Cortés, J. Impact of upwelling events on the sea water carbonate chemistry and dissolved oxygen concentration in the Gulf of Papagayo (Culebra Bay), Costa Rica: Implications for coral reefs. *RBT* **2015**, *60*, 187–195. [CrossRef]
162. Kim, S.T.; Cai, W.; Jin, F.-F.; Santoso, A.; Wu, L.; Guilyardi, E.; An, S.-I. Response of El Niño sea surface temperature variability to greenhouse warming. *Nat. Clim. Change* **2014**, *4*, 786–790. [CrossRef]
163. Cai, W.; Wang, G.; Dewitte, B.; Wu, L.; Santoso, A.; Takahashi, K.; Yang, Y.; Carréric, A.; McPhaden, M.J. Increased variability of eastern Pacific El Niño under greenhouse warming. *Nature* **2018**, *564*, 201–206. [CrossRef]
164. Beyer, H.L.; Kennedy, E.V.; Beger, M.; Chen, C.A.; Cinner, J.E.; Darling, E.S.; Eakin, C.M.; Gates, R.D.; Heron, S.F.; Knowlton, N.; et al. Risk-sensitive planning for conserving coral reefs under rapid climate change. *Conserv. Lett.* **2018**, *11*, e12587. [CrossRef]
165. Kim, C.J. Coral Reef Management and Tara Bandu on Ataúro Island: An Ecologist's Perspective. In Proceedings of the 2020 Timor-Leste Studies Association-Portugal Conference, TLSA-PT, Dili, Timor-Leste, 8 April 2021.
166. Lamb, J.B.; Willis, B.L.; Fiorenza, E.A.; Couch, C.S.; Howard, R.; Rader, D.N.; True, J.D.; Kelly, L.A.; Ahmad, A.; Jompa, J.; et al. Plastic waste associated with disease on coral reefs. *Science* **2018**, *359*, 460–462. [CrossRef]
167. Lamb, J.B.; van de Water, J.A.J.M.; Bourne, D.G.; Altier, C.; Hein, M.Y.; Fiorenza, E.A.; Abu, N.; Jompa, J.; Harvell, C.D. Seagrass ecosystems reduce exposure to bacterial pathogens of humans, fishes, and invertebrates. *Science* **2017**, *355*, 731–733. [CrossRef] [PubMed]
168. Sabdono, A.; Radjasa, O.K.; Trianto, A.; Sarjito; Munasik; Wijayanti, D.P. Preliminary study of the effect of nutrient enrichment, released by marine floating cages, on the coral disease outbreak in Ka-rimunjawa, Indonesia. *Reg. Stud. Mar. Sci.* **2019**, *30*, 100704. [CrossRef]
169. Raymundo, L.J.; Licuanan, W.L.; Kerr, A.M. Adding insult to injury: Ship groundings are associated with coral disease in a pristine reef. *PLoS ONE* **2018**, *13*, e0202939. [CrossRef]
170. Raymundo, L.J.; Diaz, R.; Miller, A.; Reynolds, T. *Baseline Surveys of Proposed and Established Marine Sanctuaries on Bantayan Island, Northern Cebu*; University of Guam: Mangilao, Guam, 2011; p. 64.
171. Kaczmarsky, L. Coral disease dynamics in the central Philippines. *Dis. Aquat. Org.* **2006**, *69*, 9–21. [CrossRef] [PubMed]
172. Raymundo, L.J.; Rosell, K.B.; Reboton, C.T.; Kaczmarsky, L. Coral diseases on Philippine reefs: Genus Porites is a dominant host. *Dis. Aquat. Org.* **2005**, *64*, 181–191. [CrossRef] [PubMed]
173. Raymundo, L. Porites ulcerative white spot disease: Description, prevalence, and host range of a new coral disease affecting Indo-Pacific reefs. *Dis. Aquat. Org.* **2003**, *56*, 95–104. [CrossRef] [PubMed]
174. Miller, J.; Sweet, M.; Wood, E.; Bythell, J. Baseline coral disease surveys within three marine parks in Sabah, Borneo. *PeerJ* **2015**, *3*, 1391. [CrossRef]
175. Green, E.P.; Bruckner, A.W. The significance of coral disease epizootiology for coral reef conservation. *Biol. Conserv.* **2000**, *96*, 347–361. [CrossRef]
176. Williams, G.J.; Work, T.M.; Aeby, G.S.; Knapp, I.S.; Davy, S.K. Gross and microscopic morphology of lesions in Cnidaria from Palmyra Atoll, Central Pacific. *J. Invertebr. Pathol.* **2011**, *106*, 165–173. [CrossRef]
177. Yasuda, N.; Nakano, Y.; Yamashiro, H.; Hidaka, M. Skeletal structure and progression of growth anomalies in Porites australiensis in Okinawa, Japan. *Dis. Aquat. Org.* **2012**, *97*, 237–247. [CrossRef]

178. Aeby, G.S. The Potential Effect on the Ability of a Coral Intermediate Host to Regenerate Has Had on the Evolution of Its Association with a Marine Parasite. In Proceedings of the 7th International Coral Reef Symposium, Guam, FSM, USA, 22–27 June 1992; pp. 809–815.
179. Aeby, G.S. Corals in the genus Porites are susceptible to infection by a larval trematode. *Coral Reefs* **2003**, *22*, 216. [CrossRef]
180. Bergsma, G.S. Tube-dwelling coral symbionts induce significant morphological change in Montipora. *Symbiosis* **2009**, *49*, 143–150. [CrossRef]
181. Floros, C.; Samways, M.; Armstrong, B. Polychaete (*Spirobranchus giganteus*) loading on South African corals. *Aquat. Conserv. Mar. Freshw. Ecosyst.* **2005**, *15*, 289–298. [CrossRef]
182. Scott, P.J.B.; Risk, M.J. The effect of Lithophaga (Bivalvia: Mytilidae) boreholes on the strength of the coral Porites lobata. *Coral Reefs* **1988**, *7*, 145–151. [CrossRef]
183. Hoeksema, B.W.; Farenzena, Z.T. Tissue loss in corals infested by acoelomorph flatworms (*Waminoa* sp.). *Coral Reefs* **2012**, *31*, 869. [CrossRef]
184. Haapkylä, J.; Seymour, A.S.; Barneah, O.; Brickner, I.; Hennige, S.; Suggett, D.; Smith, D. Association of *Waminoa* sp. (Acoela) with corals in the Wakatobi Marine Park, South-East Sulawesi, Indonesia. *Mar. Biol.* **2009**, *156*, 1021–1027. [CrossRef]
185. Rodríguez-Villalobos, J.; Work, T.; Calderón-Aguilera, L.; Reyes-Bonilla, H.; Hernández, L. Explained and unexplained tissue loss in corals from the Tropical Eastern Pacific. *Dis. Aquat. Org.* **2015**, *116*, 121–131. [CrossRef]
186. Keats, D.W.; Chamberlain, Y.M.; Baba, M. *Pneophyllum conicum* (Dawson) comb. nov. (Rhodophyta, Corallinaceae), a Widespread Indo-Pacific Non-Geniculate Coralline Alga that Overgrows and Kills Live Coral. *Bot. Mar.* **1997**, *40*, 263–280. [CrossRef]
187. Finckh, A.E. Biology of the Reef-Forming Organisms at Funafuti Atoll. In *The Atoll of Funafuti*; Royal Society: London, UK, 1904; pp. 125–150.
188. Smith, T.; Nemeth, R.; Blondeau, J.; Calnan, J.; Kadison, E.; Herzlieb, S. Assessing coral reef health across onshore to offshore stress gradients in the US Virgin Islands. *Mar. Pollut. Bull.* **2008**, *56*, 1983–1991. [CrossRef]
189. Kuffner, I.B.; Walters, L.J.; Becerro, M.A.; Paul, V.J.; Ritson-Williams, R.; Beach, K.S. Inhibition of coral recruitment by macroalgae and cyanobacteria. *Mar. Ecol. Prog. Ser.* **2006**, *323*, 107–117. [CrossRef]
190. Cetz-Navarro, N.P.; Espinoza-Avalos, J.; Hernández-Arana, H.A.; Carricart-Ganivet, J.P. Biological Responses of the Coral Montastraea annularis to the Removal of Filamentous Turf Algae. *PLoS ONE* **2013**, *8*, e54810. [CrossRef]
191. Loh, T.-L.; Pawlik, J. Friend or foe? No evidence that association with the sponge Mycale laevis provides a benefit to corals of the genus Montastraea. *Mar. Ecol. Prog. Ser.* **2012**, *465*, 111–117. [CrossRef]
192. Rützler, K.; Muzik, K. Terpios Hoshinota, a New Cyanobacteriosponge Threatening Pacific Reefs. *Sci. Mar.* **1993**, *57*, 395–403.
193. Bak, R.; Lambrechts, D.; Joenje, M.; Nieuwland, G.; Van Veghel, M. Long-term changes on coral reefs in booming populations of a competitive colonial ascidian. *Mar. Ecol. Prog. Ser.* **1996**, *133*, 303–306. [CrossRef]
194. Littler, M.M.; Littler, D.S. A Colonial Tunicate Smothers Corals and Corraline Algae on the Great Astrolabe Reef, Fiji. *Coral Reefs* **1995**, *14*, 148–149. [CrossRef]
195. Vargas-Ángel, B.; Godwin, L.S.; Asher, J.; Brainard, R.E. Invasive didemnid tunicate spreading across coral reefs at remote Swains Island, American Sāmoa. *Coral Reefs* **2009**, *28*, 53. [CrossRef]
196. Ravindran, J.; Raghukumar, C. Pink-line syndrome, a physiological crisis in the scleractinian coral Porites lutea. *Mar. Biol.* **2006**, *149*, 347–356. [CrossRef]
197. Eyre, B. Nutrient Biogeochemistry in the Tropical Moresby River Estuary System North Queensland, Australia. *Estuar. Coast. Shelf Sci.* **1994**, *39*, 15–31. [CrossRef]
198. Mitchell, A.W.; Bramley, R.; Johnson, A.K.L. Export of nutrients and suspended sediment during a cyclone-mediated flood event in the Herbert River catchment, Australia. *Mar. Freshw. Res.* **1997**, *48*, 79–88. [CrossRef]
199. Robertson, A.I.; Dixon, P.; Alongi, D.M. The Influence of Fluvial Discharge on Pelagic Production in the Gulf of Papua, Northern Coral Sea. *Estuar. Coast. Shelf Sci.* **1998**, *46*, 319–331. [CrossRef]
200. Alongi, D.M.; Boto, K.G.; Robertson, A.I. Nitrogen and phosphorus cycles. In *Ecosystems at the Land-Sea Margin: Drainage Basin to Coastal Sea*; Robertson, A.I., Alongi, D.M., Eds.; American Geophysical Union: Washington, DC, USA, 1992; Volume 41, pp. 251–292.

*Article*

# Mapping Sub-Metre 3D Land-Sea Coral Reefscapes Using Superspectral WorldView-3 Satellite Stereoimagery

**Antoine Collin [1,2,*], Mark Andel [3], David Lecchini [2,4] and Joachim Claudet [2,5]**

[1] Coastal GeoEcological Lab, Ecole Pratique des Hautes Etudes (EPHE), PSL University, 35800 Dinard, France
[2] Laboratoire d'Excellence "CORAIL", 66100 Perpignan, France; david.lecchini@ephe.psl.eu (D.L.); joachim.claudet@cnrs.fr (J.C.)
[3] Digitalglobe Foundation, Westminster, CO 80234, USA; mark.andel@digitalglobe.com
[4] CRIOBE, EPHE-UPVD-CNRS, PSL University, 98729 Moorea, French Polynesia
[5] Maison des Océans, CRIOBE, EPHE-UPVD-CNRS, PSL University, 75005 Paris, France
* Correspondence: antoine.collin@ephe.psl.eu

**Abstract:** Shallow coral reefs ensure a wide portfolio of ecosystem services, from fish provisioning to tourism, that support more than 500 million people worldwide. The protection and sustainable management of these pivotal ecosystems require fine-scale but large-extent mapping of their 3D composition. The sub-metre spaceborne imagery can neatly produce such an expected product using multispectral stereo-imagery. We built the first 3D land-sea coral reefscape mapping using the 0.3 m superspectral WorldView-3 stereo-imagery. An array of 13 land use/land cover and sea use/sea cover habitats were classified using sea-, ground- and air-truth data. The satellite-derived topography and bathymetry reached vertical accuracies of 1.11 and 0.89 m, respectively. The value added of the eight mid-infrared (MIR) channels specific to the WorldView-3 was quantified using the classification overall accuracy (OA). With no topobathymetry, the best combination included the eight-band optical (visible + near-infrared) and the MIR8, which boosted the basic blue-green-red OA by 9.58%. The classes that most benefited from this MIR information were the land use "roof" and land cover "soil" classes. The addition of the satellite-derived topobathymetry to the optical+MIR1 produced the best full combination, increasing the basic OA by 9.73%, and reinforcing the "roof" and "soil" distinction.

**Keywords:** satellite; superspectral; VHR; topobathymetry; LULC; SUSC; Moorea Island

**Citation:** Collin, A.; Andel, M.; Lecchini, D.; Claudet, J. Mapping Sub-Metre 3D Land-Sea Coral Reefscapes Using Superspectral WorldView-3 Satellite Stereoimagery. *Oceans* **2021**, *2*, 315–329. https://doi.org/10.3390/oceans2020018

Academic Editor: Rupert Ormond

Received: 29 October 2020
Accepted: 25 March 2021
Published: 2 April 2021

**Publisher's Note:** MDPI stays neutral with regard to jurisdictional claims in published maps and institutional affiliations.

## 1. Introduction

Tropical shallow hard coral reef ecosystems provide numerous and valuable services to local socio-economies, such as fish and seafood provisioning, coastal protection, or wealthy recreational activities [1]. These services have been estimated to support more than 500 million people worldwide [2]. Even though coral reefs cover only 0.1% of the oceans, they host 25% of all marine identified species [3]. However, anthropocenic changes, embodied by both sea level, sea temperature and sea acidification rises and also sedimentation related to watershed deforestation and land claiming, are strongly threatening these pivotal ecosystems [4].

The protection and sustainable management of these ecosystems requires us to adopt an integrated view of the seamless land- and seascape at a high spatial resolution, adequate to meet local stakeholders' expectations [5]. Even if the global, thus coarse (>1 m pixel size), products are insightful for assessing coral reef trends, the very high spatial (i.e., <1 m pixel size) mapping of the land use land cover (LULC) and the sea use sea cover (SUSC) constitutes a fitting response to needs of local users, managers and decision-makers. Either passive or active, airborne imagery can successfully provide some spectro-spatial combinations able to generate coastal topography and bathymetry using unmanned airborne vehicles [6,7], and map sub-metre LULC and SUSC using hyper-/multi-spectral camera [8] or multi-spectral light detection and ranging (LiDAR) system [9]. However, the manned or unmanned airborne limitations, tied to the elevation-specific flight planning and the relatively small surveyed area, impede their utilization for regional mapping [10]. The sub-metre spaceborne imagery has emerged as a tool of interest given its capability to capture large extents with a very high spatial resolution, despite its purchase cost [11]. Around 2000, IKONOS (1999) and QuickBird-2 (2001) became the first satellite sensors collecting imagery at 1 m pixel size across regional scales. These civilian and commercial pioneers were followed by sub-metre United States WorldView-1, -2, -3, -4 (2007, 2009, 2014, 2016), GeoEye-1 (2008), SkySat series (2013–2017), French Pleiades-1A and -1B (2011 and 2012), Korean Kompsat-3 and -3A (2012 and 2015), United Kingdom TripleSat (2015), and Chinese Gaofen-2 (2014), Jilin-1 (2015), Superview-1 (2018) [12]. In addition to their spatial resolution capability inherent to the panchromatic band, most of these sensors acquire four spectral bands: the visible (VIS) blue, green, red (BGR), and the optical near-infrared (NIR). Three outliers thereupon appear: the panchromatic WorldView-1, the optical 8-band WorldView-2, and the optical+mid-infrared (MIR) 16-band WorldView-3. The WorldView-2 improved the bathymetry mapping [5], the coral cover and health mapping [13,14], and the seamless LULC/SUSC mapping [11]. The WorldView-3 augmented the bathymetry [15], mineral [16], hydrocarbon [17], lithological [18], salt marsh [19], tropical forest [20], coral reef [8], and even urban plastic [21] mapping.

Furthermore, the spaceborne sub-metre LULC mapping was significantly enhanced by the (tri-)stereo-acquisition of the same scene, offering the opportunity to produce seamless land-sea digital surface models (DSMs), using the photogrammetry for land and the ratio transform for sea [22]. Horizontal and vertical accuracies of the land DSM-derived stereo-Pleiades-1 have been quantified at 0.53 and 0.65 m, respectively [12]. The addition of the topographic band to the spectral information has been shown to significantly improve spaceborne sub-metre LULC mapping [12]. Even if the novelty of the latter work relied on the sole use of a spaceborne stereo-imagery, the bathymetry and the SUSC mapping were not examined. An integration of the terrestrial and marine DSM into the spaceborne sub-metre spectral dataset was elsewhere useful in mapping the seamless coral reefscape in Japan using Google Earth imagery [23], but it was not derived from a sole spaceborne by-product. To our knowledge, a unique study has focused on the land-sea coral reefscape mapping using a sole spaceborne sub-metre stereo-imagery [24].

Despite the use of the WorldView-3 stereo-imagery to produce land-sea DSM, the authors had not previously investigated the added value of the 16-band superspectral dataset to map LULC and SUSC, simultaneously. In this paper, we innovatively propose to classify sub-metre LULC and SUSC of a coral reefscape using a sole spaceborne stereo-imagery, from which the topographic, bathymetric and superspectral information are derived. The scene studied was acquired over the complex coral reefscape of Moorea Island (French Polynesia, South Pacific) using a WorldView-3 stereo-imagery (Figure 1). The chosen area exhibits representative eight LULC and five SUSC classes, and encompasses steep volcanic vegetated watersheds, flat rural coastal areas, and a reef-dominated lagoon. An set of five issues will be considered: (1) the added value of the Coastal and yellow bands to the basic BGR classification accuracy; (2) the added value of the Red Edge (RE), NIR1 and NIR2 bands to the basic BGR classification accuracy; (3) the added value of the MIR bands to the basic BGR classification accuracy; (4) the influence of the topobathymetry (i.e., land-sea DSM) on the basic, visible, optical and optical+MIR datasets' classification performance; and (5) all four previous questions considered at the class-level.

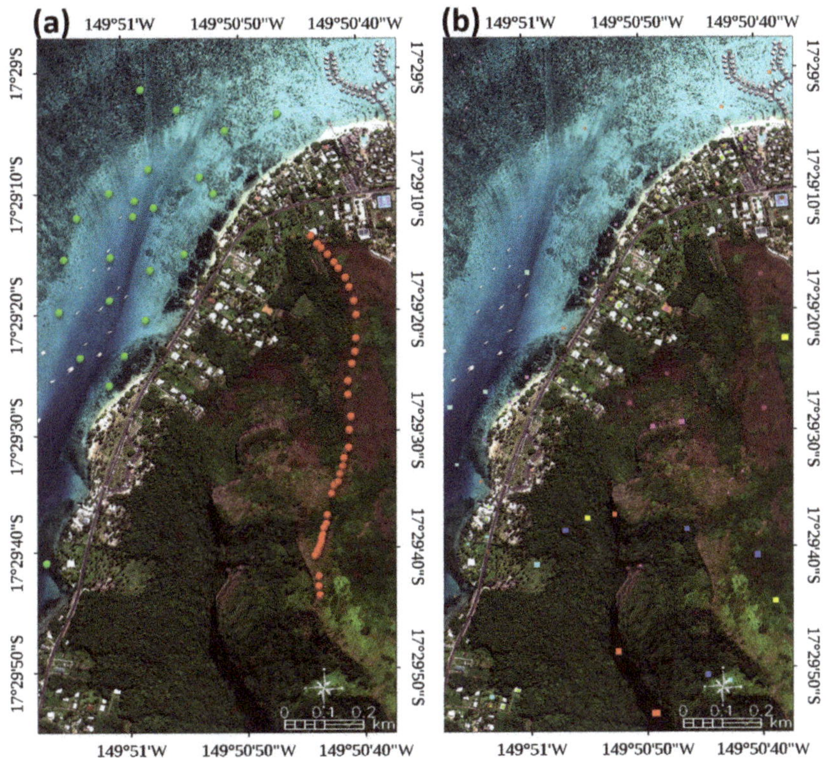



**Figure 1.** Natural-coloured WorldView-3 imagery (0.3 m × 0.3 m, 3017 × 5937 pixels) of the study area on Moorea Island (French Polynesia). (**a**) The red and green spheres represent 32 topographic and 35 bathymetric calibration/validation datasets; (**b**) the array of 105 multi-colour rectangles represents 78,000 pixels of 13 habitats, each one composed by 3000 calibration and 3000 validation pixels.

## 2. Materials and Methods

### 2.1. Study Site

The study site is located on the north shore of Moorea Island (17°32′ S, 149°50′ W) in French Polynesia (Figure 1). Moorea is a 1.6 million-year-old volcanic island which at its highest point reaches 1207 m and extends over 187 km$^2$, divided into 134 km$^2$ and 53 km$^2$ of land and sea areas, respectively. While being in the vicinity of Tahiti, the capital of French Polynesia, Moorea, is considered as a life-size laboratory, given its large array of land-sea spatial patterns and multi-scale socio-ecological processes [25]. The spatial features include rain and dry forests, volcanic and laterite soils, coconut and banana crops, urban infrastructures, coralligeneous sand, reef pavement, fringing and barrier reefs. The territory is changing rapidly due to the doubling of the local population in 40 years [26], the conversion of forest to pineapple crops and the urban growth. The lagoon hosts traditional fishing activities and is experiencing an increase in tourism activities. The test area, extending over 1.61 km$^2$, is composed of a complex land-sea coral reefscape, selected to embrace all the previously mentioned components.

### 2.2. Land-, Sea- and Air-Truth Data

The topobathymetry extraction requires XYZ control points for both land and sea realms [24]. The topographic DSM was calibrated and validated by 20 and 12 ground control points (see red spheres in Figure 1a), surveyed in September 2018 with a Mobile Mapper 120 provided with a 20 Hz Differential-GPS+GLONASS position output, ensuring a maintained 0.3 m accuracy. The sampling distribution was not optimal because it was constrained by the need to use a single pathway that ran along a crest surrounded by steep ravines. The bathymetric DSM was calibrated and validated by 20 and 15 sea control points, retrieved from the digitized French Navy chart (identified as 6657, based on a 1966-to-1972 hydrographic campaign originally referenced to the lowest astronomical tide, see green spheres in Figure 1a). Despite the gap in timing between the waterborne soundings and the spaceborne acquisition, the absence of major local events in the seascape enabled the freely available hydrographic soundings to be used, for the sake of transferability. Both topographic and bathymetric control points were then horizontally referenced to the UTM 6 South projection into the WGS-84 datum, and vertically zeroed to the mean sea level.

A suite of 13 habitats (Figure 1b) were inspected using geolocated handborne photoquadrats for land classes and geolocated airborne for sea classes. Photoquadrats were taken with an Olympus Tough TG-4 provided with BGR bands (16 million pixels each), while aerial pictures were monitored using a DJI Mavic Pro Platinum collecting BGR bands (12 million pixels) at 35 m altitude (height above the mean sea level). A series of 90 photoquadrats and 90 aerial photographs were orthorectified [8] to distinguish 10 classes, representative of an average land-sea coral reefscape (Table 1). Five land and five sea classes were each constituted of 30 and 30 seed pixels neighbourly and evenly grown to 3000 calibration and 3000 validation pixels, respectively. The three remaining classes, namely forest, roof and shadow, were straightforwardly characterized by 3000 calibration and 3000 validation pixels visually selected on the satellite imagery.

**Table 1.** Description of the 13 land use/land cover and sea use/sea cover classes.

| Class Name | Class Description | Class Colour |
| --- | --- | --- |
| Forest | Wet arborescent stratum | |
| Wood | Wet arbustive stratum | |
| Grass | Wet herbaceous stratum | |
| Dry vegetation | Wind and sun exposed forest, wood or grass | |
| Soil | Bare volcanic or lateritic substratum | |
| Roof | Wooden or metallic house covering | |
| Road | Tarmac way | |
| Shadow | Tree or house shading | |
| Backshore | Emerged coral sand | |
| Foreshore | Shallow coral sand | |
| Nearshore | Mid coral sand | |
| Offshore | Deep coral sand | |
| Coral reefs | Scleractinian and coralline algae | |

*2.3. Spaceborne Dataset*

Launched on 13 August 2014, the commercial WorldView-3 spearheads the choice of sub-metre civilian satellite, given its hyperspatial and superspectral capabilities, namely one panchromatic band at 0.3 m, five VIS + three NIR bands at 1.2 m, and eight MIR bands at 3.7 m. The pansharpening technique can successfully produce 16 spectral bands at 0.3 m spatial resolution ([27], Figure 2).

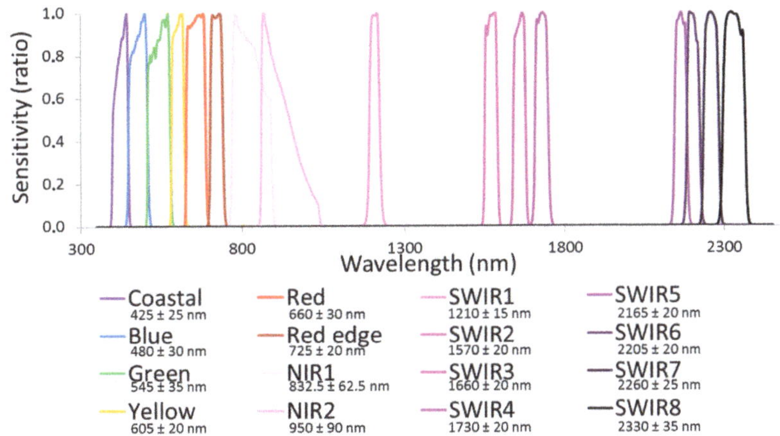

Figure 2. Lineplot of the superspectral (16 bands) WorldView-3 sensor's sensitivity as a function of their wavelengths.

Provided with a daily revisit, this sensor leverages a swath width of 13.1 km and length of 112 km. The WorldView-3 dataset used here is a stereo-imagery acquired on 18 July 2018 at 20:35:38 UTC (Figure 3a, Table 2) and 20:36:39 UTC, respectively (Figure 3b, Table 2).

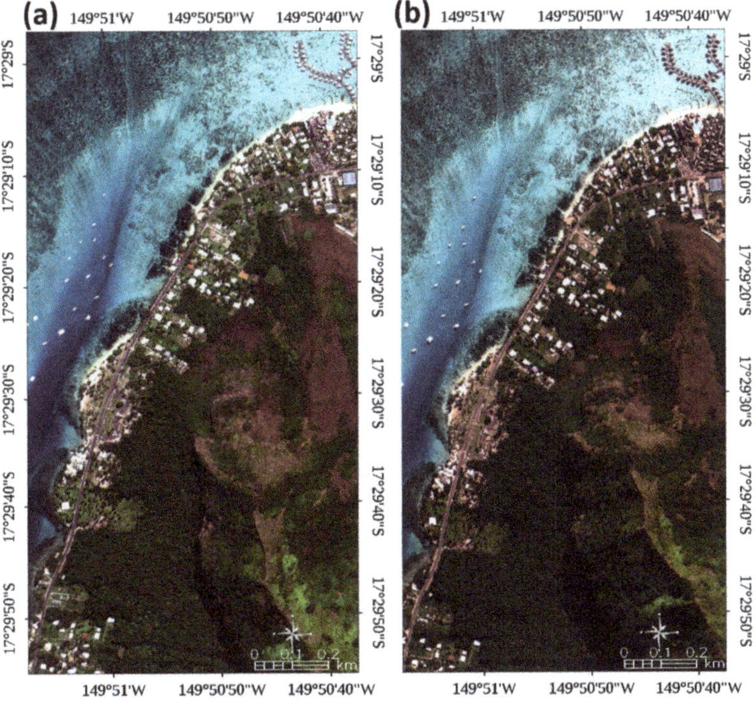

Figure 3. Natural-coloured WorldView-3 imageries of the study area, taken on 18 July 2018, over the north shore of Moorea Island (French Polynesia). (a) Imagery #1 taken with 26.4° off-nadir viewing angle; (b) Imagery #2 taken with 16.5° off-nadir viewing angle.

**Table 2.** WorldView-3 specifications related to the stereo-imagery acquisition over the study site.

| Parameters | Imagery #1 | Imagery #2 |
|---|---|---|
| Date | 12 July 2018 | 12 July 2018 |
| Time (UTC) | 20:35:38 | 20:36:39 |
| Mean viewing angle | | |
| In-track (in °) | 23.6 | −12.6 |
| Cross-track (in °) | 12.2 | 10.8 |
| Off-nadir (in °) | 26.4 | 16.5 |
| Satellite azimuth (in °) | 34.3 | 148.6 |
| Satellite elevation (in °) | 60.7 | 71.9 |
| Sun azimuth (in °) | 29.9 | 29.7 |
| Sun elevation (in °) | 45.0 | 45.1 |

### 2.4. Spaceborne Topographic DSM

The building of the satellite-based topographic DSM requires both panchromatic imageries to be radiometrically converted from digital number to top-of-atmosphere radiance, then to bottom-of-atmosphere reflectance values by considering the calibration factors (.IMD file), the atmosphere composition and sun irradiance (see [5] for details). The reflectance imageries were used to retrieve a 3D point cloud using a dense point matching algorithm [28]. The matching algorithm seeks for the pairwise pixels of two imageries by shortening the epipolar 2D to 1D, based on the rational polynomial coefficients (RPCs), then by reducing the length of the epipolar line with the global multi-resolution terrain elevation data 2010 dataset. The point cloud was then gridded at 0.3 m by converting the XY coordinates into the WGS84 datum, UTM zone 6S, and referencing in Z to the mean sea level (Figure 4). The topographic validation accuracy was estimated by the mean absolute error (MAE) and the root mean square error (RMSE) between the modeled and observed values ($N$ = 12). The MAE and RMSE attained 0.84 and 1.11 m, which corroborates the results from previous WorldView-3 works [24].

### 2.5. Spaceborne Bathymetric DSM

The creation of the satellite-based bathymetric DSM relies on the use of a single multispectral imagery that needs to be both radiometrically and geometrically corrected, as well as pansharpened. The imagery #2 was selected for this purpose since it displayed the lowest off-nadir viewing angle (Table 2). Following the radiometric correction (see Section 2.4), the VIS+NIR multispectral reflectance imagery was orthorectified using the RPCs and the 20 ground control points, as well as the corresponding panchromatic reflectance imagery. Among seven sharpening methods, the Gram-Schmidt pansharpening procedure yielded the best visual results, and it was then implemented so as to produce a VIS+NIR dataset at 0.3 m pixel size (see [5] for details). The resulting sub-metre eight-band dataset was subjected to the radiative transfer model, called the ratio transform [29]. This standard bathymetric model makes use of the fact that light absorption by water varies with wavebands. It can thereafter determine the bathymetry ($z$), as follows:

$$z = a_1 \left( \frac{ln[R_w(\lambda_i) - R_\infty(\lambda_i)]}{ln[R_w(\lambda_j) - R_\infty(\lambda_j)]} \right) - a_0 \tag{1}$$

where $a_0$ is the intercept to match the mean sea level, $a_1$ is the slope converting the relative to absolute bathymetry (20 calibration sea control points), $R_w$ is the reflectance related to the waveband $\lambda_i$, and $R_\infty$ is the reflectance over deep water. The MAE and RMSE bathymetric validation ($N$ = 15) accuracy reached 0.74 and 0.89 m, echoing the findings from previous WorldView-3 studies [15,24].

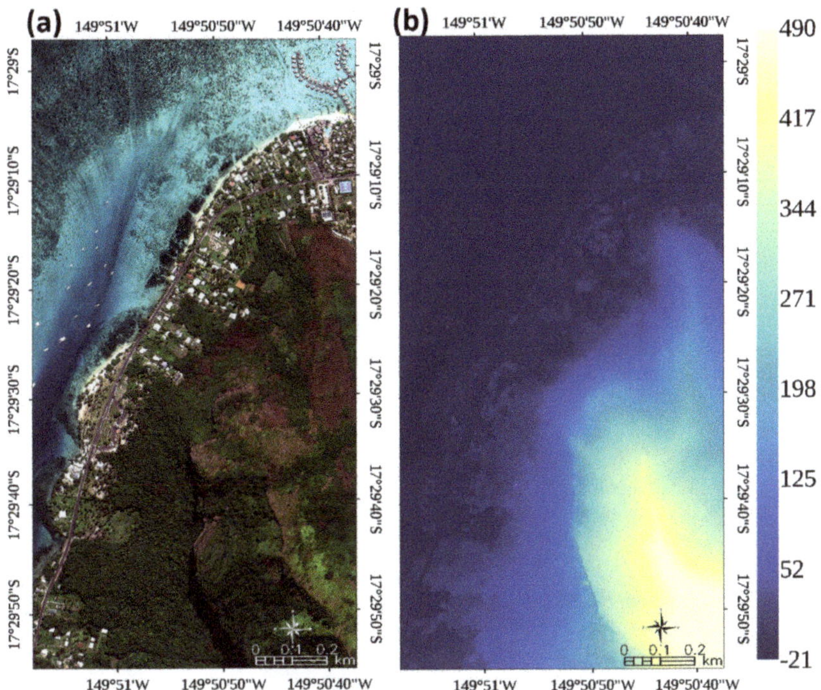

**Figure 4.** (**a**) Natural-coloured orthorectified WorldView-3 imagery of the study area; (**b**) topo-bathymetric digital surface model (in m) derived from the combination of a photogrammetry-based stereo-panchromatic imagery for land and a ratio transform model for sea, both derived from the same WorldView-3 stereo-imagery. The colour scale on the right indicates the estimated height and depth below sea-level of different parts of the image.

### 2.6. Habitat Classification

The habitat mapping stemmed from the superspectral capabilities of the WorldView-3 sensor. In addition to the pansharpened optical (VIS+NIR) reflectance, the eight-band MIR contribution to landscape (and not seascape given its water absorption) mapping was tested, which necessitated radiometric and geometric corrections purposed to the pansharpening enhancement (see [19] for WorldView-3 pansharpening). An output of 16 spectral bands at 0.3 m was used as input predictors for a supervised classification based on the commonly used probabilistic maximum likelihood (ML) algorithm. This learner assumes that the statistics for each class in each spectral band are normally distributed, enabling the probability that a given pixel belongs to a specific class to be estimated.

The 3000 calibration pixels per class were used to build the ML model, while the 3000 validation pixels per class were intended for computing the confusion matrix (CM), from which the omission, commission misclassification (or error) and overall accuracy (OA) were drawn. These accuracy metrics were based only on the multi-colour rectangular regions of interest (see Figure 1b), and not on the whole scene. The omission misclassification corresponded to the rate at which sites were erroneously omitted from the correct class in the classified map, while the commission misclassification embodied the rate at which sites were correctly classified as ground-truth sites but were erroneously omitted from the correct class in the classified map. The OA and CM were used to analyze gain patterns at the scene and the class scale, respectively. First, the VIS, NIR and MIR spectral contributions were assessed by, respectively, adding the Coastal and yellow, RE-NIR1-NIR2, and MIR1-MIR2-MIR3-MIR4-MIR5-MIR6-MIR7-MIR8 bands, to the basic BGR combination.

Second, the spatial contribution of the land-sea DSM was evaluated for the three spectral blendings. Third, the best predictions were further analyzed at the class level.

## 3. Results

Firstly, the contributions of the WorldView-3 spectral VIS, NIR and MIR were quantified for the LULC-SUSC classification accuracy. Secondly, the WorldView-3 spatial topobathymetric DSM was evaluated. Thirdly, the highest spectral and topobathymetric effects were assessed at the class scale.

### 3.1. WorldView-3 Superspectral Land-Sea Habitat Mapping

The classification performance of the BGR basic dataset provided a satisfactory OA of 89.15% (Figure 5a). On the one hand, the addition of the coastal and yellow predictors, yielding the VIS combination, increased the basic OA by 2.69% (Figure 5b). On the other hand, the addition of the RE, NIR1 and NIR2 to the VIS combination, so as to produce the optical dataset, augmented the basic OA by 8.79% (Figure 5c).

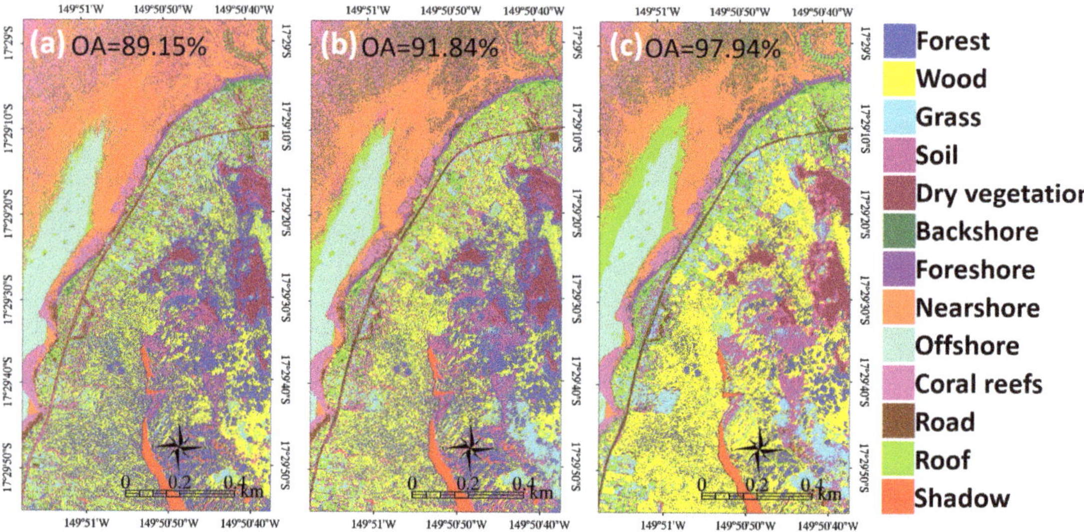


**Figure 5.** Maximum likelihood classification maps of the land-sea coral reefscape using WorldView-3: (**a**) basic dataset (blue-green-red); (**b**) visible dataset (Coastal-blue-green-yellow-red); (**c**) optical dataset (Coastal-blue-green-yellow-red-Red Edge-Near-InfraRed1-Near-InfraRed2).

The addition of the eight MIR spectral bands, individually, to the optical dataset showed a high mean contribution of 9.52% (Figure 6). However, some subtle patterns could be highlighted: the MIR8, MIR7, MIR2, MIR5 contributed to a gain of the basic OA (9.58%, 9.57%, 9.53% and 9.51%, respectively); the MIR3, MIR4 and MIR6 contributed to the mean increase of 9.5%; and the MIR1 contributed to an enhancement of 9.47%. Given that the optical boosting to the basic OA was 8.79%, all MIR bands brought novel information to discriminate LULC-SUSC.

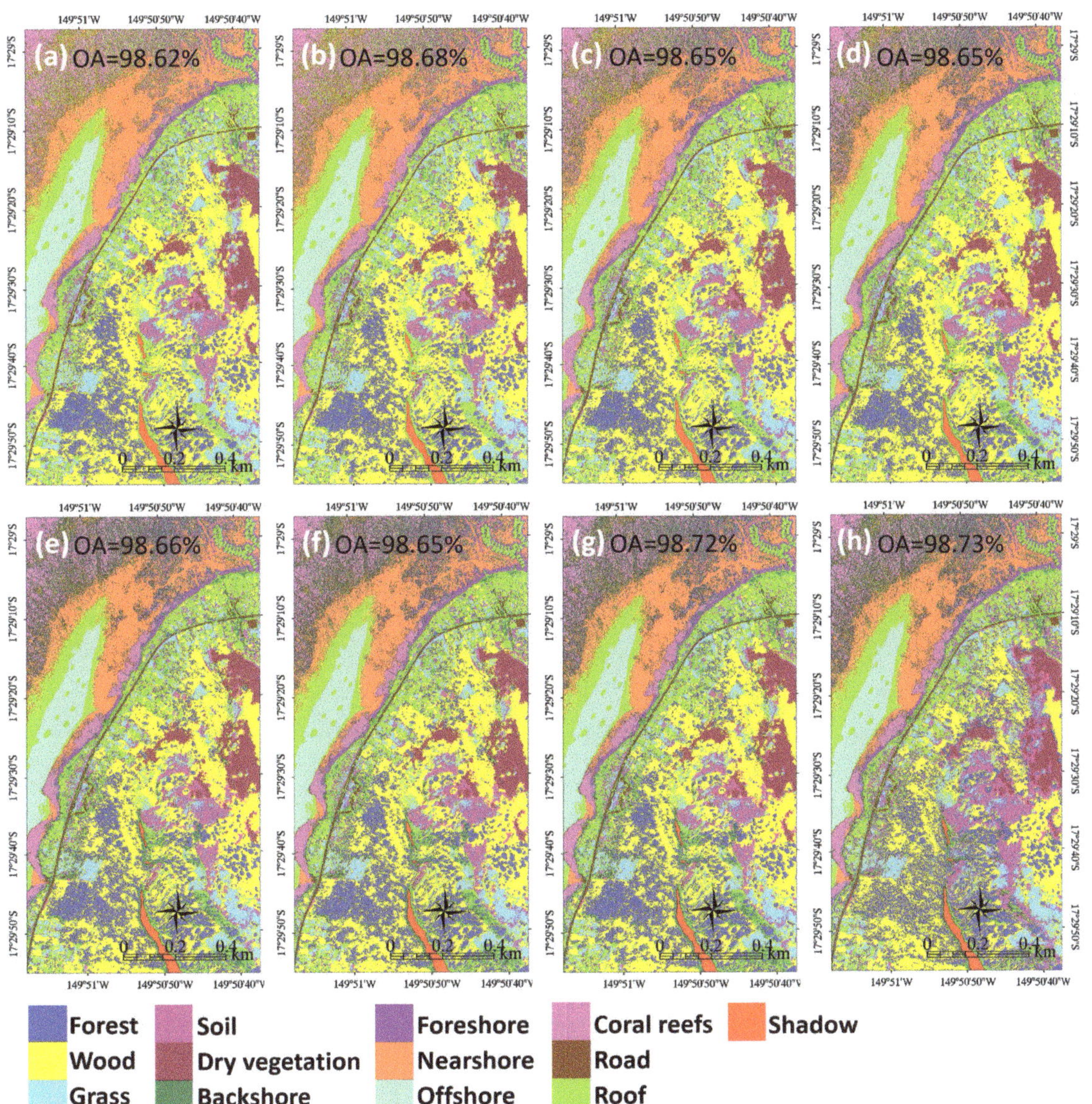



**Figure 6.** Maximum likelihood classification maps of the land-sea coral reefscape using WorldView-3 optical dataset with: (**a**) Mid-InfraRed1 (MIR1); (**b**) MIR2; (**c**) MIR3; (**d**) MIR4; (**e**) MIR5; (**f**) MIR6; (**g**) MIR7; (**h**) MIR8.

### 3.2. *WorldView-3 Topobathymetry into Land-Sea Habitat Mapping*

The evaluation of the influence of the WorldView-3-derived topobathymetry followed the same procedure as in 3.1. The addition of the DSM to the BGR basic dataset increased the OA by 2.74% (Figure 7a). The addition of the DSM to the VIS combination augmented the basic OA by 5.44% (Figure 7b). The addition of the DSM to the optical dataset enhanced the basic OA by 9.15% (Figure 7c).

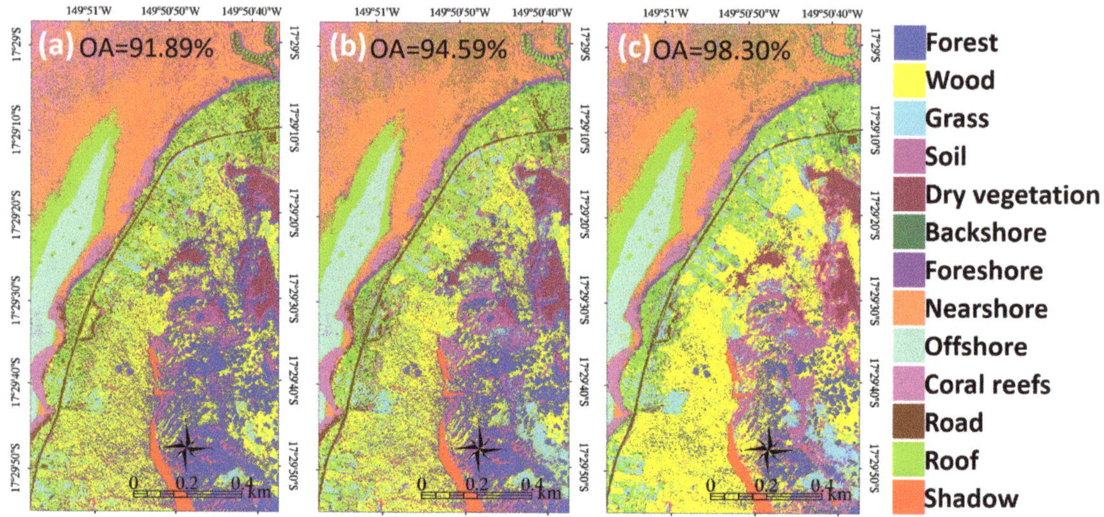



**Figure 7.** Maximum likelihood classification maps of the land-sea coral reefscape using WorldView-3 topobathymetric digital surface model with: (**a**) basic dataset (blue-green-red); (**b**) visible dataset (Coastal-blue-green-yellow-red); (**c**) optical dataset (Coastal-blue-green-yellow-red-Red Edge-Near-InfraRed1-Near-InfraRed2).

The addition of the DSM to the eight MIR spectral bands (individually added to the optical dataset) showed a high mean contribution of 9.55% (Figure 8), slightly better than the MIR bands with no DSM (Figure 6). Some variations still appeared: the MIR1, MIR8, and MIR7 contributed to the best gains of the basic OA (9.73%, 9.69%, and 9.60%, respectively); the MIR5 and MIR6 contributed to the mean boost of 9.53% and 9.51%; and the MIR4, MIR2 and MIR3 contributed to the least increases of 9.47%, 9.45% and 9.42%. Since the DSM-added optical contribution to the basic OA was 9.15%, all MIR bands continued to provide insights for improving LULC-SUSC mapping.

### 3.3. WorldView-3 Land-Sea Habitat Mapping at the Class Scale

The best OA for the WorldView-3 multispectral and superspectral, deprived of and provided with the inner topobathymetry, were further studied at the class level by comparing their confusion matrices with that for the basic BGR (Figure 9).

Concerning the multispectral level, the optical dataset strongly improved the discrimination of the grass (27.7%) and wood (19.27%), as well as the road (15.93%). The coral reefs also benefited from a better differentiation (1.2%), partly due to the decline in omission misclassification with the offshore (−0.93%). The addition of the topobathymetric DSM confirmed the greater distinction between previous classes (grass, 27.5%; wood, 26.27%; and road, 19.27%), but also improved the ability to detect the roof (23.87%) and soil (8.53%) classes. The coral reefs' separability was also heightened (1.87%) owing to the decrease in omission misclassification with the nearshore (−0.57%).

Regarding the superspectral degree, the optical dataset enhanced with the MIR8 empowered to distinguish grass (27.6%), wood (25.53%), roof (22.87%), road (20.03%), and soil (14.6%) classes. The coral reefs were also better discerned (2.07%) given the reduction in omission misclassification with the nearshore (−0.93%). The integration of the DSM information to the optical+MIR1 dataset still upgraded the classification of the grass (27.33%), wood (27.1%), roof (24.5%), road (20.67%), and soil (12.97%). Nevertheless, the coral reefs' classification remained rigorously constant (2.07%).

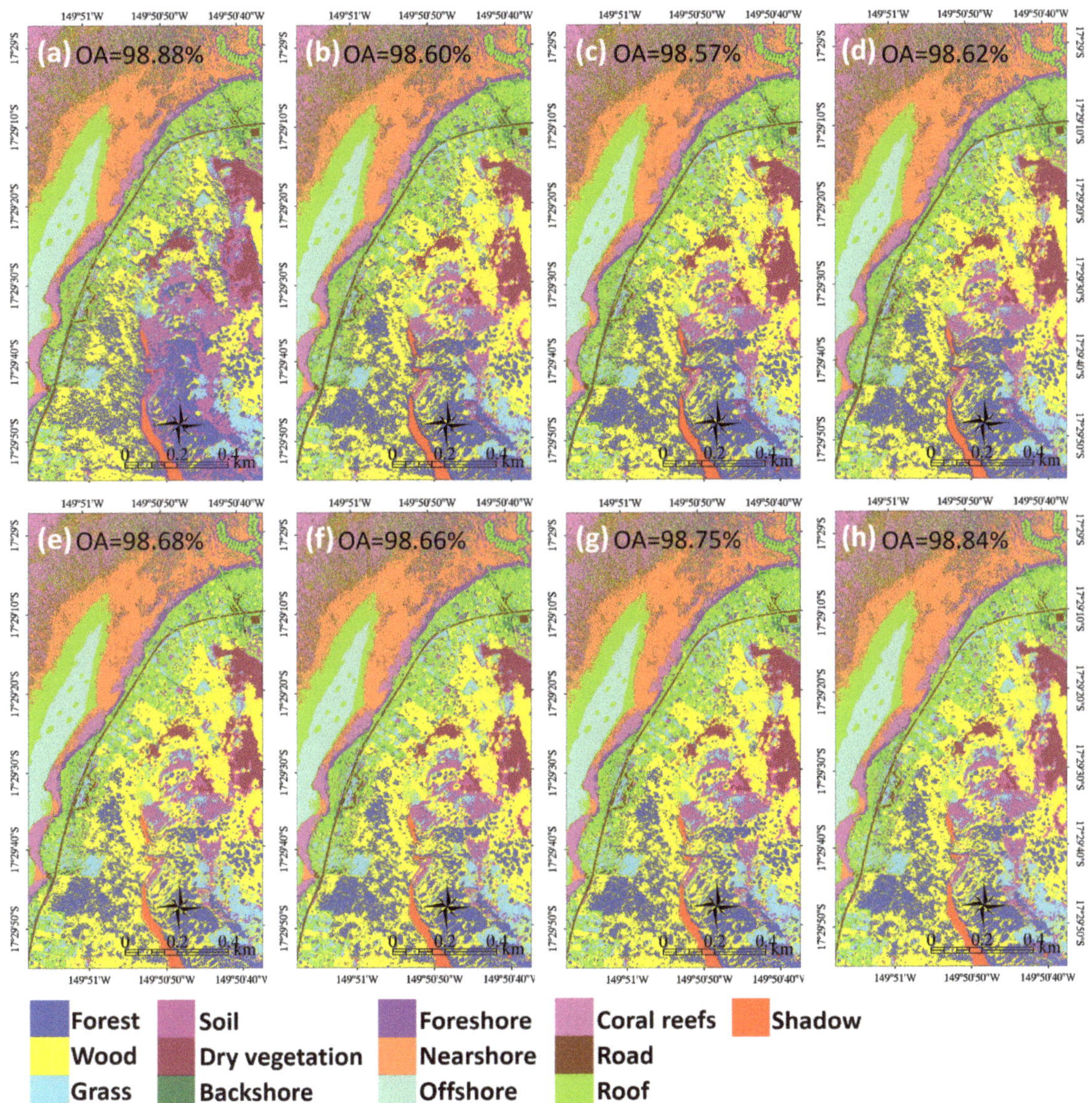



**Figure 8.** Maximum likelihood classification maps of the land-sea coral reefscape using WorldView-3 topobathymetric digital surface model and optical dataset with: (**a**) Mid-InfraRed1 (MIR1); (**b**) MIR2; (**c**) MIR3; (**d**) MIR4; (**e**) MIR5; (**f**) MIR6; (**g**) MIR7; (**h**) MIR8.

**Optical — No topobathymetry DSM**

| | Forest | Wood | Grass | Soil | Dry vegetation | Backshore | Foreshore | Nearshore | Offshore | Coral reefs | Road | Roof | Shadow |
|---|---|---|---|---|---|---|---|---|---|---|---|---|---|
| Forest | 0.23 | -1.1 | -1.9 | -0.93 | -1.07 | 0 | 0 | 0 | 0 | 0 | -0.2 | 0 | 0 |
| Wood | 0.23 | 19.27 | -25.87 | -1.47 | 0 | 0 | 0 | 0 | 0 | -0.07 | 0 | 0 | 0 |
| Grass | -0.4 | -18.17 | 27.7 | 1.43 | 0 | 0 | 0 | 0 | 0 | 0 | 0.03 | 0 | 0 |
| Soil | -0.03 | 0 | 0 | 2.07 | -1.8 | 0 | 0 | 0 | 0 | -0.37 | -12.8 | 0.17 | 0 |
| Dry vegetation | 0 | 0 | 0 | -0.6 | 3.53 | 0 | 0 | 0 | 0 | 0 | -0.23 | -0.03 | 0 |
| Backshore | 0 | 0 | 0 | 0.07 | 0 | -0.47 | -0.67 | 0 | 0 | 0 | 0 | -0.2 | 0 |
| Foreshore | 0 | 0 | 0 | 0 | 0 | -0.3 | 0.43 | -0.07 | 0 | 0 | 0 | -1.03 | 0 |
| Nearshore | 0 | 0 | 0 | 0 | 0 | 0 | -0.07 | -0.1 | 0 | -0.5 | 0 | 0 | 0 |
| Offshore | 0 | 0 | 0 | 0 | 0 | 0 | 0 | 0 | 0.07 | -0.93 | 0 | 0 | 0 |
| Coral reefs | 0 | 0 | 0 | -0.07 | -0.37 | 0 | 0 | -0.07 | 0 | 1.2 | -0.03 | 0.03 | -0.13 |
| Road | 0 | 0 | 0 | -0.5 | -0.43 | 0 | -0.07 | 0.2 | 0 | 0.03 | 15.93 | 0.33 | 0 |
| Roof | -0.03 | 0 | 0.07 | 0 | 0.13 | 0.77 | 0.37 | 0.03 | -0.07 | 0.63 | -2.7 | 0.73 | 0 |
| Shadow | 0 | 0 | 0 | 0 | 0 | 0 | 0 | 0 | 0 | 0 | 0 | 0 | 0.13 |

**Optical — Topobathymetry DSM**

| | Forest | Wood | Grass | Soil | Dry vegetation | Backshore | Foreshore | Nearshore | Offshore | Coral reefs | Road | Roof | Shadow |
|---|---|---|---|---|---|---|---|---|---|---|---|---|---|
| Forest | 5.43 | -7.93 | -1.9 | -0.93 | -1.07 | 0 | 0 | 0 | 0 | 0 | -0.2 | 0 | 0 |
| Wood | -5 | 26.27 | -25.87 | -1.37 | 0 | -0.07 | 0 | 0 | 0 | -0.07 | 0 | 0 | 0 |
| Grass | -0.43 | -18.33 | 27.5 | 0.67 | 0 | 0 | 0 | 0 | 0 | 0 | 0 | 0 | 0 |
| Soil | 0.03 | 0 | 0 | 8.53 | -2.1 | 0 | 0 | 0 | 0 | -0.43 | -13.6 | -5.23 | 0 |
| Dry vegetation | 0 | 0 | 0 | -8.93 | 3.97 | 0 | 0 | 0 | 0 | 0 | -0.23 | -0.03 | 0 |
| Backshore | 0 | 0 | 0 | 3.13 | 0 | 1.2 | -0.73 | 0 | 0 | 0.07 | 0.1 | -2.53 | 0 |
| Foreshore | 0 | 0 | 0 | 0 | 0 | -0.3 | 0.53 | -0.13 | 0 | 0.03 | 0 | -1.03 | 0 |
| Nearshore | 0 | 0 | 0 | 0 | 0 | 0 | -0.07 | 0.27 | 0 | -0.57 | 0 | 0 | 0 |
| Offshore | 0 | 0 | 0 | 0 | 0 | 0 | 0 | 0 | 0.07 | -0.93 | 0 | 0 | 0 |
| Coral reefs | 0 | 0 | 0 | -0.07 | -0.37 | 0 | 0 | -0.07 | 0 | 1.87 | -0.03 | 0 | -0.13 |
| Road | 0 | 0 | 0 | -2.13 | -0.43 | 0 | -0.07 | -0.07 | 0 | -0.1 | 19.27 | -15.03 | 0 |
| Roof | -0.03 | 0 | 0.27 | 1.1 | 0 | -0.83 | 0.33 | 0 | -0.07 | 0.13 | -5.3 | 23.87 | 0 |
| Shadow | 0 | 0 | 0 | 0 | 0 | 0 | 0 | 0 | 0 | 0 | 0 | 0 | 0.13 |

**Best Optical+MIR — MIR8**

| | Forest | Wood | Grass | Soil | Dry vegetation | Backshore | Foreshore | Nearshore | Offshore | Coral reefs | Road | Roof | Shadow |
|---|---|---|---|---|---|---|---|---|---|---|---|---|---|
| Forest | 5.43 | -7.23 | -1.9 | -0.93 | -1.07 | 0 | 0 | 0 | 0 | 0 | -0.2 | 0 | 0 |
| Wood | -4.93 | 25.53 | -25.87 | -1.5 | 0 | -0.07 | 0 | 0 | 0 | -0.07 | 0 | 0 | 0 |
| Grass | -0.43 | -18.3 | 27.6 | -0.23 | 0 | 0 | 0 | 0 | 0 | 0 | 0 | 0 | 0 |
| Soil | -0.03 | 0 | 0.13 | 14.6 | -1.63 | 0 | 0 | 0 | 0 | -0.43 | -15.13 | -1.9 | 0 |
| Dry vegetation | 0 | 0 | 0 | -10.4 | 3.5 | 0 | 0 | 0 | 0 | 0 | -0.23 | -0.03 | 0 |
| Backshore | 0 | 0 | 0 | 0.47 | 0 | 1.97 | -0.73 | 0 | 0 | 0 | 0.67 | -5.03 | 0 |
| Foreshore | 0 | 0 | 0 | 0 | 0 | -0.3 | 0.5 | -0.13 | 0 | 0.03 | 0 | -1.03 | 0 |
| Nearshore | 0 | 0 | 0 | 0 | 0 | 0 | -0.07 | 0.27 | 0 | -0.93 | 0 | 0 | 0 |
| Offshore | 0 | 0 | 0 | 0 | 0 | 0 | 0 | 0 | 0.07 | -0.93 | 0 | 0 | 0 |
| Coral reefs | 0 | 0 | 0 | -0.07 | -0.37 | 0 | 0 | -0.07 | 0 | 2.07 | -0.03 | 0 | -0.13 |
| Road | 0 | 0 | 0 | -2.13 | -0.43 | 0 | -0.07 | -0.1 | 0 | -0.13 | 20.03 | -14.87 | 0 |
| Roof | -0.03 | 0 | 0.03 | 0.2 | 0 | -1.6 | 0.37 | 0.03 | -0.07 | 0.4 | -5.1 | 22.87 | 0.03 |
| Shadow | 0 | 0 | 0 | 0 | 0 | 0 | 0 | 0 | 0 | 0 | 0 | 0 | 0.1 |

**Best Optical+MIR — MIR1**

| | Forest | Wood | Grass | Soil | Dry vegetation | Backshore | Foreshore | Nearshore | Offshore | Coral reefs | Road | Roof | Shadow |
|---|---|---|---|---|---|---|---|---|---|---|---|---|---|
| Forest | 5.9 | -8.77 | -1.9 | -0.93 | -1.07 | 0 | 0 | 0 | 0 | 0 | -0.2 | 0 | 0 |
| Wood | -5.47 | 27.1 | -25.87 | -1.4 | 0 | -0.07 | 0 | 0 | 0 | -0.07 | 0 | 0 | 0 |
| Grass | -0.43 | -18.33 | 27.33 | 0.57 | 0 | 0 | 0 | 0 | 0 | 0 | 0 | 0 | 0 |
| Soil | 0.03 | 0 | 0 | 12.97 | -1.93 | 0 | 0 | 0 | 0 | -0.43 | -15.17 | -5.73 | 0 |
| Dry vegetation | 0 | 0 | 0 | -9.77 | 3.8 | 0 | 0 | 0 | 0 | 0 | -0.23 | -0.03 | 0 |
| Backshore | 0 | 0 | 0 | 0 | 0 | 1.07 | -0.73 | 0 | 0 | 0.1 | 0.23 | -2.77 | 0 |
| Foreshore | 0 | 0 | 0 | 0 | 0 | -0.3 | 0.57 | -0.13 | 0 | 0.03 | 0 | -1.03 | 0 |
| Nearshore | 0 | 0 | 0 | 0 | 0 | 0 | -0.07 | 0.27 | 0 | -0.97 | 0 | 0 | 0 |
| Offshore | 0 | 0 | 0 | 0 | 0 | 0 | 0 | 0 | 0.07 | -0.93 | 0 | 0 | 0 |
| Coral reefs | 0 | 0 | 0 | -0.07 | -0.37 | 0 | 0 | -0.07 | 0 | 2.07 | -0.03 | 0 | -0.13 |
| Road | 0 | 0 | 0 | -2.13 | -0.43 | 0 | -0.07 | -0.07 | 0 | -0.1 | 20.67 | -14.93 | 0 |
| Roof | -0.03 | 0 | 0.43 | 0.77 | 0 | -0.7 | 0.3 | 0 | -0.07 | 0.3 | -5.27 | 24.5 | 0 |
| Shadow | 0 | 0 | 0 | 0 | 0 | 0 | 0 | 0 | 0 | 0 | 0 | 0 | 0.13 |

**Figure 9.** Confusion matrices of the percent differences between the WorldView-3 basic blue-green-red classification validation and the best results for multispectral and superspectral combinations, deprived of and provided with WorldView-3 topobathymetric DSM.

## 4. Discussion

It is necessary to bear in mind that the mapping assessment was based on the surveyed calibration/validation rectangles (see Figure 1b). The use of a reliable and detailed ground-truth image is important for providing a comprehensive estimation of the area, thus preventing misclassifications detected by visual discrepancies as seen for the roof class rectangles that were mapped into the sea.

### 4.1. Land-Sea Coral Reefscape Mapping with a Multispectral WorldView-3 Stereo-Imagery

Together with its predecessor the 2009 WorldView-2, the 2014 WorldView-3 remains the state-of-the-science sub-metre spaceborne optical sensor, leveraging two extra spectral bands in the VIS and two supplementary spectral bands in the NIR gamut, compared to all other sub-metre competitors (see Introduction). The use of the five-band VIS and eight-band optical data showed a greater classification performance of the land-sea coral reefscape than did the basic BGR, namely a gain of 2.69% and 8.79%. These results are in strong agreement with the WorldView-2 multispectral Moorea land-sea mapping project [11].

Even if the satellite-based topography, on the one hand, and bathymetry, on the other hand, are commonly studied for sub-metre optical sensors [12,15], it is still innovative to create satellite-based merged topobathymetry: Pleiades-1 [25], WorldView-3 [24]. The RMSE validation accuracies of the photogrammetry-based topography and ratio-transformed bathymetry of 1.11 m ($R^2$ = 0.99, min = 20.26 m, max = 370.25 m, $N$ = 12) and 0.89 m ($R^2$ = 0.73, min = −12.49 m, max = 0 m, $N$ = 15), computed here, are also in concordance with the stereo-WorldView-3 multispectral Moorea land-sea mapping [24]. Even if the topographic accuracy was lower than the bathymetric one, an in-depth examination showed that the accuracy rapidly diminished with depth, with a break at -5 m. In contrast, the accuracy of the altitude estimation remains constant even for higher values around 400 m. The bathymetry modeling could be ameliorated by using more soundings from a more recent survey, such as the topobathymetric LiDAR campaign locally operated in 2015. Further investigation will assess the influence of various pansharpening methods on those relief accuracies. Like the current work, this previous research highlighted that the addition of the topobathymetric DSM augmented the classification accuracy derived from the WorldView-3 basic BGR (2.74%), VIS (5.44%), and optical (9.15%) datasets. It is worth emphasizing that the DSM contribution:

- To the basic BGR neared the sole VIS performance ($\approx$2.7%);
- To the VIS prediction equaled 2.7%;
- To the optical accuracy approximated 6.4%.

*4.2. Land-Sea Coral Reefscape Mapping with a Superspectral WorldView-3 Stereo-Imagery*

The greatest novelty of the WorldView-3 resides in the collection of the eight-band MIR spectral bands, doubling the spectral bands of the WorldView-2 and extending the spectrum to 2365 nm [27]. The individual contributions of the eight MIR bands all contributed to the gain in classification accuracy of the land-sea coral reefscape compared to the basic mapping performance, ranging from 9.47% (MIR1) to 9.58% (MIR8), through 9.57% (MIR7). These significant inputs corresponded to a boosting of the optical (VIS+NIR) prediction, ranging from 0.68% to 0.79%.

The topobathymetric DSM reinforced the classification accuracy derived from the WorldView-3 optical datasets provided with the eight MIR bands. The best performances were here ensured by the MIR1 (9.73%), then the MIR8 (9.69%), while the least gain was attributed to the MIR3 (9.42%). Referenced to the optical dataset, the DSM therefore ameliorated the classification of:

- 0.94% with MIR1;
- 0.90% with MIR8;
- 0.63% with MIR3.

*4.3. Land-Sea Coral Reefscape Mapping at the Class Scale*

As might be logically expected, the best multispectral combination relied on the full optical dataset, from Coastal to NIR2. Relatively to the BGR discrete prediction, the discrimination between grass, wood and road classes were significantly refined. These findings might easily be explained if the NIR enhancement can capture the higher reflectance of both the chlorophyll-laden and the tar/asphalt -made classes better than the VIS one [30].

The DSM fusion with the optical dataset strongly improved the roof and the soil detection. This boosting was correlated with the decrease in misclassification with topographically lower road and topographically higher dry vegetation, respectively. The knowledge of the elevation component within the landscape therefore helped separate human-made, on the one hand, and natural features, on the other hand, that are spectrally similar in the optical range [30].

The best superspectral combination relied on the merge of the optical dataset with the MIR8 (2295–2365 nm). The positive effect at the class scale was tangible with grass, wood, roof, road and soil classes. Compared to the optical dataset, the roof and soil classes were better isolated. The roof outcome might stem from the MIR8 spectral fitting with the higher reflectance of the Polynesian roof made of oxidized-galvanized steel metal (0.34 reflectance) than the optical one (0.19 reflectance) [30]. The soil gain might also be due to MIR8 matching more closely with the higher reflectance (0.44) than the optical one (0.28) (see "brown loam" in [28]). These explanations were also supported by the decline in both roof and soil misclassifications with road, that displays a low MIR8 reflectance of 0.08 [30].

The DSM influence showed a better OA with the combination of the optical dataset with the MIR1 (1195–1225 nm). This optimum simply reinforced its positive effect on the same previous classes, suggesting that the additional elevation information was relatively redundant to this coming from the MIR.

Concerning the coral reefs, the successive integration of the optical bands and the DSM to the BGR dataset, slightly but consistently, strengthened their detection. The addition of the Coastal and yellow bands favoured the coral reefs' separability among other benthic features given the refinement in spectral signatures [11]. The benthic terrain information was also profitable due to the robustness of the depth proxy for delineating benthos' ecophysiological belts [23]. In view of the neat classification of the coral reefs along the lagoon width (Figures 5–8), further research should divide the coral reefs' current class according to their landscape position (fringing, barrier and outer reefs), and their morphology (encrusting, branching, massive, tabular, columnar, etc.).

## 5. Conclusions

The superspectral WorldView-3 providing 16 bands, from 400 to 2365 nm, pansharpened at 0.3 m, was acquired in the form of stereo-imagery. Both topographic and bathymetric DSMs were built using a handful of ground and sea control points, enabling us to calibrate/validate the land-based photogrammetry and the sea-based radiative transfer model, provided with 1.11 and 0.89 m vertical accuracy, respectively. The best superspectral combination for enhancing the land-sea mapping of 13 habitats relied on the merging of the optical dataset (VIS+NIR) and the MIR8, which enhanced the basic BGR classification accuracy by 9.58%, thus reaching an OA of 98.73%. The classes that most benefited from this were the land use "roof" and land cover "soil" classes. The "coral reefs" consisted of the sea class that was the most favoured. The addition of the satellite-derived topobathymetric DSM to the optical+MIR1 was the best full combination, increasing the basic BGR classification accuracy by 9.73%, thus reaching an OA of 98.88%. The discrimination of the "roof" and "soil" classes was also strengthened, but the "coral reefs" remained constant.

**Author Contributions:** Conceptualization, A.C. and M.A.; methodology, A.C., D.L. and J.C.; software, A.C. and M.A.; validation, A.C.; formal analysis, A.C.; investigation, A.C. and J.C.; resources, A.C. and M.A.; data curation, A.C. and M.A.; writing—original draft preparation, A.C.; writing—review and editing, A.C., M.A., D.L. and J.C.; visualization, A.C. and M.A.; supervision, A.C. and J.C.; project administration, A.C. and J.C.; funding acquisition, M.A. All authors have read and agreed to the published version of the manuscript.

**Funding:** This research was partly funded by the DigitalGlobe Foundation and the CNRS PEPS called "Reef I Were".

**Conflicts of Interest:** The authors declare no conflict of interest.

## References


1. Costanza, R.; De Groot, R.; Sutton, P.; van der Ploeg, S.; Anderson, S.J.; Kubiszewski, I.; Farber, S.; Turner, R.K. Changes in the global value of ecosystem services. *Glob. Environ. Chang.* **2014**, *26*, 152–158. [CrossRef]
2. Spalding, M.; Burke, L.; Wood, S.A.; Ashpole, J.; Hutchison, J.; zu Ermgassen, P. Mapping the global value and distribution of coral reef tourism. *Mar. Policy* **2017**, *82*, 104–113. [CrossRef]
3. Spalding, M.; Spalding, M.D.; Ravilious, C.; Green, E.P. *World Atlas of Coral Reefs*; UNEP-WCMC: Cambridge, UK, 2001.
4. Hughes, T.P.; Barnes, M.L.; Bellwood, D.R.; Cinner, J.E.; Cumming, G.S.; Jackson, J.B.; Kleypas, J.; Van De Leemput, I.A.; Lough, J.M.; Morrison, T.H.; et al. Coral reefs in the Anthropocene. *Nature* **2017**, *546*, 82–90. [CrossRef] [PubMed]
5. Collin, A.; Hench, J.L. Towards Deeper Measurements of Tropical Reefscape Structure Using the WorldView-2 Spaceborne Sensor. *Remote Sens.* **2012**, *4*, 1425–1447. [CrossRef]
6. Mancini, F.; Dubbini, M.; Gattelli, M.; Stecchi, F.; Fabbri, S.; Gabbianelli, G. Using Unmanned Aerial Vehicles (UAV) for High-Resolution Reconstruction of Topography: The Structure from Motion Approach on Coastal Environments. *Remote Sens.* **2013**, *5*, 6880–6898. [CrossRef]
7. Agrafiotis, P.; Karantzalos, K.; Georgopoulos, A.; Skarlatos, D. Correcting Image Refraction: Towards Accurate Aerial Image-Based Bathymetry Mapping in Shallow Waters. *Remote Sens.* **2020**, *12*, 322. [CrossRef]
8. Collin, A.M.; Andel, M.; James, D.; Claudet, J. The Superspectral/Hyperspatial Worldview-3 as The Link between Spaceborne Hyperspectral and Airborne Hyperspatial Sensors: The Case Study of the Complex Tropical Coast. *Int. Arch. Photogramm. Remote Sens. Spat. Inf. Sci.* **2019**, *XLII-2/W13*, 1849–1854. [CrossRef]

9. Collin, A.; Long, B.; Archambault, P. Merging land-marine realms: Spatial patterns of seamless coastal habitats using a multispectral LiDAR. *Remote Sens. Environ.* **2012**, *123*, 390–399. [CrossRef]
10. Mury, A.; Collin, A.; James, D. Morpho–Sedimentary Monitoring in a Coastal Area, from 1D to 2.5D, Using Airborne Drone Imagery. *Drones* **2019**, *3*, 62. [CrossRef]
11. Collin, A.; Archambault, P.; Planes, S. Bridging Ridge-to-Reef Patches: Seamless Classification of the Coast Using Very High Resolution Satellite. *Remote Sens.* **2013**, *5*, 3583–3610. [CrossRef]
12. James, D.; Collin, A.; Mury, A.; Costa, S. Very high resolution land use and land cover mapping using pleiades-1 stereo imagery and machine learning. *Int. Arch. Photogramm. Remote Sens. Spat. Inf. Sci.* **2020**, *XLIII-B2-2*, 675–682. [CrossRef]
13. Collin, A.; Planes, S. Enhancing Coral Health Detection Using Spectral Diversity Indices from WorldView-2 Imagery and Machine Learners. *Remote Sens.* **2012**, *4*, 3244–3264. [CrossRef]
14. Collin, A.; Hench, J.L.; Planes, S. A novel spaceborne proxy for mapping coral cover. In Proceedings of the 12th International Coral Reef Symposium, Cairns, Australia, 9–13 July 2012.
15. Collin, A.; Etienne, S.; Feunteun, E. VHR coastal bathymetry using WorldView-3: Colour versus learner. *Remote Sens. Lett.* **2017**, *8*, 1072–1081. [CrossRef]
16. Kruse, F.A.; Baugh, W.M.; Perry, S.L. Validation of DigitalGlobe WorldView-3 Earth imaging satellite shortwave infrared bands for mineral mapping. *J. Appl. Remote Sens.* **2015**, *9*, 96044. [CrossRef]
17. Asadzadeh, S.; Filho, C.R.D.S. Investigating the capability of WorldView-3 superspectral data for direct hydrocarbon detection. *Remote Sens. Environ.* **2016**, *173*, 162–173. [CrossRef]
18. Ye, B.; Tian, S.; Ge, J.; Sun, Y. Assessment of WorldView-3 Data for Lithological Mapping. *Remote Sens.* **2017**, *9*, 1132. [CrossRef]
19. Collin, A.; Lambert, N.; Etienne, S. Satellite-based salt marsh elevation, vegetation height, and species composition mapping using the superspectral WorldView-3 imagery. *Int. J. Remote Sens.* **2018**, *39*, 5619–5637. [CrossRef]
20. Ferreira, M.P.; Wagner, F.H.; Aragão, L.E.; Shimabukuro, Y.E.; Filho, C.R.D.S. Tree species classification in tropical forests using visible to shortwave infrared WorldView-3 images and texture analysis. *ISPRS J. Photogramm. Remote Sens.* **2019**, *149*, 119–131. [CrossRef]
21. Guo, X.; Li, P. Mapping plastic materials in an urban area: Development of the normalized difference plastic index using WorldView-3 superspectral data. *ISPRS J. Photogramm. Remote Sens.* **2020**, *169*, 214–226. [CrossRef]
22. Collin, A.; Hench, J.L.; Pastol, Y.; Planes, S.; Thiault, L.; Schmitt, R.J.; Holbrook, S.J.; Davies, N.; Troyer, M. High resolution topobathymetry using a Pleiades-1 triplet: Moorea Island in 3D. *Remote Sens. Environ.* **2018**, *208*, 109–119. [CrossRef]
23. Collin, A.; Nadaoka, K.; Nakamura, T. Mapping VHR Water Depth, Seabed and Land Cover Using Google Earth Data. *ISPRS Int. J. Geo Inform.* **2014**, *3*, 1157–1179. [CrossRef]
24. Collin, A.; Andel., M.; Lecchini, D.; Claudet, J. Submeter 3D ridge-to-reef classification using a WorldView-3 satellite stereoimagery. In Proceedings of the 14th International Coral Reef Symposium, Bremen, Germany, 5–10 July 2020.
25. Davies, N.; IDEA Consortium; Field, D.; Gavaghan, D.; Holbrook, S.J.; Planes, S.; Troyer, M.; Bonsall, M.; Claudet, J.; Roderick, G.; et al. Simulating social-ecological systems: The Island Digital Ecosystem Avatars (IDEA) consortium. *GigaScience* **2016**, *5*, 14. [CrossRef]
26. Institut de la Statistique de Polynesie Française. *Le Recensement de la Population en Polynésie Française en 2017*; ISPF: Papeete, French Polynesia, 2018.
27. Kwan, C.; Budavari, B.; Bovik, A.C.; Marchisio, G. Blind Quality Assessment of Fused WorldView-3 Images by Using the Combinations of Pansharpening and Hypersharpening Paradigms. *IEEE Geosci. Remote Sens. Lett.* **2017**, *14*, 1835–1839. [CrossRef]
28. Xu, F.; Woodhouse, N.; Xu, Z.; Marr, D.; Yang, X.; Wang, Y. Blunder elimination techniques in adaptive automatic terrain extraction. *ISPRS J.* **2008**, *29*, 21.
29. Stumpf, R.P.; Holderied, K.; Sinclair, M. Determination of water depth with high-resolution satellite imagery over variable bottom types. *Limnol. Oceanogr.* **2003**, *48*, 547–556. [CrossRef]
30. Baldridge, A.; Hook, S.; Grove, C.; Rivera, G. The ASTER spectral library version 2.0. *Remote Sens. Environ.* **2009**, *113*, 711–715. [CrossRef]

*Review*

# A Review of Current and New Optical Techniques for Coral Monitoring

Jonathan Teague [1,*], David A. Megson-Smith [1], Michael J. Allen [2,3], John C.C. Day [1] and Thomas B. Scott [1]

[1] Interface Analysis Centre (IAC), HH Wills Physics Laboratory, Bristol University, Tyndall Avenue, Bristol BS8 1TL, UK; david.megson-smith@bristol.ac.uk (D.A.M.-S.); J.C.C.Day@bristol.ac.uk (J.C.C.D.); T.B.Scott@bristol.ac.uk (T.B.S.)

[2] Plymouth Marine Laboratory (PML), Prospect Place, The Hoe, Plymouth PL1 3DH, UK; mija@pml.ac.uk

[3] College of Life and Environmental Sciences, University of Exeter, Geoffrey Pope Building, Stocker Road, Exeter EX4 4QD, UK

* Correspondence: jt16874@bristol.ac.uk

**Abstract:** Monitoring the health of coral reefs is essential to understanding the damaging impacts of anthropogenic climate change as such non-invasive methods to survey coral reefs are the most desirable. Optics-based surveys, ranging from simple photography to multispectral satellite imaging are well established. Herein, we review these techniques, focusing on their value for coral monitoring and health diagnosis. The techniques are broadly separated by the primary method in which data are collected: by divers and/or robots directly within the environment or by remote sensing where data are captured above the water's surface by planes, drones, or satellites. The review outlines a new emerging technology, low-cost hyperspectral imagery, which is capable of simultaneously producing hyperspectral and photogrammetric outputs, thereby providing integrated information of the reef structure and physiology in a single data capture.

**Keywords:** coral reef monitoring; reef health; review; hyperspectral imaging; marine optics

**Citation:** Teague, J.; Megson-Smith, D.A.; Allen, M.J.; Day, J.C.; Scott, T.B. A Review of Current and New Optical Techniques for Coral Monitoring. *Oceans* **2022**, *3*, 30–45. https://doi.org/10.3390/oceans3010003

Academic Editor: Michael W. Lomas

Received: 1 May 2021
Accepted: 13 January 2022
Published: 20 January 2022

**Publisher's Note:** MDPI stays neutral with regard to jurisdictional claims in published maps and institutional affiliations.

## 1. Introduction

Coral reefs are strategically important for coastal nations in tropical and subtropical regions due to the valuable ecosystem goods and services they provide [1]. However, reef systems are now being impacted on a global scale by a variety of conditions, namely, "coral bleaching" [2,3], diseases [4,5], nutrient pollution [6] and algal overgrowth [7], coastal engineering [8] and sedimentation [9], crown-of-thorns [10] and sea-urchin predation [11]. Corals are particularly susceptible to environmental changes as they have low tolerance to variations in temperature, salinity, and solar radiation [12]. Sustained periods of stress can lead to coral colony death, and, in some instances, to whole reef collapse [13]. This represents a significant concern for the 275 million people who live within 30 km of these ecosystems, and who rely on these reefs for their livelihoods and food security [14].

While coral bleaching is most commonly associated with changes in sea surface temperature (SST) [3,15], it can also be a response to other external factors or triggers such as ocean acidification [16], bacterial infection [17] or shading caused by extreme turbidity [18]. The term 'bleaching' refers to the loss of the symbiont algal cells of the family Symbiodiniaceae, which are normally the main provider of coral colour [19,20]. The white or bleached appearance of the coral results from the calcium carbonate exoskeleton becoming visible, since the coral tissue itself is translucent [19], or as a result of polyp death.

Coral disease is another of the main causes of reef degradation and has been increasing worldwide since first studied in the 1970s in the Red Sea [21]. It has become particularly prevalent in the Caribbean [22,23], but has also been increasingly recorded in other reef systems such as the Great Barrier Reef [4,22,24,25]. Coral can become more susceptible to disease due to factors such as a decline in water quality and fish stocks, heat stress and, more recently identified, to ocean acidification driven by anthropogenic activity [26–28]. In some cases, specific pathogens have been identified as a primary contributor [22,29]. A diverse array of diseases have now been observed, with approximately 30 diseases and syndromes affecting the health of 150 different species worldwide [5,30]. The term 'disease' is used to describe symptoms arising from a known pathogen, while 'syndrome' refers to effects arising from an unknown causative agent, whether it be a pathogen, pollutant, or climate condition such as ocean warming [19].

Visible changes associated with coral ill-health can provide a valuable metric for the monitoring of colonies. Affected corals are frequently characterised (and named accordingly) by abnormal or decreased pigmentation in compromised tissue [31]. To date, white, brown, pink, yellow, and black line diseases have all been described [26–28,32].

Bleaching or disease is often identifiable by the absence of specific pigments in individual corals [33–36]. These pigments have specific wavelength peaks characteristic of their optical reflectance or fluorescence spectra. These pigments include chlorophyll-a absorption (676 nm) [37], chlorophyll-a fluorescence (685 nm), peridinin (574 nm) [38], diatoxanthin (607 nm) [38], and green fluorescent protein (GFP) (511 nm) [39].

The photosynthetic pigment chlorophyll-a and accessory pigments peridinin and diatoxanthin provide a direct insight into the symbiont density [40,41]. As these pigments are only found within the corals' symbiotic partner Symbiodiniaceae, they provide a direct bleaching indicator. GFP fluorescence provides different insights as it is highly responsive to thermal fluctuations [15,42,43] and is often spectrally more distinctive, namely in its intensity when compared to chlorophyll fluorescence [33].

This observed specificity lends itself to utilising automated image-based techniques that can objectively quantify and monitor compromised colonies based on spectroscopic optical measurements. Current observations by human divers are unavoidably subjective and provide less robust and detailed data.

## 2. Current Optic Based Methods for Monitoring Coral Health

There currently exist many optical methods that make up the majority of coral health assessments ranging from underwater data collection via divers or robots to remote sensing techniques using satellites (Figure 1).

**Figure 1.** An overview of the current optical techniques from satellites to underwater systems such as unmanned underwater vehicles (UUVs).

There are many factors relevant to the accurate monitoring of coral reefs. These may be biological including an abundance of coral predators, coral species composition and distribution; chemical such as pH (acidity) or the presence of nutrients; physical parameters such as temperature and turbidity; and socio-economic parameters such as marine protected areas and fishing communities [44]. This review will focus on the biological parameters.

Many current reef health survey methods employ divers as 'observers'. The advantages of this method are that they offer a versatile set of skills for coral monitoring, being highly manoeuvrable, adaptable, and able, with training, to deliver reasonably precise results. Observers generally record basic data such as 'percentage cover of live coral', which is the most widely used metric of coral reef condition and is commonly used in studies that record coral reef decline and recovery across local spatial scales [45]. A drawback is that such surveys are time-consuming and so are often not able to prioritise disease identification and assessment [4,23,46,47]. A standardised survey for the assessment of coral health requires detailed examination of all coral colonies within a designated sample area (e.g., transects or quadrats), and so also involves lengthy and expensive person-intensive field time [48].

Underwater diver-generated photographic surveys employing photo-quadrats or video transects comprise the bulk of modern reef monitoring activities and may be used to record the reef substrate over hundreds of square meters. Due to difficulties associated with conducting frequent surveys over large areas by divers, there is an increasing use of 'robotic observers' (i.e., unmanned underwater vehicles (UUVs)) that can be used to cover thousands of square meters [49,50]. The increased area coverage and data generation offered by UUVs allows for the time and human effort otherwise required for physical measurements to be better used for processing, interpretation, and analysis of the often semi-quantitative digital data.

A major advantage of image-based surveys is that the images created provide a permanent record of the habitat at the time they were taken. This reduces the dependency on in-field coral experts and provides data that can be re-analysed retrospectively or compared directly between repeat surveys. The following section describes diver-based optical survey techniques that represent the 'classical' approach to coral surveyance. Many of these techniques could readily be employed using UUVs, but as yet, have found limited application in this manner. This is principally due to the higher costs associated with the UUVs and the additional training required to operate them.

### 2.1. Diver-Based Survey Techniques

### 2.1.1. RGB Imaging

Digital cameras used in current photography surveys detect light in three broad colour channels: red, green, blue (RGB), centred at approximately 660 nm, 520 nm, and 450 nm, respectively. This enables similar images to be taken to those perceived by the human eye, which also sees using only these RGB channels [51]. Cameras can be employed in many coral reef surveying techniques such as photo-quadrat sampling in which quadrats are imaged via high resolution digital cameras. This secures a visual record suitable for subsequent laboratory analysis as opposed to manual in situ coverage estimates. Imaging thus allows for a reduction in diver 'bottom time' compared with non-image based surveyance. The advantage of laboratory analysis is that it allows for images to be run through machine learning software such as CoralNet [38], which attempt to fully or partially automate classification and benthic cover estimates. However, classification can only be made after sufficient training data are provided to the algorithm [52,53]. These types of machine learning systems are still in their infancy and, to date, can only effectively estimate the cover of common coral genera.

Images can also be interpreted to provide rudimentary colour analysis. This yields outputs similar to diver assessments undertaken qualitatively by eye using colour charts or wheels. The comparison of corals against known colour hues corresponding to different concentrations of symbionts [54] enables divers to quickly identify the extent of any coral bleaching, though other aspects of coral health may be overlooked. Performing such surveys using traditional digital camera (RGB) images is an improvement, since the intensity of individual colour channels can be recorded and interpreted. For example, the intensity of red wavelengths indicates the extent of chlorophyll absorption [55]. However, the value of the method is limited as only three broad colour bands are recorded by the camera sensor. The method can aid in preliminary assessments of bleaching, but by the point bleaching is RGB-detectable, typically around 70% or more of symbionts have already been expelled [56]. A general point to remember is that these types of survey typically only cover an individual coral colony and may not be representative of the whole reef system. They also only provide a snapshot of an environment and cannot be extrapolated to an understanding of the ongoing population dynamics of the whole reef system.

An emerging and more advanced utilisation of RGB imaging is via the creation of three-dimensional (3D) reconstructions of coral systems using photogrammetry. With recent advances in photogrammetric processing, this is now relatively quick and easy to conduct [57]. Photogrammetry uses a set of overlapping images collected either by video or still photography of a target area. Ideally adjacent images should have 60–80% overlap [58]. Physical parameters can be obtained from 3D reconstruction models of coral reefs such as surface topography, estimations of rugosity and surface area as well as coral cover and distribution [59]. Crucially, a wide range of additional information can be extracted from the same original dataset, making it a useful analysis technique when image sets are recorded. However, when using standard digital cameras, the method is still limited. Resultant images are only relevant to changes in coral colour or fluorescent emissions as detected by the three RGB channels.

### 2.1.2. Underwater Spectroscopic Techniques

Spectroscopic techniques can image in numerous, narrower wavelength bands across the whole visible light spectrum and mark an improvement over simple RGB imaging. The use of spectral data enables a more definitive discrimination between live coral, macroalgae, and other photoactive organisms by using the specific spectral "signature" or "fingerprints" associated with a certain organism or type of organism [60,61]. It also identifies whether corals are displaying a decline in 'normal health' by measuring the relative intensity of the spectral signatures arising from specific pigments associated with health such as chlorophyll. This can be achieved by using reference targets to correct for incident light variations, thereby normalising spectra so they can be compared between datasets to track changes in pigment intensity and thus bleaching.

Underwater spectrometry can be achieved by using laboratory spectrometers enclosed in waterproof housings with fibre optic probes to record radiance reflectance measurements. The fibre optic probes are held at an orthogonal angle to the solar incidence angle, approximately 0.5–1.0 cm from the target [36]. An accompanying reference measurement is required to normalise for variations in ambient illumination. This is achieved by taking a reflectance measurement from a well characterised, white Lambertian reflectance target such as polytetrafluorethylene (PTFE) or a Spectralon (Labsphere, USA). The reference spectrum enables a correction function to be applied to the data. Specialised spectrometers can also be employed for certain niche applications. For example, pulse amplitude modulation (PAM) fluorometers specifically look at fluorescence to determine the photosynthetic yield. Chlorophyll density can be used to determine relative electron transport rates of photosynthetic organisms to provide a measurement of photosynthetic efficiency [62]. This is a measure of how well chlorophyll converts light into energy and detects compromised tissues that are less efficient. The diving PAM I and II (Walz, Germany) are examples of underwater fluorometers and are the most commonly used devices in studies using this technique [3,63,64]. PAM devices do have limitations. Notably, a requirement for the sampling optical fibre probe to be held in near contact (<5 mm) with the sampled object for a long time (>30 s) to obtain accurate readings.

Spectrometers and fluorometers are able to generate more accurate spectral data but also suffer from many of the same pitfalls as RGB imaging. Data acquisition is typically slow when used to cover a whole reef system. This is mainly due to the small sampling area of the probes and the requirement to make point measurements. This limitation makes the technique particularly unsuitable for large-area surveys. Additionally, multiple points are often sampled on individual corals to obtain average spectra. However, the small number of measurements precludes confidence that these average spectra are truly representative of the whole organism or a whole reef system.

Conversely, spectroscopy techniques using imagers (multispectral and hyperspectral imaging) can generate spectra for every pixel in an image within one data acquisition. This makes the process of data collection quicker and more efficient, thereby facilitating the collection of datasets that are more comprehensive and representative. In turn, imagers can categorise and quantify colour. Spectral imagers generally comprise a dispersive element (either a prism or diffraction grating) or filter, which splits or filters incoming light into wavelengths, and an imaging detector such as a charged coupled device (CCD) or complementary metal–oxide–semiconductor device (CMOS).

Multispectral imagers record data across multiple spectral bands, typically between three and 15 bands [65]. Conversely, hyperspectral imaging records in hundreds of spectral bands, which means data may be collected and processed across the whole visible and/or near infrared spectrum with improved spectral resolution.

Previously, some multispectral systems have been deployed to assess specific marine monitoring cases. These included determining coral fluorescence using narrow bandpass filters [66] and filter wheel style imagers for classification via spectral discrimination [67]. Other imagers have been produced for applications such as the exploration of marine minerals and ores [68], but are not currently being used in coral monitoring surveys [69].

Underwater hyperspectral imaging (UHI) is a relatively new, emerging technology with limited published instances to date. Current diver operated hyperspectral systems such as the "HyperDiver" system [70] can generate hyperspectral and traditional RGB images simultaneously capturing synchronised high-resolution digital images, hyperspectral, and topographic data [70]. The system utilises a push-broom hyperspectral imager (Pika 2, Resonon Inc., Bozeman, MT, USA) with a spectral range of 400–900 nm sampled at ~1.5 nm resolution with 480 fixed bands and 640 spatial pixels [70].

Push broom or line scanning imaging methods acquire full spectral data one spatial line at a time. The line is imaged onto the entrance slit of a spectrometer, which disperses the light into its spectral components before reaching the sensor array. The composite image is constructed by either moving the slit across the image plane or by moving the entire system across the scene [71]. This is advantageous as spectral data can be gathered whilst the imager is moving, which provides both full spectral and spatial data. Other hyperspectral systems such as 'Full data cube snapshot' imagers work from fixed viewpoints, similar to traditional RGB imagers. In this case, a push broom effect is achieved by optically scanning a linear field of view across the hyperspectral detector within the device. The need for a stable platform and the delicate nature of the optics involved make them generally unsuitable for use in UHI.

UHI presents additional potential applications using an 'objects of interest' (OOI) identification technique, as described by Johnsen [72], which includes mapping and monitoring of seafloor habitats for minerals or soft versus hard bottom; seafloor pipeline inspections to determine type of material, cracks, rust, and leakage; shipwrecks (type and state of wood, nails, rust, and artefacts); deep-water coral reefs and sponge fields for species identification, area coverage and physiological state, and kelp forests (species identification, area coverage, physiological state, and growth rates of benthic organisms).

Current UHI technologies (outlined in Table 1) are generally bulky systems that are difficult to deploy and manoeuvre. For example, the "Hyperdiver" system [70] including all its additional sensors and payloads weighs 32 kg in air. Other sensors, specifically the tunable LED-based underwater multispectral imaging system (TuLUMIS) and ocean vision (UHI OV), are designed to be mounted on a UUV. The UUV provides the interface system to operate the camera as well as a translation platform. These are not easily deployed by a diver.

The use of UUVs does, however, eliminate the limitations imposed by diver reliance. For example, dive surveys require substantial amounts of time as there is a finite period a diver can spend underwater; on dives, this is usually dependent on air-tank capacity and depth. Subsequent dives can be achieved through the use of multiple air-tanks but ultimately, a diver will fatigue. The corresponding issue on UUV based surveys is battery life, although multiple batteries can be used to extend the survey time. Crucially, UUVs do not suffer fatigue and can be deployed longer than their human counterparts. A UUV can also cover a larger distance in a shorter time. For example, a 120 m squared area may take two scuba divers up to 2.5 h [4], equating to a surveying rate of 0.13 $m^2/s$. Comparatively, a low-cost remotely operated vehicle (ROV) such as BlueROV2 can achieve survey rates of 1 $m^2/s$. Key limitations to both divers and UUVs are repeatability and accuracy when surveying reefs because global positioning system(s) (GPS) do not work underwater. Acoustic transponder networks designed for UUVs create a way of translating GPS coordinates underwater and thus improve the repeatability and accuracy by recording accurate georeferenced data [57].

The use of UHI on UUVs is currently limited with only a few studies having been reported. One such study [73] used a prototype UHI system for mapping the seafloor for the automated identification of seabed, habitat, and OOI in coral reefs. Other studies [61], specifically using hyperspectral imaging with corals, have mainly focused on coverage and benthic discrimination with machine learning to classify corals and have not focused on assessing health or disease. The current generation of commercially available hyperspectral imagers are often cost prohibitive to both acquire and insure for marine studies. Consequently, there exists a need for technology development and application to study marine environments such as the surveyance of coral health.

### 2.1.3. The Potential of 'New' Underwater Hyperspectral Imagers

A new type of hyperspectral imaging technology that utilises linear variable filter(s) (LVF) has recently emerged. A LVF is an optical filter whose bandpass windows varies continuously across its surface [74]. LVFs allow for lower cost imagers to be produced [74]. For example, LVFs have been integrated with consumer grade digital cameras to convert them into hyperspectral imagers [74–76]. The Bi-Frost [74] DSLR reduces the financial burden of spectral imagers by up to 75%. Using DSLR cameras offers key advantages as they are already well implemented in underwater photography (in both scientific and hobbyist applications), making them affordable and accessible in both supply and use. Non-specialised personnel can access and use the technology with relatively little training, thus reducing dependency on highly skilled divers and marine scientists for surveying.

The Bi-Frost DSLR can be implemented in two operational modes: hyperspectral reflectance imaging (HyRi) and fluorescence imaging (HyFi). These modes use the same Bi-Frost DSLR with the only difference being the lighting conditions a used for imaging; HyRi images under sunlight/white light; and HyFi under ultraviolet (UV) produced by light emitting diodes (LEDs).

The methodology for gathering HyRi/HyFi data is similar to that of underwater photogrammetry. A delivery platform such as a diver or UUV translates the imager across a target scene in a single line for measurements on a colony scale, or a 'lawn mower' pattern for a reef scale [77], ensuring sufficient overlap between line intersects [57] (Figure 2).

Spectral data can be extracted from the raw images generated from Bi-Frost cameras using photogrammetric approaches. Specifically, software was developed to interface with commercially available photogrammetric solutions (Photoscan 1.3.4, Agisoft, Russia). This software can produce hypercubes of intensity data having three spatial and one wavelength coordinate. In this way, its use enables the construction of 3D models with full spectral information for each 3D surface point. The sizes of the datasets are ultimately limited by the computing power required to run the reconstruction software. This approach generates camera positions in 3D space relative to the scene for each image taken. This, in turn, allows for the accurate derivation of the correction factors needed to compensate for optical attenuation between the source and the detector. A single survey with a HyRi system therefore gathers spectral data (enabling for coral identification, zonation, physiological assessments) and 3D data (reef structure, rugosity), giving it twice the value.

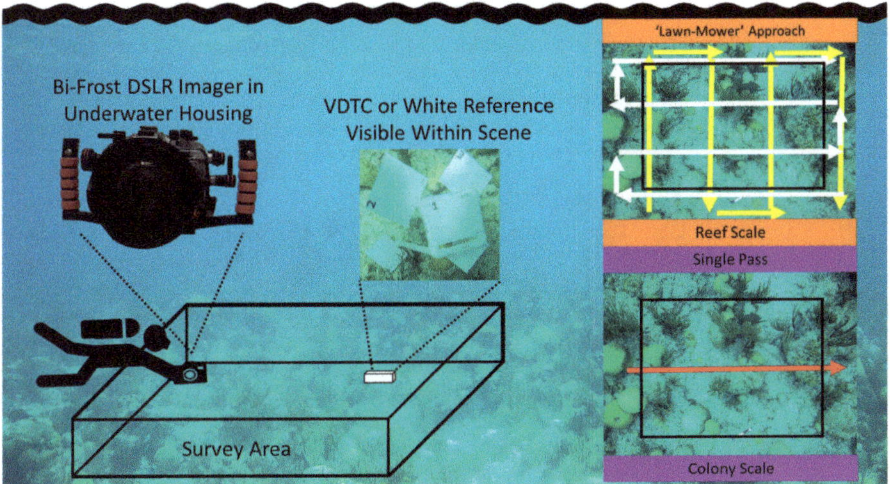


**Figure 2.** The survey method for the Bi-Frost DSLR hyperspectral imager can either be conducted in a 'lawn mower' approach [77] for reef scale surveys or a single line pass for colony scale surveys. To ensure sufficient coverage data, the recorded area must exceed the target survey area. Water attenuations effects are corrected for by imaging white reference objects at various depths. These are variable depth target calibrators (VDTCs) that comprise a series of white references at different depths or a single white reference (only for correcting against incident light where variable depth is not considered).

### 2.2. Remote Sensing Techniques

Data on coral can also be gained remotely using satellite or aerial imagery combined with spectral discrimination. However, this approach is usually only able to discriminate between coral colour types and broad benthic communities, and not between coral species [33]. This section describes the remote sensing techniques, which represent the more modern approach to global coral surveyance.

### 2.2.1. Airborne Multi/Hyperspectral Imaging

Imaging techniques can be deployed above the water's surface using multispectral/hyperspectral imagers. These surveys primarily look at coral distribution. However, they are limited by low spectral and spatial resolution and are only able to distinguish between coral, algae, and sand [61]. Aerial surveys can be undertaken using light aircraft or helicopters flying at an altitude of approximately 150 m. Aircraft and satellites can be equipped with multispectral and hyperspectral imagers. Aircraft are able to use benthic reflectance signatures to map the composition and condition of shallow water ecosystems in higher spatial resolution than their satellite counterparts, albeit at the cost of lower spatial scale [78]. In bleaching surveys, each reef is typically assigned a number from zero to four, these categories are associated with bleaching severity. The category classifications are as follows: CAT 0, <1% of corals bleached; CAT 1, 1–10%; CAT 2, 10–30%; CAT 3, 30–60%; and CAT 4, >60% of corals bleached [79]. Correction algorithms are required to account for loss of light through factors such as atmospheric scattering and the attenuation coefficient of water. As such, airborne surveys still require underwater ground-truthing to compare and validate these correction procedures [80].

The spatial resolution and cost of remote hyperspectral observations can be further improved by using unmanned aerial vehicles (UAVs). Lightweight hyperspectral cameras deployed on UAVs typically produce images with a spatial resolution of around 15 cm/pixel, allowing for the identification and monitoring of individual corals [81]. In a set period, this method can cover larger areas than diver or UUV solutions. Compared to manned aircraft, UAVs achieve higher resolution primarily due to the lower flight altitude (30–100 m), but at a smaller spatial scale. This technique allows for rapid data of areas of reef for preliminary assessments. For example, a 2017 study by Queensland University of Technology demonstrated that a UAV could photograph 40 hectares of coral reef in approximately 30 min to enable the study of coral bleaching [82].

### 2.2.2. Satellite Multi/Hyperspectral Imaging

Global programs such as the Coral Reef Watch by the National Oceanographic and Atmospheric Administration (NOAA) use satellite technology to observe and monitor reef conditions across all visible reefs. In practice, this is mainly limited to shallow reefs that are less than 25 m deep. Satellites are used to estimate SSTs and predict the potential extent of coral reef bleaching [83]. Temporal data can be used to monitor the effect of SST anomalies on coral [84]. During the warmest months of the year, often a 1 °C elevation above the monthly mean maximum can be associated with bleaching events [85]. Coral Reef Watch's HotSpot program uses these satellite observations to provide a "Satellite Bleaching Alert" or SBA [86]. Coral Reef Watch issues four levels of SBA for 24 reef sites in the tropics [87] based on satellite near-real-time HotSpot levels. This provides an early warning system for vulnerable coral reef systems determined by the change in SST from the norm. The technique, however, is largely speculative as there is no actual data taken directly from the corals themselves and should therefore be considered as a top-level predictive tool for bleaching events.

The loss of pigmented Symbiodiniaceae from corals during mass bleaching events results in an optical signal that can be strong enough for detection by remote sensing satellites in low-Earth orbit. Multispectral satellite systems such as the Landsat satellites allow the surveys to cover vast areas quickly with around 30,000 km$^2$ acquired in a 5-h period, with a spatial resolution of 30–60 m [88]. Other satellites such as the European Space Agency's Sentinel-2, are able to capture data with a 290 km field of view with spatial resolution varying from 10 m to 60 m depending on the spectral band [89].

Satellites equipped with multispectral cameras are able to provide data on coral conditions as outlined below.

The Landsat Thematic Mapper (TM) carried by Landsats 4 and 5 has mapped the geomorphology of Australia's Great Barrier Reef [90]. Landsat TM and Enhanced Thematic Mapper Plus (ETM+) have also been used to monitor changes in groups of coral reefs [91]. More recently, a detailed survey of the geological features and spectral characteristics of reefs near the Nansha Islands in the South China Sea was conducted using the Landsat 8 operational land imager (OLI) [92]. Specialised, marine focused remote sensors have also been deployed. In 2009, the Hyperspectral Imager for the Coastal Ocean (HICO) was installed on the International Space Station [93]. HICO focused on selected coastal regions and imaged them with full spectral coverage (380 to 960 nm sampled at 5.7 nm intervals). During its five years in operation, HICO collected over 10,000 scenes from around the world [94], collecting data on water clarity, bottom types, bathymetry, and on-shore vegetation maps.

Both airborne systems and instruments deployed in low-Earth orbit provide the ability to conduct large area reconnaissance of coral reef health, albeit with a relatively poor spatial resolution. These systems can image most global shallow reefs but are depth limited and struggle to map deeper reefs [95] whereas light absorption precludes the recognition of features below a critical depth threshold of approximately 20 m water depth, dependent on water clarity [96,97]. Any spectral data taken above the water's surface require a correction for the attenuation of light through the atmosphere and the water, which are wavelength specific. These corrections vary due to daily conditions and water types, each producing variability of the spectral diffuse attenuation coefficient in coral reefs and adjacent waters [98]. Again, ground truthing is required to validate the spectra used for these corrections.

## 2.3. Limitations of Non-Invasive Monitoring

As outlined previously, a variety of data can be collected using these different methods of assessment to indicate coral 'health'. Each technique generally gives multiple metrics that can be used, as summarised in Figures 3 and 4.

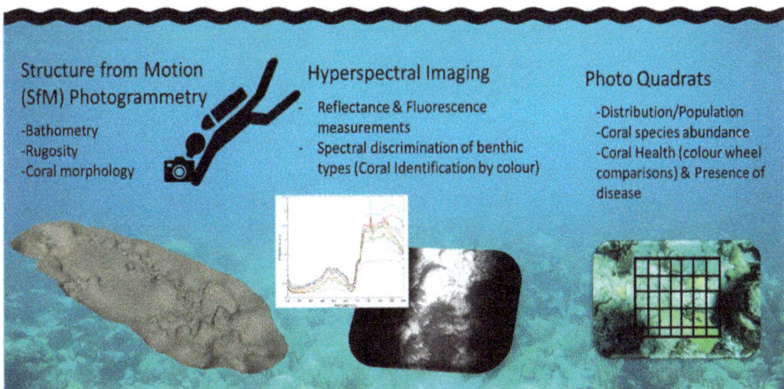


**Figure 3.** Diver image based techniques and examples of the types of data produced from the main techniques.

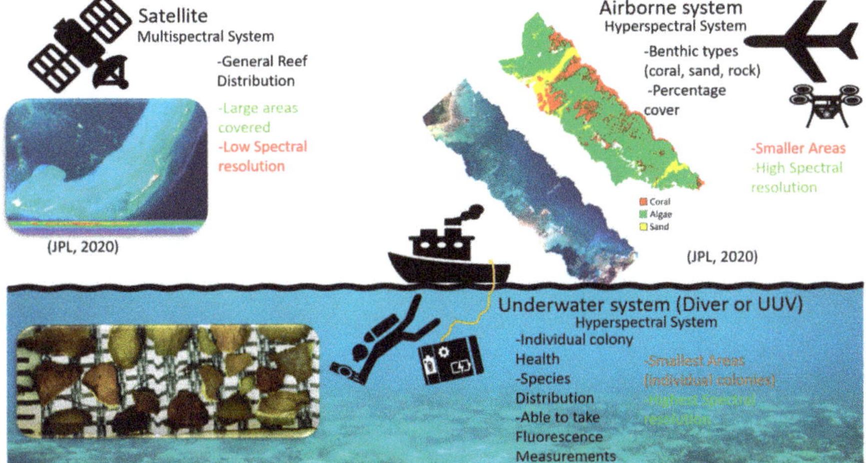


**Figure 4.** The types of data from each type of spectral system at the different levels from satellites to underwater systems. Images from the Jet Propulsion Lab (JPL) (2020) and Eric J. Hochberg, Bermuda Institute of Ocean Sciences. UUV: Unmanned underwater vehicle.

The use of spectral techniques to look at other parameters such as distinguishing between coral species and identifying diseases and bleaching suffers from limiting factors that include phenotypic plasticity in the host, differing symbiont pigment compositions [99], and changing physical parameters of the water column [100] such as turbity.

In order to derive meaning from colour and its relation to bleaching, bar a simple presence vs. absence, the exact symbiont density and chlorophyll-a content needs to be quantified to link to their corresponding wavelengths. Unfortunately, this procedure is destructive. Typically, whole corals or tissue samples are removed for laboratory analysis. Coral health is assessed based on the composition of extracted Symbiodiniaceae and chlorophyll content [101]. Laboratory bleaching experiments allow links to be made between the effects of bleaching and pigment intensity. These experiments link the levels of symbiont density to spectral peaks and approximate the stage of bleaching that can be determined using in situ hyperspectral imagery [33]. Effective coral monitoring in this way can thus become a victim of itself; effective identification in a plethora of sites featuring the signatures of coral bleaching creates an insurmountable requirement for downstream laboratory based truthing.

Nevertheless, as outlined in Tables 2 and 3, the techniques described have many features that contribute to their overall suitability for coral monitoring including cost, spatial scale, spectral resolution, and any additional data taken alongside sample acquisition. The choice of instrumentation is normally governed by research requirements [102,103]. The cost of the instrumentation is often the most prohibitive factor in the adoption of improved sensing strategies. For instance, to determine the global extent of coral bleaching, airborne and satellite imagers are best suited despite their cost and problems assessing deep reefs. However, as stated, ground truthing data would be required to validate and correct for atmospheric and aqueous attenuation correction algorithms. For in situ surveys that require high spectral resolution, an underwater hyperspectral system is best suited. This provides the desired spectral resolution at a non-prohibitive cost whilst simultaneously gathering related supporting measurements (see Table 4).

Therefore, the choice of the underwater spectroscopy tool is also crucial. Each imaging technique has a number of defining qualities and trade-offs regarding spectral range, resolution, scale of operation, depth rating, and cost. These determine their suitability and effectiveness in any given situation.

Looking towards the future, new spectral tools will likely emerge to address these issues. A range of underwater imagers identified by Liu (2020) [69] (Table 1) were compared as well as the new "Bi-Frost" digital single-lens reflex (DSLR) hyperspectral camera system presented in Section 2.1. Whilst spectral performance and cost are key factors, other considerations should be considered. Of particular importance are how the device is interfaced and how portable it is.

**Table 1.** A comparison of the selection of underwater spectral imaging systems as outlined by Liu et al., 2020 [69] including the HyRi/HyFi ordered by cost.

| Model | Developer | Spectral Range/Bands | Resolution | Spatial Imaging | Depth Rating | Cost (£) | Notes |
|---|---|---|---|---|---|---|---|
| Bi-Frost DSLR | Bristol University | 339–789 nm/192 | 18 nm@450 nm | Push-broom | 60 m | ~5000 | |
| TuLUMIS | Liu et al., 2018 [104] | 400–700 nm/8 | >10 nm | Staring array | 2000 m | ~5210.00 ($6730) | UUV mounted |
| HyperDiver | Chennu et al., 2017 [70] | 400–900 nm/480 | 1.5 nm | Push-broom | 50 m | ~20,040 (22,000 €) | Air weight ~32 kg |
| LUMIS 2 | Zawada et al., 2010 [105] | 460,522,582,678 nm/4 | 12.0–42.1 nm | Staring array | 20 m | ~46,470 ($60,000) | 4 imagers used |
| U185 | Cubert Gmbh | 450–950 nm/125 | 8 nm @ 532 nm | Snapshot | 5 m | ~49,850 (54,900 €) | |
| WaterCam | Sphere Optics | 450–950 nm/138 | 8 nm @ 532 nm | Snapshot | 5 m | ~49,850 (54,900 €) | |
| UMSI | Wu et al., 2019 [106] | 400–700 nm/31 | 10 nm | Staring | 50 m | ~56,170 ($72,800) | |
| UHI OV | Ecotone | 380–750 nm/150–200 | 2.2–5.5 nm | Push-broom | 2000 m | ~57,800 ($75,000) | UUV mounted |

LUMIS: Low-light-level underwater multispectral imaging system; UMSI: Underwater spectral imaging system; HyRi/HyFi: Hyperspectral reflectance/fluorescence imager; TuLUMIS: Tunable LED-based underwater multispectral imaging system; UHI OV: Ocean Vision (costs derived from manufacturers quotes with exchange rate applied on 8 October 2020.

**Table 2.** Optical techniques rated by cost, spatial scale, spatial resolution, and additional data gathered.

| Technique | Cost | Spatial Scale | Spatial Resolution | Additional Data Gathered | Notes |
|---|---|---|---|---|---|
| RGB imaging (Based on GoPro) | Very Low | Moderate | Very High | Photogrammetry | Limited spectral data obtained |
| Spectrometers (Waltz Diving PAM) | Moderate | Very Low | Very High | N/A | N/A |
| Bi-Frost DSLR | Low | Moderate | Very High | Photogrammetry, Fluorescence (HyFi) | Night-time imaging required for Fluorescence |
| Current UHI systems (See Table 4) | Moderate to High | Moderate | Very High to Moderate | N/A | Often large and cumbersome or designed specifically for UUVs |
| Drone multi/hyperspectral imaging [81] | Moderate to High | Moderate | Moderate | N/A | Requires Ground truthing |
| Aeroplane multi/hyperspectral imaging [61,107] | High | High | Moderate to Low | N/A | Requires Ground truthing |
| Satellite multi/hyperspectral imaging [108] | * Very High | Very High | Low to Very Low | SST, RGB images (Dependant on additional sensors equipped) | Requires Ground truthing |

HyFi: hyperspectral fluorescence imaging, UUV's: unmanned underwater vehicles, SST: Sea surface temperatures.
* Based on total cost of building and launching into orbit.

**Table 3.** Guide to rankings for each classification outlined in Table 2.

| Classification | Cost | Spatial Scale | Spatial Resolution |
|---|---|---|---|
| Very Low | <£1000 | mm–cm | km |
| Low | <£5000 | m | <100 m |
| Moderate | <£10,000 | <100 m | <10 m |
| High | >£25,000 | km | m |
| Very High | >£100,000 | 100+ km | mm-cm |

**Table 4.** An outline of the ability of the different optical techniques to assess important criteria used in coral surveyance as outlined by Leujak and Ormond, 2007 [103].

| Criteria | RGB Imaging | Spectrometers | Bi-Frost DSLR | Drone Multi/Hyperspectral Imaging | Aeroplane Multi/Hyperspectral Imaging | Satellite Multi/Hyperspectral Imaging |
|---|---|---|---|---|---|---|
| Damage (Disease/Bleaching) | Yes | Yes (Local Scale) | Yes | Yes | Yes (Mass Scale) | Yes (Mass Scale) |
| Recruits | Yes (modified camera) | No | Yes (HyFi) | No | No | No |
| Number of Colonies | Yes | No | Yes | Yes | Yes (Height Dependent) | No |
| Growth Measurements | Yes | No | Yes | No | No | No |
| Repeatability * | No | No | Yes (Reef Scale) | Yes | Yes | Yes |

* Repeatability was defined as the possibility of returning to exactly the same sampling unit in future monitoring.

## 3. Concluding Remarks

Current optical monitoring methods use a range of different approaches to answer the overly simplified question of 'Is this coral reef healthy?'. Due to coral reefs being physically and ecologically complex ecosystems, each technique offers a different piece of the puzzle, shedding light on parameters pertaining to overall reef health and status.

The Earth's coral reefs face a difficult and challenging future. If they are to survive the impending onslaught, effective monitoring and assessment will be fundamental to aiding in their recovery. A wealth of technological know-how is already being applied to study coral reefs from the colony to global scale. Consequently, our understanding of the problem and its severity continues to improve. New tools are being developed to help facilitate all survey requirements across all scales. On the larger scale, satellites and aerial multispectral/hyperspectral imaging provide the greatest spatial coverage with the trade-off of reduced spatial resolution. For colony or reef scale surveys, HyRi can produce high spatial resolution but cannot replace remote systems. Rather, it is a tool to complement these assessments and provide ground truth data to look at environments in closer detail. The use of in situ imagers also enables areas to be studied where aerial imaging is not applicable such as deep reefs or reefs with complex structures (i.e., steep reef faces/walls and overhangs). Aerial based imaging is unsuitable for deep coral studies, but tends to work best in mapping shallow lagoons and reef flats as these are within the depth threshold for this type of imaging (20 m) [96].

In the near future, Bi-Frost DSLRs are predicted to be a powerful new addition in the diagnostic toolbox for coral health. The benefits of enabling rapid non-destructive and repeatable measurements [33] address many of the shortcomings of the current generation of UHI instruments. The 3D data obtained with a Bi-Frost DSLR can be used in many ways, making this single tool capable of recording data for several different diagnostic needs from health to population surveys.

**Author Contributions:** Conceptualization, J.C.C.D. and J.T.; Methodology, J.T.; Software, D.A.M.-S.; Validation, J.C.C.D., J.T. and D.A.M.-S.; Investigation, J.T.; Writing—original draft preparation, J.T.; Writing—review and editing, M.J.A., J.C.C.D., D.A.M.-S. and T.B.S.; Supervision, J.C.C.D. and T.B.S.; Funding acquisition, T.B.S. All authors have read and agreed to the published version of the manuscript.

**Funding:** Pervioli Foundation, Roddenberry Foundation, and Bristol Alumni Foundation.

**Institutional Review Board Statement:** Not applicable.

**Informed Consent Statement:** Not applicable.

**Data Availability Statement:** Not applicable.

**Acknowledgments:** We would like to acknowledge the editor Rupert Ormond for his guidance and wisdom to better focus the manuscript. We would also like to thank James Alexandroff for his continued support without which this project would not be possible.

**Conflicts of Interest:** The authors declare no conflict of interest.

## References

1. Moberg, F.; Folke, C. Ecological goods and services of coral reef ecosystems. *Ecol. Econ.* **1999**, *29*, 215–233. [CrossRef]
2. Douglas, A.E. Coral bleaching—How and why? *Mar. Pollut. Bull.* **2003**, *46*, 385–392. [CrossRef]
3. Ralph, P.J.; Gademann, R.; Larkum, A. Zooxanthellae expelled from bleached corals at 33 °C are photosynthetically competent. *Mar. Ecol. Prog. Ser.* **2001**, *220*, 163–168. [CrossRef]
4. Willis, B.L.; Page, C.A.; Dinsdale, E.A. Coral Disease on the Great Barrier Reef. *Coral Health Dis.* **2002**, 69–104. [CrossRef]
5. Sutherland, K.P.; Porter, J.W.; Torres, C. Disease and immunity in Caribbean and Indo-Pacific zooxanthellate corals. *Mar. Ecol. Prog. Ser.* **2004**, *266*, 273–302. [CrossRef]
6. Zaneveld, J.R.; Burkepile, D.E.; Shantz, A.; Pritchard, C.E.; McMinds, R.; Payet, J.P.; Welsh, R.; Correa, A.M.S.; Lemoine, N.P.; Rosales, S.; et al. Overfishing and nutrient pollution interact with temperature to disrupt coral reefs down to microbial scales. *Nat. Commun.* **2016**, *7*, 11833. [CrossRef]
7. Mumby, P.J. The Impact of Exploiting Grazers (Scaridae) on the Dynamics of Caribbean Coral Reefs. *Ecol. Appl.* **2006**, *16*, 747–769. [CrossRef]
8. Mora, C. A clear human footprint in the coral reefs of the Caribbean. *Proc. R. Soc. B Boil. Sci.* **2008**, *275*, 767–773. [CrossRef] [PubMed]
9. Wenger, A.S.; Fabricius, K.E.; Jones, G.P.; Brodie, J.E. Effects of sedimentation, eutrophication, and chemical pollution on coral reef fishes. In *Ecology of Fishes on Coral Reefs*; Mora, C., Ed.; Cambridge University Press: Cambridge, UK, 2015; pp. 145–153.
10. Kayal, M.; Vercelloni, J.; De Loma, T.L.; Bosserelle, P.; Chancerelle, Y.; Geoffroy, S.; Stievenart, C.; Michonneau, F.; Penin, L.; Planes, S.; et al. Predator Crown-of-Thorns Starfish (Acanthaster planci) Outbreak, Mass Mortality of Corals, and Cascading Effects on Reef Fish and Benthic Communities. *PLoS ONE* **2012**, *7*, e47363. [CrossRef]
11. Bak, R.P.M.; van Eys, G. Predation of the sea urchin Diadema antillarum Philippi on living coral. *Oecologia* **1975**, *20*, 111–115. [CrossRef]
12. Hoegh-Guldberg, O. Climate change, coral bleaching and the future of the world's coral reefs. *Mar. Freshw. Res.* **1999**, *50*, 839–866. [CrossRef]
13. Hoegh-Guldberg, O.; Pendleton, L.; Kaup, A. People and the changing nature of coral reefs. In *Regional Studies in Marine Science*; Elsevier: Amsterdam, The Netherlands, 2019; Volume 30, p. 100699. [CrossRef]
14. Lamb, J.B.; True, J.D.; Piromvaragorn, S.; Willis, B.L. Scuba diving damage and intensity of tourist activities increases coral disease prevalence. *Biol. Conserv.* **2014**, *178*, 88–96. [CrossRef]

15. Roth, M.S.; Deheyn, D.D. Effects of cold stress and heat stress on coral fluorescence in reef-building corals. *Sci. Rep.* **2013**, *3*, 1421. [CrossRef]
16. Hoegh-Guldberg, O.; Mumby, P.J.; Hooten, A.J.; Steneck, R.S.; Greenfield, P.; Gomez, E.; Harvell, C.D.; Sale, P.F.; Edwards, A.J.; Caldeira, K.; et al. Coral Reefs Under Rapid Climate Change and Ocean Acidification. *Science* **2007**, *318*, 1737–1742. [CrossRef] [PubMed]
17. Glynn, P.W. Coral reef bleaching: Ecological perspectives. *Coral Reefs* **1993**, *12*, 1–17. [CrossRef]
18. Rogers, C.S. The effect of shading on coral reef structure and function. *J. Exp. Mar. Biol. Ecol.* **1979**, *41*, 269–288. [CrossRef]
19. Sheppard, C.; Davy, S.; Pilling, G.; Graham, N. *The Biology of Coral Reefs*; Oxford University Press: Oxford, UK, 2017.
20. Field, S.F.; Bulina, M.Y.; Kelmanson, I.V.; Bielawski, J.P.; Matz, M.V. Adaptive Evolution of Multicolored Fluorescent Proteins in Reef-Building Corals. *J. Mol. Evol.* **2006**, *62*, 332–339. [CrossRef] [PubMed]
21. Antonius, A. The 'band' diseases in coral reefs. In Proceedings of the Fourth International Coral Reef Symposium, Manila, Philippines, 18–22 May 1981.
22. Galloway, S.B.; Bruckner, A.W.; Woodley, C.M. *The Global Perspective on Incidence and Prevalence of Coral Diseases*; NOAA Technical Memorandum NOS NCCOS 97 and CRCP 7; National Oceanic and Atmospheric Administration: Silver Spring, MD, USA, 2009.
23. Ruiz-Moreno, D.; Willis, B.L.; Page, A.C.; Weil, E.; Croquer, A.; Vargas-Angel, B.; Jordan-Garza, A.G.; Jordán-Dahlgren, E.; Raymundo, L.; Harvell, C.D. Global coral disease prevalence associated with sea temperature anomalies and local factors. *Dis. Aquat. Org.* **2012**, *100*, 249–261. [CrossRef]
24. Dinsdale, A.E. Abundance of black-band disease on corals from one location on the Great Barrier Reef: A comparison with abundance in the Caribbean region. In Proceedings of the 9th International Coral Reef Symposium, Bali, Indonesia, 23–27 October 2000.
25. Miller, I. Black band disease on the Great Barrier Reef. *Coral Reefs* **1996**, *15*, 58.
26. Cervino, J.M.; Thompson, F.L.; Gomez-Gil, B.; Lorence, E.A.; Goreau, T.J.; Hayes, R.L.; Winiarski-Cervino, K.B.; Smith, G.W.; Hughen, K.; Bartels, E.; et al. The Vibrio core group induces yellow band disease in Caribbean and Indo-Pacific reef-building corals. *J. Appl. Microbiol.* **2008**, *105*, 1658–1671. [CrossRef]
27. Bongiorni, L.; Rinkevich, B. The pink-blue spot syndrome in Acropora eurystoma (Eilat, Red Sea): A possible marker of stress? *Zoology* **2005**, *108*, 247–256. [CrossRef]
28. Ravindran, J.; Raghukumar, C. Pink-line syndrome, a physiological crisis in the scleractinian coral Porites lutea. *Mar. Biol.* **2006**, *149*, 347–356. [CrossRef]
29. Antonius, A. Black band disease infection experiments on hexacorals and octocorals. In Proceedings of the Fifth International Coral Reef Congress, Tahiti, French, 27 May–1 June 1985.
30. Green, E.P.; Bruckner, A.W. The significance of coral disease epizootiology for coral reef conservation. *Biol. Conserv.* **2000**, *96*, 347–361. [CrossRef]
31. Palmer, C.V.; Modi, C.K.; Mydlarz, L.D. Coral Fluorescent Proteins as Antioxidants. *PLoS ONE* **2009**, *4*, e7298. [CrossRef] [PubMed]
32. Muller, E.M.; Van Woesik, R. Caribbean coral diseases: Primary transmission or secondary infection? *Glob. Chang. Biol.* **2012**, *18*, 3529–3535. [CrossRef]
33. Teague, J.; Willans, J.; Allen, M.; Scott, T.; Day, J. Hyperspectral imaging as a tool for assessing coral health utilising natural fluorescence. *J. Spectr. Imaging* **2019**, *8*, a7. [CrossRef]
34. Holden, H.; LeDrew, E. Spectral Discrimination of Healthy and Non-Healthy Corals Based on Cluster Analysis, Principal Components Analysis, and Derivative Spectroscopy. *Remote Sens. Environ.* **1998**, *65*, 217–224. [CrossRef]
35. Myers, M.R.; Hardy, J.T.; Mazel, C.H.; Dustan, P. Optical spectra and pigmentation of Caribbean reef corals and macroalgae. *Coral Reefs* **1999**, *18*, 179–186. [CrossRef]
36. Leiper, I.A.; Siebeck, U.E.; Marshall, N.J.; Phinn, S.R. Coral health monitoring: Linking coral colour and remote sensing techniques. *Can. J. Remote Sens.* **2009**, *35*, 276–286. [CrossRef]
37. Brown, J.S. Absorption and fluorescence spectra of chlorophyll-proteins isolated from Euglena gracilis. *Biochim. Biophys. ActaBioenergy* **1980**, *591*, 9–21. [CrossRef]
38. Falkowski, P.; Knoll, A.H. *Evolution of Primary Producers in the Sea*; Academic Press; Elsevier: New York, NY, USA, 2007.
39. Alieva, N.O.; Konzen, K.A.; Field, S.F.; Meleshkevitch, E.A.; Hunt, M.E.; Beltran-Ramirez, V.; Miller, D.J.; Wiedenmann, J.; Salih, A.; Matz, M.V. Diversity and Evolution of Coral Fluorescent Proteins. *PLoS ONE* **2008**, *3*, e2680. [CrossRef] [PubMed]
40. Gil, T.S.; Corredor, J. Studies of photosynthetic pigments of zooxanthellae in Caribbean hermatypic corals. In Proceedings of the Fourth International Coral Reefs Symposium, Manila, Philippines, 18–22 May 1981.
41. Venn, A.A.; Wilson, M.A.; Trapido-Rosenthal, H.G.; Keely, B.J.; Douglas, A.E. The impact of coral bleaching on the pigment profile of the symbiotic alga, Symbiodinium. *Plant Cell Environ.* **2006**, *29*, 2133–2142. [CrossRef] [PubMed]
42. Rodriguez-Lanetty, M.; Harii, S.; Hoegh-Guldberg, O. Early molecular responses of coral larvae to hyperthermal stress. *Mol. Ecol.* **2009**, *18*, 5101–5114. [CrossRef]
43. Desalvo, M.K.; Voolstra, C.R.; Sunagawa, S.; Schwarz, J.A.; Stillman, J.H.; Coffroth, M.A.; Szmant, A.M.; Medina, M. Differential gene expression during thermal stress and bleaching in the Caribbean coralMontastraea faveolata. *Mol. Ecol.* **2008**, *17*, 3952–3971. [CrossRef]
44. Hill, J.; Wilkinson, C. *Methods for Ecological Monitoring of Coral Reefs*; Australian Institute of Marine Science: Townsville, Australia, 2004.

45. Bruno, J.F.; Selig, E.R. Regional Decline of Coral Cover in the Indo-Pacific: Timing, Extent, and Subregional Comparisons. *PLoS ONE* **2007**, *2*, e711. [CrossRef]

46. Page, C.A.; Baker, D.M.; Harvell, C.D.; Golbuu, Y.; Raymundo, L.; Neale, S.J.; Rosell, K.B.; Rypien, K.L.; Andras, J.P.; Willis, B.L. Influence of marine reserves on coral disease prevalence. *Dis. Aquat. Organ.* **2009**, *87*, 135–150. [CrossRef] [PubMed]

47. Page, C.A.; Field, S.N.; Pollock, F.J.; Lamb, J.; Shedrawi, G.; Wilson, S.K. Assessing coral health and disease from digital photographs and in situ surveys. *Environ. Monit. Assess.* **2016**, *189*, 18. [CrossRef]

48. Raymundo, L.J.; Couch, C.S.; Bruckner, A.W.; Harvell, C.D. *Coral Disease Handbook: Guidelines for Assessment, Monitoring & Management*; Coral Reef Targeted Research & Capacity Building for Management (CRTR); Currie Communications: Melbourne, Australia, 2008.

49. Patterson, M.R.; Relles, N.J. Autonomous Underwater Vehicles resurvey Bonaire: A new tool for coral reef management. In Proceedings of the 11th International Coral Reef Symposium, Fort Lauderdale, FL, USA, 7–11 July 2008.

50. Williams, S.B.; Pizarro, O.; Jakuba, M.V.; Johnson, C.; Barrett, N.; Babcock, R.; Kendrick, G.; Steinberg, P.D.; Heyward, A.J.; Doherty, P.; et al. Monitoring of Benthic Reference Sites: Using an Autonomous Underwater Vehicle. *IEEE Robot. Autom. Mag.* **2012**, *19*, 73–84. [CrossRef]

51. Bokolonga, E.; Hauhana, M.; Rollings, N.; Aitchison, D.; Assaf, M.; Das, S.R.; Biswas, S.N.; Groza, V.; Petriu, E.M. A compact multispectral image capture unit for deployment on drones. In Proceedings of the Conference Record-IEEE Instrumentation and Measurement Technology Conference, Taipei, Taiwan, 23–26 May 2016.

52. Beijbom, O.; Edmunds, P.J.; Roelfsema, C.; Smith, J.; Kline, D.I.; Neal, B.P.; Dunlap, M.J.; Moriarty, V.; Fan, T.Y.; Tan, C.J.; et al. Towards Automated Annotation of Benthic Survey Images: Variability of Human Experts and Operational Modes of Automation. *PLoS ONE* **2015**, *10*, e0130312. [CrossRef]

53. Williams, I.D.; Couch, C.; Beijbom, O.; Oliver, T.; Vargas-Angel, B.; Schumacher, B.; Brainard, R. Leveraging Automated Image Analysis Tools to Transform Our Capacity to Assess Status and Trends of Coral Reefs. *Front. Mar. Sci.* **2019**, *6*, 222. [CrossRef]

54. Siebeck, U.E.; Marshall, N.J.; Klüter, A.; Hoegh-Guldberg, O. Monitoring coral bleaching using a colour reference card. *Coral Reefs* **2006**, *25*, 453–460. [CrossRef]

55. Winters, G.; Holzman, R.; Blekhman, A.; Beer, S.; Loya, Y. Photographic assessment of coral chlorophyll contents: Implications for ecophysiological studies and coral monitoring. *J. Exp. Mar. Bio. Ecol.* **2009**, *380*, 25–35. [CrossRef]

56. Fitt, W.K.; McFarland, F.K.; Warner, M.E.; Chilcoat, G.C. Seasonal patterns of tissue biomass and densities of symbiotic dinoflagellates in reef corals and relation to coral bleaching. *Limnol. Oceanogr.* **2000**, *45*, 677–685. [CrossRef]

57. Teague, J.; Scott, T.B. Underwater Photogrammetry and 3D Reconstruction of Submerged Objects in Shallow Environments by ROV and Underwater GPS. *J. Mar. Sci. Res. Technol.* **2017**, *1*, 1.

58. Colomina, I.; Molina, P. Unmanned aerial systems for photogrammetry and remote sensing: A review. *ISPRS J. Photogramm. Remote Sens.* **2014**, *92*, 79–97. [CrossRef]

59. Bayley, D.T.I.; Mogg, A.O.M. A protocol for the large-scale analysis of reefs using Structure from Motion photogrammetry. *Methods Ecol. Evol.* **2020**, *11*, 1410–1420. [CrossRef]

60. Hedley, J.D.; Mumby, P.J. Biological and remote sensing perspectives of pigmentation in coral reef organisms. *Adv. Mar. Biol.* **2002**, *43*, 277–317. [CrossRef]

61. Hochberg, E.J.; Atkinson, M.J. Spectral discrimination of coral reef benthic communities. *Coral Reefs* **2000**, *19*, 164–171. [CrossRef]

62. Jones, R.J.; Kildea, T.; Hoegh-Guldberg, O. PAM Chlorophyll Fluorometry: A New in situ Technique for Stress Assessment in Scleractinian Corals, used to Examine the Effects of Cyanide from Cyanide Fishing. *Mar. Pollut. Bull.* **1999**, *38*, 864–874. [CrossRef]

63. Chauka, L.J.; Steinert, G.; Mtolera, M.S.P. Influence of local environmental conditions and bleaching histories on the diversity and distribution of *Symbiodinium* in reef-building corals in Tanzania. *Afr. J. Mar. Sci.* **2016**, *63*, 57–64. [CrossRef]

64. Kurihara, H.; Takahashi, A.; Reyes-Bermudez, A.; Hidaka, M. Intraspecific variation in the response of the scleractinian coral Acropora digitifera to ocean acidification. *Mar. Biol.* **2018**, *165*, 38. [CrossRef]

65. Chang, C.I. *Hyperspectral Imaging: Techniques for Spectral Detection and Classification*; Springer Science & Business Media: Berlin, Germany, 2003.

66. Zawada, D.G. The Application of a Novel Multispectral Imaging System to the in vivo Study of Flourescent Compounds in Selected Marine Organisms. Ph.D. Thesis, University of California, San Diego, CA, USA, 2002.

67. Gleason, A.C.R.; Reid, R.P.; Voss, K.J. Automated classification of underwater multispectral imagery for coral reef monitoring. In Proceedings of the Oceans Conference Record (IEEE), Charleston, WV, USA, 29 September–4 October 2007.

68. Sture, O.; Ludvigsen, M.; Soreide, F.; Aas, L.M.S. Autonomous underwater vehicles as a platform for underwater hyperspectral imaging. In Proceedings of the OCEANS 2017, Aberdeen, UK, 19–22 June 2017.

69. Liu, B.; Liu, Z.; Men, S.; Li, Y.; Ding, Z.; He, J.; Zhao, Z. Underwater Hyperspectral Imaging Technology and Its Applications for Detecting and Mapping the Seafloor: A Review. *Sensors* **2020**, *20*, 4962. [CrossRef]

70. Chennu, A.; Färber, P.; De'Ath, G.; De Beer, D.; Fabricius, K.E. A diver-operated hyperspectral imaging and topographic surveying system for automated mapping of benthic habitats. *Sci. Rep.* **2017**, *7*, 7122. [CrossRef] [PubMed]

71. Maglione, P. Very High Resolution Optical Satellites: An Overview of the Most Commonly used. *Am. J. Appl. Sci.* **2016**, *13*, 91–99. [CrossRef]

72. Johnsen, G.; Ludvigsen, M.; Sørensen, A.; Sandvik Aas, L.M. The use of underwater hyperspectral imaging deployed on remotely operated vehicles—methods and applications. *IFAC-Papers* **2016**, *49*, 476–481. [CrossRef]

73. Johnsen, G.; Volent, Z.; Dierssen, H.; Pettersen, R.; Ardelan, M.; Søreide, F.; Fearns, P.; Ludvigsen, M.; Moline, M. Underwater hyperspectral imagery to create biogeochemical maps of seafloor properties. In *Subsea Optics and Imaging*; Elsevier: Amsterdam, The Netherlands, 2013.

74. Pust, O. Innovative Filter Solutions for Hyperspectral Imaging. *Opt. Photonik* **2016**, *11*, 24–27. [CrossRef]

75. Renhorn, I.G.E.; Bergström, D.; Hedborg, J.; Letalick, D.; Möller, S. High spatial resolution hyperspectral camera based on a linear variable filter. *Opt. Eng.* **2016**, *55*, 114105. [CrossRef]

76. Song, S.; Gibson, D.; Ahmadzadeh, S.; Chu, H.O.; Warden, B.; Overend, R.; Macfarlane, F.; Murray, P.; Marshall, S.; Aitkenhead, M.; et al. Low-cost hyper-spectral imaging system using a linear variable bandpass filter for agritech applications. *Appl. Opt.* **2020**, *59*, A167–A175. [CrossRef]

77. Burns, J.; Delparte, D.M.; Gates, R.D.; Takabayashi, M. Integrating structure-from-motion photogrammetry with geospatial software as a novel technique for quantifying 3D ecological characteristics of coral reefs. *PeerJ* **2015**, *3*, e1077. [CrossRef] [PubMed]

78. Thompson, D.R.; Hochberg, E.; Asner, G.P.; Green, R.O.; Knapp, D.E.; Gao, B.-C.; Garcia, R.; Gierach, M.; Lee, Z.; Maritorena, S.; et al. Airborne mapping of benthic reflectance spectra with Bayesian linear mixtures. *Remote Sens. Environ.* **2017**, *200*, 18–30. [CrossRef]

79. Berkelmans, R.; De'Ath, G.; Kininmonth, S.; Skirving, W.J.S. A comparison of the 1998 and 2002 coral bleaching events on the Great Barrier Reef: Spatial correlation, patterns, and predictions. *Coral Reefs* **2004**, *23*, 74–83. [CrossRef]

80. Hughes, T.P.; Kerry, J.T.; Álvarez-Noriega, M.; Álvarez-Romero, J.G.; Anderson, K.D.; Baird, A.H.; Babcock, R.C.; Beger, M.; Bellwood, D.R.; Berkelmans, R.; et al. Global warming and recurrent mass bleaching of corals. *Nature* **2017**, *543*, 373–377. [CrossRef]

81. Boyd, J. Drones survey the great barrier reef: Aided by AI, hyperspectral cameras can distinguish bleached from unbleached coral-[News]. *IEEE Spectr.* **2019**, *56*, 7–9. [CrossRef]

82. Queensland University of Technology (QUT). Queensland's Own Rapid Response Tool for Monitoring Coral Bleaching. 2017. Available online: https://www.qut.edu.au/news?news-id=122198 (accessed on 12 June 2018).

83. Sully, S.; Burkepile, D.E.; Donovan, M.K.; Hodgson, G.; Van Woesik, R. A global analysis of coral bleaching over the past two decades. *Nat. Commun.* **2019**, *10*, 1264. [CrossRef]

84. Brown, B.E.; Dunne, R.P.; Ambarsari, I.; Le Tissier, M.D.A.; Satapoomin, U. Seasonal fluctuations in environmental factors and variations in symbiotic algae and chlorophyll pigments in four Indo-Pacific coral species. *Mar. Ecol. Prog. Ser.* **1999**, *191*, 53–69. [CrossRef]

85. Strong, A.E.; Barrientos, C.S.; Duda, C.; Sapper, J. Improved satellite techniques for monitoring coral reef bleaching. In Proceedings of the 8th International Coral Reef Symposium, Panama City, Panama, 24–29 June 1996.

86. Liu, G.; Strong, A.E.; Skirving, W.J.; Arzayus, L.F. Overview of NOAA Coral Reef Watch Program's Near-Real-Time Satellite Global Coral Bleaching Monitoring Activities. In Proceedings of the 10th International Coral Reef Symposium, Okinawa, Japan, 28 June–2 July 2004.

87. Liu, G.; Heron, S.F.; Eakin, C.M.; Muller-Karger, F.E.; Vega-Rodriguez, M.; Guild, L.S.; De La Cour, J.L.; Geiger, E.F.; Skirving, W.J.; Burgess, T.F.R.; et al. Reef-Scale Thermal Stress Monitoring of Coral Ecosystems: New 5-km Global Products from NOAA Coral Reef Watch. *Remote Sens.* **2014**, *6*, 11579–11606. [CrossRef]

88. Clark, M.L. Comparison of simulated hyperspectral HyspIRI and multispectral Landsat 8 and Sentinel-2 imagery for multi-seasonal, regional land-cover mapping. *Remote Sens. Environ.* **2017**, *200*, 311–325. [CrossRef]

89. Drusch, M.; Del Bello, U.; Carlier, S.; Colin, O.; Fernandez, V.; Gascon, F.; Hoersch, B.; Isola, C.; Laberinti, P.; Martimort, P.; et al. Sentinel-2: ESA's Optical High-Resolution Mission for GMES Operational Services. *Remote Sens. Environ.* **2012**, *120*, 25–36. [CrossRef]

90. Joyce, K.E.; Phinn, S.R.; Roelfsema, C.M.; Neil, D.T.; Dennison, W.C. Combining Landsat ETM+ and Reef Check classifications for mapping coral reefs: A critical assessment from the southern Great Barrier Reef, Australia. *Coral Reefs* **2004**, *23*, 21–25. [CrossRef]

91. Palandro, D.; Andréfouët, S.; Muller-Karger, F.E.; Dustan, P.; Hu, C.; Hallock, P. Detection of changes in coral reef communities using Landsat-5 TM and Landsat-7 ETM+ data. *Can. J. Remote Sens.* **2003**, *29*, 201–209. [CrossRef]

92. Duan, Y.; Liu, Y.; Li, M.; Zhou, M.; Yang, Y. Survey of reefs based on Landsat 8 operational land imager (OLI) images in the Nansha Islands, South China Sea. *Acta Oceanol. Sin.* **2016**, *35*, 11–19. [CrossRef]

93. Keith, D.J.; Schaeffer, B.A.; Lunetta, R.S.; Gould, R.W., Jr.; Rocha, K.; Cobb, D.J. Remote sensing of selected water-quality indicators with the hyperspectral imager for the coastal ocean (HICO) sensor. *Int. J. Remote Sens.* **2014**, *35*, 2927–2962. [CrossRef]

94. O. S. University. What Is HICO? 2015. Available online: http://hico.coas.oregonstate.edu/ (accessed on 12 June 2018).

95. Stumpf, R.P.; Holderied, K.; Sinclair, M. Determination of water depth with high-resolution satellite imagery over variable bottom types. *Limnol. Oceanogr.* **2003**, *48*, 547–556. [CrossRef]

96. Foo, S.A.; Asner, G.P. Scaling Up Coral Reef Restoration Using Remote Sensing Technology. *Front. Mar. Sci.* **2019**, *6*, 79. [CrossRef]

97. Casella, E.; Collin, A.; Harris, D.; Ferse, S.; Bejarano, S.; Parravicini, V.; Hench, J.L.; Rovere, A. Mapping coral reefs using consumer-grade drones and structure from motion photogrammetry techniques. *Coral Reefs* **2017**, *36*, 269–275. [CrossRef]

98. Hochberg, E.J.; Peltier, S.A.; Maritorena, S. Trends and variability in spectral diffuse attenuation of coral reef waters. *Coral Reefs* **2020**, *39*, 1377–1389. [CrossRef]

99. Takabayashi, M.; Hoegh-Guldberg, O. Ecological and physiological differences between two colour morphs of the coral Pocillopora damicornis. *Mar. Biol.* **1995**, *123*, 705–714. [CrossRef]

100. Anderson, D.A.; Armstrong, R.A.; Weil, E. Hyperspectral Sensing of Disease Stress in the Caribbean Reef-Building Coral, Orbicella faveolate—Perspectives for the Field of Coral Disease Monitoring. *PLoS ONE* **2013**, *8*, e81478. [CrossRef]

101. Johannes, R.E.; Wiebe, W.J. Method for Derterminations of Coral Tissue Biomass and Composition. *Limnol. Oceanogr.* **1970**, *15*, 822–824. [CrossRef]

102. Jokiel, P.L.; Rodgers, K.S.; Brown, E.K.; Kenyon, J.C.; Aeby, G.; Smith, W.R.; Farrell, F. Comparison of methods used to estimate coral cover in the Hawaiian Islands. *PeerJ* **2015**, *3*, e954. [CrossRef]

103. Leujak, W.; Ormond, R.F.G. Comparative accuracy and efficiency of six coral community survey methods. *J. Exp. Mar. Biol. Ecol.* **2007**, *351*, 168–187. [CrossRef]

104. Liu, H.; Sticklus, J.; Köser, K.; Hoving, H.-J.T.; Song, H.; Chen, Y.; Greinert, J.; Schoening, T. TuLUMIS—A tunable LED-based underwater multispectral imaging system. *Opt. Express* **2018**, *26*, 7811–7828. [CrossRef] [PubMed]

105. Zawada, D.G.; Jaffe, J.S. Changes in the fluorescence of the Caribbean coral Montastraea faveolata during heat-induced bleaching. *Limnol. Oceanogr.* **2010**, *48*, 412–425. [CrossRef]

106. Wu, C.; Shentu, Y.; Chaofan, C.; Guo, Y.; Zhang, Y.; Wei, H.; Yang, P.; Huang, H.; Song, H. Development of an underwater multispectral imaging system based on narrowband color filters. In Proceedings of the OCEANS 2018 MTS/IEEE, Charleston, SC, USA, 22–25 October 2019.

107. Kobryn, H.T.; Wouters, K.; Beckley, L.E.; Heege, T. Ningaloo Reef: Shallow Marine Habitats Mapped Using a Hyperspectral Sensor. *PLoS ONE* **2013**, *8*, e70105. [CrossRef] [PubMed]

108. Hedley, J.D.; Roelfsema, C.M.; Chollett, I.; Harborne, A.R.; Heron, S.F.; Weeks, S.; Skirving, W.J.; Strong, A.E.; Eakin, C.M.; Christensen, T.R.L.; et al. Remote Sensing of Coral Reefs for Monitoring and Management: A Review. *Remote Sens.* **2016**, *8*, 118. [CrossRef]

MDPI
St. Alban-Anlage 66
4052 Basel
Switzerland
Tel. +41 61 683 77 34
Fax +41 61 302 89 18
www.mdpi.com

*Oceans* Editorial Office
E-mail: oceans@mdpi.com
www.mdpi.com/journal/oceans